D1482888

R. G. Wilkins

Kinetics and Mechanism of Reactions of Transition Metal Complexes

© VCH Verlagsgesellschaft mbH, D-6940 Weinheim (Federal Republic of Germany), 1991

Distribution:

VCH, P. O. Box 101161, D-6940 Weinheim (Federal Republic of Germany)

Switzerland: VCH, P. O. Box, CH-4020 Basel (Switzerland)

United Kingdom and Ireland: VCH (UK) Ltd., 8 Wellington Court, Cambridge CB1 1HZ (England)

USA and Canada: VCH, Suite 909, 220 East 23rd Street, New York, NY 10010–4606 (USA)

ISBN 3-527-28389-7 (VCH, Weinheim)   ISBN 1-56081-198-6 (VCH, New York)

Ralph G. Wilkins

Kinetics and Mechanism of Reactions of Transition Metal Complexes

2nd Thoroughly Revised Edition

VCH

Weinheim · New York · Basel · Cambridge

Prof. Ralph G. Wilkins
University of Warwick
Dept. of Chemistry
Coventry CV4 7AL
Great Britain

1st edition 1974
2nd edition 1991

Published jointly by
VCH Verlagsgesellschaft mbH, Weinheim (Federal Republic of Germany)
VCH Publishers, Inc., New York, NY (USA)

Editorial Director: Karin von der Saal
Production Manager: Elke Littmann

Library of Congress Card No. applied for.

A CIP catalogue record for this book is available
from the British Library.

Deutsche Bibliothek Cataloguing-in-Publication Data:
Wilkins, Ralph G.:
Kinetics and mechanism of reactions of transition metal
complexes / Ralph G. Wilkins. – 2., thoroughly rev. ed. –
Weinheim ; New York ; Basel ; Cambridge : VCH, 1991
ISBN 3-527-28389-7 (Weinheim ...) brosch.
ISBN 3-527-28253-X (Weinheim ...) Gb.
ISBN 1-56081-198-6 (New York) brosch.
ISBN 1-56081-125-0 (New York) Gb.

Composition: Filmsatz Unger und Sommer GmbH, D-6940 Weinheim. Printing and bookbinding: Konrad Triltsch, Graphischer Betrieb, D-8700 Würzburg. Cover design: TWI, Herbert J. Weisbrod, D-6943 Birkenau.
Printed in the Federal Republic of Germany

Preface

Seventeen years is a long time between editions of a book. In order to add some of the vast amount of new material which has been published in that time, I have needed to abridge the older edition and in so doing apologise to oldtimers (myself included!) whose work may have been removed or modified. Nevertheless, the approach used is unchanged. In the first three chapters I have dealt with the acquisition of experimental data and discussed use for building up the rate law and in the deduction of mechanism. In the second part of the book, the mechanistic behavior of transition metal complexes of the Werner type is detailed, using extensively the principles and concepts developed in the first part.

There are noticeable changes from the first edition. The past decade or so has seen a marked increase in the use of photolytic and nmr methods, pressure effects and so on, to discover the intimate details of mechanism. These developments have been incorporated. So too, the growing interest in the biological aspects of inorganic chemistry has not been ignored. On the other hand, photochemical behavior and organometallic compounds are used for illustrative purposes only. I have succumbed to the use of SI units for energy, recognizing that much of the past (and a sizeable amount of the present) literature do not employ these. I shall never forget the factor 4.184 which converts old to new. I have increased the number of problems threefold and provided a section containing hints on their solution.

I would like to thank Dave Pennington for his careful examination of Chapter 5, Steve Davies for producing the structures, Ellen Foley and Merlin Callaway for valient typing and the staff of VCH, especially Michael Weller and Karin von der Saal, for their kind help in the production of the book. Many copyright holders gave permission to reproduce material. All these people have my sincere thanks. At the end of course, errors that remain are my responsibility alone.

My especial thanks go to my wife, Pat. She critically read the whole manuscript and gave much encouragement and understanding from the beginning. The book would never have appeared without her support and I dedicate it to her, with much love.

Ralph G. Wilkins, June 1991
Wappenbury, England

Contents

Ligands and Complexes

Ligand Abbreviations

$NH_2(CH_2)_xNH_2$	en ($x=2$); tn ($x=3$)
$NH_2(CH_2)_2NH(CH_2)_2NH_2$	dien
$NH_2(CH_2)_2NH(CH_2)_2NH(CH_2)_2NH_2$	trien sal$_2$trien
$N[(CH_2)_xNH_2]_3$	tren ($x=2$); trpn ($x=3$)
$NH(CH_2CO_2)_2^{2-}$	ida^{2-}
$N(CH_2CO_2)_3^{3-}$	nta^{3-}
$[^-OCOCH_2]_2N-X-N[CH_2COO^-]_2$	edta^{4-} (X=(CH$_2$)$_2$); pdta^{4-} (X=CH(CH$_3$)CH$_2$); cydta^{4-} (X=1,2-cyclohexyl)

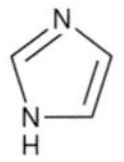

imid

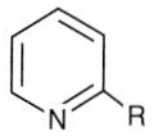

py (R=H); pic$^-$ (R=CH$_2$CO$_2^-$)

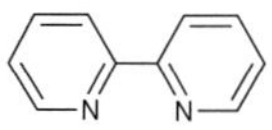

2,2'-bpy

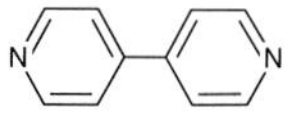

4,4'-bpy

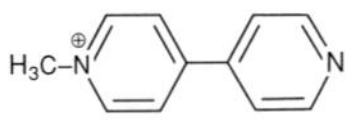

mbpy$^+$

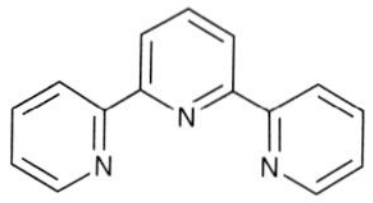

tpy

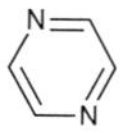

pz

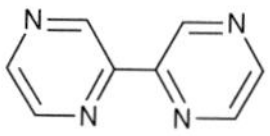

bpz

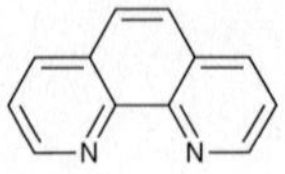

phen

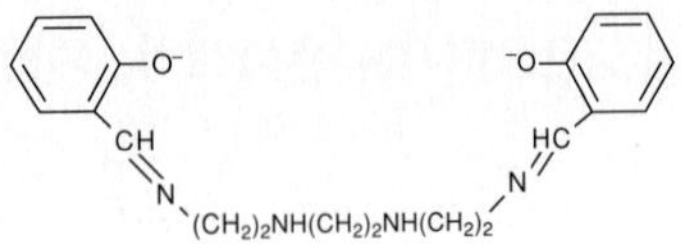

$CH_3COCH_2COCH_3$ acacH

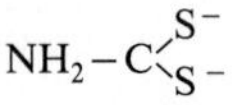

$NH_2-C\underset{S^-}{\overset{S^-}{\diagdown}}$ dtc^{2-}

$(CH_2)_2NH(CH_2)_2NH(CH_2)_2$

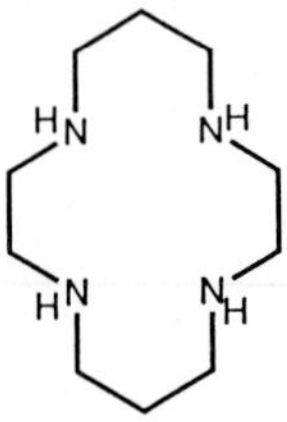

[14]ane-1,4,8,11-N$_4$
[14]aneN$_4$ or cyclam

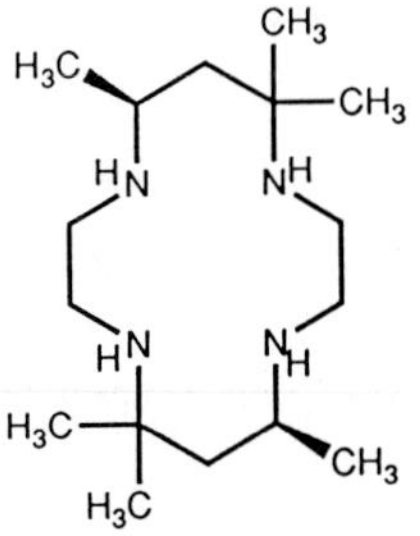

rac-(5,12)-Me$_6$[14]aneN$_4$
tet b

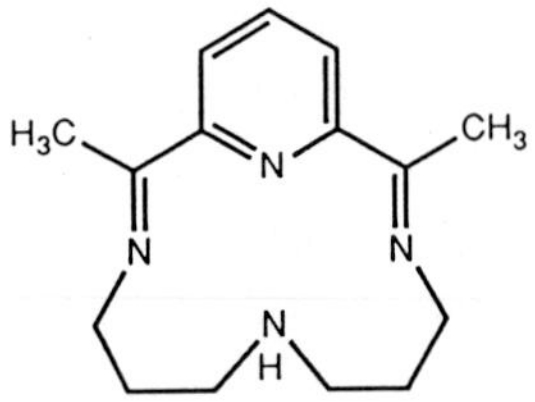

Me$_2$pyo[14]trieneN$_4$

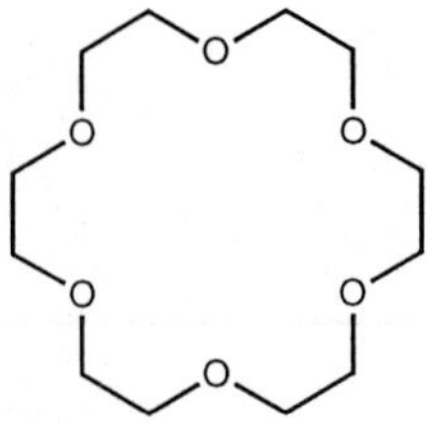

[18]crown-6

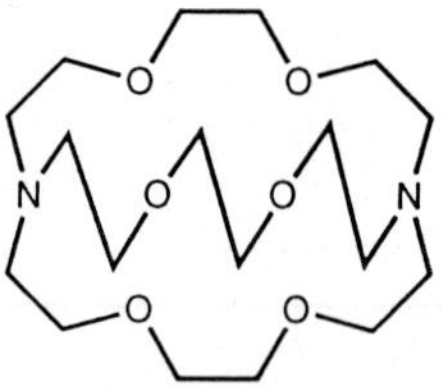

cryptand 2$_o$2$_o$2$_o$
(the numbers indicate
the oxygens on each
strand of the
macrocyclic ligand)

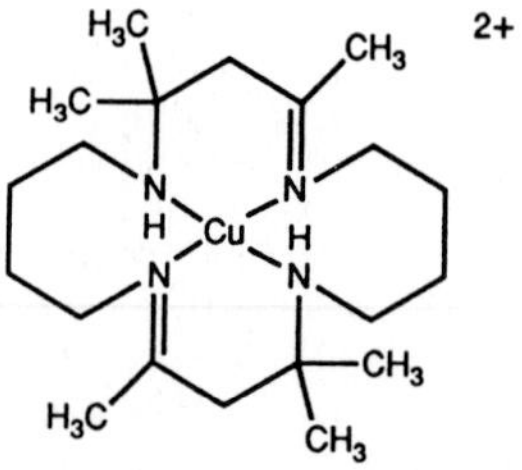

Cu(*trans*-(CH$_3$)$_6$[18]dieneN$_4$)$^{2+}$

Some commonly used macrocycles. − The number in brackets indicates the size of the ring. The terms ane and ene denote saturated and unsaturated rings respectively. The number of ligating atoms is indicated by a subscript. A trivial name is often used (G. A. Melson in Coordination Chemistry of Macrocyclic Compounds, Ed. G. A. Melson, Plenum, New York, 1979, Chapter 1; Comprehensive Coordination Chemistry, Ed. G. Wilkinson, Pergamon, Oxford, 1987. Several chapters in Vol. 2 (Ligands)).

R = H, tpypH$_2$
R = CH$_3$, tmpypH$_2$

X = SO$_3^-$, tppsH$_2$
X = N(CH$_3$)$_3^+$, tapH$_2$

Some Useful Water-Soluble Porphyrins

Co(sep)$^{2+/3+}$ Co(sar)$^{2+/3+}$ Co(azamesar)$^{2+/3+}$ Co(dinosar)$^{2+/3+}$

Some Caged Cobalt Complexes — The synthesis of these macrocycles is based on both the metal template effect as well as metal-ion activation of the imine moiety to nucleophilic attack, which initiates encapsulation (Chap. 6). A large variety of groups can replace NO$_2$ in Co(dinosar)$^{3+}$ yielding complexes with a wide range of properties. Again, trivial abbreviations are employed e.g. sep replaces 1,3,6,8,10,13,16,19-octaazabicyclo[6.6.6]eicosane. A. M. Sargeson, Pure Appl. Chem. 56, 1603 (1984).

Chapter 1

The Determination of the Rate Law

The single most important factor that determines the rate of a reaction is concentration – primarily the concentrations of the reactants, but sometimes of other species that may not even appear in the reaction equation. The relation between the rate of a reaction and the concentration of chemical species is termed the rate law; it is the cornerstone of reaction mechanisms. The rate law alone allows much insight into the mechanism. This is usually supplemented by an examination of other factors which can also be revealing. (For these, see Chap. 2)

1.1 The Rate of a Reaction and the Rate Law

The rate, or velocity, of a reaction is usually defined as the change with time t of the concentration (denoted by square brackets) of one of the reactants or of one of the products of the reaction; that is,

$$\text{rate} = V = -d[\text{reactant}]/dt = n \times d[\text{product}]/dt \qquad (1.1)$$

The negative sign arises because there is a loss of reactant. The value of n is often 1, but a value other than unity arises when one molecule of the reactant produces other than one molecule of the product. Rates are usually expressed in moles per liter per second, which we shall designate $M\ s^{-1}$, although $dm^{-3}\ mol\ s^{-1}$ is also a popular abbreviation. The rate law expresses the rate of a reaction in terms of the concentrations of the reactants and of any other species in solution, including the products, that may affect the rate.[1]

Suppose that the rate of a reaction depends only on the concentrations of A and B. The proportionality factor k relating rate to the concentrations of [A] and [B] in the rate law, or rate expression,

$$V = k\,[\text{A}]^a\,[\text{B}]^b \qquad (1.2)$$

is usually termed the *rate constant,* although it is sometimes referred to as the *specific rate,* or *rate coefficient.* Although the latter terms are in some respects preferable, since the proportionality factor is rarely invariant, the term *rate constant* is used in most of the literature and it is unlikely that it will be replaced. The values of a and b in (1.2) determine the *order of the reaction.* If $a = 1$, the reaction is termed *first-order* in A, and if $a = 2$, the reaction is *second-*

order in A. These are the most frequently encountered orders of reaction. The overall order of the reaction is $(a + b)$. The rate may be independent of the concentration of A, even if it participates in the overall stoichiometry. In this case, $a = 0$, and the reaction is *zero-order* in A. The concentration of A does not then feature in the rate law.

The formation of a number of chromium(III) complexes $Cr(H_2O)_5 X^{2+}$ from their constituent ions $Cr(H_2O)_6^{3+}$ and X^-,

$$Cr(H_2O)_6^{3+} + X^- \rightarrow Cr(H_2O)_5 X^{2+} + H_2O \tag{1.3}$$

where X^- represents a unidentate ligand, obeys the two-term rate law (the coordinated water being usually omitted from the formula)

$$V = d[CrX^{2+}]/dt = k_1[Cr^{3+}][X^-] + k_2[Cr^{3+}][X^-][H^+]^{-1} \tag{1.4}$$

over a wide range of concentration of reactants and acid.[2] Symbols such as a and b instead of k_1 and k_2 may be used in (1.4), since these quantitites are often composite values, made up of rate and equilibrium constants. To maintain both sides of (1.4) dimensionally equivalent,

$$V = M\ s^{-1} = k_1 \times M^2 = k_2 \times M \tag{1.5}$$

k_1 and k_2 must obviously be expressed in units of $M^{-1}\ s^{-1}$ and s^{-1} respectively. This simple application of dimension theory is often useful in checking the correctness of a complex rate law (see 1.121 and Probs. 1, Chap. 1 and Chap. 2).

If we consider specifically the formation of $CrBr^{2+}$, the rate law

$$d[CrBr^+]/dt = 3.0 \times 10^{-8}[Cr^{3+}][Br^-] + 3.6 \times 10^{-9}[Cr^{3+}][Br^-][H^+]^{-1} \tag{1.6}$$

holds at 25.0°C, and an ionic strength of 1.0 M.[2] The majority of kinetic determinations are carried out at a constant ionic strength by the addition of an "unreactive" electrolyte which is at a much higher concentration than that of the reactants. At high acid concentrations (>1 M), the first term in (1.6) is larger than the second, and the reaction rate is virtually acid-independent. At lower acid concentrations ($\sim 10^{-2}$ M), the second term dominates and the rate of reaction is now inversely proportional to $[H^+]$. This emphasizes the importance of studying the rate of a reaction over as wide a range of concentrations of species as possible so as to obtain an extensive rate law.

1.2 The Rate Law Directly from Rate Measurements

If the rate of a reaction can be measured at a time for which the concentrations of the reactants are known, and if this determination can be repeated using different concentrations of reactants, it is clear that the rate law (1.2) can be deduced directly. It is not often obtained in this manner, however, despite some distinct advantages inherent in the method.

1.2.1 Initial-Rate Method

The rate of reaction is measured at the commencement of the reaction, when the concentrations of the reactants are accurately known, indeed predetermined.

The decomposition of H_2O_2 (1.7)

$$2 H_2O_2 \rightarrow 2 H_2O + O_2 \tag{1.7}$$

is catalyzed by many metal complexes and enzymes.[3] The rate law for catalysis by an iron(III) macrocycle complex **1** has been deduced by measuring the initial rates V_0 of oxygen production.[3] A selected number from many data are shown in Table 1.1.

1

Table 1.1. Initial Rate Data for the Catalyzed Decomposition of H_2O_2 by an Fe(III) Macrocycle at 25°C[3]

Run	$[H_2O_2]_0$	Fe(III)	$[H^+]$	$[OAc]_T$	$10^3 \times$ Initial Rate	$10^3 k$[a]
	M	mM	10^5 M	M	10^3 M s^{-1}	s^{-1}
1	0.18	0.41	2.45	0.05	54	5.0
2	0.18	0.74	2.45	0.05	85	4.4
3	0.18	1.36	2.45	0.05	160	4.5
4	0.12	0.66	2.45	0.05	37	5.0
5	0.18	0.66	2.45	0.05	78	4.5
6	0.36	0.66	2.45	0.05	306	4.5
7	0.18	0.66	1.0	0.05	191	4.9
8	0.18	0.66	2.45	0.05	86	4.8
9	0.18	0.66	11.5	0.05	24	6.5
10	0.18	0.66	2.45	0.02	233	5.2
11	0.18	0.66	2.45	0.05	86	4.8
12	0.18	0.66	2.45	0.17	26	4.9

[a] Obtained from $k = V_0[H^+][OAc]_T[Fe(III)]^{-1}[H_2O_2]_0^{-2}$, a rearrangement of (1.8). Note the units of k.

It is apparent from these data that within experimental error the initial rate is proportional to the initial concentration of iron(III) complex, (Runs 1–3) to the square of the initial H_2O_2 concentration $[H_2O_2]_0$ (4–6) and inversely dependent on both H^+ (7–9) and total acetate $[OAc]_T$ (10–12) concentrations. A rate law involving the initial rate V_0,

$$V_0 = +d[O_2]/dt = k[Fe(III)][H_2O_2]_0^2[H^+]^{-1}[OAc]_T^{-1} \tag{1.8}$$

applies therefore and values of k calculated on this basis are reasonably constant. They are shown in the last column of Table 1.1. Raw initial rate data are also included in papers dealing with the Cr(VI)-I$^-$ reaction,[4] and the reaction of pyridoxal phosphate with glutamate in the presence of copper ions.[5] See Prob. 2.

Initial reaction rates are often estimated from the steepest tangents to absorbance/time traces. These initial gradients (absorbance units/sec) are easily converted into M s^{-1} units by using the known molar absorptivities of products and reactants. Alternatively, absorbance (D)/Time (T) plots can be fit to a function (1.9) leading to (1.10) and (1.11)

$$F(D) \quad = C1 + C2T + C3T^2 + C4T^3 \ldots \tag{1.9}$$

$$d[D]/dt = C2 + 2C3T + 3C4T^2 \tag{1.10}$$

$$\text{At } T \quad = 0, \quad d[D]/dt = C2 \tag{1.11}$$

as illustrated in the study of the reaction of H_2O_2 with 1,3-dihydroxybenzene in the presence of Cu(II) ions,[6(a)] and in the disproportionation of Mn(VI).[6(b)]

1.2.2 Critique of the Initial-Rate Method

The merits and difficulties in the use of this method are summarized:

(a) The initial-rate method is useful in the study of reactions complicated by side reactions or subsequent steps. The initial step of hydrolytic polymerization of Cr(III) is dimerization of the monomer. As soon as the dimer is formed however, it reacts with the monomer or with dimer to form trimer or tetramer, respectively. Initial rates (using a pH stat, Sec. 3.10.1 (a)) are almost essential in the measurement of dimer formation.[7]

The reverse reaction in (1.12)

$$Cr(H_2O)_6^{3+} + NCS^- \rightleftharpoons Cr(H_2O)_5 NCS^{2+} + H_2O \tag{1.12}$$

is unimportant during the early stages of the forward reaction when the products have not accumulated. A small initial loss of free thiocyanate ion can be accurately monitored by using the sensitive Fe(III) colorimetric method. The reverse reaction can be similarly studied for the initial rate of thiocyanate ion appearance by starting with $Cr(H_2O)_5 NCS^{2+}$ ions.[8]

(b) Complicated kinetic expressions and manipulations may be thus avoided. The initial-rate method is extensively employed for the study of enzyme kinetics and in enzyme assay (Prob. 3), simplifying the kinetics (see Selected Bibliography). Obviously the method cannot cope with the situation of induction periods or initial burst behavior. Furthermore a complex rate law, giving a wealth of mechanistic information, may now be disguised (for an example see Ref. 9). Despite this the initial-rate approach can be useful for distinguishing between two similar reaction schemes e.g. (1.13) and (1.14) (Sec. 1.6.4(d)).

$$A + B \rightleftharpoons C \rightarrow D \tag{1.13}$$

$$C \rightleftharpoons A + B \rightarrow D \tag{1.14}$$

in which A plus B yields D either *via* C or directly. The resolution otherwise requires very accurate relaxation data.[10]

(c) Two obvious disadvantages of the method are that many individual runs must be carried out to build up a rate law and a sensitive monitoring technique is required in order to obtain accurate concentration/time data for the first few percentage of the reaction. While nothing can be done about the first point, the advent of very sensitive computer-linked monitoring devices is ameliorating the sensitivity problem.

1.2.3 Steady-State Approach

In the steady-state approach to determining the rate law, solutions containing reactants are pumped separately at a constant flow rate into a vessel ("reactor"), the contents of which are vigorously stirred. After a while, products and some reactants will flow from the reactor at the same total rate of inflow and a steady state will be attained, in which the reaction will take place in the reactor with a *constant concentration of reactants,* and therefore a *constant rate.* This is the basis of the stirred-flow reactor, or capacity-flow method. [11] Although the method has been little used, it has the advantage of a simplified kinetic treatment even for complex systems.

1.3 Integrated Forms of the Rate Expression

Consider a reaction that proceeds to completion in which the concentration of only one reactant, A, changes appreciably during the reaction. This may arise because (1) there is only one reactant A involved — for example, in a stereochemical change; (2) all the other possible reactants are in much larger ($\geqslant$ tenfold) concentration than A; or (3) the concentration of one of the other reactants may be held constant by buffering, or be constantly replenished, as it would be if it were acting in a catalytic role. Attention needs to be focused therefore only on the change of concentration of A as the reaction proceeds, and we can, for the present, forget about the other reactants. Now,

$$-d[A]/dt = k[A]^a \tag{1.15}$$

The manner in which [A] varies with time determines the order of the reaction with respect to A. Since it is usually much easier to measure a concentration than a rate, the form (1.15) is integrated. [12] The three situations $a = 0$, 1, and 2 account for the overwhelming number of kinetic systems we shall encounter, with $a = 1$ by far the most common behavior.
$a = 0$, zero-order in A:

$$-d[A]/dt = k \tag{1.16}$$

$$[A]_t = [A]_0 - kt \tag{1.17}$$

$a = 1$, first order in A:

$$-d[A]/dt = k[A] \tag{1.18}$$

$$\log[A]_t = \log[A]_0 - kt/2.3 \tag{1.19}$$

or now increasingly common,

$$\ln[A]_t = \ln[A]_0 - kt \tag{1.19a}$$

$a = 2$, second-order in A:

$$-d[A]/dt = k[A]^2 \tag{1.20}$$

$$\frac{1}{[A]_t} = \frac{1}{[A]_0} + kt \tag{1.21}$$

The differential (rate) forms are (1.16), (1.18) and (1.20), and the corresponding integrated forms are (1.17), (1.19) (or (1.19a)) and (1.21). The designations $[A]_0$ and $[A]_t$ represent the concentrations of A at zero time and time t. Linear plots of $[A]_t$, $\ln[A]_t$ or $[A]_t^{-1}$ vs time therefore indicate zero-, first, or second order dependence on the concentration of A. The important characteristics of these order reactions are shown in Fig. 1.1. Notwithstanding the appearance of the plots in 1.1(b) and 1.1(c), it is not always easy to differentiate between first- and second-order kinetics.[13,14] Sometimes a second-order plot of kinetic data might be mistaken for successive first-order reactions (Sec. 1.6.2) with similar rate constants.[15]

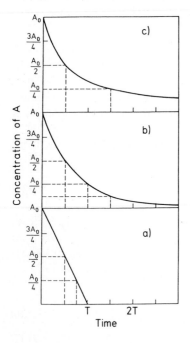

Fig. 1.1 The characteristics of (a) zero- (b) first- and (c) second-order reactions. In (a) the concentration of A decreases linearly with time until it is all consumed at time T. The value of the zero-order rate constant is given by A_0/T. In (b) the loss of A is exponential with time. The plot of $\ln[A]_t$ vs time is linear, the slope of which is k, the first-order rate constant. It obviously does not matter at which point on curve (b) the first reading is taken. In (c) the loss of A is hyperbolic with time. The plot of $[A]_t^{-1}$ vs time is linear with a slope equal to k, the second-order rate constant.

In the unlikely event that a in (1.15) is a non-integer, the appropriate function must be plotted. In general, for (1.15), $a \neq 1$[12] (see Prob. 8)

$$\frac{1}{[A]_t^{a-1}} - \frac{1}{[A]_0^{a-1}} = (a - 1)kt \tag{1.22}$$

It should be emphasized that in the case that A is in deficiency over other reactants, B etc., then k, relating to loss of A, is a pseudo rate constant and the effect of the other reactants on the rate must still be assessed separately (Sec. 1.4.4).

1.4 Monophasic Unidirectional Reactions

We shall first consider some straightforward kinetics, in which the loss of A, in the treatment referred to above, is monophasic and the reaction is unidirectional, that is, it leads to $\geq 95\%$ loss of A.

1.4.1 Zero-Order Dependence

It is impossible to conceive of a reaction rate as being independent of the concentration of *all* the species involved in the reaction. The rate might, however, very easily be independent of the concentration of one of the reactants. If this species, say A, is used in deficiency, then a pseudo zero-order reaction results. The rate $-d[\text{A}]/dt$ will not vary as [A] decreases, and will *not* depend on the initial concentration of A.

In the substitution reaction

$$\text{Ni(POEt}_3)_4 + \text{C}_6\text{H}_{11}\text{NC} \rightarrow \text{Ni(POEt}_3)_3(\text{C}_6\text{H}_{11}\text{NC}) + \text{POEt}_3 \qquad (1.23)$$

the loss of $\text{C}_6\text{H}_{11}\text{NC}$ has been followed in the presence of excess Ni complex.[16] The linear plot of absorbance, which is proportional to isonitrile concentration, vs time indicates a reaction zero-order in isonitrile (Fig. 1.2).

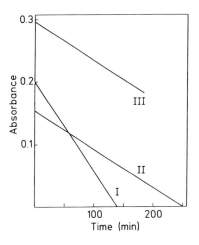

Fig. 1.2 Zero-order kinetic plots for reaction (1.23). Concentrations of $\text{Ni(POEt}_3)_4$ are 95 mM(I), 48 mM(II) and 47 mM(III). Those of $\text{C}_6\text{H}_{11}\text{NC}$ are 6.5 mM(I), 5.0 mM(II) and 9.7 mM(III).[16]

The slope of the zero-order plot (when absorbance is converted into concentration) is $k, \text{M s}^{-1}$. The value of k is found to be proportional to the concentration of $Ni(POEt_3)_4$, which is used in excess (Fig. 1.2),[17]

$$-d\,[Ni(POEt_3)_4]/dt = -1/2\,d\,[C_6H_{11}NC]/dt = k = k_1\,[Ni(POEt_3)_4] \qquad (1.24)$$

The reaction is therefore overall first-order, with a first-order rate constant $k_1\,(\text{s}^{-1})$. The zero-order situation is not often encountered (Prob. 4). A number of examples are compiled in Ref. 18 and one is shown in Fig. 8.3.

1.4.2 First-Order Dependence

First-order reactions are extremely common and form the bulk of reported kinetic studies. The rate of loss of the reactant A decreases as the concentration of A decreases. The differential form (1.18) leads to a number of equivalent integrated expressions, in addition to (1.19) and (1.19a):

$$[A]_t = [A]_0 \exp{(-kt)} \qquad (1.25)$$

$$\ln{([A]_0/[A]_t)} = kt \qquad (1.26)$$

$$-d\ln{[A]_t}/dt = k \qquad (1.27)$$

Important quantities characteristic of a first-order reaction are $t_{1/2}$, the half-life of the reaction, which is the value of t when $[A]_t = [A]_0/2$, and τ, the relaxation time, or mean lifetime, defined as k^{-1}.

$$k = 0.693/t_{1/2} = 1/\tau \qquad (1.28)$$

The latter is invariably used in the relaxation or photochemical approach to rate measurement (Sec. 1.8), and is the time taken for A to fall to $1/e$ $(1/2.718)$ of its initial value. Half-lives or relaxation times are constants over the complete reaction for first-order or pseudo first-order reactions. The loss of reactant A with time may be described by a single exponential but yet may hide two or more concurrent first-order and/or pseudo first-order reactions.

The change in the absorbance at 450 nm when $cis\text{-}Ni([14]aneN_4)(H_2O)_2^{2+}$ (for structure see "Ligands and Complexes") is plunged into 1.0 M $HClO_4$ is first-order (rate constant $= k$). It is compatible with concurrent isomerization and hydrolysis (1.29)

$$
cis\text{-}Ni([14]aneN_4)(H_2O)_2^{2+}
\begin{array}{c}
\xrightarrow{k_{isom}} trans\text{-}Ni([14]aneN_4)(H_2O)_2^{2+} \\
\mathbf{3} \\
\xrightarrow{k_{hyd}} Ni^{2+} + \text{protonated } [14]aneN_4 \\
\mathbf{4}
\end{array}
\qquad (1.29)
$$

$\mathbf{2}$

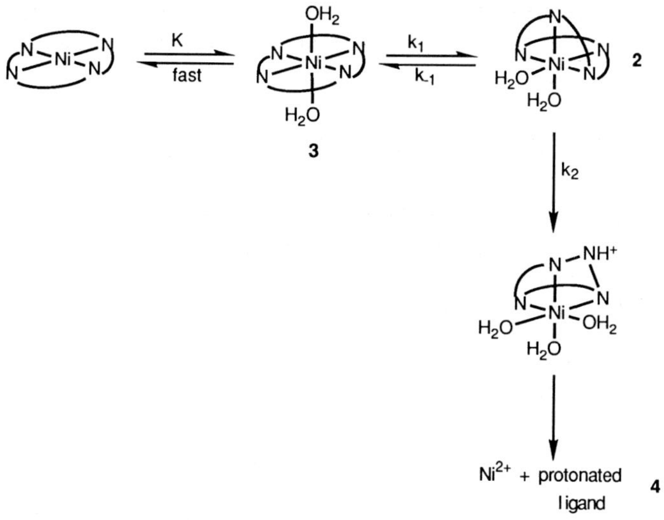

For this system

$$-d[\mathbf{2}]/dt = k[\mathbf{2}] = (k_{isom} + k_{hyd})[\mathbf{2}] \tag{1.30}$$

Estimation of the amounts of **3** and **4** produced from the final absorbance change[19] allows determination of both k_{isom} and k_{hyd}.[20] See also Ref. 21.

$$k_{isom}/k_{hyd} = [\mathbf{3}]/[\mathbf{4}] \tag{1.31}$$

Concurrent first-order changes invariably arise from the decay of an excited state *A which can undergo a number of first-order changes. In the presence of an excess of added B, pseudo first-order transformations can also occur:

$$\mathrm{A} \xrightarrow{h\nu} {}^*\mathrm{A} \rightarrow \begin{cases} \text{photochemical reaction products} \\ \mathrm{A} + h\nu' \text{ (luminescence, includes fluorescence and phosphorescence)} \\ \mathrm{A} + \text{heat (radiationless deactivation)} \\ \xrightarrow{(\mathrm{B})} \mathrm{A} \text{ and/or products (quenching process)} \end{cases} \tag{1.32}$$

The analysis of the immediate products or the isolation of one of the steps is essential in order to resolve the complexity.[22]

The reaction of a mixture of species A and A_1 which interconvert rapidly compared with the reaction under study, can also lead to a single first-order process. In order to resolve the kinetic data, information on the $A \rightleftharpoons A_1$ equilibrium is essential. When the relative amounts of A and A_1 are pH-controlled however (Sec. 1.10.1) or when the products of reaction of A and A_1 differ and do not interconvert readily, resolution is also in principle possible.

1.4.3 Second-Order Dependence

Second-order kinetics play an important role in the reactions of complex ions. Two identical reactants may be

$$2A \rightarrow \text{products} \tag{1.33}$$

involved. Eqn (1.21) is modified slightly to (1.34). By convention, the

$$\frac{1}{[A]_t} - \frac{1}{[A]_0} = 2kt \tag{1.34}$$

second-order rate constant for (1.33) is designated $2k$, because two identical molecules disappear for each encounter leading to reaction.[23] Reaction (1.33) is a key step in the process of dimerization or disproportionation of A and is often observed when a transient radical A is produced by photolysis or pulse radiolysis. Flash photolysis of hexane solutions of CO and $M_2(CO)_{10}$ (M = Mn or Re) produces $M(CO)_5^{\bullet}$ radicals. At the termination of

$$M_2(CO)_{10} \underset{2k}{\overset{h\nu}{\rightleftharpoons}} 2M(CO)_5^{\bullet} \tag{1.35}$$

the flash the radicals recombine in a second-order manner to regenerate $M_2(CO)_{10}$ with a rate law[24,25]

$$-d[M(CO)_5^{\bullet}]/dt = 2k[M(CO)_5^{\bullet}]^2 \tag{1.36}$$

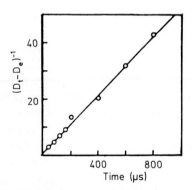

Fig. 1.3 Second-order kinetic plot for decay of Re(CO)$_5^{\bullet}$ obtained by flash photolysis of 120 μM Re$_2$(CO)$_{10}$ with 10 mM CO in isooctane. (CO slows further reactions of Re(CO)$_5^{\bullet}$) At λ = 535 nm, the molar absorbance coefficient, ε_M of Re(CO)$_5^{\bullet}$ is $1.0 \times 10^3 M^{-1} cm^{-1}$. Since therefore from Beer-Lambert law $(D_t - D_e) = [A]\varepsilon_m l$, where D's are optical absorbances and l is the path length (10 cm), the slope of Fig. 1.3 is $2k/\varepsilon_m l$ and $2k$ therefore equals $(5.3 \times 10^5)(10^3)(10) = 5.3 \times 10^9 M^{-1} s^{-1}$. Ref. 25. Reproduced with permission of the Journal of the American Chemical Society. © 1982, American Chemical Society.

The appropiate absorbance function vs time is linear (Fig. 1.3).[25] The values of k are $9.5 \times 10^8 M^{-1} s^{-1}$ (M = Mn) and $3.7 \times 10^9 M^{-1} s^{-1}$ (M = Re).[24,25] The second-order decay of Cr(V) affords another example, of many, of the type (1.33).[26]

Second-order reactions between two dissimilar molecules A and B are invariably studied under pseudo first-order conditions (Sec. 1.4.4) because this is by far the simpler procedure. If however this condition cannot be used because, for example, both reactants absorb heavily or low concentrations of *both* reactants must be used because of high rates, then for the reaction

$$A + B \rightarrow C \quad k_1 \tag{1.37}$$

$$-d[A]/dt = k_1[A][B] \tag{1.38}$$

Integration results in

$$\frac{1}{[B]_0 - [A]_0} \frac{\ln [A]_0[B]_t}{[B]_0[A]_t} = k_1 t \tag{1.39}$$

with subscripts t and 0 representing time t and 0, respectively. Applications of this equation are given in Refs. 27–29 and include the treatment of flow traces for the rapid second-order reaction between Eu(II) and Fe(III).[27] A simplified form of (1.39) equivalent to (1.21) arises when the starting concentrations of A and B are equal. This condition $[A]_0 = [B]_0$ must be set up experimentally with care[30] although in rare cases equal concentrations of A and B may be imposed by the conditions of their generation.[31] (See Prob. 9.)

1.4.4 Conversion of Pseudo to Real Rate Constants

Having obtained the exponent a in (1.15) by monitoring the concentration of A in deficiency we may now separately vary the concentration of the other reactants, say B and C, still keeping them however in excess of the concentration of A. The variation of the pseudo rate-constant k with [B] and [C] will give the order of reaction b and c with respect to these species,[32] leading to the expression

$$k = k_1[B]^b[C]^c \tag{1.40}$$

and therefore the full rate law

$$-d[A]/dt = k_1[A]^a[B]^b[C]^c \tag{1.41}$$

We can use this approach also to examine the effects on the rate, of reactants that may not be directly involved in the stoichiometry (for example, H^+) or even of products. It is the most popular method for determining the rate law, and only rarely cannot be used.

Considering just one other reactant B, we generally find a limited number of observed variations of $k(= V/[A]^a)$ with $[B]^b$. These are shown as (a) to (c) in Fig. 1.4. Nonlinear plots of k vs $[B]^b$ signify complex multistep behavior (Sec. 1.6.3).

(a) The rate and value of k may be independent of the concentrations of other reactants (Fig. 1.4(a)). This situation may occur in a dimerization, rearrangement or conformational change involving A. It may also arise in the solvolysis of A since the reaction order with respect to solvent is indeterminable, because its concentration cannot be changed.

(b) The value of k may vary linearly with $[B]^b$, and have a zero intercept for the appropriate plot (Fig. 1.4(b)). This conforms to a single-term rate law

$$-d[A]/dt = k[A]^a = k_3[A]^a[B]^b \tag{1.42}$$

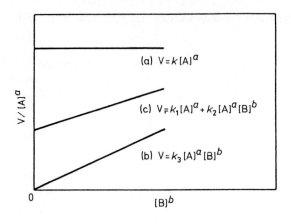

Fig. 1.4 Common variations of $V/[A]^a$ with $[B]^b$.

In the graph:
(a) $V = k[A]^a$
(c) $V = k_1[A]^a + k_2[A]^a[B]^b$
(b) $V = k_3[A]^a[B]^b$

y-axis: $V/[A]^a$
x-axis: $[B]^b$

4+

5

In the oxidation of Fe(II)P^{4+} (**5**) by excess O_2 there is a second-order loss of Fe(II). The second-order rate constant k is linearly dependent on $[O_2]$ with a zero intercept for the $k/[O_2]$ plot. An overall third-order rate law therefore holds.[14]

$$-d\,[Fe(II)P^{4+}]/dt = k_1\,[Fe(II)P^{4+}]^2[O_2] \tag{1.43}$$

When $a = b = 1$ in (1.42) the overall reaction is second-order. Even a quite small excess of one reagent (here B) can be used and pseudo first-order conditions will still pertain.[33] As the reaction proceeds, the ratio of concentration of the excess to that of the deficient reagent progressively increases so that towards the end of the reaction, pseudo first-order conditions certainly hold. Even if [B] is maintained in only a two-fold excess over [A], the error in the computed second-order rate constant is $\leqslant 2\%$ for 60% conversion.[34]

(c) The value of k may be linear with respect to $[B]^b$ but have a residual value at zero [B] (Fig. 1.4(c)). This behavior is compatible with concurrent reactions of A. The oxidation of Fe(phen)$_3^{2+}$ by excess Tl(III) conforms to this behavior leading to the rate law

$$-d\,[\text{Fe(phen)}_3^{2+}]/dt \;=\; k\,[\text{Fe(phen)}_3^{2+}] \;=\; k_1\,[\text{Fe(phen)}_3^{2+}] \;+\; k_2\,[\text{Fe(phen)}_3^{2+}]\,[\text{Tl(III)}]$$

$$(1.44)$$

The second-order redox reaction, giving rise to the rate constant k_2, is accompanied also by loss of the iron(II) complex by hydrolysis, which leads to the k_1 term. The latter can be more accurately measured in the absence of Tl(III).[35] The kinetics of substitution of many square-planar complexes conform to behavior (c), see Sec. 4.6. It is important to note that an intercept might be accurately defined and conclusive only if low concentrations of B are used. In the base catalyzed conversion

$$\text{Co(NH}_3)_5\text{ONO}^{2+} \;\rightarrow\; \text{Co(NH}_3)_5\text{NO}_2^{2+} \tag{1.45}$$

$$-d\,[\text{Co(NH}_3)_5\text{ONO}^{2+}]/dt \;=\; (k_1 + k_2[\text{OH}^-])\,[\text{Co(NH}_3)_5\text{ONO}^{2+}] \tag{1.46}$$

the k_1 term can be established only if low OH^- concentrations of 0.01–0.1 M are used.[36]

1.5 Monophasic Reversible Reactions

The only reactions considered so far have been those that proceed to all intents and purposes ($>95\%$) to completion. The treatment of *reversible* reactions is analogous to that given above, although now it is even more important to establish the stoichiometry and the thermodynamic characteristics of the reaction. A number of reversible reactions are reduced to pseudo first-order opposing reactions when reactants or products or both are used in excess

$$\text{A} \rightleftharpoons \text{X} \qquad k_1, k_{-1}, K \tag{1.47}$$

of A and X. The order with respect to these can then be separately determined. The approach to equilibrium for (1.47) is still first-order, but the derived first-order rate constant k is the sum of k_1 (the forward rate constant) and k_{-1} (the reverse rate constant):

$$\ln \frac{[\text{A}]_0 - [\text{A}]_e}{[\text{A}]_t - [\text{A}]_e} \;=\; kt \;=\; (k_1 + k_{-1})\,t \tag{1.48}$$

This equation resembles (1.26) but includes $[\text{A}]_e$, the concentration of A at equilibrium, which is not now equal to zero. The ratio of rate constants, $k_1/k_{-1} = K$, the so-called *equilibrium constant,* can be determined independently from equilibrium constant measurements. The value of k, or the relaxation time or half-life for (1.47), will all be independent of the direction from which the equilibrium is approached, that is, of whether one starts with pure A or X or even a nonequilibrium mixture of the two. A first-order reaction that hides concurrent first-order reactions (Sec. 1.4.2) can apply to reversible reactions also.

The scheme

$$\text{A} + \text{B} \rightleftharpoons \text{X} \qquad k_1, k_{-1} \tag{1.49}$$

can be reduced to (1.47) by using B in excess, and creating thereby a pseudo first-order reversible reaction. The rate law that arises is[37]

$$V = k[A] = (k_1[B] + k_{-1})[A] \qquad (1.50)$$

where $k_1[B]$ and k_{-1} are the forward and reverse first-order rate constants. Such a situation arises in the interaction of $V(H_2O)_6^{3+}$ with SCN^- ions,

$$V(H_2O)_6^{3+} + NCS^- \rightleftharpoons V(H_2O)_5NCS^{2+} + H_2O \qquad k_1, k_{-1}, K \qquad (1.51)$$

which is studied using a large excess of V^{3+} ions, although an equilibrium position is still attained. A plot of the pseudo first-order rate constant k for the approach to equilibrium vs $[V^{3+}]$ is linear (Fig. 1.5).[38] The slope is k_1 and the intercept is k_{-1}:

$$k = k_1[V^{3+}] + k_{-1} \qquad (1.52)$$

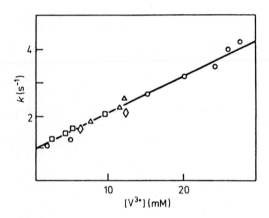

Fig. 1.5 Plot of $k\,(s^{-1})$ vs $[V^{3+}]$ for reaction (1.51) at 25°C. $[H^+] = 1.0$ M (circles); 0.50 M (triangles); 0.25 M (diamonds) and 0.15 M (squares).[38]

It is not always easy to obtain an accurate value for k_{-1} from such a plot. However, combination of k_1 with $K(= k_1/k_{-1})$ obtained from spectral measurements, yields a meaningful value for k_{-1}. The plots in Fig. 1.5 show the independence of the values of k_1 and k_{-1} on the acid concentrations in the range 0.15 to 1.0 M. There are slight variations to this approach, which have been delineated in a number of papers.[39]

Since the k vs [B] plot illustrated in Fig. 1.5 is identical to that obtained with unidirectional concurrent first- and second-order reactions of A (Fig. 1.4(c)) confusion might result if the equilibria characteristics are not carefully assessed. The pseudo first-order rate constant k for the reaction

$$Pt(dien)H_2O^{2+} + Cl^- \rightleftharpoons Pt(dien)Cl^+ + H_2O \qquad k_1, k_{-1} \qquad (1.53)$$

has been determined using excess $[Cl^-]$. The variation of k with $[Cl^-]$ resembles that shown in Fig. 1.4(c), i.e.

$$k = a + b[Cl^-] \qquad (1.54)$$

The results were interpreted as a two-term rate law for an irreversible reaction, b representing the second-order rate constant for attack by Cl^- ion and a an unusual dissociative path (Sec. 4.6). More recent work indicates that at the low $[Cl^-]$ concentrations used, reaction (1.53) is reversible and Eqn. (1.54) is better interpreted in terms of a reversible reaction as depicted in (1.53) in which in (1.54), $b = k_1$ and $a = k_{-1}$.[40] As a check the value of $b/a = K$ is close to that estimated for reaction (1.53).

The conversion of second-order reversible reactions to reversible first-order kinetics by using all but one of the reactants and all but one of the products in excess is a valuable ploy. The reversible reaction

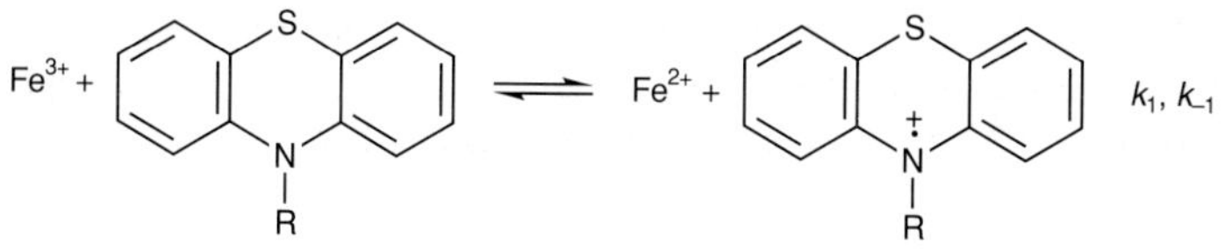

$$Fe^{3+} + \qquad \rightleftharpoons \qquad Fe^{2+} + \qquad\qquad k_1, k_{-1} \qquad (1.55)$$

is studied by using excess Fe^{3+} *and* Fe^{2+}. The plot of $\ln(D_e - D_t)$ vs time is linear, where D_e and D_t represent the absorbance of the highly colored radical at equilibrium and time t, respectively. The slope of this plot is k_{obs}, which from (1.48) is given by

$$k_{obs} = k_1[Fe^{3+}] + k_{-1}[Fe^{2+}] \qquad (1.56)$$

Thus a plot of $k_{obs}/[Fe^{2+}]$ vs $[Fe^{3+}]$ is linear with slope k_1 and intercept k_{-1}.[41] For other examples of this approach see Refs. 42 and 43. The full treatment for

$$A + B \rightleftharpoons X + Y \qquad (1.57)$$

is quite tedious[41,44,45] as it is for (1.58) Prob. 10[46,47]

$$A + B \rightleftharpoons X \qquad (1.58)$$

and (1.59)[48,49]

$$A \rightleftharpoons X + Y \qquad (1.59)$$

but simplified approaches,[45,49,50] relaxation treatment and computers[51] have largely removed the pain.

1.5.1 Conversion of Reversible to Unidirectional Reactions

An often useful approach is to eliminate the elements of reversibility from a reaction and force it to completion, either by the use of a large excess of reactant or by rapid removal of one of the products. A good illustration is afforded by the study of

$$PtCl(PPh_3)_2CO^+ + ROH \rightleftharpoons PtCl(PPh_3)_2COOR + H^+ \qquad k_1, k_{-1}, K_1 \qquad (1.60)$$

Since the equilibrium quotient K_1 is small, a nonnucleophilic base is added to the reaction mixture to react with liberated protons and drive the reaction to completion (left to right). Using an excess of ROH then ensures simple unidirectional pseudo first-order (rate constant k_f) kinetics:

$$-d \ln [PtCl(PPh_3)_2CO^+]/dt = k_f = k_1[ROH] \tag{1.61}$$

The reverse reaction also gives simple first-order (rate constant k_r) kinetics when studied with excess $HClO_4$:

$$-d \ln [PtCl(PPh_3)_2COOR]/dt = k_r = k_{-1}[H^+] \tag{1.62}$$

It was verified that k_1/k_{-1} equaled K_1, determined in a separate experiment.[52] For another example, see Ref. 53.

The reversible first-order reaction (1.47) can be converted into an irreversible A → X process by scavenging X rapidly and preventing its return to A. Thus the intramolecular reversible electron transfer in modified myoglobin (Sec. 5.9)

$$\overbrace{\text{PROTEIN}}^{} \qquad \overbrace{\text{PROTEIN}}^{} \qquad \overbrace{\text{PROTEIN}}^{}$$
$$Ru^{2+} \qquad Fe^{3+} \underset{k_{-1}}{\overset{k_1}{\rightleftharpoons}} Ru^{3+} \qquad Fe^{2+} \xrightarrow{CO} Ru^{3+} \qquad Fe^{2+}CO \tag{1.63}$$

can be converted into a unidirectional process controlled by k_1, by carrying out the experiments in the presence of CO which binds strongly to the ferrous heme.[54]

Care should be exercised that excess of one reactant does in fact promote irreversible reaction if this is the desired object, otherwise invalid kinetics and mechanistic conclusions will result. Consideration of the reduction potentials for cytochrome-c Fe(III) and $Fe(CN)_6^{3-}$ (0.273 V and 0.420 V respectively) indicates that even by using a 10^2–10^3 fold excess of $Fe(CN)_6^{4-}$, reduction of cytochrome-c Fe(III) will still not be complete. An equilibrium kinetic treatment is therefore necessary.[45]

1.6 Multiphasic Unidirectional Reactions

Attention is now directed to reactions that show a nonlinear plot of the appropriate function or that have rate laws that are altered with changes in the concentration of the species involved in the reaction. Such deviations are usually associated with concurrent and consecutive reactions.

1.6.1 Concurrent Reactions

A single species A (produced for example by radiolytic or photolytic means) may often disappear by concurrent first- (or pseudo first-) order (k_1) as well as by a second order process $(2 k_2)$ already alluded to (Sec. 1.4.3). Thus

$$-d[A]/dt = k_1[A] + 2 k_2[A]^2 \qquad (1.64)$$

At higher concentrations of A, the second-order process is more important and loss of A is second-order. As the concentration of A decreases, so the first term in (1.64) becomes dominant and decay of A is a purely first-order process (it may, for example, represent decomposition by solvent[24]) (Fig. 1.6).[55] The rate curve may be analyzed by a relatively straightforward linearizing method.[23]

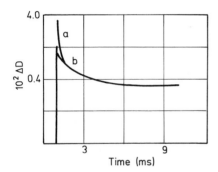

Fig. 1.6 Decay of $(NH_3)_5Co(mbpy^\bullet)^{3+}$ at pH 7.2 and 25 °C. The transient was generated by using equivalent amounts (10 μM) of $(NH_3)_5Co(mbpy)^{4+}$ and CO_2^-. The decay of the transient was nicely second-order up to 85% reaction (a). After this, when the Co(III) radical concentration is small (1–2 μM), there is a *slight* deviation for the expected second-order plot (not shown) and a first-order reaction (b) remains $(k = 5.4 \times 10^2 s^{-1})$.[55] The difference between the (steep) second-order decay and the (extrapolated) first-order loss is apparent.

The plot axes read: $10^2 \Delta D$ on the vertical axis (values 0.4 and 4.0), Time (ms) on the horizontal axis (values 3 and 9), with curves labeled a and b.

If a mixture of A and B undergoes parallel first-order or pseudo first-order reactions to give a common product C, and A and B do not interconvert readily compared with the reaction under study,

$$[C]_e - [C]_t = [A]_0 \exp(-k_1 t) + [B]_0 \exp(-k_2 t) \qquad (1.65)$$

where k_1 and k_2 are the first-order rate constants for conversion of A and B, respectively.[12] The resultant semilog plot of [C] vs time will in general be curved, and can be disected algebraically or by a computer program.[56] The hydrolysis of the anions in the complexes $Na[Co(medta)Br]$ and $K[Co(medta)Br]$ has been examined (medta = N-methyl-ethylene-diaminetriacetate). For the semilog plots the Na salt gives marked curvature, whereas the plot of the K salt is linear over four half-lives (Fig. 1.7). It is considered that the sodium salt is a mixture of isomers, in which the bromine is either in an equatorial or an axial position of the cobalt(III) octahedron. The biphasic plot can be separated into a fast component and a slow one. Significantly, the fast portion matches exactly the semilog plot for the K salt, which is considered isomerically pure.[57] A similar concurrent hydrolysis pattern has been observed with other complex ions.[58] It is not always easy to distinguish concurrent from consecutive reactions, as we shall see in the next section.

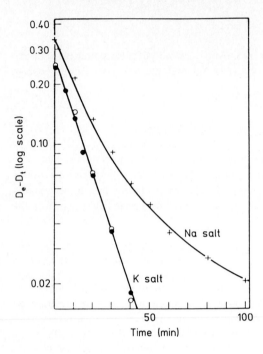

Fig. 1.7 Semilog plot of optical absorbance at 540 nm vs time for hydrolysis of Co(medta)Br$^-$ at 78.8 °C and pH 2.6. Mixture of isomers (+); fast component derived from mixture by subtraction of slow component (o); experimental points for fast isomer (•).[57]

1.6.2 Consecutive Reactions with no Elements of Reversibility

Consecutive reactions figure prominently in Part II. Since complex ions have a number of reactive centers, the product of one reaction may very well take part in a subsequent one. The simplest and very common, but still surprisingly involved, sequence is that of two irreversible first-order (1.66) or pseudo first-order (with X and Y in large excess) (1.67) reactions,

$$A \xrightarrow{k_1} B \xrightarrow{k_2} C \tag{1.66}$$

$$A \xrightarrow{(+X)k_1'} B \xrightarrow{(+Y)k_2'} C \tag{1.67}$$

$$k_1 = k_1'[X]; \quad k_2 = k_2'[Y] \tag{1.68}$$

If $k_1 \gg k_2$ then both steps in (1.66) can be analyzed separately as described previously. If $k_2 \gg k_1$, then only the first step is observed and

$$-d[A]/dt = d[C]/dt = k_1[A] \tag{1.69}$$

The only way for determining k_2 will be through isolation and separate examination of B (see later). If B is not isolable however and its properties are unknown, real difficulties might arise.[59]

The rate equations for (1.66) are

$$-d[A]/dt = k_1[A] \tag{1.70}$$

$$d[B]/dt = k_1[A] - k_2[B] \tag{1.71}$$

$$d[C]/dt = k_2[B] \tag{1.72}$$

Integrating these equations, and assuming $[A] = [A]_0$, and $[B] = [C] = 0$ at $t = 0$, we obtain the concentrations of A, B and C at any time t in terms of the concentration, $[A]_0$:

$$[A] = [A]_0 \exp(-k_1 t) \tag{1.73}$$

$$[B] = \frac{[A]_0 k_1}{k_2 - k_1} [\exp(-k_1 t) - \exp(-k_2 t)] \tag{1.74}$$

$$[C] = [A]_0 \left[1 - \frac{k_2}{k_2 - k_1} \exp(-k_1 t) + \frac{k_1}{k_2 - k_1} \exp(-k_2 t) \right] \tag{1.75}$$

Often k_1 will be comparable in value to k_2 and in this event the distribution of A, B and C with time is illustrated for example by the stepwise hydrolysis of a cobalt(III) complex **6**, $(X = -OC_6H_3(NO_2)_2)$

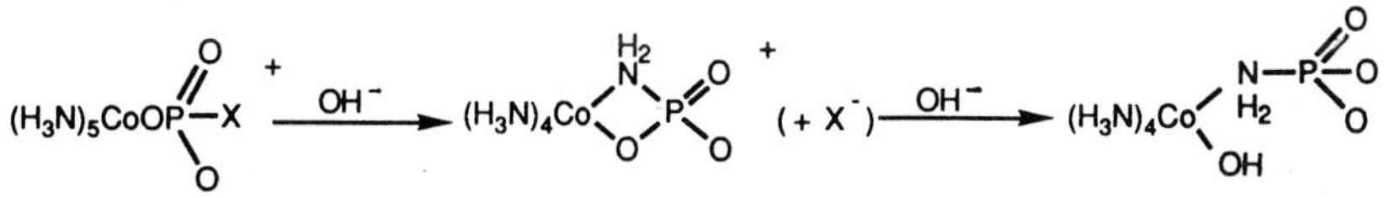

$$\tag{1.76}$$

6   **7**   **8**

Stepwise Hydrolysis of a Cobalt (III) Phosphate Complex (X = $^-OC_6H_4(NO_2)_2$)

The concentrations of **6**, **7** and **8** are determined by integration of their characteristic ^{31}P nmr spectra (Sec. 3.9.5) and their variations with time are shown in Fig. 1.8.[60] These curves illustrate some general features of the system (1.66). At the maximum concentration of **7**,

$$d[7]/dt = 0 \tag{1.77}$$

and

$$k_1[6] = k_2[7] \tag{1.78}$$

Since the relative concentrations of **6** and **7** are easily assessed at the maximum concentration of **7**, the ratio k_1/k_2 can be determined even if this is quite close to unity. It can be shown[61] that the time for the concentration of **7** to reach a maximum (t_{max}) is given by (1.79), Prob. 11,

$$t_{max} = \frac{\ln(k_1/k_2)}{k_1 - k_2} \tag{1.79}$$

(1.78) and (1.79) can be used to determine rate constants for consecutive reactions.[62,63] There is a lag period in the buildup of **8**, the inflection in the [**8**]/time curve corresponding to $[7]_{max}$. For another example where A, B and C in (1.66) are all monitored, see Ref. 64.

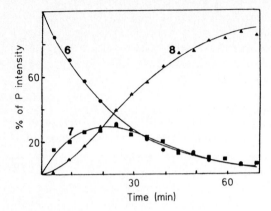

Fig. 1.8 Distribution of **6, 7** and **8** vs time from integrated ^{31}P nmr spectra in 0.35 M OH$^-$, $\mu = 1.0$ M at 5°C. Signals (ppm) are at 6.5 (**6**), 22–31 (**7**, dependent on [OH$^-$]) and 7.1 ppm (**8**).[60] (Reprinted with permission from P. Hendry and A. M. Sargeson, Inorg. Chem. **25**, 865 (1986). © 1986, American Chemical Society)

Such systems as (1.66) or (1.67) appear to have been examined mainly by spectral methods, and so discussion will center around this monitoring method. The occurrence of steps with similar rates is indicated by the lack of isosbestic points (Sec. 3.9.1) over some portion of the reaction. In addition the appropriate kinetic plots are biphasic although perhaps only slightly so, or only at certain wavelengths. Data obtained at a wavelength that monitors only the concentration of A will give a perfect first-order plot (or pseudo first-order in the presence of excess X) rate constant k_1 (1.73). Considering the optical absorbance D_t of the reacting solution at time t, in a cell of path length 1 cm,

$$D_t = \varepsilon_A [A] + \varepsilon_B [B] + \varepsilon_C [C] \tag{1.80}$$

where ε_A is the molar absorptivity of A and so on (Sec. 3.9.1). By substituting (1.73) through (1.75) into (1.80) and rearranging terms, it is not difficult to derive the expression[65,66]

$$D_t - D_e = a_1 \exp(-k_1 t) + a_2 \exp(-k_2 t) \tag{1.81}$$

where $D_e = \varepsilon_C [A]_0$ and a_1 and a_2 are composed of rate constants and molar absorptivities:[67]

$$a_1 = \varepsilon_A [A]_0 + \frac{\varepsilon_B [A]_0 k_1}{k_2 - k_1} + \frac{\varepsilon_C [A]_0 k_2}{k_1 - k_2} \tag{1.82}$$

$$a_2 = \frac{k_1 [A]_0 (\varepsilon_B - \varepsilon_C)}{k_1 - k_2} \tag{1.83}$$

It is not always easy by inspection to be certain that two reactions are involved. The use of a semi-log plot helps[68-70] since it shows better the deviation from linearity that a biphasic reaction demands but a computer treatment of the data is now the definitive approach.[71-77] Less apparent is the fact that having resolved the curve into two rate constants k_{fast} and k_{slow} we cannot simply assign these to k_1 and k_2 respectively (in 1.66) without additional information since (1.81) is derived without specifying a k_{fast}/k_1 or k_{fast}/k_2 condition.[71,78] The most popular method for resolving this ambiguity is by resorting to spectral considerations. The spectrum of the intermediate B (which is a collection of ε_b at various wavelengths) can be calculated from (1.81), (1.82) and (1.83) on the two premises that (a) $k_{fast} = k_1$ and $k_{slow} = k_2$ or (b) $k_{fast} = k_2$ and $k_{slow} = k_1$. One of the two spectra estimated by this means will usually be much more plausible than the other (which may even have negative molar absorptivities!) and the sequence which leads to the unreasonable spectrum for B can be discarded. An approach of this type was first made[79] in a study of

$$\text{Cr(en)}_3^{3+} \xrightarrow{\text{H}^+} cis-\text{Cr(en)}_2(\text{enH})(\text{H}_2\text{O})^{4+} \xrightarrow{\text{H}^+} cis-\text{Cr(en)}_2(\text{H}_2\text{O})_2^{3+} + \text{enH}_2^{2+} \tag{1.84}$$

and has since been adopted on numerous occasions.[70,73,76,80] In the majority of cases the fast and slow rate constants are assigned to k_1 and k_2 indicating fast formation of a relatively weakly absorbing intermediate.[70,73,76,80] A few instances are known however[61,81] where the reverse pertains i.e. a highly absorbing intermediate arises which decays rapidly compared with its formation. Now $k_{fast} = k_2$.

An intense purple-blue species forms within a few seconds when Cr^{2+}, in excess, is added to 4,4'-bipyridinium ion, bpyH_2^{2+}. The color fades slowly over many minutes. The formation and disappearance of the intermediate, k_{fast} and k_{slow} respectively, are assigned to k_2 and k_1 respectively from considerations of the assessed spectrum of the intermediate.[81] The reactions involved are

$$\text{HN}\!\!-\!\!\langle\text{bpy}\rangle\!\!-\!\!\text{NH}^{2+} + \text{Cr}^{2+} \longrightarrow \text{HN}\!\!-\!\!\langle\text{bpy}\rangle\!\!-\!\!\text{NH}^{\cdot+} + \text{Cr}^{3+} \tag{1.85}$$

highly colored

$$\text{HN}\!\!-\!\!\langle\text{bpy}\rangle\!\!-\!\!\text{NH}^{\cdot+} + \text{Cr}^{2+} \longrightarrow \text{HN}\!\!=\!\!\langle\text{bpy}\rangle\!\!=\!\!\text{NH} \tag{1.86}$$

colorless

and the rate equations are

$$V = a\,[\text{Cr}^{2+}]\,[\text{bpyH}_2^{2+}] \tag{1.87}$$

$$V = \{b + c\,[\text{Cr}^{2+}]\,[\text{H}^+]\}\,[\text{bpyH}_2^{\cdot+}] \tag{1.88}$$

The radical intermediate is always less than 1% of bpyH_2^{2+} but shows up because of its high absorbance.

If the spectrum of an intermediate is known, a choice of observation wavelength may allow isolation of each stage. In the reaction (A) → (C) using concentrated HCl:

$$\text{Re(en)}_2(\text{OH})_2^{3+} \xrightarrow[\text{2HCl}]{k_1} \text{Re(en)Cl}_2(\text{OH})_2^+ + \text{enH}_2^{2+} \tag{1.89}$$
$$\text{(A)} \qquad\qquad\qquad \text{(B)}$$

$$\text{Re(en)Cl}_2(\text{OH})_2^+ \xrightarrow[\text{2HCl}]{k_2} \text{ReCl}_4(\text{OH})_2^- + \text{enH}_2^{2+} \tag{1.90}$$
$$\text{(C)}$$

examination at 395 or 535 nm (where $\varepsilon_B = \varepsilon_C$) gives $-d[A]/dt$ and hence a value for k_1. Observation at 465 nm (where $\varepsilon_A = \varepsilon_B$) allows a value for k_2 to be obtained after a short induction period. Absorbance changes at 685 nm ($\varepsilon_A = \varepsilon_C$) reflect $d[B]/dt$, and specifically allow an estimate of t_{max}, the time for the concentration of B to reach a maximum. A confirmation of k_1 and k_2 is then possible by using (1.79).[82]

Finally we return to the situation mentioned at the beginning of the section of a system of two consecutive reactions still yielding a single linear plot of $\ln(D_t - D_e)$ vs time. A condition which is not obvious is (1.91) which emerges from (1.81)[65,78,83-85]

$$\frac{k_1}{k_2} = \frac{\varepsilon_C - \varepsilon_A}{\varepsilon_B - \varepsilon_A} \tag{1.91}$$

This may appear to be an unlikely situation to encounter until one recalls that there are a number of reactions involving two isolated and independently reacting centers. One might then anticipate that statistically $k_1 = 2k_2$ and that $\varepsilon_B = 1/2(\varepsilon_A + \varepsilon_C)$. These are precisely the conditions demanded by (1.91). It is worth noting that in this case the observed first-order rate constant is k_2.

Eight-iron ferredoxin contains two 4-Fe-clusters separated by about 12 Å. Each cluster can undergo a one-electron redox reaction (r and o represent the reduced and oxidized forms)

$$\text{8-Fe(rr)} \xrightarrow{k_1} \text{8-Fe(or)} \tag{1.92}$$

$$\text{8-Fe(or)} \xrightarrow{k_2} \text{8-Fe(oo)} \tag{1.93}$$

Oxidation of 8-Fe(rr) to 8-Fe(oo) by a number of one-electron oxidants gives a single first-order process. After ruling out the more obvious reasons for this observation, it is concluded that condition (1.91) holds,[86] one often referred to as "statistical kinetics". For other examples, see Refs. 87 and 88. Even if (1.91) is not strictly satisfied, linear plots may still be obtained.[84,88]

Further complexities may be anticipated. For example in the consecutive reaction (1.66), direct conversion of A to C may occur as well as conversion that proceeds via B

$$\text{A} \xrightarrow{k_1} \text{B} \xrightarrow{k_2} \text{C} \tag{1.94}$$
$$\underset{k_3}{\overline{\phantom{\text{A}\xrightarrow{k_1}\text{B}\xrightarrow{k_2}}}}$$

The SCN^- anation of $Co(NH_3)_5H_2O^{3+}$ has only recently been recognized to form $(NH_3)_5CoSCN^{2+}$ in parallel with the stable N-bonded $Co(NH_3)_5NCS^{2+}$. The system has been fully analyzed and illustrates well the difficulties in detecting biphasic behavior.[89]

$$Co(NH_3)_5H_2O^{3+} \underset{\searrow}{\overset{\nearrow}{}} \begin{matrix} Co(NH_3)_5SCN^{2+} \\ \downarrow \\ Co(NH_3)_5NCS^{2+} \end{matrix} \tag{1.95}$$

For another example, see Ref. 90.

As the number of reaction steps increases so, of course, does the complexity. A novel example is the six consecutive steps for hydrolysis of a cyanobridged polynuclear complex **9**[91]

$$[Ru(bpz)_3^{2+}(Fe(CN)_5^{3-})_6]^{16-} + 6H_2O \rightarrow Ru(bpz)_3^{2+} + 6Fe(CN)_5H_2O^{3-} \quad (1.96)$$

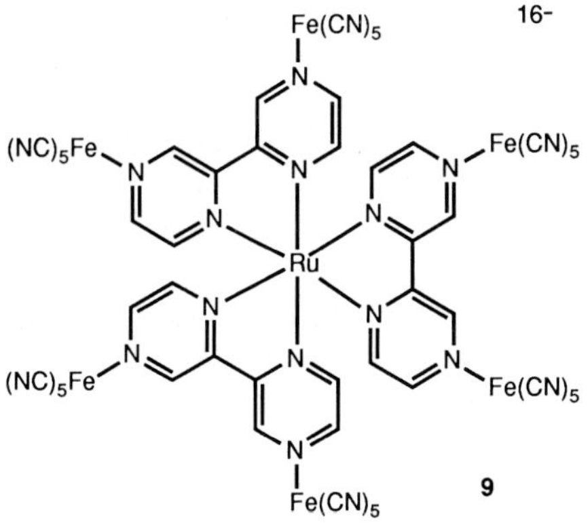

The successive rate constants vary only from 1.3×10^{-3} s^{-1} to 4.6×10^{-4} s^{-1} at $\mu = 0.1$ M and the multistep kinetics are analyzed with a computer. A large excess of dmso is used to aid the dissociation of the polynuclear complex by reacting irreversibly with the product, $Fe(CN)_5H_2O^{3-}$

$$Fe(CN)_5H_2O^{3-} + dmso \rightarrow Fe(CN)_5(dmso)^{3-} + H_2O \quad (1.97)$$

The ambiguities which have been alluded to in this section may sometimes be circumvented by changing the conditions, the concentrations of excess reactants, temperature, pH and so on.

1.6.3 Two-Step Reactions with an Element of Reversibility

Suppose that an irreversible reaction between A and B leading to $>95\%$ product or products, designated D, is examined in the usual way. One of the reactants, B, is held in excess and the loss of A monitored. It is likely that the loss will be a first-order process (rate constant k). At low concentrations of B (but still $\gg$ [A]), the value of k may be proportional to the concentration of B. At higher concentrations of B however this direct proportionality may disappear and eventually k will become independent of [B]. Obviously, a second-order reaction at low reactant concentrations has lost its simplicity at higher reactant concentrations and eventually turned over to first-order in A alone. Such a situation is accommodated by a rate law of the form

$$V = \frac{-d[A]}{dt} = k[A] = \frac{a[A][B]}{1 + b[B]} \quad (1.98)$$

This behavior is sometimes referred to as saturation kinetics. When $b[B] < 1$, the observed second-order is easily understood (rate constant $= a$). When $b[B] \sim 1$ there is a mixed-order and eventually, when $b[B] > 1$, the reaction is first-order in A, rate constant $= a/b$. The k vs $[B]$ plot is described by a hyperbola and (1.98) can be treated directly by a computer program. A favorite approach in the past, and still useful, is to convert (1.98) into a linear form (1.99). A plot of k^{-1} vs $[B]^{-1}$

$$\frac{1}{k} = \frac{1}{a[B]} + \frac{b}{a} \tag{1.99}$$

yields $1/a$ (slope) and b/a (intercept). An example of the rate law (1.98) is shown in the redox reaction (1.100).[92] $Co(edta)^{2-}$ $(= B)$ is used in excess and $a = 4.5\,M^{-1}s^{-1}$ and $b = 831\,M^{-1}$ at 25°C.

$$Co^{II}(edta)^{2-} + Fe(CN)_6^{3-} \rightarrow Co^{III}(edta)^- + Fe(CN)_6^{4-} \tag{1.100}$$

1.6.4 Reaction Schemes Associated with (1.98)

There are a number of possible schemes which may explain the rate behavior associated with (1.98). A single step can be ruled out. At least two consecutive or competitive reactions including one reversible step must be invoked.

(a) Consider the scheme

$$A + B \underset{k_{-1}}{\overset{k_1}{\rightleftharpoons}} C \overset{k_2}{\longrightarrow} D \tag{1.101}$$

which is a very important one in chemistry.[93] If we do not see deviations from a single first-order process it is likely that the first reversible step is much more rapid than the second. At higher concentrations of B the rapid formation of substantial amounts of C may be discernable. If the first step is the more rapid one, C will be in equilibrium with A and B throughout the reaction and

$$\frac{[C]}{[A][B]} = \frac{k_1}{k_{-1}} = K_1 \tag{1.102}$$

will be continually maintained. We shall be monitoring the loss of both A and C or the equivalent gain in one of the products (D).

$$d[D]/dt = k([A] + [C]) = k_2[C] = k_2K_1[A][B] \tag{1.103}$$

From (1.102) and (1.103),

$$k = \frac{k_2K_1[B]}{1 + K_1[B]} \tag{1.104}$$

which is of the form (1.98) with $a = k_2K_1$ and $b = K_1$. In reaction (1.100), C would represent some adduct, now believed to be a cyano bridged Co(III)Fe(II) species, with a formation constant $K_1 = b = 831\,M^{-1}$. This adduct would break down to products with a rate constant $k_2 = a/b = 5.4 \times 10^{-3}\,s^{-1}$

$$Co(edta)^{2-} + Fe(CN)_6^{3-} \underset{}{\overset{K_1}{\rightleftharpoons}} (edta)Co^{III}NCFe^{II}(CN)_5^{5-} \tag{1.105}$$

$$(edta)Co^{III}NCFe^{II}(CN)_5^{5-} \underset{k_{-2}}{\overset{k_2}{\rightleftharpoons}} Co(edta)^- + Fe(CN)_6^{4-} \tag{1.106}$$

The rate constant k_{-2} can be ignored at present.

(b) A related scheme to that of (a) is one in which A and B react directly to form D, but are also in a rapid "dead-end" or nonproductive equilibrium to give C (competitive reactions)

$$A + B \rightleftharpoons C \quad K_1 \tag{1.107}$$

$$A + B \rightleftharpoons D \quad k_3, k_{-3} \text{ (negligible)} \tag{1.108}$$

$$d[D]/dt = k([A] + [C]) = k_3[A][B] \tag{1.109}$$

whence

$$k = \frac{k_3[B]}{1 + K_1[B]} \tag{1.110}$$

This is again equivalent to (1.98). In the reaction (1.100), the reactants still form the adduct, K_1 remains $831\,M^{-1}$, but the reactants interact separately to give products with a rate constant $k_3 = a = 4.5\,M^{-1}\,s^{-1}$.

(c) There is yet another possible explanation for the observed data namely the sequence

$$A \rightleftharpoons C \quad k_4, k_{-4} \tag{1.111}$$

$$C + B \rightarrow D \quad k_5 \tag{1.112}$$

A very useful simplification that can often be made in these systems is to assume that the intermediate C is in a small "steady-state" concentration. Therefore

$$\pm d[C]/dt = 0 \tag{1.113}$$

$$k_4[A] \text{ (gain of C)} = k_{-4}[C] + k_5[C][B] \text{ (loss of C)} \tag{1.114}$$

$$d[D]/dt \simeq k[A] = k_5[C][B] = \frac{(k_5 k_4/k_{-4})[A][B]}{1 + (k_5/k_{-4})[B]} \tag{1.115}$$

which is (1.98) with $a = (k_5 k_4/k_{-4})$ and $b = (k_5/k_{-4})$. In (1.100), $Fe(CN)_6^{4-}$ would rearrange to a reactive form with a rate constant, $k_4 = a/b = 5.4 \times 10^{-3}\,s^{-1}$.

The ratio $k_5/k_{-4} = 831\,M^{-1}$ but the analysis can be taken no further without knowledge of the equilibrium constant for (1.111) (k_4/k_{-4}).

(d) Distinguishing Schemes (a), (b) and (c)

It is relatively easy to spot behavior (c). Plots of k vs [B] are non-linear with [B] $\gg$ [A] but will always remain linear with [A] $\gg$ [B].[94] Saturation kinetics will arise with (a) and (b) whether A or B is the reagent used in excess. Secondly, the value of k_4 calculated with scheme (c) will be independent of the nature of B. Finally the value of chemical intuition cannot be underestimated in resolving these problems. Two different reactive forms of $Fe(CN)_6^{4-}$ are unreasonable (although not impossible) thus tending to rule out Scheme (c) for reaction (1.100). The scheme (c) is more likely to operate for example in the reactions of proteins. The $A \rightleftharpoons C$ transformation in (1.111) would represent a conformational change. Such a mechanism is favored in the oxidation of blue copper proteins. [94]

It is often however very difficult to distinguish between schemes (a) and (b). [95] With scheme (a) there will be an induction period in the appearance of products as C is being built-up, see also Fig. 1.8. This will not be the case with scheme (b) since production of D starts directly from A and B. By examining very carefully the reaction progress at very early times while (1.107) is being set up, it may be possible to distinguish between the two schemes. [96-99] Once the rapid equilibrium has been established however the steady-state kinetics are identical for (a) and (b). Arguments over which of the two schemes is preferred must then be based on chemical or rate considerations, and these are usually equivocal. However they can be used to distinguish (a) and (b) in the reactions (1.100) using the following reasoning. [100] The overall equilibrium constant for (1.100) can be estimated from oxidation potential data. The value (20) is equal to $K_1 k_2 / k_{-2}$ for (a) and k_3 / k_{-3} for (b) where k_{-2} and k_{-3} are second-order rate constants for the steps

or

$$Co(edta)^- + Fe(CN)_6^{4-} \xrightarrow{k_{-2}} (edta)Co^{III}NCFe^{II}(CN)_5^{5-} \tag{1.116}$$

$$Co(edta)^- + Fe(CN)_6^{4-} \xrightarrow{k_{-3}} Co(edta)^{2-} + Fe(CN)_6^{3-} \tag{1.117}$$

This means that either k_{-2} or k_{-3} is $0.21\,M^{-1}\,s^{-1}$. This is a reasonable value for a second-order redox process which (1.117) represents, but is very unlikely for k_{-2} since formation of the bridged adduct must involve Co^{III}–O bond cleavage in $Co(edta)^-$ and such a process would be expected to be much slower (Ch. 4). For this, and other reasons [101, 102] mechanism (b) is strongly preferred.

Since the propensity to form adducts in chemistry is high and these adducts undergo a variety of reactions, the rate law (1.98) is quite common. This is particularly true in enzyme kinetics. [103] In reality, these reaction schemes give biphasic first-order plots but because the first step is usually more rapid, for example between A and B in (1.101) we do not normally, nor do we need to, examine this step in the first instance. [104-106] The value of K_1 in (1.107) obtained kinetically can sometimes be checked directly by examining the rapid preequilibrium before reaction to produce D occurs. In the reactions of Cu(I) proteins with excited Cr and Ru polypyridine complexes, it is considered that (a) and (b) schemes may be operating concurrently. [107]

Finally, we consider a straightforward example of three consecutive steps as in the scheme

$$A + B \rightleftharpoons C + D \qquad k_1, k_{-1} \tag{1.118}$$

$$C + B \rightleftharpoons E + D \qquad k_2, k_{-2} \tag{1.119}$$

$$E + B \rightarrow F + D \qquad k_3 \tag{1.120}$$

$$V = k_3[E][B] = \frac{k_1 k_2 k_3[A][B]^3}{k_2 k_3[B]^2 + k_{-1} k_3[B][D] + k_{-1} k_{-2}[D]} \tag{1.121}$$

This rate equation is derived by assuming a steady-state treatment in which

$$d\,[C]/dt = d\,[E]/dt = 0 \tag{1.122}$$

or alternatively the Christiansen formulation is applied. [108] The inclusion of the concentration of product D in (1.121) indicates that D features in a reversible step that occurs prior to or at the rate-determining stage. Deliberate addition of the product D and observation of retardation of the rate will show whether this scheme is plausible.

This oxidation of inorganic reductants by Cr(VI) show a variety of behaviors based on scheme (1.118–1.120) (A = Cr(VI); C = Cr(V); E = Cr(IV); F = Cr(III); B = reduced form and D = oxidized form of reductant). With B = Ag(I) or Ce(III), the rate law (1.121) is obeyed, indicating that the rate-determining step (1.120) involves Cr(IV). [109, 110]

1.6.5 Two-Step Reactions with Total Reversibility

The scheme

$$A \underset{k_{-1}}{\overset{k_1}{\rightleftharpoons}} B \underset{k_{-2}}{\overset{k_2}{\rightleftharpoons}} C \tag{1.123}$$

forms an important basis for understanding the kinetics of many reactions. [111–113] Although the addition of the reverse terms k_{-1} or k_{-2} (or even worse both) to (1.66) complicates the treatment considerably, some complexity can usually be removed by various subterfuges. The two steps are unlikely to be both purely first-order and any or even all of the rate constants may be functions of the concentrations of other reagents D − G. Provided these are maintained in

$$A \underset{k'_{-1}[E]}{\overset{k'_1[D]}{\rightleftharpoons}} B \underset{k'_{-2}[G]}{\overset{k'_2[F]}{\rightleftharpoons}} C \tag{1.124}$$

excess of A, B or C, the treatment of (1.124) is based on (1.123) with $k_1 = k'_1[D]$, $k_{-1} = k'_{-1}[E]$ etc. and k'_1, k'_{-1} etc. the second-order rate constants for A reacting with D etc. We shall see that with the system (1.123) or (1.124) identical results are obtained whether pure A, pure C, or a mixture is the starting condition, or an equilibrium mixture is perturbed (Sec. 1.8.2).

Biphasic kinetics will in principle be observed with (1.123). The associated first-order rate constants k_I and k_{II} are related to the rate constants of (1.123) by the relationships (Sec. 1.8.2)

$$k_I + k_{II} = k_1 + k_{-1} + k_2 + k_{-2} \tag{1.125}$$

$$k_I k_{II} = k_1 (k_2 + k_{-2}) + k_{-1} k_{-2} \tag{1.126}$$

Only if $K_1 (= k_1/k_{-1})$ and $K_2 (= k_2/k_{-2})$ are known, are all the rate constants calculable. This necessity is removed if one of the steps is pseudo first-order, e.g. A reacts with D in the first step, as we shall now discover.

The interaction of excess concanavalin A [114] (P) with the chromogenic disaccharide p-nitro-phenyl-2-O-α-D-mannopyranosyl-α-D-mannopyranoside(M) **10**(a) displays a spectral-time course shown in Fig. 1.9. [115] It is clearly biphasic and

Concanavalin A (Con A) – There is a binding site for a single D-mannopyranosyl (manp) group not far from the Ca, Mn binuclear site in Con A. Carbohydrate binding is the basis of important biological properties of the protein. The interaction can be probed by attaching a nitrophenyl or a methylumbelliferyl group (Meumb) to the sugar and using spectral or fluorescence monitoring, respectively. A single phase attends the Con A reaction with Meumb-manp, but with the disaccharide Meumb(manp)$_2$ interaction is biphasic. The two mechanisms may be represented as either

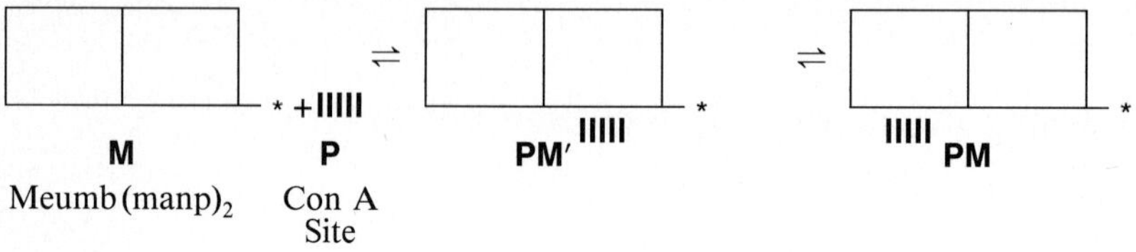

in which binding is followed by an isomerization, or by:

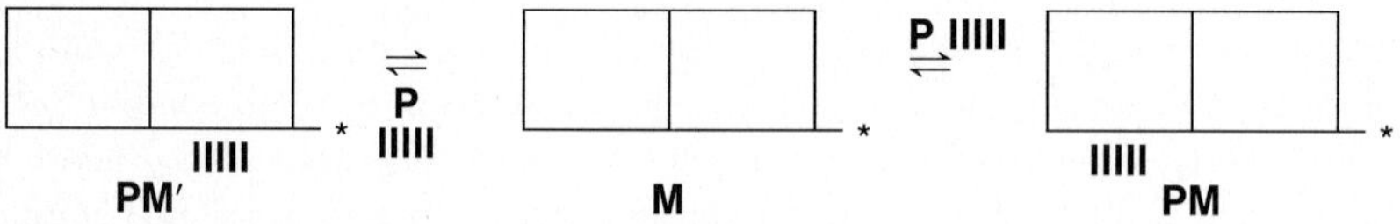

now binding occurs simultaneously with groups on both internal and terminal residues. The asterisk represents the chromophoric group of the disaccharide.

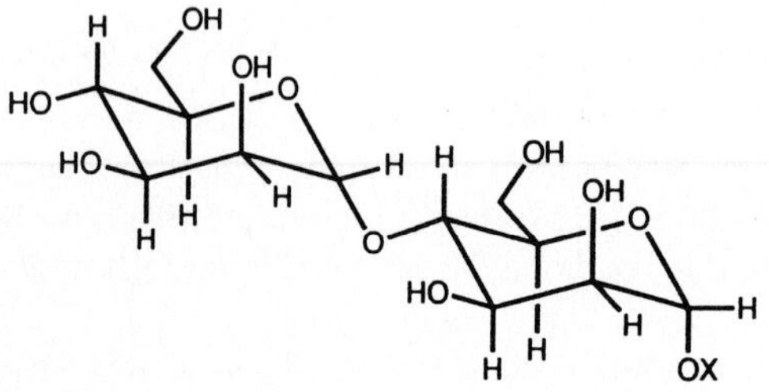

10

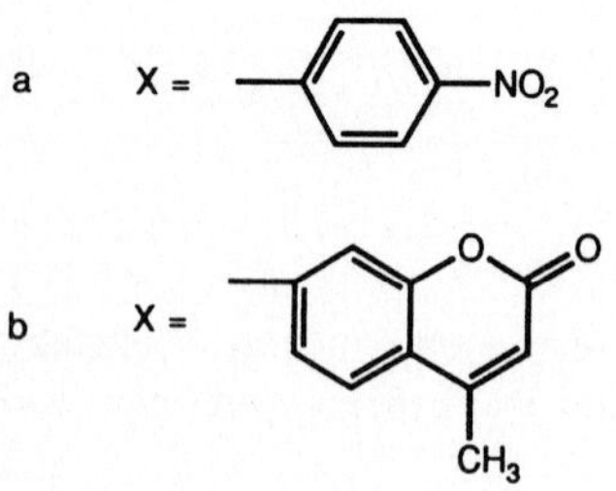

a   X =

b   X =

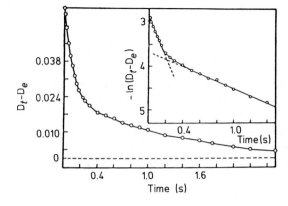

Fig. 1.9 Time dependence of absorbance obtained after mixing Con A (200 μM) and **10** (a) (20 μM). The semi-log plot of the data (inset) shows even clearer the biphasic nature of the reaction. [115] Reprinted with permission from T. J. Williams, J. A. Shafer, I. T. Goldstein and T. Adamson, J. Biol. Chem. **253**, 8538 (1978).

is analyzed by

$$D_t - D_e = \alpha \exp(-k_I t) + \beta \exp(-k_{II} t) \tag{1.127}$$

where α and β ($= D_0 - D_e - \alpha$) are composite parameters. One reaction scheme related to (1.123) with A reacting with D in the first step (therefore k_1 is a second-order rate constant) is

$$M + P \underset{k_{-1}}{\overset{k_1}{\rightleftharpoons}} PM' \underset{k_{-2}}{\overset{k_2}{\rightleftharpoons}} PM \tag{1.128}$$

in which there is a conformational change of the protein-sugar adduct PM' subsequent to its more rapid formation. It is easy to see from (1.125) and (1.126) that

$$k_I + k_{II} = k_1[P] + k_{-1} + k_2 + k_{-2} \tag{1.129}$$

$$k_I k_{II} = k_1(k_2 + k_{-2})[P] + k_{-1}k_{-2} \tag{1.130}$$

From the slopes and intercepts of the two plots of $(k_I + k_{II})$ vs [P] and $k_I k_{II}$ vs [P] in excess, all four rate constants can, in principle, be extracted.

However as was seen in Sec. 1.6.4 another scheme is possible

$$A + B \rightleftharpoons C \tag{1.131}$$

$$A + B \rightleftharpoons D \tag{1.132}$$

which applied to the present system yields

$$PM' \underset{k_{-3}}{\overset{k_3}{\rightleftharpoons}} P + M \underset{k_{-4}}{\overset{k_4}{\rightleftharpoons}} PM \tag{1.133}$$

This is a fundamentally different mechanism. The two adducts PM and PM' are initially formed at relative rates equal to k_4/k_{-3}. In the slow phase PM and *PM equilibrate and their relative concentrations are controlled by the equilibrium constant $k_3 k_4/k_{-3}k_{-4}$. For the scheme (1.133) by substituting $k_3 + k_{-4}$ for $k_1 + k_{-2}$ and $(k_{-3} + k_4)[P]$ for $k_{-1} + k_2$ in (1.125) and (1.126) one obtains

$$k_I + k_{II} = k_3 + k_{-4} + (k_{-3} + k_4)\,[P] \tag{1.134}$$

$$k_I k_{II} = k_3 k_{-4} + (k_3 k_4 + k_{-3} k_{-4})\,[P] \tag{1.135}$$

Comparison of (1.129) and (1.130) with (1.134) and (1.135) shows that the schemes cannot easily be distinguished kinetically. Considerations of spectral changes accompanying the two phases and values of α and β for the two schemes, as well as chemical considerations, strongly support the second interpretation. Both spectral and fluorescence monitoring (using **10(b)**) and the techniques of stopped-flow (starting with P and M) and temperature-jump (starting with an equilibrium mixture of P, M, *PM and PM) have been applied with similar results to this important interaction. The references 115–118 should be consulted for detailed analyses of the system.

The reaction of metal ion M^{n+} with the keto, enol tautomeric mixture of acetylacetone (acacH) in acidic aqueous solution has been treated by a similar approach to that outlined above (see Prob. 16).

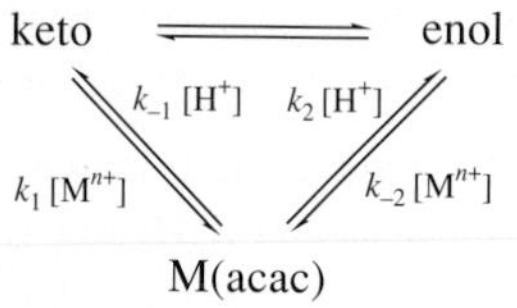

$$
\begin{array}{ccc}
\text{keto} & \rightleftharpoons & \text{enol} \\
& k_{-1}\,[H^+] \quad k_2\,[H^+] & \\
k_1\,[M^{n+}] & & k_{-2}\,[M^{n+}] \\
& M(\text{acac}) &
\end{array}
\tag{1.136}
$$

Simplifications ease the extraction of accurate values for the rate constants. For example, the keto, enol interconversion may sometimes be ignored, and $k_{-2} \gg k_1$ and $k_2 \gg k_{-1}$ are justifiable assumptions. [119–121]

A very important scheme in transition metal chemistry is illustrated by

$$AB \rightleftharpoons A + B \tag{1.137}$$

$$A + C \rightleftharpoons AC \tag{1.138}$$

which arises if the mechanism for interchange of B and C on A is dissociative in character (Sec. 4.2.2). Specifically we might consider the interchange of O_2 and CO coordinated to an iron respiratory protein e.g. myoglobin (PFe)

$$PFe(O_2) \underset{k_{-1}}{\overset{k_1}{\rightleftharpoons}} PFe + O_2 \tag{1.139}$$

$$PFe + CO \underset{k_{-2}}{\overset{k_2}{\rightleftharpoons}} PFe(CO) \tag{1.140}$$

The steady-state assumption is widely applied since the concentration of adduct-free myoglobin, PFe is very small.

$$\pm d\,[PFe]/dt = 0 \tag{1.141}$$

A single first-order reaction is observed and

$$d\,[PFe(CO)]/dt = -d\,[PFe(O_2)]/dt = k_{\text{obs}}\,([PFe(CO)]_e - [PFe(CO)]_t) \tag{1.142}$$

where

$$k_{obs} = \frac{k_1 k_2 [CO] + k_{-1} k_{-2} [O_2]}{k_2 [CO] + k_{-1} [O_2]} \tag{1.143}$$

Further simplification may be possible, dependent on the relative values of the four terms in (1.143). [122] Care must be taken to ensure that both $[O_2]$ and $[CO] \gg [PFe(O_2)]$, so that the concentrations of O_2 and CO remain approximately constant throughout the reaction. The interchange

$$Fe(CN)_5 X^{3-} + Y \rightleftharpoons Fe(CN)_5 Y^{3-} + X \tag{1.144}$$

may be similarly treated. [123]

The schemes considered are only a few of the variety of combinations of consecutive first-order and second-order reactions possible including reversible and irreversible steps. [124] Exact integrated rate expressions for systems of linked equilibria may be solved with computer programs. Examples other than those we have considered are rarely encountered however except in specific areas such as oscillating reactions or enzyme chemistry, and such complexity is to be avoided if at all possible.

1.7 Recapitulation

At this stage we ought to restate briefly the sequences necessary in the construction of the rate law.

1. Decide the reactant (A) whose concentration is most conveniently monitored. Use A in deficiency and determine whether it is totally ($>95\%$) consumed in all the experimental conditions envisaged. If it is not, the reaction is reversible, and should be allowed for in the kinetic treatment.

2. Determine the order of the reaction with respect to A. The value of $[A]$, $\ln [A]$, or $[A]^{-1}$ will probably be linearly related to time, indicating zero-, first-, or second-order, respectively. All other reactants are maintained in constant concentration.

3. Repeat the experiments with different concentrations of the other reactants B, and so on, varied one at a time. Thus determine the order with respect to these also. Plots of $\log k_{obs}$ vs $\log [B]$, and so on, are sometimes useful in giving the reaction orders as slopes (Prob. 5).

4. If complexity is suspected from the kinetic behavior, the effect of products and of possible impurities and the occurrence of side reactions should be considered. Later we shall see that medium composition (Sec. 2.9), temperature and pressure (Sec. 2.3) are other important parameters that affect rate. The rate law incorporating these effects is obtained by further experiments of the type indicated in step 3.

1.8 Relaxation Kinetics

With the availability of perturbation techniques for measuring the rates of rapid reactions (Sec. 3.4), the subject of relaxation kinetics − rates of reaction near to chemical equilibrium − has become important in the study of chemical reactions.[125] Briefly, a chemical system at equilibrium is perturbed, for example, by a change in the temperature of the solution. The rate at which the new equilibrium position is attained is a measure of the values of the rate constants linking the equilibrium (or equilibria in a multistep process) and is controlled by these values.

1.8.1 Single-Step Reactions

Consider a simple equilibrium, second-order in the forward direction and first-order in the reverse:

$$A + B \rightleftharpoons C \qquad k_1, k_{-1} \tag{1.145}$$

After the perturbation, let the final equilibrium concentrations be represented by A, B, and C. At any time t after the perturbation is imposed, and before the final equilibrium is reached, let the concentrations of A, B, and C be $(A - a)$, $(B - b)$, and $(C - c)$. Thus a, b, and c represent deviations from the final equilibrium concentrations. It is apparent from the stoichiometry of the system that

$$a = b = -c \tag{1.146}$$

At time t,

$$d(C - c)/dt = k_1(A - a)(B - b) - k_{-1}(C - c) \tag{1.147}$$

At final equilibrium,

$$dC/dt = 0 = k_1 AB - k_{-1} C \tag{1.148}$$

Since the perturbations are small, the term ab can be neglected for (1.147); then combination of (1.146), (1.147) and (1.148) gives

$$-dc/dt = [k_1(A + B) + k_{-1}] c \tag{1.149}$$

Similarly,

$$-da/dt = [k_1(A + B) + k_{-1}] a \tag{1.150}$$

The shift to the new equilibrium as a result of the perturbation, the *relaxation* of the system, is therefore a first-order process with a first-order rate constant $k = \tau^{-1}$ (1.28) made up of $k_1(A + B) + k_{-1}$.

A treatment similar to that above can be applied to other single equilibria. If the stoichiometry condition akin to (1.146), the zero net rate condition at final equilibrium as in (1.148), and the neglect of squared terms in the deviation concentrations are applied to the rate equation similar to (1.147), it is found that there is always a linear relation of the form

$$-da/dt = ka \qquad (1.151)$$

with a value for k characteristic of the system (Table 1.2), Prob. 12.

Table 1.2. Values of Relaxation Rate Constants (k) for Various Single Equilibria

System	k
$A \rightleftharpoons B$	$k_1 + k_{-1}$ [a]
$2A \rightleftharpoons B$	$4 k_1 [A] + k_{-1}$
$A + B \rightleftharpoons C$	$k_1 ([A] + [B]) + k_{-1}$
$A + C \rightleftharpoons B + C$	$(k_1 + k_{-1}) [C]$
$A + B \rightleftharpoons C + D$	$k_1 ([A] + [B]) + k_{-1} ([C] + [D])$

[a] the symbols k_1, k_{-1} represent the forward and reverse rate constants for all systems; [A], [B], [C] and [D] represent the final equilibrium concentration of these species.

Thus the determination of the relaxation times for a number of different reactant concentrations (estimated *in situ* or from a knowledge of the equilibrium constant) will give both the reaction order and the associated rate constants. It should be noted that the concentrations in (1.149) are equilibrium ones. Algebraic manipulation allows the determination of the rate constants for some systems in Table 1.2 without a knowledge of equilibrium constants. For $2A \rightleftharpoons B$, it can be easily deduced that a plot of k^2 vs $[A]_0$ is linear with a slope of $8 k_1 k_{-1}$ and an intercept k_{-1}^2. The value $[A]_0$ is the stoichiometric concentration $= [A] + 2 [B]$, [126-128] see Prob. 13. Normally, changes in concentration from the perturbation are held to $<10\%$ of the total concentration, an amount which is usually easily monitored. It is perhaps surprising that perturbations imposed can be larger than this for single- and multistep reactions, virtually without loss of the first-order relaxation features. [14,129-132] In any event, analysis of the last portion of a trace arising from a large perturbation will be a valid procedure.

If in the relaxation systems listed in Table 1.2 one of the reactants A or B and one of the products C or D is in large excess, that is if pseudo first-order conditions obtain, the relaxation expression is identical with the rate law obtained starting from pure reactants (1.148). For conditions other than these however, the simplified treatment with relaxation conditions is very evident, as can be seen, for example, in the simple expression for the first-order relaxation rate constant for the $A + B \rightleftharpoons C + D$ scheme compared with the treatment starting from only A and B, and when pseudo first-order conditions cannot be imposed. [41]

1.8.2 Multistep Reactions

One does not often encounter the simple scheme involving only first-order reactions such as (1.152)

$$A \rightleftharpoons B \rightleftharpoons C \qquad (1.152)$$

However, consider the very common and important two-step mechanism (1.153).

$$A + B \underset{k_{-1}}{\overset{k_1}{\rightleftharpoons}} C \underset{k_{-2}}{\overset{k_2}{\rightleftharpoons}} D \tag{1.153}$$

This can be reduced to (1.152) when $[B] \gg [A]$. The difficult situation to analyze arises when the rates associated with the two steps in (1.153) are similar and in addition $[A]_0 \sim [B]_0$ and the reduction to (1.152) cannot be made. This case will be treated first. The objective is to express da/dt and dc/dt each in terms of a and c, which are the deviations from equilibrium concentrations symbolized A, B, C, and D. These provide the basis for the two relaxation times observable with the system. Now

$$a = b = -(c + d) \tag{1.154}$$

$$d(A - a)/dt = -k_1(A - a)(B - b) + k_{-1}(C - c) \tag{1.155}$$

$$-da/dt = k_1(A + B)a - k_{-1}c \tag{1.156}$$

$$d(C - c)/dt = k_1(A - a)(B - b) - k_{-1}(C - c) - k_2(C - c) + k_{-2}(D - d) \tag{1.157}$$

$$k_{-2}D = k_2 C \tag{1.158}$$

$$-dc/dt = -k_1(A + B)a + k_{-1}c + k_2 c + k_{-2}(a + c) \tag{1.159}$$

$$-dc/dt = -(k_1(A + B) - k_{-2})a + (k_{-1} + k_2 + k_{-2})c \tag{1.160}$$

Equations (1.156) and (1.160) are of the forms

$$-da/dt = \alpha_{11} a + \alpha_{12} c \tag{1.161}$$

$$-dc/dt = \alpha_{21} a + \alpha_{22} c \tag{1.162}$$

Making the substitution $a = Xe^{-kt}$ and $c = Ye^{-kt}$ gives

$$kXe^{-kt} = \alpha_{11} Xe^{-kt} + \alpha_{12} Ye^{-kt} \tag{1.163}$$

$$kYe^{-kt} = \alpha_{21} Xe^{-kt} + \alpha_{22} Ye^{-kt} \tag{1.164}$$

or [133]

$$(\alpha_{11} - k)a + \alpha_{12} c = 0 \tag{1.165}$$

$$(\alpha_{22} - k)c + \alpha_{21} a = 0 \tag{1.166}$$

Solving for k by eliminating a and c gives

$$k^2 - (\alpha_{11} + \alpha_{22})k + \alpha_{11} \alpha_{22} - \alpha_{12} \alpha_{21} = 0 \tag{1.167}$$

The two first-order rate constants k_I and k_{II} associated with this scheme are given by

$$2k_I = \alpha_{11} + \alpha_{22} + [(\alpha_{11} + \alpha_{22})^2 - 4(\alpha_{11} \alpha_{22} - \alpha_{12} \alpha_{21})]^{1/2} \tag{1.168}$$

and

$$2 k_{II} = \alpha_{11} + \alpha_{22} - [(\alpha_{11} + \alpha_{22})^2 - 4(\alpha_{11}\alpha_{22} - \alpha_{12}\alpha_{21})]^{1/2} \qquad (1.169)$$

and

$$\alpha_{11} = k_1 (A + B) \qquad (1.170)$$

$$\alpha_{12} = -k_{-1} \qquad (1.171)$$

$$\alpha_{22} = k_{-1} + k_2 + k_{-2} \qquad (1.172)$$

$$\alpha_{21} = k_{-2} - k_1 (A + B) \qquad (1.173)$$

These are cumbersome equations to use but (1.168) and (1.169) can be computer treated. Alternatively, it is easily deduced that

$$k_I + k_{II} = k_1 (A + B) + k_{-1} + k_2 + k_{-2} \qquad (1.174)$$

and

$$k_I k_{II} = k_1 (k_2 + k_{-2})(A + B) + k_{-1}k_{-2} \qquad (1.175)$$

From plots of $(k_I + k_{II})$ vs [A + B] and $k_I k_{II}$ vs [A + B], all rate constants can be determined (see Sec. 1.6.5). [134]

This treatment yields the time course of the relaxation which is of most concern to us, but ignores the relative magnitudes of the relaxations (contained in the X and Y terms in (1.163)). These latter are complex functions of reaction enthalpies and absorbance coefficients, but can yield equilibrium constants for the two steps. [135,136] However, relaxation data are much less used for thermodynamic than kinetic information.

A common simplification arises when the bimolecular step in (1.153) equilibrates rapidly compared with the unimolecular step (it may, for example, be a proton-base reaction). This means that the change in concentrations of A, B, and C due to the first process in (1.153) will have occurred before D even starts to change. The relaxation time τ_I associated with it will therefore be the same as if it were a separated equilibrium:

$$\tau_I^{-1} = k_I = k_1 (A + B) + k_{-1} \qquad (1.176)$$

The changes of concentration of C and D resulting from the second equilibrium are however coupled to the first, and the associated relaxation time τ_{II} might be expected to be a more complex function. It is however fairly easily derived.

$$d(D - d)/dt = k_2(C - c) - k_{-2}(D - d) \qquad (1.177)$$

from which

$$dd/dt = k_2 c - k_{-2} d \qquad (1.178)$$

Now we must express c in terms of d, so that an equation relating dd/dt and d only may be obtained. Since the first equilibrium is always maintained, compared with the second,

$$k_1(A - a)(B - b) = k_{-1}(C - c) \qquad (1.179)$$

$$k_1(Ab + Ba) = k_{-1}c \qquad (1.180)$$

Since

$$-a = -b = c + d \tag{1.181}$$

$$-k_1(A + B)(c + d) = k_{-1}c \tag{1.182}$$

$$c = \frac{-k_1(A + B)d}{k_1(A + B) + k_{-1}} \tag{1.183}$$

$$\frac{dd}{dt} = \frac{-k_1k_2(A + B)d}{k_1(A + B) + k_{-1}} - k_{-2}d \tag{1.184}$$

$$\tau_{II}^{-1} = k_{II} = k_{-2} + \frac{k_1k_2(A + B)}{k_1(A + B) + k_{-1}} \tag{1.185}$$

Equations (1.176) and (1.185) can be derived from the general expression (1.167) by using the relationship $k_1(A + B) + k_{-1} \gg k_{-2} + k_2$ and making the approximation $(1 - x)^{1/2} \sim 1 - x/2$. The approximate approach is often used when the direct product of a bimolecular reaction undergoes a slower change (isomerization or conformational reaction). Both fast and slow relaxations are analyzed by (1.176) and (1.185) using Fig. 1.10(a) which indicates the variation of k_I and k_{II} with [A] + [B] on the basis that $k_I > k_{II}$ ($K_1 = k_1/k_{-1}$). Examples of its occurrence are in the reactions[137]

$$\text{Cu(II)bleomycin + DNA} \underset{7.5 \times 10^3 \text{ M}^{-1}\text{s}^{-1}}{\overset{2.6 \times 10^6 \text{ M}^{-1}\text{s}^{-1}}{\rightleftharpoons}} \text{outside-bound complex}$$

$$\underset{8.4 \text{ s}^{-1}}{\overset{45 \text{ s}^{-1}}{\rightleftharpoons}} \text{intercalated complex} \tag{1.186}$$

and[138]

$$\text{E + S} \underset{50 \text{ M}^{-1}}{\overset{K_1 =}{\rightleftharpoons}} \text{*ES} \underset{1.0 \times 10^3 \text{ s}^{-1}}{\overset{3.3 \times 10^3 \text{ s}^{-1}}{\rightleftharpoons}} \text{ES} \tag{1.187}$$

(fast)

$$\text{E + P} \underset{100 \text{ M}^{-1}}{\overset{K_1 =}{\rightleftharpoons}} \text{*EP} \underset{1.7 \times 10^3 \text{ s}^{-1}}{\overset{1.2 \times 10^3 \text{ s}^{-1}}{\rightleftharpoons}} \text{EP} \tag{1.188}$$

(fast)

Reaction (1.187) represents the binding of the substrate (S), Gly-L-Phe and (1.188) is binding of the product (P), L-Phe, to a colored derivative of carboxypeptidase (Chap. 8. Zn) ((E), arsanilazotyrosine-248 labelled) **11**. Only the slow step has been analyzed (starting in either direction) and it conformed to (1.185). Values for K_1 (which equals k_1/k_{-1} in (1.153)), k_2 and k_{-2} for the formation of both ES and EP are obtained. The two directions are isolable because the conversion of ES to EP is relatively slow ($k_{cat} = 0.01$ s^{-1}).[138]

The relaxation approach has played an important role in our understanding of the mechanisms of complex formation in solution (Chap. 4)[139,140] The use of computer programs has now eased the study of multiple equilibria. For example, four separate relaxation effects with τ's ranging from 100 μs to 35 ms are observed in a temperature-jump study of the reactions of Ni^{2+} with flavin adenine dinucleotide (fad) (Eqn. (8.121)). The complex relaxation

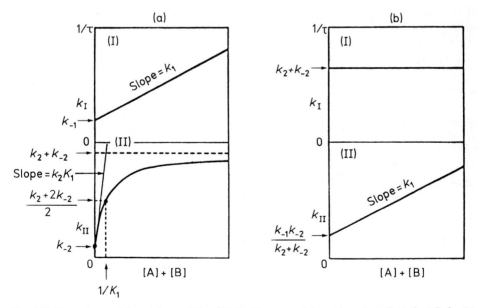

11

equations may be computer-analyzed and the associated four steps rationalized in terms of initial phosphate attachment to Ni^{2+} followed by intramolecular binding of the other two ligand sites of the coenzyme. [141, 142]

Only rarely is the situation encountered when the first step in (1.153) is *slower* than the second transformation. It is easily recognized from the dependence of k_I and k_{II} on [A] + [B], Fig. 1.10(b). [143] The difficulties of distinguishing Scheme (1.153) from (1.189):

$$C \rightleftharpoons A + B \rightleftharpoons D \qquad (1.189)$$

have been already alluded to and the relaxation approach to this problem thoroughly examined. [144] Comprehensive tables of rate expressions for a variety of schemes both possible and unlikely in practice are collected in the texts by Bernasconi and Hiromi (Prob. 14).

Fig. 1.10 Dependence of k_I and k_{II} on [A] + [B] for the sequential reactions $A + B \rightleftharpoons C \rightleftharpoons D$ for (a) $A + B \rightleftharpoons C$ the more rapid and (b) $C \rightleftharpoons D$ the more rapid step. [143]

1.9 Exchange Kinetics

Somewhat in the same vein as relaxation kinetics, there is a simplicity about the manipulation of isotopic exchange results that makes the method an important and useful tool for studying mechanism. When AX and BX, both containing a common atom or group of atoms X, are mixed, there will be a continual interchange of X between the two environments that may range from extremely rapid to negligibly slow. This exchange will go undetected unless we tag AX or BX with some labeled X, which we denote by *X:

$$AX + B*X \rightleftharpoons A*X + BX \tag{1.190}$$

Consider a mixture of AX and BX at chemical equilibrium. When, for example, radioisotopes are used as tracers, they are injected into the equilibrium mixture in the form of a very small amount of B*X. At various times, either (BX + B*X) or (AX + A*X) is separated from the mixture and analyzed. When nmr line broadening is used to monitor the exchange the tracer is already present e. g. ^{1}H or ^{17}O (or an additional amount can be added) and the exchange is monitored *in situ* and assessed from the shape of the nmr signals (Sec. 3.9.6). If the concentration of (AX + A*X) is a and the concentration of (BX + B*X) is b, and the fraction of exchange at time t is F, it is not difficult to show that the gross or overall rate of X transfer between AX and BX, V_{exch} (M s^{-1}) is given by[145]

$$V_{exch} = \frac{-ab \ln(1 - F)}{(a + b)\, t} = k_{obs} \frac{ab}{(a + b)} \tag{1.191}$$

F will equal x/x_e where x and x_e represent the mole fractions of A*X at sampling and equilibrium times. The rate of exchange will be identical in both directions in (1.190) and be always first-order, comparable to the situation with relaxation kinetics, a relationship which has been explored.[146] The equation (1.191) is modified slightly when isotope effects are included.[147] The rate of exchange will depend on the concentrations a, b, [H$^+$], and so on, in a manner that determines the rate expression. The exchange rate is measured with different concentrations of AX, BX, [H$^+$] etc., and the rate law is constructed exactly as in the initial-rate or stationary-state methods (Prob. 17). A popular method for treating (1.191) has been by the use of a plot of $-\ln(1 - F)$ vs time. For a single exchange process this will be linear, from the slope of which k_{obs} is obtained.

The exchange of Mn between MnO$_4^-$ and MnO$_4^{2-}$ has been followed using the ^{54}Mn radioisotope and quenched-flow methods (Sec. 3.3.2).[148] The results are shown in Table 1.3, from which it is apparent that

$$V_{exch} = k\, [MnO_4^-]\, [MnO_4^{2-}] \tag{1.192}$$

Even complex rate laws may be easily constructed by examining the dependence of V_{exch} on the concentrations of the various species in solution. The rate of exchange of Ni between Ni^{2+} and Ni(edta)$^{2-}$ obeys the rate law

$$V_{exch} = k_1 [Ni^{2+}]\, [Ni(edta)^{2-}] + k_2 [Ni^{2+}]\, [Ni(edta)^{2-}]\, [H^+] + k_3 [Ni(edta)^{2-}]\, [H^+]$$

$$+ k_4 [Ni(edta)^{2-}]\, [H^+]^2 + k_5 [Ni(edta)^{2-}]\, [H^+]^3 \tag{1.193}$$

Table 1.3. Dependence of ^{54}Mn Exchange Rate on Concentrations of MnO_4^{2-} and MnO_4^- at 0.1 °C in 0.16 M NaOH

$10^5 \times [MnO_4^{2-}]$ M	$10^5 \times [MnO_4^-]$ M	$t_{1/2}$ exch s	$10^6 \times V_{exch}$ M s^{-1}	$10^{-2} \times k$ M^{-1} s^{-1}
4.3	4.8	10.6	1.5	7.2
4.1	4.8	11.2	1.4	7.0
4.6	9.7	6.6	3.3	7.3
4.5	14.6	5.3	4.5	6.8
4.3	19.4	4.3	5.7	7.6
4.2	24.3	3.2	7.8	7.3
4.1	34.0	2.5	10.2	6.5
1.0	9.7	9.2	0.69	7.3
2.3	9.5	9.0	1.4	6.5
4.6	9.7	6.6	3.3	7.3
10.1	9.7	4.9	6.9	7.1
19.7	9.7	3.1	15	7.9
33.0	337	0.25	830	7.5
195	188	0.26	2550	7.0
29	381	0.25	750	6.8
			Average	7.1 ± 0.3

The five terms simply represent paths through which exchange can occur (Sec. 4.4.3).[149]

If there is more than one exchanging atom of X in the interacting molecules, for example AX_n exchanging with BX_m, the rate expression (1.191) is modified accordingly, with a and b replaced by na and mb respectively. This applies only when the nX or mX atoms are equivalent. In basic solution the vanadium(V) ion, VO_4^{3-} exchanges oxygen with solvent H_2O. The plot of $\ln(1 - F)$ is linear with time for at least four half-lives and it can be shown that all four oxygens in VO_4^{3-} exchange, from the distribution of ^{18}O between VO_4^{3-} and H_2O at equilibrium. They are thus equivalent.[150]

$$V_{exch} = k_{obs} \frac{4 [VO_4^{3-}] [H_2O]}{4 [VO_4^{3-}] + [H_2O]} \tag{1.194}$$

and since $[H_2O] \gg 4 [VO_4^{3-}]$

$$V_{exch} = 4 k_{obs} [VO_4^{3-}] \tag{1.195}$$

A special and important type of exchange arises when one of the exchanging species is the solvent. The rate of exchange of solvent molecules, S, between free and metal-coordinated solvent has been studied for a large variety of metal ions and solvents.[151] A good deal of confusion has arisen over the definition and meaning of the exchange rate constant in such systems. It has been recently shown,[152,153] surprisingly, that the rates of isotopically labelled solvent exchange of all n ligands in MS_n^{2+} and of a particular one of these n ligands are the same. Previous divison of k_{exch} by a statistical factor of n should not have been carried out.[154] This can be illustrated by reference to the exchange of H_2O between $Pt(H_2O)_4^{2+}$ and free solvent, Fig. 1.11.[155] From (1.191)

$$V_{exch} = k_{obs}\frac{4[Pt(H_2O)_4^{2+}][H_2O]}{4[Pt(H_2O)_4^{2+}] + H_2O} \sim 4k_{obs}[Pt(H_2O)_4^{2+}] = k[Pt(H_2O)_4^{2+}] \quad (1.196)$$

k is the rate constant for exchange of any H_2O molecule

$$Pt(H_2O)_4^{2+} + H_2{}^*O \xrightarrow{k} Pt(H_2O)_3(H_2{}^*O)^{2+} + H_2O \qquad (1.197)$$

while k_{obs} is the rate constant for exchange of a particular

$$Pt(H_2{}^*O)_3(H_2O) + H_2{}^*O \xrightarrow{k_{obs}} Pt(H_2{}^*O)_4 + H_2O \qquad (1.198)$$

ligand and equals τ_{exch}^{-1} ($\tau_{exch} = 116$ s from Fig. 1.11). k therefore equals $4k_{obs}$.

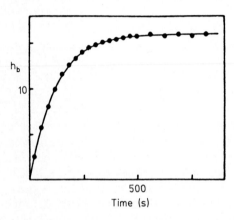

Fig. 1.11 Increase with time of height h_b (arbitrary units) of the ^{17}O nmr signal from coordinated water in $Pt(H_2O)_4^{2+}$ when treated with ^{17}O-enriched water. Fast injection of 0.7 g of $Pt(H_2O)_4(ClO_4)_2$ (0.58 M) and $HClO_4$ (3.5 M) into 0.55 g of 20% ^{17}O-enriched water at 50.6°C was employed.[155] Reprinted with permission from L. Helm, L. I. Elding and A. E. Merbach, Inorg. Chem. **24**, 1719 (1985). © 1985, American Chemical Society.

When the molecule contains more than one type of exchanging atom, but the associated exchange rates differ widely, there is no problem in treating this system as separate single exchange steps (see the markedly different oxygen exchange types in $Mo_3O_4(H_2O)_9^{4+}$, **6** in Chap. 8[156]). However when the exchange rates are similar, their resolution and treatment is much more complex. This has been a vexing problem in the study of oxygen exchange between H_2O and metal-coordinated oxalate. There are two kinetically distinct types of coordinated oxygen in $Rh(en)_2C_2O_4^+$ (**12**). The slower exchange is attributed to inner/outer oxygen interchange which is presumed to occur before inner oxygens can exchange with solvent.

12

The rate processes differ only slightly and the treatment resembles that for Sec. 1.8.2.[157]

In recent years there have been relatively few studies of isotopic exchange using radioisotopes,[158] many more using ^{18}O labelling[159] and a large number probed, both qualitatively and quantitatively by using the nmr method,[151] which has the decided advantage that *in situ* monitoring can be used (Chap. 3). Two overriding values of exchange reactions cannot be overemphasized. They *must* take place if a kinetic path exists (i.e. thermodynamics are not a consideration) and the associated very small driving force makes them easier to interpret than net chemical changes.

1.10 The Inclusion of [H$^+$] Terms in the Rate Law

So far only the reactants directly involved in a reaction have been considered in their contribution to the rate law. Added "inert" cations and anions can sometimes contribute in a profound way by modifying the major reactants (e.g. by ion pairing) but usually the effects of their concentrations on the rates of reactions are best accommodated by the general theories of the effect of ionic strength on the reaction rate (Sec. 2.9.1).

Many reaction rates are affected by the pH of the solution. The modification of the rate as the pH is changed can be ascribed to formation of different species with different reactivities. Thus, the oxygen exchange between oxions and water is faster in acid because of the enhanced reactivity of protonated oxions. It is therefore essential that as wide a range of pH as possible be studied so as to detail a full reaction scheme, and thus delineate the reactive forms of the reactants. In no area is this recognized as important as in enzyme kinetics.[103] Assigning [H$^+$] terms in the rate law presents little problem, the rate constant for the reaction being simply measured at a number of hydrogen ion concentrations. The [H$^+$] may be in excess over that of other reagents, or alternatively the solutions may be buffered. In both cases, no change of pH occurs during the reaction. Since this is such an important parameter in its effect on rates, we shall discuss in some detail the most common types of behavior (rate/pH profiles) encountered. The determination of rate constants at even a few pH values can often indicate the extent and type of H$^+$ (or OH$^-$) involvement. It may in certain cases be necessary to separate a "medium" from a "mechanistic" effect of [H$^+$] on the rate (Sec. 2.9.2).

1.10.1 One Monoprotic Reactant, One Acid-Base Equilibrium

If the profile of the observed or the intrinsic rate constant plotted against pH resembles the profile for an acid-base titration curve, this strongly suggests that one of the reactants is involved in an acid-base equilibrium in that pH range. Such behavior is fairly common and is illustrated by the second-order reaction between the Co(II)-trien complex and O$_2$ (Fig. 1.12).[160] The limiting rate constants at the higher and low acidities correspond to the acidic and basic forms of the Co(II) reactant, probably,

$$\text{Co(trien)(H}_2\text{O)}_2^{2+} \;\rightleftharpoons\; \text{Co(trien)(H}_2\text{O)OH}^+ + \text{H}^+$$
$$\quad + \qquad\qquad\qquad\qquad +$$
$$\quad \text{O}_2 \qquad\qquad\qquad\qquad \text{O}_2 \qquad\qquad\qquad\qquad\qquad (1.199)$$
$$\quad \downarrow \qquad\qquad\qquad\qquad \downarrow$$
$$\text{products} \qquad\qquad\quad \text{products}$$

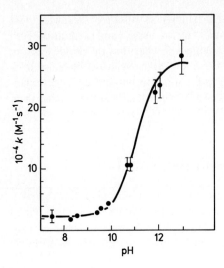

Fig. 1.12 The pH dependence of the first stage of the reaction between Co(II)-trien complex and O_2. The solid line represents Eqn. (1.207) with $K_{AH} = 6 \times 10^{-12}$ M, which is a spectrally determined value. [160]

The rate law and the general reaction scheme associated with this system are easily derived. Consider the reaction of an acid AH and its conjugate base A with a substrate B (any charges are omitted from these for convenience):

$$AH \rightleftharpoons A + H^+ \qquad K_{AH} \tag{1.200}$$

$$A + B \rightarrow \text{products} \qquad k_A \tag{1.201}$$

$$AH + B \rightarrow \text{products} \qquad k_{AH} \tag{1.202}$$

Invariably, the total concentration of A and AH is monitored, and so the rate expression is formulated in terms of $([A] + [AH])$

$$-d([AH] + [A])/dt = k([AH] + [A])[B]^n \tag{1.203}$$

where k is the experimental rate constant, first-order if $n = 0$, second-order if $n = 1$, and so on. The rate is also the sum of the contributions of AH and A.

$$-d([AH] + [A])/dt = k_{AH}[AH][B]^n + k_A[A][B]^n \tag{1.204}$$

Combining (1.203) and (1.204), we obtain

$$k = (k_{AH}[AH] + k_A[A])/([AH] + [A]) \tag{1.205}$$

Using

$$K_{AH} = [A][H^+]/[AH] \tag{1.206}$$

yields

$$k = (k_{AH}[H^+] + k_A K_{AH})/([H^+] + K_{AH}) \tag{1.207}$$

The full curve of Fig. 1.12 is drawn in accordance with (1.207). Considering limits, we find $k = k_{AH}$ when $[H^+] \gg K_{AH}$ and $k = k_A$ with $[H^+] \ll K_{AH}$. The full S-shaped profile will be observed in reactions of acid/base pairs where both forms are attainable and show different reactivities. Examples are hydroxy and aqua complexes as typified by the example shown in (1.199), O_2^- and HO_2, [161, 162] CN^- and HCN, $Au(NH_3)_4^{3+}$ and $Au(NH_3)_3NH_2^{2+}$, [163] and others. In rare cases, an unknown pK_{AH} may be determined kinetically (Prob. 20).

When one of the forms predominates over the pH range of investigation, yet the other form is much more reactive, only one limiting rate constant is obtained.

$[A] \gg [AH]; [H^+] \ll K_{AH},$

$$k = k_A + (k_{AH}[H^+])/K_{AH} \tag{1.208}$$

$[AH] \gg [A]; [H^+] \ll K_{AH},$

$$k = k_{AH} + (k_A K_{AH})/[H^+] \tag{1.209}$$

Such behavior is more common than the full rate/pH profile of (1.207). Equation (1.208) is observed in acid catalysis [164-166] and (1.209) in base catalysis. [167] The rate constant for the reaction of only one of the two forms can be obtained directly, that is, k_A in (1.208) and k_{AH} in (1.209). Ancillary information on K_{AH} is required to assess the rate constant of the acid-base partner. The absence of reliable data on K_{AH} can pose a problem in assessing the missing rate constant. [167]

Observation of the semblance of an S-shaped profile has been used to estimate the pK_{AH} value for an acid/base pair which may be difficult to obtain directly. This may be as the result of one of the pairs being unstable (polynuclear formation or hydrolysis, for example). [168, 169] This approach must however be used with care since the errors are likely to be larger than when the whole, or a goodly portion, of the curve is defined. [164, 165] See Problem 18.

A serious ambiguity in the interpretation of the rate law,

$$V = k[A][B][H^+]^n \tag{1.210}$$

exists when A and B are both basic and n is one or greater (acid hydrolysis). The actual species involved in the rate determining step, AH with B or BH with A in the case of $n = 1$ cannot usually be assessed on the basis of kinetics but may sometimes be differentiated by resort to plausibility (Sec. 2.1.7). A similar problem arises when we consider reaction between AH and BH and $n = -1$ in (1.210) (base hydrolysis).

1.10.2 Two Acid-Base Equilibria

When the sigmoidal shape of the rate constant/pH profile associated with (1.207) or the simpler derivatives (1.208) or (1.209) give way to a bell-shape or inverted bell-shape plot, the reactions of at least three acid-base-related species (two equilibria) have to be considered. This may involve acid-base forms of (a) one reactant or (b) two different reactants.

(a) Diprotic reactant A. Consider the scheme

$$AH_2 \rightleftharpoons AH + H^+ \qquad K_{AH_2} \qquad (1.211)$$

$$AH \rightleftharpoons A + H^+ \qquad K_{AH} \qquad (1.212)$$

$$A \rightarrow products \qquad k_A \qquad (1.213)$$

$$AH \rightarrow products \qquad k_{AH} \qquad (1.214)$$

$$AH_2 \rightarrow products \qquad k_{AH_2} \qquad (1.215)$$

in which the rate constants may be first-order or pseudo first-order. Usually the products from the three steps are identical or at least pH-related. The observed rate constant k at any $[H^+]$ can be shown by reasoning similar to that used in developing (1.207) to be:

$$k = \frac{k_{AH_2}[H^+]^2 + k_{AH}K_{AH_2}[H^+] + k_A K_{AH}K_{AH_2}}{[H^+]^2 + K_{AH_2}[H^+] + K_{AH}K_{AH_2}} \qquad (1.216)$$

If the species AH reacts more rapidly or more slowly than either A or AH_2, a bell shape or inverted bell shape respectively results for the k/pH profile.

The ruthenium(III) complex of edta in which the ligand acts only as a five-coordinate species and in which an acetate arm remains free, exists in three pH-related forms:

$$
\begin{array}{lll}
k_{AH_2} & (AH_2) & Ru(edtaH)H_2O \xrightarrow[<10\ M^{-1}s^{-1}]{SCN^-} Ru(edtaH)(SCN)^- \\
 & & \quad\quad 10^{-2.4}\updownarrow \\
k_{AH} & (AH) & Ru(edta)H_2O^- \xrightarrow[270\ M^{-1}s^{-1}]{SCN^-} Ru(edta)(SCN)^{2-} \\
 & & \quad\quad 10^{-7.6}\updownarrow \qquad\qquad \xrightarrow[<10\ M^{-1}s^{-1}]{SCN^-} \\
k_A & (A) & Ru(edta)OH^{2-}
\end{array}
\qquad (1.217)
$$

The second-order rate constants for thiocyanate anation vs pH are shown in Fig. 1.13. The full line represents (1.216) with the values shown in scheme (1.217).[170] This profile had been earlier recognized in the ring closure of the three analogous pH-related forms of Co(III)-edta to give Co(edta)$^-$ in which the edta is completely coordinated.[171] In the Co(III) case the reactivities of the three forms are much closer. A plot of $k\{[H^+]^2 + K_{AH_2}[H^+] + K_{AH}K_{AH_2}\}$ vs $[H^+]$ is a quadratic curve from which k_{AH_2}, K_{AH} and k_A can be obtained.[172,173]

If the predominant form of the reactant is AH in the pH region under examination, that is $K_{AH_2} \gg [H^+] > K_{AH}$, then only the middle term of the denominator in (1.216) is important. Thus

$$k = (k_{AH_2}[H^+]/K_{AH_2}) + k_{AH} + (k_A K_{AH}/[H^+]) \qquad (1.218)$$

This describes the behavior of a number of complexes towards hydrolysis, for example CrX^{2+} (Sec. 4.3.1).

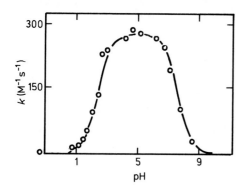

Fig. 1.13 The pH dependence of the reaction between the Ru(III)-edta complex and SCN⁻ at 25 °C ($\mu = 0.2$ M).[170] Reprinted with permission from T. Matsabura and C. Creutz, *Inorg. Chem.* **18**, 1956 (1979). © (1979) American Chemical Society.

(b) Two protic reactants. Suppose now that we have two reactants both of which may be involved in an acid-base equilibrium

$$AH \rightleftharpoons A + H^+ \qquad K_{AH} = [A][H^+]/[AH] \qquad (1.219)$$

$$BH \rightleftharpoons B + H^+ \qquad K_{BH} = [B][H^+]/[BH] \qquad (1.220)$$

In principle (but not often in practice) four reactions are possible, their importance dependent on pH

$$A + B \rightarrow products \qquad k_A^B \qquad (1.221)$$

$$AH + B \rightarrow products \qquad k_{AH}^B \qquad (1.222)$$

$$A + BH \rightarrow products \qquad k_A^{BH} \qquad (1.223)$$

$$AH + BH \rightarrow products \qquad k_{AH}^{BH} \qquad (1.224)$$

All four products are likely to be the same or at least pH-related. The apparent second-order association rate constant k is given by

$$k(A + AH)(B + BH) = k_A^B[A][B] + k_{AH}^B[AH][B] + k_A^{BH}[A][BH] + k_{AH}[AH][BH] \qquad (1.225)$$

By using (1.219) and (1.220) and a little algebraic manipulation one obtains (1.226)

$$k = \frac{k_A^B}{[1 + ([H^+]/K_{AH})][1 + ([H^+]/K_{BH})]} + \frac{k_{AH}^B K_{BH}/K_{AH}}{[1 + [H^+]/K_{AH})][1 + (K_{BH}/[H^+])]} +$$

$$\frac{k_A^{BH}}{[1 + ([H^+]/K_{AH})][1 + (K_{BH}/[H^+])]} + \frac{k_{AH}^{BH}}{[1 + (K_{AH}/[H^+])][1 + (K_{BH}/[H^+])]} \qquad (1.226)$$

The full equation (1.226) reduces to (1.207) when K_{BH} is small, that is when B is aprotic. The observed rate constant k approaches limiting values of k_{AH}^{BH} at high [H⁺] and k_A^B at low [H⁺]. The total profile resembles a bell-shape or inverted bell-shape which may be symmetrical with either a maximum or a minimum at $k \sim k_A^{BH}$ or $k_{AH}^B K_{BH}/K_{AH}$ and at [H⁺] $= (K_{AH}K_{BH})^{1/2}$. The bell-shape behavior is beautifully illustrated in Fig. 1.14.[174] It is clear from the kinetic form that it is not easy to distinguish the reaction of AH with B from A with BH or a mixture of both. The calculated values of k_A^{BH} and k_{AH}^B from (1.226) do however differ by a factor of K_{BH}/K_{AH} and this ratio may be sufficiently different from unity to transform the calculated value of either k_A^{BH} or k_{AH}^B above the diffusion-controlled limit, and therefore unacceptable (Prob. 19).[174,175] Similarly we can guess that the combination of aquacobalamin with CNO^- is the correct reactive pair (rather than hydroxocobalamin with HCNO) since the hydroxo form is much less reactive than the aqua form with an aprotic ligand such as SCN^- ion.[176] For a final example see Ref. 163.

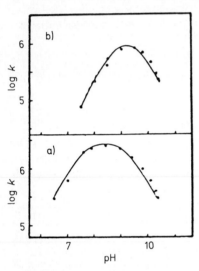

Fig. 1.14 (a) The pH dependence of the rate constants for the association of carbonic anhydrase B (E) and p-nitrobenzenesulfonamide (S). The reaction is monitored by using stopped-flow and the quenching of a tryptophan fluorescence in the protein which occurs when sulfonamides bind. The full line fits Eqn. (1.226) with $k = 3.5 \times 10^6 M^{-1}s^{-1}$, $pK_E = 7.5$ and $pK_s = 9.3$.[174] The plot in (b) refers to a similar interaction with the carboxymethylated derivative of carbonic anhydrase (Prob. 19). Aromatic sulfonamides are powerful inhibitors of the action of the enzyme and are useful probes of the site characteristics.

An interesting situation arises when the molecules A and B are identical. Now we are considering two protic species reacting with one another leading perhaps to disproportionation[162,177] or dimerization[178]

$$HO_2^{\bullet} \rightleftharpoons H^+ + O_2^{\bar{\bullet}} \qquad\qquad K_{HO_2^{\bullet}} = K_{AH} \qquad (1.227)$$

$$HO_2^{\bullet} + HO_2^{\bullet} \rightarrow H_2O_2 + O_2 \qquad\qquad k_{HO_2^{\bullet}} = k_{AH}^{AH} \qquad (1.228)$$

$$HO_2^{\bullet} + O_2^{\bar{\bullet}} + H_2O \rightarrow H_2O_2 + O_2 + OH^- \qquad\qquad k_{HO_2^{\bullet}}^{O_2^{\bar{\bullet}}} = k_{AH}^A \qquad (1.229)$$

$$O_2^{\bar{\bullet}} + O_2^{\bar{\bullet}} + H_2O \rightarrow HO_2^- + O_2 + OH^- \qquad\qquad k_{O_2^{\bar{\bullet}}} = k_A^A \qquad (1.230)$$

The k/pH profile is shown in Fig. 1.15 and is consistent with (1.231)[177]

$$k = \frac{k_{HO_2^{\bullet}} + k_{HO_2^{\bullet}}^{O_2^{\bar{\bullet}}}(K_{HO_2^{\bullet}}/[H^+])}{(1 + K_{HO_2^{\bullet}}/[H^+])^2} + k_{O_2^{\bar{\bullet}}} \qquad (1.231)$$

with $k_{HO_2^{\bullet}} = 11.0 \times 10^5 M^{-1}s^{-1}$, $k_{HO_2^{\bullet}}^{O_2^{\bar{\bullet}}} = 18 \times 10^7 M^{-1}s^{-1}$, $k_{O_2^{\bar{\bullet}}} = 5.0 M^{-1}s^{-1}$ and $K_{HO_2^{\bullet}} = 2.2 \times 10^{-5} M$.[177,179] The expression (1.231) is derived from (1.225) assuming A = B, $k_A^A \approx 0$, and $K_{AH} = K_{BH}$. For another example, examine the pH profile for dimerization of the $Fe(III) - tppsH_2^{4-}$ complex. The most effective combination again is of the acid form $(Fe(tpps)H_2O^{3-})$ with the basic one, $(Fe(tpps)(OH)^{4-})$, giving a bell shaped k/pH profile.[178]

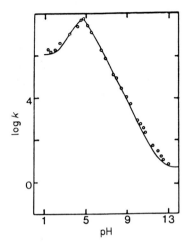

Fig. 1.15 Second-order superoxide disproportionation constant vs pH at 25°C. Potassium superoxide (~ 1 mM) in pH ≥ 12 was mixed in a stopped-flow apparatus with buffers at various pH's and the change in absorbance at 250 nm monitored. The decays were second-order and data were treated in a similar manner to that described in Fig. 1.3.[177] The full line fits Eqn. (1.231) using the parameters given in the text. Reprinted with permission from Z. Bradić and R. G. Wilkins, J. Am. Chem. Soc. **106**, 2236 (1984). © (1984) American Chemical Society.

Finally, we cite an interesting example of the care which must be shown in interpreting pH effects, probably more relevant to protein studies, but a lesson for all kineticists. Investigation of the oxidation of the reduced form of the iron protein, HIPIP_R *(Chromatium vinosum)* by Fe(CN)_6^{3-} indicated a near invariance of the second-order rate constant in the pH range 5–10 at $\mu = 0.2$ M. However, there is a changing charge on the HIPIP as the pH changes and this will by itself likely lead to rate changes (Sec. 2.9.6). The valid procedure is to determine the rate constants at each pH at various ionic strengths and use the value extrapolated to zero ionic strength to assess the pH effect. More marked changes with pH are then observed.[180] The problem appears less important in the corresponding oxidation of HIPIP_R by $\text{Co}^{III}(\text{bpds})^{3-}$, **13**, because the negative charge of the oxidant is more diffuse and electrostatic and ionic strength effects appear much less important.[181]

13

1.10.3 The Effect of High Acid Concentration

In higher acidity, the rate constant may correlate better with h_0 (the Hammet-Deyrup acidity scale) than with the stoichiometric concentration of H^+. Since nearly all the studies involve hydrolysis reactions, the depletion of the reagent water may be an important consideration also.[182]

These points are illustrated nicely in the study of the aquation of CrN_3^{2+} in 1-11 M $HClO_4$.[183] The loss of CrN_3^{2+} monitored at 270 nm is first-order (rate constant = k). The $-\log k$ vs $-H_0$ ($\log h_0$) profile is reproduced in Fig. 1.16. It is marked by a linear dependence at lower acidities, a short plateau, and a decrease in k with increasing acidity at $[HClO_4] > 8.0$ M, at which point there is decreasing value for a_w, the activity of water. This behavior conforms to a rate law of the form

$$-d\ln [CrN_3]_T/dt = k = \frac{h_0}{K_1 + h_0}\,(k_1 a_w + k_2) \tag{1.232}$$

where $[CrN_3]_T = [CrN_3^{2+}] + [CrN_3H^{3+}]$, and the $h_0[K_1 + h_0]^{-1}$ term allows for substantial protonation of CrN_3^{2+}:

$$K_1 = [CrN_3^{2+}]\,h_0/[CrN_3H^{3+}] \tag{1.233}$$

The k_2 term must be included since a plot of $k\,(K_1 + h_0)\,h_0^{-1}$ vs a_w, although linear (slope k_1), has a positive intercept (k_2) at $a_w = 0$. A simple mechanism consistent with this rate law is

$$Cr(H_2O)_5N_3H^{3+} + H_2O \xrightarrow{k_1} Cr(H_2O)_6^{3+} + HN_3 \tag{1.234}$$

$$Cr(H_2O)_5N_3H^{3+} \xrightarrow{k_2} Cr(H_2O)_5^{3+} + HN_3 \tag{1.235}$$

$$Cr(H_2O)_5^{3+} + H_2O \rightarrow Cr(H_2O)_6^{3+} \tag{1.236}$$

with $k_1 = k_1^0 f_{3+}/f_{\neq}$ and $k_2 = k_2^0 f_{3+}/f_{\neq}$. The ratios of activity coefficients may remain constant even when the reaction medium changes.

Information on the kinetics of complex ion reactions in high acid is sparse, partly because of the instability problems, but also because of the difficulties in the interpretation of results in such a complex medium.

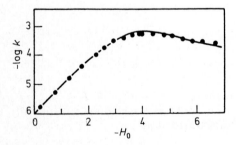

Fig. 1.16 The variation of $-\log k$ with $-H_0$ for the hydrolysis of CrN_3^{2+} in 1–11 M $HClO_4$. The line is calculated from Eqn. (1.232).[183]

1.11 Kinetics and Thermodynamics

We have seen how a comparison of the equilibrium constant estimated from kinetic data for the forward and reverse directions (i.e. $K = k_f/k_r$) with that obtained by measurements on the equilibrated system, may be used to provide strong support (or otherwise) for a particular reaction scheme (see also Chap. 8 Pd(II)). The kinetic approach may be useful also for providing information on thermodynamic data not otherwise easily available.

The interaction

$$IrCl_6^{2-} + N_3^- \rightleftharpoons IrCl_6^{3-} + N_3^\bullet \qquad k_1, k_{-1}, K_1 \tag{1.237}$$

has been used to assess the oxidation potential for the $N_3^\bullet$, N_3^- couple. Reaction (1.237) can be studied in the forward direction by stopped-flow mixing of the reactants in the presence of the spin-trap pbn, **14** or dmpo, **15**, which forces the reaction to completion ($2 N_3^\bullet \rightarrow 3 N_2$) and obviates complicating slow reactions of $N_3^\bullet$. The value of $2 k_1$ is $1.6 \times 10^2 M^{-1}s^{-1}$ at $\mu = 1.0M$ and 25 °C (the factor 2 enters because the pbn, $N_3^\bullet$ adduct is probably rapidly oxidized by $IrCl_6^{2-}$). Reaction (1.237) is studied in the reverse direction by measuring the rate of the reaction of pulse-radiolytically generated $N_3^\bullet$ with added $IrCl_6^{3-}$ (Sec. 3.5.2). The value of k_{-1} is $5.5 \times 10^8 M^{-1}s^{-1}$ with the same conditions as for k_1. This means that K_1 is 1.5×10^{-7} and since E^0 for the $IrCl_6^{2-/3-}$ couple is 0.93 V that for $N_3^\bullet/N_3^-$ is 1.33 V.[184] This is in good agreement with values obtained by cyclic voltametry and by examining an equilibrium involving $Br_2^\bullet$ and N_3^- using pulse radiolysis.[185] The oxidation potentials for a number of couples involving radicals have been determined by kinetic methods.[186-188]

Me₃CN=CHC₆H₅
|
O

14 (PBN)

H_3C
H_3C
N
|
O

15 ((DMPO)

Spin-Trapping Reagents — These can scavenge short-lived radicals to produce free radical nitroxides of longer lifetime. The short-lived radical can thus be prevented from interfering kinetically (as in this case) or be characterised by the epr of the nitroxide.

Non-statistical successive binding of O_2 and CO to the four heme centers of hemoglobin ("cooperativity") has been thoroughly documented. It is difficult to test for a similar effect for NO since the equilibrium constants are very large ($\approx 10^{12}M^{-1}$) and therefore difficult to measure accurately. It is found that the four successive formation rate constants for binding NO to hemoglobin are identical. In contrast, the rate constant for dissociation of the first NO from $Hb(NO)_4$ is at least 80 times less than that for removal of NO from the singly bound entity Hb(NO). This demonstrates cooperativity for the system, and shows that it resides in the dissociation process.[189] The thermodynamic implications of any kinetic data should therefore always be assessed.

1.12 Concluding Remarks

It is important in building up the rate law that a wide range of concentrations of species involved in the reaction be examined so that a complete picture can be obtained. By extending the concentrations of reactants used, additional rate terms have been revealed in the reaction of Fe(II) with Cr(VI),[190] and Fe(II) with $Co(C_2O_4)_3^{3-}$ Ref. 191, and the rate law confirmed in the reaction of $Co(CN)_5^{3-}$ with H_2 (by increasing the concentration with high pressure).[192] In addition, a thorough study of the effect on the rate of reactants such as electrolytes, products and so forth is essential. Other variables, temperature and pressure particularly, will be explored in the next chapter.

It would be blatantly untrue to suggest that the rate behavior of all complex-ion reactions could be fitted into categories contained in this chapter. There are many complicated reaction schemes that require solution by computer and this is becoming increasingly straightforward. It is also true that many complicated reactions can be reduced in complexity by judicious choice of reaction conditions and by so doing become amenable to the type of treatment outlined above.

References

1. The concentration of a species is usually determined *in situ* using some physical property which is linearly dependent on its concentration. By far the most utilized is the uv/vis absorbance (see (1.80) and (3.21)).
2. F.-C. Xu, H. R. Krouse and T. W. Swaddle, Inorg. Chem. **24**, 267 (1985); J. H. Espenson, Inorg. Chem. **8**, 1554 (1969) for Table and References.
3. A. C. Melnyk, N. K. Kildahl, A. R. Rendina and D. H. Busch, J. Amer. Chem. Soc. **101**, 3232 (1979). Full details are given for converting O_2-pressure-time profiles into initial rates of O_2 evolution in terms of Ms^{-1}.
4. K. E. Howlett and S. Sarsfield, J. Chem. Soc. A, 683 (1968); D. C. Gaswick and J. H. Krueger, J. Amer. Chem. Soc. **91**, 2240 (1969).
5. M. E. Farago and T. Matthews, J. Chem. Soc. A, 609 (1969).
6. (a) R. H. Dinius, Inorg. Chim. Acta **99**, 217 (1985); (b) D. G. Lee and T. Chen, J. Amer. Chem. Soc. **111**, 7534 (1989).
7. F. P. Rotzinger, H. Stunzi and W. Marty, Inorg. Chem. **25**, 489 (1986).
8. C. Postmus and E. L. King, J. Phys. Chem. **59**, 1216 (1955).
9. K. W. Hicks and J. R. Sutter, J. Phys. Chem. **75**, 1107 (1971).
10. R. O. Viale, J. Theor. Biol. **31**, 501 (1971).
11. J. E. Taylor, J. Chem. Educ. **46**, 742 (1969).
12. No attempt will be made to derive the integrated form for these and subsequent differential equations. This is found in a number of places including Moore and Pearson, Harris and particularly in Capellos and Bielski. Neither do we intend to say much about computer treatment of raw kinetic data except that this is largely replacing their algebraic manipulation (K. B. Wiberg, B5, Chap. 16).
13. R. A. Scott and H. B. Gray, J. Amer. Chem. Soc. **102**, 3219 (1980).
14. G. A. Tondreau and R. G. Wilkins, Inorg. Chem. **25**, 2745 (1986).
15. F. G. Halaka, G. T. Babcock and J. L. Dye, J. Biol. Chem. **256**, 1084 (1981).
16. M. Meier, F. Basolo and R. G. Pearson, Inorg. Chem. **8**, 795 (1969).

17. The factor $1/2$ enters into (1.24) because a second substitution which consumes another molecule of isonitrile, rapidly follows each first stage. The rate of loss of $C_6H_{11}NC$ is therefore twice that of $Ni(POEt_3)_4$.

18. H. Ogino, M. Shimura, N. Yamamoto and N. Okubo, Inorg. Chem. **27**, 172 (1988).

19. The product **3** is in rapid equilibrium with planar $Ni([14]aneN_4]^{2+}$ and it is the changing absorbance of the planar complex at 450 nm ($\varepsilon = 50M^{-1}cm^{-1}$) which monitors the change in concentration of **3**. $k_{hyd} = k_2$; $k_{isom} \sim k_{-1}$.

20. E. J. Billo, Inorg. Chem. **23**, 236 (1984).

21. D. A. Kamp, R. L. Wilder, S. C. Tang and C. S. Garner, Inorg. Chem. **10**, 1396 (1971).

22. A. Juris, V. Balzani, F. Barigelletti, S. Campagna, P. Belser and A. von Zelewsky, Coordn. Chem. Revs. **84**, 85 (1988).

23. M. S. Matheson and L. M. Dorfman, Pulse Radiolysis, M.I.T. Press, Cambridge, Mass. 1969, Chap. 5.

24. R. W. Wegman, R. J. Olsen, D. R. Gard, L. R. Faulkner and T. L. Brown, J. Amer. Chem. Soc. **103**, 6089 (1981).

25. W. K. Meckstroth, R. T. Walters, W. L. Waltz, A. Wojcicki and L. M. Dorfman, J. Amer. Chem. Soc. **104**, 1842 (1982).

26. M. Krumpolc and J. Rocek, Inorg. Chem. **24**, 617 (1985).

27. D. W. Carlyle and J. H. Espenson, J. Amer. Chem. Soc. **90**, 2272 (1968).

28. J. Barrett and J. H. Baxendale, Trans. Faraday Soc. **42**, 210 (1956).

29. S. C. Chan and S. F. Chan, J. Chem. Soc. A, 202 (1969).

30. B12, p. 44.

31. R. C. Young, T. J. Meyer and D. G. Whitten, J. Amer. Chem. Soc. **98**, 286 (1974); R. Berkoff, K. Krist and H. D. Gafney, Inorg. Chem. **19**, 1 (1980).

32. The values of b and c may sometimes be assessed by using log-log plots. Thus for (1.40)

$$\log k = \log k_1 + b \log [B] + c \log [C]$$

the slope of the $\log k/\log [B]$ plot ([C], constant) is b. Although hardly used, the method is useful when the rate law contains several terms involving $[H^+]$ raised to various powers, J. P. Birk, J. Chem. Educ. **53**, 704 (1976); Inorg. Chem. **17**, 504 (1978). (See Prob. 5 and Fig. 8.8.)

33. J. L. Jensen, R. C. Kanner and G. R. Shaw, Int. J. Chem. Kinetics *22* 1211 (1990).

34. J. Rawlings, S. Wherland and H. B. Gray, J. Amer. Chem. Soc. **99**, 1968 (1977).

35. J. Burgess, J. Chem. Soc. A, 3123 (1968).

36. W. G. Jackson, G. A. Lawrance, P. A. Lay and A. M. Sargeson, J. Chem. Educ. **58**, 734 (1981).

37. D. Meyerstein, Inorg. Chem. **14**, 1716 (1975).

38. B. R. Baker, N. Sutin and T. J. Welch, Inorg. Chem. **6**, 1948 (1967).

39. F. Below, Jr., R. E. Connick and C. P. Coppel, J. Amer. Chem. Soc. **80**, 2961 (1958); J. H. Espenson and D. F. Dustin, Inorg. Chem. **8**, 1760 (1969); W. F. Pickering and A. McAuley, J. Chem. Soc. A, 1173 (1968).

40. J. K. Beattie, Inorg. Chim. Acta **76**, L69 (1983).

41. E. Pelizzetti and E. Mentasti, Inorg. Chem. **18**, 583 (1979).

42. R. F. Pasternack and E. G. Spiro, J. Amer. Chem. Soc. **100**, 968 (1978).

43. D. H. Macartney and N. Sutin, Inorg. Chem. **22**, 3530 (1983).

44. B. Bosnich, F. P. Dwyer and A. M. Sargeson, Aust. J. Chem. **19**, 2051, 2213 (1966).

45. J. Butler, D. M. Davies and A. G. Sykes, J. Inorg. Biochem. **15**, 41 (1981); N. Ohno and M. A. Cusanovich, Biophys. J. **36**, 589 (1981).

46. D. B. Rorabacher, T. S. Turan, J. A. Defever and W. G. Nickels, Inorg. Chem. **8**, 1498 (1969).

47. M. Moriyasu and Y. Hashimoto, Bull. Chem. Soc. Japan **53**, 3590 (1980).

48. C. K. Poon and M. L. Tobe, J. Chem. Soc. A, 2069 (1967).

49. E. L. King, Int. J. Chem. Kinetics **14**, 1285 (1982).

50. D. Cummins and H. B. Gray, J. Amer. Chem. Soc. **99**, 5158 (1977).

51. T. R. Crossley and M. A. Slifkin, Prog. React. Kinetics **5**, 409 (1970).

52. J. E. Byrd and J. Halpern, J. Amer. Chem. Soc. **93**, 1634 (1971).

53. T. W. Kallen and J. E. Earley, Inorg. Chem. **10**, 1149 (1971).

54. R. J. Crutchley, W. R. Ellis, Jr. and H. B. Gray, J. Amer. Chem. Soc. **107**, 5002 (1985).

55. K. Tsukahara and R. G. Wilkins, Inorg. Chem. **28**, 1605 (1989).

56. S. W. Provencher, J. Chem. Phys. **64**, 2772 (1976); S. W. Provencher, Biophys. J. **16**, 27 (1976).

57. M. H. Evans, B. Grossman and R. G. Wilkins, Inorg. Chim. Acta **14**, 59 (1975).

58. D. H. Busch, J. Phys. Chem. **63**, 340 (1959); P. Krumholz, Inorg. Chem. **4**, 609 (1965).

59. These problems may be circumvented if one or both of the steps is multi-order, since the associated pseudo first-order rate constant is modified by changes in concentrations of the excess reagent.

60. P. Hendry and A. M. Sargeson, Inorg. Chem. **25**, 865 (1986).

61. E. T. Gray, Jr., R. W. Taylor and D. W. Margerum, Inorg. Chem. **16**, 3047 (1977).

62. A. Haim and N. Sutin, J. Amer. Chem. Soc. **88**, 5343 (1966).

63. R. W. Hay and L. J. Porter, J. Chem. Soc. A, 127 (1969); R. G. Pearson, R. E. Meeker and F. Basolo, J. Amer. Chem. Soc. **78**, 709 (1956).

64. Z. Bradić, K. Tsukahara, P. C. Wilkins and R. G. Wilkins in Bioinorganic Chemistry 85, A. V. Xavier, ed. VCH, Weinheim, 1985, p. 337.

65. D. A. Buckingham, D. J. Francis and A. M. Sargeson, Inorg. Chem. **13**, 2630 (1974).

66. B8, p. 66.

67. Generally, $P_t - P_e = A_1 \exp(-a_1 t) + A_2 \exp(-a_2 t)$ where P is a measured property, and A_1, A_2, a_1 and a_2 are constants.

68. K. R. Ashley and R. E. Hamm, Inorg. Chem. **5**, 1645 (1966).

69. J. H. Espenson and T.-H. Chao, Inorg. Chem. **16**, 2553 (1977).

70. N. E. Dixon, W. G. Jackson, M. J. Lancaster, G. A. Lawrance and A. M. Sargeson, Inorg. Chem. **20**, 470 (1981).

71. N. W. Alcock, D. J. Benton and P. Moore, Trans. Faraday Soc. **66**, 2210 (1970).

72. J. C. Pleskowicz and E. J. Billo, Inorg. Chim. Acta **99**, 149 (1985).

73. J. Hoch and R. M. Milburn, Inorg. Chem. **18**, 886 (1979).

74. G. M. Clore, A. N. Lane and M. R. Hollaway, Inorg. Chim. Acta, **46**, 139 (1980).

75. S. A. Jacobs and D. W. Margerum, Inorg. Chem. **23**, 1195 (1984).

76. M. Shimura and J. H. Espenson, Inorg. Chem. **23**, 4069 (1984).

77. A. Ekstrom, L. F. Lindoy, H. C. Lip, R. J. Smith, H. J. Goodwin, M. McPartlin and P. A. Tasker, J. Chem. Soc. Dalton Trans. 1027 (1979).

78. W. G. Jackson, J. MacB. Harrowfield and P. D. Vowles, Int. J. Chem. Kinetics, **9**, 535 (1977).

79. E. Jørgensen and J. Bjerrum, Acta Chem. Scand. **13**, 2075 (1959).

80. A. N. Singh, E. Gelerinter and E. S. Gould, Inorg. Chem. **21**, 1232; 1236 (1982).

81. A. J. Miralles and A. Haim, Inorg. Chem. **19**, 1158 (1980).

82. J. H. Beard, J. Casey and R. K. Murmann, Inorg. Chem. **4**, 797 (1965).

83. D. B. Vanderheiden and E. L. King, J. Amer. Chem. Soc. **95**, 3860 (1973).

84. J. R. Chipperfield, J. Organomet. Chem. **137**, 355 (1977).

85. K. Jackson, J. H. Ridd and M. L. Tobe, J. Chem. Soc. Perkins II 611 (1979).

86. F. A. Armstrong, R. A. Henderson and A. G. Sykes, J. Amer. Chem. Soc. **102**, 6545 (1980).

87. W. Marty and J. H. Espenson, Inorg. Chem. **18**, 1246 (1979).

88. M. C. Pohl and J. H. Espenson, Inorg. Chem. **19**, 235 (1980).

89. W. G. Jackson, S. S. Jurisson and B. C. McGregor, Inorg. Chem. **24**, 1788 (1985).

90. D. J. MacDonald and C. S. Garner, Inorg. Chem. **1**, 20 (1962).

91. H. E. Toma and A. B. P. Lever, Inorg. Chem. **25**, 176 (1986).

92. D. Huchital and R. G. Wilkins, Inorg. Chem. **6**, 1022 (1967).

93. F. M. Beringer and E. M. Findler, J. Amer. Chem. Soc. **77**, 3200 (1955), where this scheme is fully analyzed.

94. A. G. Lappin in Metal Ions in Biological Systems, H. Sigel, ed. M. Dekker, NY, 1981, 13, p. 15–71; N. Al-Shatti, A. G. Lappin and A. G. Sykes, Inorg. Chem. **20,** 1466 (1981).

95. A. M. Martin, K. J. Grant and E. J. Billo, Inorg. Chem. **25,** 4904 (1986).

96. B11, p. 88.

97. J. Halpern, J. Chem. Educ. **45,** 372 (1968).

98. R. B. Martin, J. Chem. Educ. **62,** 789 (1985), for a useful treatment of the two schemes.

99. R. B. Martin, Inorg. Chem. **26,** 2197 (1987).

100. L. Rosenheim, D. Speiser and A. Haim, Inorg. Chem. **13,** 1571 (1974).

101. B. T. Reagor and D. H. Huchital, Inorg. Chem. **21,** 703 (1982).

102. G. C. Seaman and A. Haim, J. Amer. Chem. Soc. **106,** 1319 (1984).

103. I. H. Segel, Enzyme Kinetics, Wiley-Interscience, NY, 1975. P. C. Engel, Enzyme Kinetics, Chapman and Hall, London, 1977. M. Dixon, E. C. Webb, C. J. R. Thorne and K. F. Tipton, Enzymes, Third Ed. Academic, NY, 1979. K. F. Tipton and H. B. F. Dixon, Methods in Enzym. **63A,** 183 (1979).

104. When the two steps have comparable rates, the situation is more complex, see Refs. 98, 105, and 106.

105. R. L. Reeves and L. K. J. Tong, J. Amer. Chem. Soc. **84,** 2050 (1962).

106. R. L. Reeves, G. S. Calabrese and S. A. Harkaway, Inorg. Chem. **22,** 3076 (1983).

107. B. S. Brunschwig, P. J. DeLaive, A. M. English, M. Goldberg, H. B. Gray, S. L. Mayo and N. Sutin, Inorg. Chem. **24,** 3743 (1985).

108. J. A. Christiansen, Z. Phys. Chem. **339,** 145 (1936); **378,** 374 (1937).

109. J. H. Espenson, Acc. Chem. Res. **3,** 347 (1970).

110. R. D. Cannon, Electron Transfer Reactions, Butterworth, London, 1980, p. 63–68.

111. The kinetic analysis of (1.123) has been fully discussed in a number of texts in the Selected Bibliography.

112. First treated by T. M. Lowry and W. T. John, J. Chem. Soc. 2634 (1910).

113. F. A. Matsen and J. J. Franklin, J. Amer. Chem. Soc. **72,** 3337 (1950); E. S. Lewis and M. D. Johnson, J. Amer. Chem. Soc. **82,** 5399 (1960).

114. I. J. Goldstein and C. E. Hayes, Adv. Carbohydrate Chem. Biochem. **35,** 127 (1978).

115. T. J. Williams, J. A. Shafer, I. J. Goldstein and T. Adamson, J. Biol. Chem. **253,** 8538 (1978).

116. A. Van Landschoot, F. G. Loontiens, R. M. Clegg and T. M. Jovin, Eur. J. Biochem. **103,** 313 (1980).

117. T. J. Williams, L. D. Homer, J. A. Shafer, I. J. Goldstein, P. J. Garegg, H. Hultberg, T. Iversen and R. Johansson, Arch. Biochem. Biophys. **209,** 555 (1981).

118. F. G. Loontiens, R. M. Clegg and A. Van Landschoot, J. Bioscience, **5,** Supplement 1, 105 (1983).

119. R. G. Pearson and O. P. Anderson, Inorg. Chem. **9,** 39 (1970).

120. M. R. Jaffe, D. P. Fay and N. Sutin, J. Amer. Chem. Soc. **93,** 2878 (1971).

121. M. J. Hynes and M. T. O'Shea, Inorg. Chim. Acta **73,** 201 (1983).

122. E. Antonini and M. Brunori, Hemoglobin and Myoglobin in their Reactions with Ligands, North-Holland, Amsterdam, 1971, p. 197 for the derivation of (1.143).

123. J. M. Malin, H. E. Toma and E. Giesbrecht, J. Chem. Educ. **54,** 385 (1977).

124. J. K. Beattie, M. B. Celap, M. T. Kelso and S. M. Nesić, Aust. J. Chem. **42,** 1647 (1989) treat the interconversion of the geometrical isomers of $Co(L\text{-ala})_2(NO_2)_2^-$ in terms of the scheme $A \rightarrow B \rightleftharpoons C$; $B \rightarrow D$; $C \rightarrow D$.

125. For a general treatment of relaxation kinetics see B4; G. Schwarz in B6, Chap. II, and G. W. Gastellan, Ber. Bunsenges. Phys. Chem. **67,** 898 (1963).

126. B4, p. 15.

127. R. Koren and G. G. Hammes, Biochemistry **15,** 1165 (1976).

128. M. Krishamurthy and J. R. Sutter, Inorg. Chem. **24,** 1943 (1985).

129. B4, Chap. 5.

130. B13, p. 198.

131. R. Brouillard, J. Chem. Soc. Faraday Trans. I **76,** 583 (1980) this paper fully analyzes the degree that an equilibrium may be shifted in a chemical relaxation for a single step reaction.

132. J. P. Bertigny, J. E. Dubois, R. Brouillard, J. Chem. Soc. Faraday Trans. I **79**, 209 (1983), treat a two-step system, following up Ref. 131.

133. The most elegant way to tackle the simultaneous equations (1.165) and (1.166) is by determinants, see B4, p. 27.

134. B. Havsteen, J. Biol. Chem. **242**, 769 (1967); M. Eigen, Quart. Rev. Biophysics **1**, 1 (1968).

135. M. M. Palcic and H. B. Dunford, Arch. Biochem. Biophys. **211**, 245 (1981).

136. B4, Chap. 6.

137. L. F. Povirk, M. Hogan, N. Dattagupta and M. Buechner, Biochemistry **20**, 665 (1981). Cu(II)bleomycin can be used as a model for Fe(II)bleomycin which is an antitumor antibiotic that catalyses the degradation of DNA, J. Stubbe, Biochemistry **27**, 3893 (1988).

138. L. W. Harrison and B. L. Vallee, Biochemistry **17**, 4359 (1978).

139. H. Brintzinger and G. G. Hammes, Inorg. Chem. **5**, 1286 (1966).

140. G. G. Hammes and J. I. Steinfeld, J. Amer. Chem. Soc. **84**, 4639 (1962).

141. J. C. Thomas, C. M. Frey and J. E. Stuehr, Inorg. Chem. **19**, 501; 505 (1980).

142. J. Bidwell, J. Thomas and J. Stuehr, J. Amer. Chem. Soc. **108**, 820 (1986).

143. B13, p. 207.

144. B13, p. 211.

145. Eqn. (1.191) is usually referred to as the McKay equation (H. A. C. McKay, Nature (London) **142**, 497 (1938)) although inklings of the relationship had been recognized. J. N. Wilson and R. G. Dickenson, J. Amer. Chem. Soc. **59**, 1358 (1937).

146. E. L. King, E. B. Fleischer and R. D. Chapman, J. Phys. Chem. **86**, 4273 (1982).

147. L. Melander and W. H. Saunders, Jr., Reaction Rates of Isotopic Molecules Wiley, NY, 1980, p. 115.

148. J. C. Sheppard and A. C. Wahl, J. Amer. Chem. Soc. **79**, 1020 (1957).

149. C. M. Cook, Jr., and F. A. Long, J. Amer. Chem. Soc. A, **80**, 33 (1958).

150. R. K. Murmann, Inorg. Chem. **16**, 46 (1977).

151. A. E. Merbach, Pure Appl. Chem. **54**, 1479 (1982); **59**, 161 (1987).

152. L. Mønsted and O. Mønsted, Acta Chem. Scand. **A34**, 259 (1980).

153. A. E. Merbach, P. Moore, O. W. Howarth and C. H. McAteer, Inorg. Chim. Acta **39**, 129 (1980).

154. T. W. Swaddle, Adv. Inorg. Bioinorg. Mechs. **2**, 95 (1983).

155. L. Helm, L. I. Elding and A. E. Merbach, Inorg. Chem. **24**, 1719 (1985).

156. K. R. Rodgers, R. K. Murmann, E. O. Schlemper and M. E. Shelton, Inorg. Chem. **24**, 1313 (1985); D. T. Richens, L. Helm, P.-A. Pittet, A. E. Merbach, F. Nicolò and G. Chapuis, Inorg. Chem. **28**, 1394 (1989).

157. N. S. Rowan, R. M. Milburn and T. P. Dasgupta, Inorg. Chem. **15**, 1477 (1976) and references therein.

158. A. Nagasawa, H. Kido, T. M. Hattori and K. Saito, Inorg. Chem. **25**, 4330 (1986).

159. H. Gamsjager and R. K. Murmann, Adv. Inorg. Bioinorg. Mechs. **2**, 317 (1983).

160. F. Miller and R. G. Wilkins, J. Amer. Chem. Soc. **92**, 2687 (1970), and unpublished results.

161. P. Natarajan and N. V. Raghavan, J. Amer. Chem. Soc. **102**, 4518 (1980).

162. B. H. J. Bielski, D. E. Cabelli, R. L. Arudi and A. B. Ross, J. Phys. Chem. Ref. Data **14**, 1041 (1985).

163. B. Brønnum, H. S. Johansen and L. H. Skibsted, Inorg. Chem. **27**, 1859 (1988).

164. R. Langley, P. Hambright, K. Alston and P. Neta, Inorg. Chem. **25**, 114 (1986).

165. P.-I. Ohlsson, J. Blanck and K. Ruckpaul, Eur. J. Biochem. **158**, 451 (1986).

166. M. M. Palcic and H. B. Dunford, J. Biol. Chem. **255**, 6128 (1980).

167. I. Rapaport, L. Helm, A. E. Merbach, P. Bernhard and A. Ludi, Inorg. Chem. **27**, 873 (1988).

168. J. M. Malin and R. C. Koch, Inorg. Chem. **17**, 752 (1978).

169. J. E. Sutton and H. Taube, Inorg. Chem. **20**, 4021 (1981).

170. T. Matsubara and C. Creutz, Inorg. Chem. **18**, 1956 (1979).

171. A. W. Shimi and W. C. E. Higginson, J. Chem. Soc. 260 (1958).

172. H. Igino, K. Tsukahara and N. Tanaka, Inorg. Chem. **19**, 255 (1980).

173. The addition of a further protonated species AH_3^{3+} to the reaction scheme can be treated in a similar manner to that already considered. It is applied in the reaction of leghemoglobin with H_2O_2, D. Job, B. Zeba, A. Puppo and J. Rigaud, Eur. J. Biochem. **107**, 391 (1980).

174. P. W. Taylor, R. W. King and A. S. V. Burgen, Biochemistry **9**, 3894 (1970).

175. R. W. King and A. S. V. Burgen, Proc. Roy. Soc. London, B, **193**, 107 (1976); S. Lindskog in Zinc Enzymes, T. G. Spiro, ed. Wiley-Interscience, NY, 1983.

176. W. C. Randall and R. A. Alberty, Biochemistry **6**, 1520 (1967).

177. Z. Bradić and R. G. Wilkins, J. Amer. Chem. Soc. **106**, 2236 (1984).

178. A. A. El-Awady, P. C. Wilkins and R. G. Wilkins, Inorg. Chem. **24**, 2053 (1985).

179. These values are in excellent agreement with those collected from a compilation of previous studies[162] except for that of $k_{O_2^-}$.

180. B. A. Feinberg and W. V. Johnson, Biochem. Biophys. Res. Communs. **93**, 100 (1980).

181. I. K. Adzamli, D. M. Davies, C. S. Stanley and A. G. Sykes, J. Amer. Chem. Soc. **103**, 5543 (1981).

182. H. L. Chum and M. E. M. Helene, Inorg. Chem. **19**, 876 (1980).

183. T. C. Templeton and E. L. King, J. Amer. Chem. Soc. **93**, 7160 (1971).

184. M. S. Ram and D. M. Stanbury, Inorg. Chem. **24**, 4233 (1985); J. Phys. Chem. **90**, 3691 (1986).

185. Z. B. Alfassi, A. Harriman, R. E. Huie, S. Mosseri and P. Neta, J. Phys. Chem. **91**, 2120 (1987); see also M. R. De Felippis, M. Faraggi and M. H. Klapper, J. Phys. Chem. **94**, 2420 (1990).

186. B. W. Carlson and L. L. Miller, J. Amer. Chem. Soc. **105**, 7453 (1983).

187. S. Goldstein and G. Czapski, Inorg. Chem. **24**, 1087 (1985).

188. D. M. Stanbury, Adv. Inorg. Chem. **33**, 69 (1989).

189. E. G. Moore and Q. H. Gibson, J. Biol. Chem. **251**, 2788 (1976).

190. D. R. Rosseinsky and M. J. Nicol, J. Chem. Soc. A, 2887 (1969).

191. R. D. Cannon and J. S. Stillman, J. Chem. Soc. Dalton Trans. 428 (1976).

192. J. Halpern and M. Pribanić, Inorg. Chem. **9**, 2616 (1970).

Selected Bibliography

B1. J. D. Atwood, Inorganic and Organometallic Reaction Mechanisms, Brooks/Cole, Monterey, California, 1985.

B2. C. H. Bamford, C. F. H. Tipper and R. G. Compton eds. Comprehensive Chemical Kinetics, Vols 1–29, Elsevier, Amsterdam, 1969–1989.

B3. S. W. Benson, The Foundations of Chemical Kinetics, McGraw-Hill, NY, 1960.

B4. C. F. Bernasconi, Relaxation Kinetics, Academic, New York, 1976.

B5. C. F. Bernasconi, ed. Investigation of Rates and Mechanisms of Reactions, Fourth Edition, Part I. General Considerations and Reactions at Conventional Rates, Wiley, NY, 1986.

B6. C. F. Bernasconi, ed. Investigation of Rates and Mechanisms of Reactions, 4th Edition, Part II. Investigation of Elementary Reaction Steps in Solution and Fast Reaction Techniques, Wiley, NY, 1986.

B7. C. Capellos and B. H. J. Bielski, Kinetic Systems, Wiley-Interscience, NY, 1972.

B8. J. H. Espenson, Chemical Kinetics and Reaction Mechanisms, McGraw-Hill, NY, 1981.

B9. H. J. Fromm, Initial Rate Kinetics, Springer-Verlag, NY, 1975.

B10. G. G. Hammes, Principles of Chemical Kinetics, Academic, NY, 1978.

B11. L. P. Hammett, Physical Organic Chemistry, 2nd Edition, McGraw-Hill, NY, 1970.

B12. G. M. Harris, Chemical Kinetics, Heath, Boston, 1966.

B13. K. Hiromi, Kinetics of Fast Enzyme Reactions, Wiley, NY, 1979.

B14. D. Katakis and G. Gordon, Mechanisms of Inorganic Reactions, Wiley-Interscience, NY, 1987.

B15. K. J. Laidler, Chemical Kinetics, 3rd Edition, Harper and Row, NY, 1987.

B16. J. W. Moore and R. G. Pearson, Kinetics and Mechanism, 3rd Edition, Wiley, NY, 1981.

B17. R. van Eldik, ed., Inorganic High Pressure Chemistry. Kinetics and Mechanism, Elsevier, Amsterdam, 1986.

B18. H. Strehlow and W. Knoche, Fundamentals of Chemical Relaxation, Verlag Chemie, Weinheim, 1977.

Problems

1. The conventional equation of transition state theory is expressed as Eq. (2.119)

$$k = \mathbf{k}T/\mathbf{h} \exp\left(-\Delta G^{\ddagger}/RT\right)$$

where $\mathbf{k}$ is Boltzmann's constant and $\mathbf{h}$ is Planck's constant. Show that this is dimensionally incorrect and suggest how a correction might be made. R. D. Cannon, Inorg. Reaction Mechs. Vol. 6, p. 6; see also J. R. Murdoch, J. Chem. Educ. **58**, 32 (1981) and B14, p. 59.

2. The reaction

$$N_2H_4 + 2\,H_2O_2 \xrightarrow{\;Cu^{2+}\;} N_2 + 4\,H_2O$$

has been studied by measuring the initial rate of production of N_2 gas at 25 °C with the following results (total volume of solution = 300 ml)

N_2H_4 mM	H_2O_2 mM	Cu^{2+} μM	pH	Initial rate of N_2 production ml of N_2/min
16	65	1.23	9.5	7.3
33	65	1.23	9.7	7.4
131	65	1.23	10.0	7.4
33	33	1.23	9.8	3.6
33	131	1.23	9.0	15.0
65	131	1.23	9.7	15.0
33	65	0.33	9.7	1.95
33	65	1.3	9.7	8.3
33	65	1.64	9.7	10.4
33	65	2.46	9.7	16.2

Determine the rate law, the value of any rate constants (using M and seconds units) and after reading Chap. 2 suggest a mechanism. Is the principle of initial rate being complied with?

C. R. Wellman, J. R. Ward and L. P. Kuhn, J. Amer. Chem. Soc. **98**, 1683 (1976).

3. A method for the assay (enzyme activity) of carbonic anhydrase (Chap. 8. Zn(II)) uses the catalysis at pH 7.0 of the hydrolysis reaction

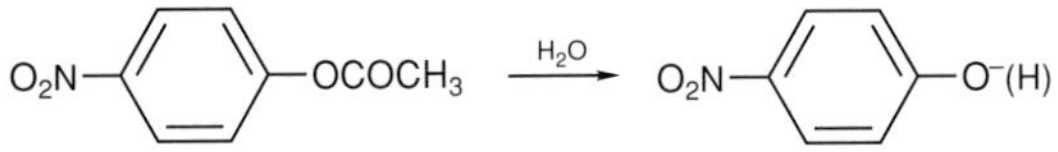

[A]

The initial rate is measured at 348 nm which is an isosbestic point for the product mixture of phenolate/phenol ($\varepsilon = 5.5 \times 10^3 \mathrm{M}^{-1}\mathrm{cm}^{-1}$; $pK_a = 7.1$). The initial rate is easily determined from the linear absorbance (D) vs. time plot, with the following results

[A]	[Enzyme]	Rate
mM	μM	$\Delta D/20$ sec in 1 cm cell
1.1	21	0.80
0.4	21	0.30
1.1	10.5	0.41

Show that $V = k\,[\text{Enzyme}]\,[\text{A}]$ and determine the values of k. Estimate the percentage of A which has reacted in the 20 seconds period and thus rationalize the linearity of this plot.

4. The Ag^+ catalyzed oxidation of VO_3^+ by $\mathrm{S}_2\mathrm{O}_8^{2-}$ can be studied by examining $d\,[\mathrm{VO}_3^+]/dt$ at 455 nm. The loss of VO_3^+ absorbance is *linear* with time, when Ag^+ and $\mathrm{S}_2\mathrm{O}_8^{2-}$ are used in large excess. The following data are obtained at 20°C and $[\mathrm{HClO}_4] = 0.95 - 1.05$ M, $[\mathrm{VO}_3^+]_0 = 1.2$ mM and $[\mathrm{VO}_2^+]_0 = 3$ mM

$[\mathrm{S}_2\mathrm{O}_8^{2-}]_0$	$[\mathrm{Ag}^+]_0$	$10^6 \times$ slope[a]
mM	mM	Ms^{-1}
9.5	23	1.35
9.5	45	2.5
9.2	89	4.7
4.6	89	2.3
19	45	5.1

[a] From $[\mathrm{VO}_3^+]$ vs time plot

Deduce the rate law and suggest a likely mechanism. See also Chap. 8, Prob. 7. R. C. Thompson, Inorg. Chem. **22**, 584 (1983).

5. The reduction of BrO_3^- by IrCl_6^{3-}

$$\mathrm{BrO}_3^- + 5\,\mathrm{IrCl}_6^{3-} + 6\,\mathrm{H}^+ \rightarrow 5\,\mathrm{IrCl}_6^{2-} + 1/2\,\mathrm{Br}_2 + 3\,\mathrm{H}_2\mathrm{O}$$

using excess BrO_3^- and H^+ conformed to the rate law

$$-d\,[\mathrm{IrCl}_6^{3-}]/dt = k\,[\mathrm{IrCl}_6^{3-}]$$

The values of k with different concentrations of BrO_3^- and H^+ are shown in the Table (selected data) at 25 °C, $\mu = 0.50$ M:

$[H^+]$	$[BrO_3^-]$	$10^2 \times k$
M	mM	s^{-1}
0.05	5.0	0.19
0.10	0.80	0.16
0.10	20.0	1.1
0.30	0.50	0.36
0.30	0.80	0.49
0.40	0.80	0.78
0.40	3.0	2.4
0.40	5.0	4.4
0.50	5.0	6.5
0.50	10.0	12.4

Try a log k/log $[BrO_3^-][H^+]^2$ plot and from the result, deduce the rate law (which turns out to be a common one for the reduction of BrO_3^- by a number of complexes). J. P. Birk, Inorg. Chem. **17**, 504 (1978).

6. The following data were obtained for the decay of methyl radicals in the presence of argon diluent in a flash photolysis experiment.

time (μs)	0	10	20	30	40	50
$[CH_3^\bullet]$ μM	1.25	0.95	0.80	0.65	0.57	0.50

Determine the order of the reaction and evaluate the rate constant.

7. The reduction of the Cu(II) protein, azurin, with excess dithionite, $S_2O_4^{2-}$ was monitored at 625 nm (first-order loss of azurin) at pH 9.2 and 25 °C with the following results

$S_2O_4^{2-}$	k_{obs}
mM	s^{-1}
1.0	5.5
2.5	10
5.0	16
7.5	20
10	24
15	32
20	39
25	43

Determine the relationship between k_{obs} and $[S_2O_4^{2-}]$ and $[S_2O_4^{2-}]^{1/2}$. Thus determine the rate law. Calculate the rate constants using the value of 1.4×10^{-9} M for the equilibrium constant for $S_2O_4^{2-} \rightleftharpoons 2 SO_2^-$.
D. O. Lambeth and G. Palmer, J. Biol. Chem. **248**, 6095 (1973).

Z. Bradić and R. G. Wilkins, J. Amer. Chem. Soc. **106**, 2236 (1984).
Generally similar results are reported by G. D. Jones, M. G. Jones, M. T. Wilson, M. Brunori, A. Colosimo and P. Sarti, Biochem. J. **209**, 175 (1983).

8(a). The Cu(II) catalyzed oxidation by O_2 of ascorbic acid (in excess) shows linear plots of $[O_2]^{1/2}$ *vs* time. What is the order of the reaction with respect to O_2?

R. F. Jameson and N. J. Blackburn, J. Chem. Soc. Dalton Trans. 9 (1982).

(b). The reaction in hexane at $0\,°C$ of $Co_2(CO)_8$ with PPh_3 (in excess) shows linear plots of $\{[Co_2(CO)_8]_t^{-1/2} - [Co_2(CO)_8]_0^{-1/2}\}$ *vs* time. What is the order of the reaction with respect to $Co_2(CO)_8$?

M. A.-Halabi, J. D. Atwood, N. P. Forbus and T. L. Brown, J. Amer. Chem. Soc. **102**, 6248 (1980).

9. Consider a reaction that has the stoichiometry

$$2\,A + B \;\longrightarrow\; products$$

but that is second-order, with a probable mechanism

$$A + B \xrightarrow{\ k\ } AB$$

$$AB + A \;\longrightarrow\; products \qquad fast$$

Show that a plot of $\ln[(2\,b - x)/a - x)]$ against t should be linear with the characteristics

$$Slope = (2\,b - a)\,k \qquad Intercept = \ln(2\,b/a)$$

where a and b are the initial concentrations of A and B and x is the amount of A that has been consumed at time t.

The situation is encountered in the second-order reaction of Mn(III) $=$ A with p-$C_6H_4(OH)_2$ $=$ B.

G. Davies and K. Kustin, Trans. Faraday Soc. **65**, 1630 (1969).

10. The reversible reaction ($K_1 = 4.0$)

$$MA_2 + MB_2 \;\rightleftharpoons\; 2\,MAB \qquad k_1, k_{-1}, K_1$$

was studied by using $[MA_2]_0 = [MB_2]_0 = A_0$ and following $-d[MA_2]/dt$ or $d[MAB]/dt$. Show

$$-\ln(1 - (2[MA_2]/[MA_2]_0)) = k_1[MA_2]_0\,t$$

and

$$-\ln(1 - ([MAB]/[MA_2]_0)) = k_1[MA_2]_0\,t$$

How are the equations modified when $K \neq 4$ and when K approaches an infinite value (i.e. the reaction is irreversible)? Equal concentrations of MA_2 and MB_2 are still assumed.

M. Moriyasu and Y. Hashimoto, Bull. Chem. Soc. Japan **53**, 3590 (1980); **54**, 3374 (1981); J. Stach, R. Kirmse, W. Dietzsch, G. Lassmann, V. K. Belyaeva and I. N. Marov, Inorg. Chim. Acta **96**, 55 (1985).

11. In the sequence A $\xrightarrow{k_1}$ B $\xrightarrow{k_2}$ C show that a maximum of B, $[B]_{max}$, will occur at a time t_{max} given by

$$t_{max} = \frac{\ln (k_2/k_1)}{(k_2 - k_1)}$$

that $[B]_{max}$ will have a value

$$[B]_{max} = [A]_0 \exp (-k_2 t_{max})$$

and that when $k_2 \gg k_1$

$$[B]_{max} = [A]_0 k_1/k_2$$

E. T. Gray, Jr., R. W. Taylor and D. W. Margerum, Inorg. Chem. **16**, 3047 (1977).

12. A general rate equation has been derived for reactions with a single relaxation time. Consider:

$$a A + b B \rightleftharpoons d D + e E \qquad k_1, k_{-1}$$

show

$$\tau^{-1} = k_1 [A]_e^{\ a} [B]_e^{\ b} [(a^2/[A]_e) + (b^2/[B]_e)] + k_{-1} [D]_e^{\ d} [E]_e^{\ e} [(d^2/[D]_e) + (e^2/[E]_e)]$$

E. L. King, J. Chem. Educ. **56**, 580 (1979).

13. The equilibrium

$$Mo_2^V O_3 L_4 \rightleftharpoons Mo^{IV} OL_2 + Mo^{VI} O_2 L_2 \qquad k_1, k_{-1}$$

in 1,2-dichloroethane (L = $XYCNEt_2$, XY = SSe or SeSe) has been perturbed by a 1:1 (volume) dilution with 1,2-dichloroethane in a flow apparatus. The relaxation is a single first-order process (rate constant = k). The figure shows plots of k^2 vs $[Mo_2O_3L_4]_0$ (the concentration of $Mo_2O_3L_4$ if the equilibrium were shifted completely to left). Estimate the values of k_1 and k_{-1}.

T. Matsuda, K. Tanaka and T. Tanaka, Inorg. Chem. **18**, 454 (1979).

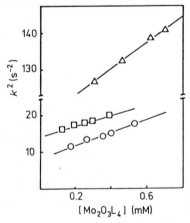

Problem 13. Plots of k^2 vs $[Mo_2O_3L_4]_0$ in 1,2-dichloroethane at 25 °C. L = $Et_2NCS_2(O)$, $Et_2NCSSe(\square)$, $Et_2NCSe_2(\triangle)$. Reprinted with permission from T. Matsuda, K. Tanaka and T. Tanaka, Inorg. Chem. **18**, 454 (1979). © (1979) American Chemical Society.

14. Derive the expression for the relaxation times for

$$A + B \rightleftharpoons C \rightleftharpoons D + E$$

This system is fully discussed and analyzed in the early pressure-jump work on

$$CO_2 + H_2O \rightleftharpoons H_2CO_3 \rightleftharpoons H^+ + HCO_3^-$$

S. Ljunggren and O. Lamm, Acta Chem. Scand. **12**, 1834 (1958).

15. For the Michaelis-Menten scheme involving interaction of enzyme (E) with substrate (S) or product (P):

$$E + S \underset{k_{-1}}{\overset{k_1}{\rightleftharpoons}} X \underset{k_{-2}}{\overset{k_2}{\rightleftharpoons}} E + P$$

show using steady-state conditions for X and $[S]_0 \gg [E]_0$ that

$$\frac{-d[S]}{dt} = \frac{[V_s/K_s][S] - [V_p/K_p][P]}{1 + [S]/K_s + [P]/K_p}$$

where
$$V_s = k_2[E]_0$$
$$V_p = k_{-1}[E]_0$$
$$K_s = (k_{-1} + k_2)/k_1$$
$$K_p = (k_{-1} + k_2)/k_{-2}$$

Convince yourself that (a) this equation reduces to (1.104) when initial-rate measurements, $[P] = 0$, are used and (b) the same form of equation results when a number of intermediates arises as in

$$E + S \rightleftharpoons X_1 \rightleftharpoons X_2 \rightleftharpoons P + E$$

which is a more realistic portrayal of enzyme behavior.
B10, p. 220–222.

16. The reaction of Cu(II) ion with acetylacetone (acacH) to form the mono complex $Cu(acac)^+$ in methanol is interpreted in terms of the scheme

$$\text{keto} \underset{k_{-1}[H^+]}{\overset{k_1[Cu^{2+}]}{\rightleftharpoons}} \begin{matrix} Cu(acac)^+ \\ + H^+ \end{matrix} \underset{k_{-2}[Cu^{2+}]}{\overset{k_2[H^+]}{\rightleftharpoons}} \text{enol}$$

where the keto and enol represent those forms of acetylacetone. The reaction is studied by mixing a solution containing Cu^{2+} and H^+ with one containing acacH and H^+ in a stopped flow apparatus. Two reactions are seen, k_I and k_{II} with the following conditions:

total [acacH] $=$ 0.1 mM, [H$^+$] $=$ 1.09 mM, 25 °C

[Cu^{2+}] mM	k_I s^{-1}	k_{II} s^{-1}
1.18	28	1.64
1.41	36	2.21
1.65	41	2.26
1.88	43	2.64
2.12	50	2.55
2.35	53	2.76
1.18	28	1.77
1.65	37	2.02
2.35	50	3.10

Treating the system as a relaxation with all pseudo first-order rate constants, deduce the expressions for $k_I + k_{II}$ and $k_I k_{II}$ in terms of the component rate constants and hence determine the values of k_1, k_{-1}, k_2 and k_{-2}.
R. G. Pearson and O. P. Anderson, Inorg. Chem. **9**, 39 (1970).

17. Determine the rate law for the exchange of Ag between Ag(I) and Ag(II) in 5.9 M HClO$_4$ at 0 °C. Use the accompanying data, obtained by the quenched-flow method. Suggest a mechanism for the exchange.

[Ag^I], mM	[Ag^{II}], mM	$t_{1/2}$ exch, s
2.2	0.64	0.77
3.7	1.4	0.35
3.6	1.5	0.34
3.9	1.2	0.42
6.8	2.2	0.26
7.6	1.9	0.34
10.0	1.3	0.49
9.7	2.0	0.29
10.6	1.8	0.32
15.5	1.4	0.45
17.7	1.2	0.51
24.1	1.3	0.54
30.3	1.2	0.60

B. M. Gordon and A. C. Wahl, J. Amer. Chem. Soc. **80**, 273 (1958).

18. The reduction of Mn(III) myoglobin by dithionite (in excess) obeys the rate law at 25 °C, $\mu = 0.45$ M (Na$_2$SO$_4$)

$$-d\,[\text{Mn(III)Myo}]/dt = k_{obs}\,[\text{Mn(III)Myo}]$$

where $k_{obs} = k\,[\text{S}_2\text{O}_4^{2-}]^{1/2}$

pH	$k, \mathrm{M}^{-1/2}\mathrm{s}^{-1}$	pH	$k, \mathrm{M}^{-1/2}\mathrm{s}^{-1}$
5.3	33	6.9	1.6
5.5	21	7.0	1.3
5.7	15	7.1	1.2
5.8	12	7.3	0.76
5.9	10	7.5	0.72
6.0	7.5	7.7	0.75
6.1	6.0	7.8	0.60
6.3	4.6	7.9	0.58
6.4	3.2	8.0	0.51
6.6	2.7	8.4	0.47
6.7	1.9	8.6	0.47

Plot k vs pH and see if there is a pK associated with the protein (assume dithionite is aprotic in this pH range). Deduce the expression relating k and $[\mathrm{H}^+]$.
R. Langley, P. Hambright, K. Alston and P. Neta, Inorg. Chem. **25**, 114 (1986).

19. Aromatic sulfonamides are specific inhibitors of carbonic anhydrase (E). The apparent second-order rate constants for association of p-nitrobenzenesulfonamide with (a) carbonic anhydrase-B and (b) the carboxymethylated derivative of the enzyme are shown against pH in Figure 1.14. Estimate using equations (1.225) and (1.226) the values for pK_E, pK_s and k_1 and k_2 for the two possible schemes shown for both carbonic anhydrase-B and the carboxymethylated derivative. (It is uncertain whether the bound inhibitor is ionized in the product.)

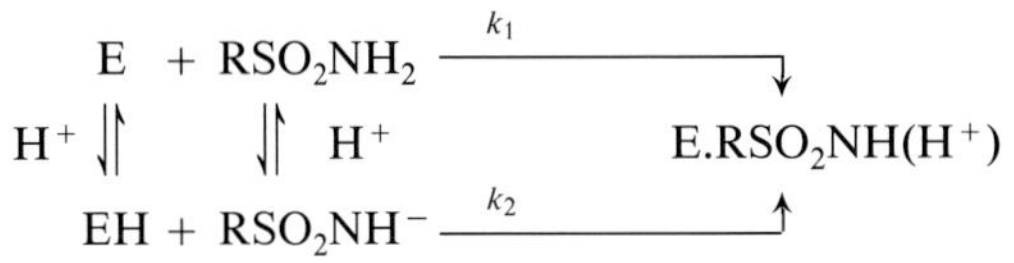

$$
\begin{array}{ccccc}
 & \mathrm{E} & + & \mathrm{RSO_2NH_2} & \xrightarrow{\;k_1\;} \\[2pt]
\mathrm{H}^+ \updownarrow & & \updownarrow \mathrm{H}^+ & & \qquad\qquad \mathrm{E.RSO_2NH(H^+)} \\[2pt]
 & \mathrm{EH} & + & \mathrm{RSO_2NH^-} & \xrightarrow{\;k_2\;}
\end{array}
$$

pK_E and pK_s are the ionization constants for the enzyme (or derivative) and sulfonamide respectively. Suggest which is the likely path.
P. W. Taylor, R. W. King and A. S. V. Burgen, Biochemistry **9**, 3894 (1970); S. Lindskog in Zinc Enzymes, T. G. Spiro, ed., Wiley-Interscience, NY 1983.

20. The second-order rate constant for oxidation of $Fe(CN)_6^-$ by $OH^•$ radicals, produced by low-intensity-pulse radiolysis of water, varies with pH as in the accompanying table. Determine the pK for acid dissociation of the $OH^•$ radical in aqueous solution. (This is difficult to obtain by any other method.)

pH	$10^{-10} \times k$ $M^{-1}s^{-1}$
neutral	1.2
11.94	0.49
12.10	0.36
12.57	0.19
13.07	0.06

J. Rabani and M. S. Matheson, J. Amer. Chem. Soc. **86**, 3175 (1964).

21. The second-order rate constant for the reaction of a hydrogen atom with a hydroxide ion to give an electron and water (hydrated electron) is $2.0 \times 10^7 M^{-1}s^{-1}$. The rate constant for the decay of a hydrated electron to give a hydrogen atom and hydroxide ion is $16 M^{-1}s^{-1}$. Both rate constants can be determined by pulse radiolytic methods. Estimate, using these values, the pK_a of the hydrogen atom. Assume the concentration of water is $55.5 M$ and that the ionization constant of water is $10^{-14} M$.
W. L. Jolly, Modern Inorganic Chemistry, McGraw Hill, NY, 1984, p. 239.

Chapter 2
The Deduction of Mechanism

We are concerned in this chapter with the mechanism of a reaction, that is, the detailed manner in which it proceeds, with emphasis on the number and nature of the steps involved. There are several means available for elucidation of the mechanism, including using the rate law, and determining the effect on the rate constant of varying the structure of reactants (linear free energy relations) and of outside parameters such as temperature and pressure. Finally chemical intuition and experiments are often of great value. These means will be analyzed.

2.1 The Rate Law and Mechanism

The kineticist should always strive to get as complete (and accurate!) a rate law as conditions will allow. The mechanism that is suggested to account for the rate law is, however, a product of the imagination, and since it may be one of several plausible mechanisms, it might very well turn out to be incorrect. Indeed, it is impossible to prove any single mechanism but so much favorable data may be amassed for a mechanism that one can be fairly certain of its validity.

Some general rules exist for deducing the mechanism from a rate law, and the subject is hardly a magical one.[1] Probably the most important single statement is that *the rate law gives the composition of the activated complex — nothing more nor less — but yields no clue about how it is assembled.* Once this is appreciated, many of the problems and ambiguities that have arisen on occasion are easily understood, though not necessarily resolved.

For the moment, we can consider the activated complex as a type of "intermediate" (although not isolatable) reached by the reactants as the highest energy point of the most favorable reaction path. The activated complex is in equilibrium with the reactants and is commonly regarded as an ordinary molecule, except that movement along the reaction coordinate will lead to decomposition. The activated complex can be assumed to have the associated properties of molecules, such as volume,[2] heat content,[3] acid-base behavior,[4] entropy,[5] and so forth. Indeed, formal calculations of equilibrium constants involving reactions of the activated complex to form another activated complex can be carried out (Sec. 5.6 (b)).[6]

Consider the formation of CrO_5 from $Cr(VI)$ and H_2O_2 in an acid medium.[7,8]

$$HCrO_4^- + 2H_2O_2 + H^+ \rightarrow CrO_5 + 3H_2O \tag{2.1}$$

The reaction is third-order:

$$d\,[CrO_5]/dt = k\,[HCrO_4^-]\,[H_2O_2]\,[H^+] \tag{2.2}$$

From the rate law, the composition of the activated complex must therefore be

$$[HCrO_4^-, H_2O_2, H^+, (H_2O)_n]^{\ddagger} \tag{2.3}$$

although we do not know how the various groups are assembled. The activated complex might arise, for example, from a rate-determining step (rds)[9] involving H_2CrO_4 reacting with H_2O_2, $HCrO_4^-$ reacting with $H_3O_2^+$, or even, in principle, CrO_4^{2-} reacting with $H_4O_2^{2+}$. Also involved in the step will be an unknown number of solvent molecules.

Any reagent that appears as part of the reaction stoichiometry but does not feature in the rate law must react in a step that follows the rate-determining one. It is clear from the stoichiometry of (2.1) that one H_2O_2 molecule must react after the rds. In light of these various points, two possible mechanisms would be

$$HCrO_4^- + H^+ \;\rightleftharpoons\; H_2CrO_4 \qquad\qquad K_1 \tag{2.4}$$

$$H_2CrO_4 + H_2O_2 \;\rightarrow\; H_2CrO_5 + H_2O \qquad k_1\ (\text{rds}) \tag{2.5}$$

$$H_2CrO_5 + H_2O_2 \;\rightarrow\; CrO_5 + 2\,H_2O \qquad \text{fast} \tag{2.6}$$

and

$$H_2O_2 \;\;+ H^+ \;\rightleftharpoons\; H_3O_2^+ \qquad\qquad K_2 \tag{2.7}$$

$$H_3O_2^+ \;\;+ HCrO_4^- \rightarrow H_2CrO_5 + H_2O \qquad k_2\ (\text{rds}) \tag{2.8}$$

$$H_2CrO_5 + H_2O_2 \;\rightarrow\; CrO_5 + 2\,H_2O \qquad \text{fast} \tag{2.9}$$

Activated complexes will be associated with all three steps of each mechanism since each step is in principle a separable reaction. The important activated complex is that produced in the rds. This is as far as we can go at present, using the rate law, but we shall return to this problem later (Sec. 2.1.7 (c)).

It is clear that many reactions, particularly those without simple stoichiometry, will have mechanisms containing several steps, one or more of which may include reversible equilibria (consider for example (1.118) to (1.120)). The number of separate terms in the rate law will indicate the number of paths by which the reaction may proceed, the relative importance of which will vary with the conditions. The complex multiterm rate laws, although tedious to characterize, give the most information on the detailed mechanism.

We shall discuss in the following sections the mechanisms that might be associated with the common rate laws. We have already referred to reaction schemes (mechanisms) in discussing rate laws in Chap. 1. Indeed, experienced kineticists often have some preconceived notions of the mechanism before they plan a kinetic study. Certainly in principle, however, the rate law can be obtained before any thoughts of mechanism arise.

2.1.1 First-Order Reactions

As well as the obvious example involving one reactant (even with this solvent may be involved) a number of reactions between A and B that might have been expected to be second-order, first-order in A and in B, turn out to be first-order only (say in A). Obviously some feature of A, not directly connected with the main reaction with B, must be determining the rate. The product of this rds, A_1, must react more readily with B than A does. It is possible to check the correctness of this idea by independent study of the A $\rightarrow$ A_1 interconversion. An isomerization within a complex may limit the rate of its reaction with another reagent.

The reaction of planar Ni ([14]aneN$_4$)$^{2+}$ represented as shown in (2.10) with a number of bidentate ligands (XY) to produce cis-octahedral Ni ([14]aneN$_4$) XY^{2+} is first-order in nickel complex and [OH$^-$] and independent of the concentration of XY.[10] In the preferred mechanism, the folding of the macrocycle (base-catalyzed trans $\rightarrow$ cis isomerization) is rate determining, and this is followed by rapid coordination of XY:

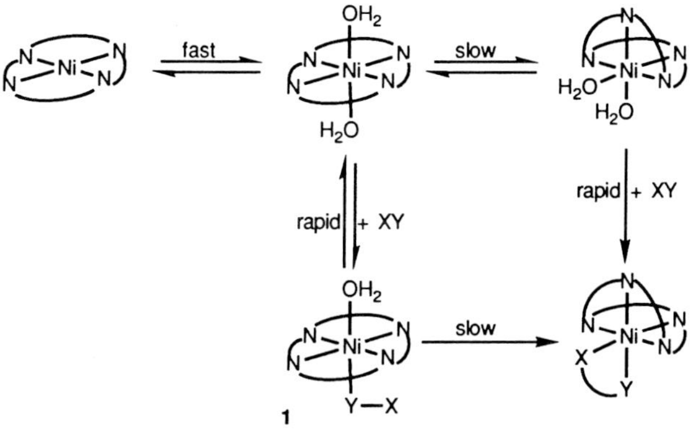

(2.10)

In an alternative mechanism a monodentate intermediate (1) is in rapid equilibrium with reactants and it undergoes at high [XY] rate-determining ring closure. Such a type of mechanism is believed to operate for Ni(trien)$^{2+}$ interacting with XY.[11] Reasons for the preferred mechanisms are given.[10] The isomerization may take the form of a conformational change in a metalloprotein.[12]

The reactivity of a dimer may be limited by its fragmentation. The rates of a number of reactions of cobalt(III) peroxo species (any charges omitted) are limited by their breakdown (rds)

$$L_5Co^{III}O_2Co^{III}L_5 \rightarrow 2Co(II)L_5 + O_2 \qquad (2.11)$$

The scavenging of Co(II)L$_5$ or O$_2$ by added reagent follows rapidly. The rate law does not therefore include the concentration of the added reagent,[13, 14] except in certain instances.[15] Finally, the first-order dominance of a reaction between A and B may only become apparent at higher concentrations of B (Sec. 1.6.3).

The reaction of Co(III) complexes with Cr^{2+} is almost universally a second-order process (Chap. 5). The reaction of Co(NH$_3$)$_5$OCOCH(OH)$_2^{3+}$ with Cr^{2+} is second-order at low concentrations of reductant, but becomes almost independent of [Cr^{2+}] when the concentration is high.[16] The glyoxalate is hydrated in the complex (>98% from nmr measurements) and the rate behavior can be understood on the basis of the scheme

$$Co(NH_3)_5OCCH(OH)_2^{2+} \quad \rightleftharpoons \quad Co(NH_3)_5OCCHO^{2+} + H_2O \quad k_1, k_1 \qquad (2.12)$$

$$Co(NH_3)_5OCCHO^{2+} + Cr^{2+} \rightarrow products \qquad\qquad k_2 \qquad (2.13)$$

for which a rate law of the form (1.115) applies:

$$V = -d\,[Co^{III}]/dt = \frac{k_1\,k_2\,[Co^{III}]\,[Cr^{2+}]}{k_{-1} + k_2\,[Cr^{2+}]} \qquad (2.14)$$

At high $[Cr^{2+}]$,

$$V = k_1\,[Co^{III}] \qquad (2.15)$$

This explanation is supported by nmr rate data for the dehydration of hydrated pyruvic acid, which is similar to glyoxalate. For this at 25°C

$$k_1 = 0.22 + 1.25\,[H^+] \qquad (2.16)$$

compared with

$$k_1 = 0.075 + 0.64\,[H^+] \qquad (2.17)$$

for the glyoxalate complex from the reduction data.[16]

2.1.2 Second-Order Reactions

These are among the most commonly encountered reactions, and can be either between two different reagents or two identical species, as in disproportionation reactions of radicals.
    The distinguishing of a single step

$$A + B \rightarrow products \qquad (2.18)$$

from a stepwise mechanism

$$A + B \rightleftharpoons intermediate \rightarrow products \qquad (2.19)$$

may prove very difficult (Sec. 5.5).
    An excellent example of a "simple" second-order reaction being far from simple is the interaction of small molecules e.g. O_2, CO or NO with deoxymyoglobin (PFe). Examination

using flow, T-jump and flash photolysis methods shows up an excellent second-order reaction between ligand dissolved in the solvent and the heme iron center. This however disguises a much more complicated mechanism:

$$\underset{\text{(solv)}}{PFe} + O_2 \underset{14\mu s^{-1}}{\overset{43\,\mu M^{-1}s^{-1}}{\rightleftharpoons}} PFe\cdots O_2 \underset{120\mu s^{-1}}{\overset{8.5\,\mu s^{-1}}{\rightleftharpoons}} PFe\cdot O_2 \underset{92\,s^{-1}}{\overset{490\,\mu s^{-1}}{\rightleftharpoons}} PFeO_2 \quad (2.20)$$

2	**3**	**4**	**5**
separated	O_2 embedded in	"Geminate" state	O_2 bound to Fe
reactants	protein but distant	$- O_2$ embedded	
in solution	from Fe	in protein near	
		heme but not	
		Fe bound	

   Irradiation of $PFeO_2$ in solutions by short, very intense, laser pulses produces transients such as 3 and 4. Absorbance changes following the production of 3 and 4 are ascribed to their decay and rate parameters can be estimated as shown in the Scheme. Note the units in (2.20) which are occasionally used for large rate constants. The mechanism shown is certainly a simplified one. [17, 18]

Geminate recombination and quantum yield. – The hemoprotein adducts are photoactive, light breaking the iron-ligand bond. The quantum yields (unliganded protein molecules formed/quanta absorbed) however vary widely being about 0.5 to 1.0 for CO, 0.05 for O_2 and quite low, 0.001 for NO adducts. These differences can be rationalized in terms of geminate recombination. This represents the return of the photolyzed ligand to the iron site from a nearby site and therefore not actually getting into solution. It is shown that a high fraction of NO, much less O_2 and very little CO, recombines with heme after short-time laser pulse photolysis. This accounts for the order of quantum yields above, which relate to the complete removal of the ligand into the solution on sustained photolysis. The rate constant for binding generally decreases NO > $O_2 \geqslant$ CO (M. P. Mims, A. G. Porras, J. S. Olson, R. W. Nobel and J. A. Peterson, J. Biol. Chem. **258**, 14219 (1983)). The rds for NO binding is diffusion of NO into the protein, whereas with the other ligands attachment to the metal is also an important rate-controlling step. These points emerge from the geminate recombination studies (Q. H. Gibson, J. Biol. Chem. **264**, 20155 (1989)).

2.1.3 Third-Order Reactions

An activated complex containing three species (other than solvent or electrolyte), which attends a third-order reaction, is not likely to arise from a single termolecular reaction involving the three species. Third- (and higher-) order reactions invariably result from the combination of a rapid preequilibrium or preequilibria with a rds, often unidirectional. Such reactions are fairly common in transition metal chemistry because of the stepwise nature of metal complex-ligand interactions. A third-order rate law,

$$V = k [A] [B] [C] \tag{2.21}$$

is usually best understood as arising from a rate-determining reaction of the binary product, say AB, of a rapid preequilibrium, with the third reactant C.

The Pt(II)-catalyzed substitution of Pt(IV) complexes was first established in 1958.[19] The rate of exchange of chloride between $Pt(en)_2Cl_2^{2+}$ and Cl^- ions is *extremely* slow, but the rate is markedly enhanced in the presence of $Pt(en)_2^{2+}$ ions. The third-order exchange law

$$V_{exch} = k [Pt(en)_2Cl_2^{2+}] [Pt(en)_2^{2+}] [Cl^-] \tag{2.22}$$

can be beautifully rationalized by the mechanism

$$Pt(en)_2^{2+} \quad + Cl^- \quad \rightleftharpoons Pt(en)_2Cl^+ \qquad K_1 \tag{2.23}$$

$$Pt(en)_2Cl_2^{2+} + Pt(en)_2Cl^+ \rightleftharpoons exchange \qquad k_2, \text{ slow} \tag{2.24}$$

for which

$$k = K_1 k_2 \tag{2.25}$$

Exchange is visualized as occurring through a symmetrical intermediate or transition state **6**, which allows for interchange of Cl between Pt(II) and Pt(IV). Breakage of the Cl bridge at *a* produces the original

$$
\overset{\displaystyle en^+}{\underset{\displaystyle en}{*Cl-Pt}} + \overset{\displaystyle en}{\underset{\displaystyle en}{Cl-Pt-Cl^{2+}}} \rightleftharpoons \overset{a \quad b}{*Cl-Pt\!\mid\!Cl\!\mid\!Pt-Cl^{3+}} \tag{2.26}
$$

$$\mathbf{6}$$

$$\rightleftharpoons \overset{\displaystyle en}{\underset{\displaystyle en}{*Cl-Pt-Cl^{2+}}} + \overset{\displaystyle en}{\underset{\displaystyle en}{Pt-Cl^+}}$$

isotopic distribution, while cleavage at *b* leads to exchange. It should be noted that this mechanism leads also to exchange of both Pt and en between Pt(II) and Pt(IV), catalyzed by Cl^- ion. All these exchanges have been studied and the existence of similar values of k from Cl^- exchange (12–15 $M^{-2}s^{-1}$), Pt exchange (11 $M^{-2}s^{-1}$), and en exchange (16 $M^{-2}s^{-1}$, all data at 25 °C) is striking evidence for the correctness of the mechanism. These original studies have led to substantial developments in the chemistry of substitution in Pt(IV),[20] and modification of the mechanism with certain systems.[21]

A third-order rate law of the form

$$V = k [A]^2 [B] \tag{2.27}$$

perhaps suggests (although there are other possibilities, as we shall see) that a dimer A_2 is rapidly formed, in small quantities, from the monomer A, and that it is this dimer that reacts with B in the rate-determining step. Such a situation may apply in the H_2 reduction of $Co(CN)_5^{3-}$ (Ref. 22)

$$2Co(CN)_5^{3-} + H_2 \rightleftharpoons 2Co(CN)_5H^{3-} \qquad k_1, k_{-1} \qquad (2.28)$$

for which the rate law

$$-d[Co(CN)_5^{3-}]/dt = 2k_1[Co(CN)_5^{3-}]^2[H_2] - 2k_{-1}[Co(CN)_5H^{3-}]^2 \qquad (2.29)$$

suggests a mechanism

$$2Co(CN)_5^{3-} \rightleftharpoons Co_2(CN)_{10}^{6-} \qquad (2.30)$$

$$Co_2(CN)_{10}^{6-} + H_2 \rightleftharpoons 2Co(CN)_5H^{3-} \qquad (2.31)$$

The rate law (2.27) is not helpful in detailing the sequence leading to the formation of the activated complex, only that it consists of two molecules of A and one of B.

Reaction of Vitamin B_{12r} (B_{12r}) with organic iodides (RI) in aqueous solution

$$2B_{12r} + RI \rightarrow B_{12a} + RB_{12} + I^- \qquad (2.32)$$

Vitamin B_{12} (cyanocobalamin) – This has a central Co(III) bound to four N's of a corrin ring, a methylbenzimidazole group (which is attached to the corrin ring) and a CN^- group. When the CN^- is replaced by H_2O or OH^-, aquacobalamin (B_{12a}) and hydroxocobalamin (B_{12b}) result. Reduced derivatives are B_{12r} also termed cob(II)alamin and B_{12s} (cob(I)alamin). The redox interconversion of the Co(III), Co(II) and Co(I) derivatives is of key importance, D. Lexa and J.-M. Saveant, Accs. Chem. Res. **16**, 235 (1983).

---

occurs with the rate law

$$V = k[B_{12r}]^2[RI] \qquad (2.33)$$

Associated mechanisms consistent with this rate law include

$$B_{12r} + RI \rightleftharpoons B_{12r} \cdot RI \qquad \text{fast} \qquad (2.34)$$

$$B_{12r} + B_{12r} \cdot RI \xrightarrow[\text{rds}]{} RB_{12} + B_{12a} + I^- \qquad (2.35)$$

or

$$2B_{12r} \xrightarrow{\text{fast}} [B_{12r}]_2 \xrightarrow[\text{rds}]{RI} RB_{12} + B_{12a} + I^- \qquad (2.36)$$

or

$$2B_{12r} \xrightarrow{\text{fast}} B_{12a} + B_{12s} \qquad (2.37)$$

$$B_{12s} + RI \xrightarrow[\text{rds}]{} RB_{12} + I^- \qquad (2.38)$$

(2.37) involves the disproportionation of Co(II) in B_{12r}. For chemical reasons, the first mechanism is preferred.[23]

2.1.4 Even Higher-Order Reactions

In the formation of the highly colored $Fe(bpy)_3^{2+}$ ion from Fe^{2+} ion and excess bipyridine (bpy) in acid solution, the following rate law has been demonstrated

$$d\,[Fe(bpy)_3^{2+}]/dt = k\,[Fe^{2+}]\,[bpy]^3 \tag{2.39}$$

with $k = 1.4 \times 10^{13}\,M^{-3}s^{-1}$ and temperature-independent from $0\,°C$ to $35\,°C$.[24] This rate law follows from (2.40)–(2.42), with K_1 and K_2 associated with rapid pre-equilibria and (2.42) the rds.

$$Fe(H_2O)_6^{2+} + bpy \quad \rightleftharpoons \quad Fe(bpy)\,(H_2O)_4^{2+} + 2\,H_2O \qquad K_1 \tag{2.40}$$

$$Fe(bpy)\,(H_2O)_4^{2+} + bpy \quad \rightleftharpoons \quad Fe(bpy)_2\,(H_2O)_2^{2+} + 2\,H_2O \qquad K_2 \tag{2.41}$$

$$Fe(bpy)_2\,(H_2O)_2^{2+} + bpy \quad \rightleftharpoons \quad Fe(bpy)_3^{2+} + 2\,H_2O \qquad\qquad k_3 \tag{2.42}$$

On this basis, k is a composite rate constant

$$k = K_1 K_2 k_3 \tag{2.43}$$

Since $K_1 K_2$ is $10^8\,M^{-2}$ at $25\,°C$, k_3 is calculated as $1.4 \times 10^5\,M^{-1}s^{-1}$. The latter is a reasonable value from our knowledge of the substitution reactions of iron(II); see Table 4.1. For another example, see Ref. 25.

2.1.5 Negative-Order Reactions

The inclusion in the rate law of a simple inverse dependence on the concentration of a species (negative-order reactions) usually indicates that this reagent features as the product of a rapid step preceding the rate-determining step. This is illustrated by the multiterm rate law that governs the reaction of Fe(III) with V(III) in acid,[26]

$$Fe^{III} + V^{III} \rightarrow Fe^{II} + V^{IV} \tag{2.44}$$

$$-d\,[Fe^{III}]/dt = -d\,[V^{III}]/dt = k_1\,[Fe^{III}]\,[V^{III}] + k_2\,[Fe^{III}]\,[V^{III}]\,[V^{IV}]\,[Fe^{II}]^{-1} \tag{2.45}$$

At high initial [Fe(II)], the term in k_2 is negligible and the k_1 term represents a straightforward second-order (acid-dependent) reaction,

$$k_1 = k + k'\,[H^+]^{-1} + k''\,[H^+]^{-2} \tag{2.46}$$

From experiments at high initial [V(IV)], the second term becomes important and is then easily measurable. For this term a possible mechanism is

$$Fe^{III} + V^{IV} \rightleftharpoons Fe^{II} + V^V \qquad k_3, k_{-3} \tag{2.47}$$

$$V^V + V^{III} \rightarrow 2\,V^{IV} \qquad\qquad k_4 \tag{2.48}$$

The stationary-state approximation, $d[V(V)]/dt = 0$, leads to

$$\frac{-d[V^{III}]}{dt} = \frac{k_3 k_4 [Fe^{III}][V^{III}][V^{IV}]}{k_{-3}[Fe^{II}] + k_4[V^{III}]}$$
(2.49)

and if $k_{-3}[Fe(II)] \gg k_4[V(III)]$, the observed rate term is obtained with $k_2 = k_3 k_4 / k_{-3}$. The value of k_3 / k_{-3} equals the equilibrium constant for reaction (2.47) and can be independently determined. From this, and the k_2 value obtained experimentally, k_4 can be calculated. A later direct determination of the rate constant for the V(III), V(V) reaction gave a value (and pH dependence)[27] in good agreement with that obtained indirectly, thus affording strong support for the correctness of the mechanism.

The lack of retardation by added $Fe(III)(tmpyp)OH^{4+}$ on the rate of the $Fe(II)tmpyp^{4+}$ (7) reaction with O_2 indicates that Fe(III) porphyrin is *not* formed in the first step, i.e. (2.50) rather than (2.51) is the better description for the first step in the overall reaction[28]

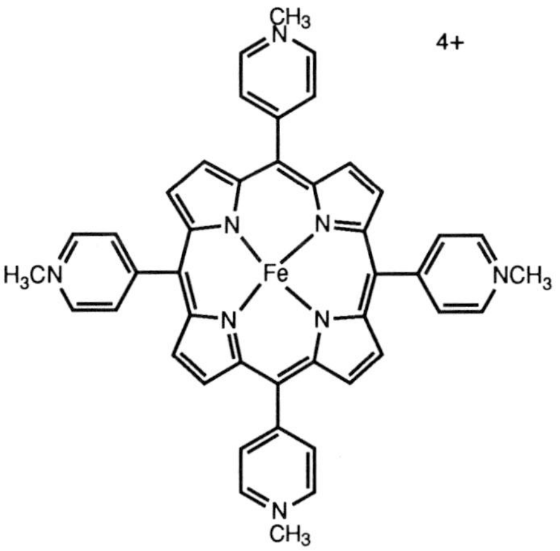

7

$$Fe(tmpyp)^{4+} + O_2 \rightleftharpoons Fe(tmpyp)O_2^{4+}$$
(2.50)

$$Fe(tmpyp)^{4+} + O_2 + OH^- \rightleftharpoons Fe(tmpyp)(OH)^{4+} + O_2^{\bar{}}$$
(2.51)

See also Chap. 8. Fe(II).

2.1.6 Fractional-Order Reactions

A fractional order may arise when a reaction with a multiterm rate law (containing no fractional orders) is examined over only a small range of concentrations (see (1.6.3)). Such an origin can be easily detected, since it disappears when the rate law is fully resolved.

Monomer, polymer equilibria can be the basis of a genuine fractional order. Many reductions by dithionite $S_2O_4^{2-}$ of oxidants (ox) to produce reductants (red) contain in the rate law a term which includes a square root dependence on the $S_2O_4^{2-}$ concentration (often this is the only term). This arises from the presence of very small amounts of the radical SO_2^{-} in equilibrium with $S_2O_4^{2-}$:

$$S_2O_4^{2-} \rightleftharpoons 2\,SO_2^{-} \qquad K = k_1/k_{-1} \tag{2.52}$$

$$ox + SO_2^{-} \xrightarrow{k_2} red + SO_2 \longrightarrow products \tag{2.52a}$$

It is not too difficult to show[29-31]

$$V = -d\,[S_2O_4^{2-}]/dt = \frac{2k_1\,[S_2O_4^{2-}]}{1 + (1 + a\,[S_2O_4^{2-}])^{1/2}} \tag{2.53}$$

where

$$a = \frac{16k_1 k_{-1}}{k_2^2\,[ox]^2} \tag{2.54}$$

Two limits are immediately obvious,

$$a\,[S_2O_4^{2-}] \ll 1, \quad V = k_1\,[S_2O_4^{2-}] \tag{2.55}$$

and

$$a\,[S_2O_4^{2-}] \gg 1, \quad V = 1/2\,k_2\,(k_1/k_{-1})^{1/2}\,[S_2O_4^{2-}]^{1/2}\,[ox] \tag{2.56}$$

The latter behavior is usually observed. Similar kinetics apply with the $Cr_2(OAc)_6^{2-}/Cr(OAc)_3^{-}$ system.[31] (Chap.8. Cr(II)).

Fractional orders such as 3/2 often hint at a chain mechanism. The autoxidation of $(H_2O)_5CrCH(CH_3)_2^{2+}$ leads to a number of products. Log (initial rate) vs log (initial concentration of organochromium cation) plots give a 3/2 slope. The rate is independent of $[H^+]$ and $[O_2]$ and the rate law is therefore

$$-d\,[CrCH(CH_3)_2^{2+}]/dt = k_{obs}\,[CrCH(CH_3)_2^{2+}]^{3/2} \tag{2.57}$$

A consistent mechanism is:

$$CrCH(CH_3)_2^{2+} \xrightarrow{k_1} Cr^{2+} + {}^{\bullet}CH(CH_3)_2 \tag{2.58}$$

$${}^{\bullet}CH(CH_3)_2 + O_2 \xrightarrow{k_2} {}^{\bullet}OOCH(CH_3)_2 \tag{2.59}$$

$$CrCH(CH_3)_2^{2+} + {}^{\bullet}OOCH(CH_3)_2 \xrightarrow{k_3} CrOOCH(CH_3)_2^{2+} + {}^{\bullet}CH(CH_3)_2 \tag{2.60}$$

$$2\,{}^{\bullet}OOCH(CH_3)_2 \xrightarrow{k_4} (CH_3)_2CO + (CH_3)_2CHOH + O_2 \tag{2.61}$$

With a steady state approximation for the chain-carrying intermediates in (2.59) and (2.60) and the assumption of long chain length[32]

$$k_{obs} = \frac{k_3 k_1^{1/2}}{(2k_4)^{1/2}} \tag{2.62}$$

see also Refs. 33 and 34.

2.1.7 The Inclusion of [H⁺] Terms in the Rate Law

There are problems in correctly ascribing $[H^+]$ terms in the rate law to a mechanism for the reaction. First, it must be decided whether a medium effect rather than a distinctive reaction pathway might be responsible for the variation of rate with $[H^+]$, particularly if this is a small contribution. This is an important point that we shall deal with later (Sec. 2.9.2). Secondly, even when it has been established that the pH term has a mechanistic basis, there may be an ambiguity in the interpretation of the rate law. On occasion, such ambiguity has been quite severe and has led to much discussion.

(a) *Positive Dependence on* $[H^+]$. Inclusion of an $[H^+]^n$, $n \geqslant 1$, term in the rate law can usually be explained by the operation of a rate-determining reaction of a protonated species. Usually there is a likely basic site on one of the reactants for protonation, and the greater reactivity of the protonated species compared with the unprotonated form can usually be rationalized (see Sec. 4.3.1). A two-term rate law for the acid hydrolysis of CrX^{n+} (see (1.208)),

$$-d \ln [CrX^{n+}]/dt = k + k' [H^+] \tag{2.63}$$

has been noted with a number of basic ligands, F^-, N_3^-, and so forth. The terms can be attributed to reactions of protonated and unprotonated forms of the complex

$$CrX^{n+} + H^+ \underset{K_1}{\rightleftharpoons} CrXH^{(n+1)+} \xrightarrow[H_2O]{k_1} products \tag{2.64}$$

$$CrX^{n+} \xrightarrow[H_2O]{k_0} products \tag{2.65}$$

with $k = k_0$ and $k' = K_1 k_1$. The depletion of the reagent water will be an important consideration in high acid concentrations if water features in the activated complex. The rate of hydrolysis might then decrease with an increase of (high) perchloric acid concentration and

$$V \sim k [complex] a_w \tag{2.66}$$

where a_w represents the activity of water. Linear rate dependencies on the activity of water are observed in the hydrolysis of $CrOCOCH_3^{2+}$ in 6–8 M $HClO_4$[35], of $CrClO_4^{2+}$ in 5–10 M $HClO_4$[36] and of $Co(NH_3)_5OPO_3H_3^{3+}$ in >5 M $HClO_4$.[37]

(b) *Negative Dependence on* $[H^+]$. Inclusion of an $[H^+]^n$ term in the rate law, where n is a negative integer, can be attributed to a proton being a product of a preequilibrium step (see Sec. 2.1.5), and therefore arising from the rate-determining reaction of a deprotonated species. It is often likely to occur when one of the reactants is acidic.

The formation of $FeCl^{2+}$ from Fe^{3+} and Cl^- ions in acid solution obeys the rate law (compare with (1.6))[38]

$$d [FeCl^{2+}]/dt = k_1 [Fe^{III}] [Cl^-] + k_2 [Fe^{III}] [Cl^-] [H^+]^{-1} \tag{2.67}$$

The first term simply represents the reaction between the fully hydrated iron(III) ion, $Fe(H_2O)_6^{3+}$, abbreviated Fe^{3+}, the predominant iron species, and chloride ion, with an associated activated complex $[FeCl(H_2O)_n]^{\ddagger}$.

$$Fe^{3+} + Cl^- \rightarrow FeCl^{2+} \qquad k_1 \tag{2.68}$$

Inclusion of the $[H^+]^{-1}$ term is reasonably ascribed to the reaction of $Fe(H_2O)_5OH^{2+}$, abbreviated $FeOH^{2+}$, with Cl^- ions in the slow step:

$$Fe^{3+} + OH^- \rightleftharpoons FeOH^{2+} \qquad K \qquad (2.69)$$

$$H_2O \rightleftharpoons H^+ + OH^- \qquad K_w \qquad (2.70)$$

$$FeOH^{2+} + Cl^- \rightarrow Fe(OH)Cl^+ \qquad k \qquad (2.71)$$

$$Fe(OH)Cl^+ + H^+ \rightarrow FeCl^{2+} \qquad \text{fast} \qquad (2.72)$$

for which

$$k_2 = kK_wK \qquad (2.73)$$

The similar rate laws for the reactions of Fe(III) with a number of aprotic ligands can be rationalized in the same manner. The rate constants for ligation of Fe^{3+} and $FeOH^{2+}$ are shown in Table 2.1. The exchange of water with Fe(III) and other tervalent transition metal ions, Cr(III), Rh(III), as well as Al(III) and Ga(III), Eqn. 4.5, proceeds via $M(H_2O)_6^{3+}$ as well as by $M(H_2O)_5OH^{2+}$, so that here also an $[H^+]^{-1}$ term features in the rate law.

Table 2.1 Rate Constants (k, $M^{-1}s^{-1}$) for Reactions of Fe^{3+} and $FeOH^{2+}$ with Ligands at 25°C

Ligand	Fe^{3+}	$FeOH^{2+}$	Ref.
Cl^-	9.4	1.1×10^4	41
NCS^-	127	1.0×10^4	41
HF	11.4	—	41
HN_3	4.0	—	41
Ambiguous			
CrO_4^{2-}	5×10^7		
$HCrO_4^-$		9.2×10^3	42
F^-	5×10^3		
HF		3.2×10^3	41
N_3^-	1.6×10^5		
HN_3		6.3×10^3	41
SO_4^{2-}	4×10^3		
HSO_4^-		2.4×10^4	41
$RCON(O^-)R_1$	$\sim 10^9$		
$RCON(OH)R_1$		$\sim 10^3$	45
$H_2PO_4^-$	6.8×10^4		
H_3PO_4		9.2×10^6	43

(c) *Proton Ambiguity.* Acid-catalyzed reaction between species A and B, *both* of which are basic, leads to interpretive difficulties

$$V = k[A][B][H^+] \qquad (2.74)$$

since we are uncertain whether A or B takes the proton into the activated complex. The rate law for the oxidation of hydrazinium ion by Cr(VI) takes the form of (2.74). The reactant pair $N_2H_5^+$ and CrO_2H^{3+} are preferred over $N_2H_6^{2+}$ and CrO_3^{2+} for chemical reasons. [39]

We can now return to the reaction considered at the beginning of this chapter. The third-order rate constant k will equal K_1k_1 on the basis of (2.4) and K_2k_2 if mechanism (2.7) is correct. It is possible to make rough estimates of the values (at 4°C) of $k_1(2.5 \times 10^4 M^{-1}s^{-1})$ and $k_2(\sim 5 \times 10^8 \ M^{-1}s^{-1})$ from the values of K_1 (0.1 M^{-1}), $K_2(\sim 2 \times 10^{-5} \ M^{-1})$ and $k(5 \times 10^3 M^{-2}s^{-1})$. The improbably high value for k_2 is one of the reasons that the first mechanism is preferred. [7,8] Deviation from a first-order dependence on $[H^+]$ occurs at high acidity since a stage is being reached where H_2CrO_4 is in significant concentration (see (1.98)). [40] Even here and in the limiting region, where the rate is independent of $[H^+]$ because H_2CrO_4 is the major chromium species and its concentration is pH-independent, it is not difficult to see that the concentration products $[H_2CrO_4][H_2O_2]$ and $[HCrO_4^-][H_3O_2^+]$ are still kinetically indistinguishable.

Diffusion-controlled reaction — The maximum rate of a reaction between two species *1* and *2* of charge z_1 and z_2 is controlled by the rate at which the reactants come together (diffusion-controlled). The second-order rate constant k_1 ($M^{-1}s^{-1}$) for this interaction is given by

$$k_1 = \frac{4 \times 10^{-3}\pi N(d_1 + d_2)\, aU(a)}{kT[\exp(U(a)/kT) - 1]}$$

The maximum rate for the reverse reaction (unimolecular dissociation k_{-1}, s^{-1}) is via separation by diffusion of the two molecules

$$k_{-1} = \frac{3(d_1 + d_2)U(a)}{kT[\exp(U(a)/kT) - 1]a^2} \cdot \left[\exp\left(\frac{U(a)}{kT}\right)\right]$$

$U(a)$ is the Debye-Hückel interionic potential

$$U(a) = \frac{z_1z_2e^2}{aD} - \frac{z_1z_2e^2\kappa}{D(1 + \kappa a)}$$

$$\kappa^2 = \frac{8\pi Ne^2\mu}{1000\,Dk\,T}$$

d_1 and d_2 are the diffusion coefficients of *1* and *2* in cm^2s^{-1}; e is the charge on an electron in esu units; D is the dielectric constant of the medium; k is Boltzmann's constant in ergs; N is Avogadro's number and a is the distance in cm of closest approach of the ions ($r_1 + r_2$), all at T (in K). The ratio k_1/k_{-1} is a diffusion-controlled equilibrium constant and equals the theoretically-deduced outer-sphere formation constant, K_0 in (4.17). See G. Q. Zhou and W. Z. Zhong, Eur. J. Biochem. **128**, 383 (1982).

Obviously, ambiguity can also arise in the interpretation of the rate law

$$V = k\,[\text{AH}]\,[\text{BH}]\,[\text{H}^+]^{-1} \tag{2.75}$$

where both AH and BH are acidic. In the reaction of Fe(III) with ligands that can take part in acid-base equilibria, interpretive difficulties arise.[41] With $HCrO_4^-$, for example, the term $k_2\,[\text{Fe(III)}]\,[HCrO_4^-]\,[\text{H}^+]^{-1}$ can arise from reaction of $Fe(H_2O)_6^{3+}$ with CrO_4^{2-},

$$HCrO_4^- \quad\rightleftharpoons\quad CrO_4^{2-} + H^+ \qquad K_{Cr} \tag{2.76}$$

$$Fe(H_2O)_6^{3+} + CrO_4^{2-} \xrightarrow{\ k_a\ } Fe(H_2O)_5OCrO_3^+ + H_2O \tag{2.77}$$

or from $Fe(H_2O)_5OH^{2+}$ interacting with $HCrO_4^-$,

$$Fe(H_2O)_6^{3+} \quad\rightleftharpoons\quad Fe(H_2O)_5OH^{2+} + H^+ \qquad K_{Fe} \tag{2.78}$$

$$Fe(H_2O)_5OH^{2+} + HCrO_4^- \xrightarrow{\ k_b\ } Fe(H_2O)_5OCrO_3^+ + H_2O \tag{2.79}$$

In Scheme (2.76, 2.77), $k_2 = K_{Cr}k_a$ and in Scheme (2.78, 2.79), $k_2 = K_{Fe}k_b$. Since $K_{Cr} = 3 \times 10^{-7}$M and $K_{Fe} = 1.7 \times 10^{-3}$M, the observed value of k_2, 15 s^{-1}, leads to values for k_a of 5×10^7M^{-1}s^{-1} and for k_b of 9.2×10^3M^{-1}s^{-1}, all at 25°C. The calculated value of k_b seems much more reasonable than the calculated value of k_a which is inordinately high. Thus FeOH^{2+}, $HCrO_4^-$ as the kinetically active pair is therefore preferred.[42] Table 2.1 shows a selected number of examples that may be resolved similarly.[41] It is exceptional for $Fe(H_2O)_6^{3+}$ to be the preferred reactant.[43,44]

The reactions of Cr(III),[46] Al(III)[47] and Ga(III)[48] have been rationalized in a manner similar to that used for the reactions of Fe(III). Although the hydroxy form MOH^{2+} is the minor species present in the acid medium used in such studies (typically 1–2%), its enhanced reactivity compared with M^{3+}, both in substitution and redox reactions, will ensure its participation in the overall rate. We encountered the problem of the interpretation of (2.75) in Chap. 1 (Sec. 1.10.2). The rate constant for reaction of the acid form of bovine carbonic anhydrase with deionized p-(salicyl-5-azo)benzenesulfonamide is calculated as 10^{10}M^{-1}s^{-1}. This appears to be slightly too large a value for such a reaction. For this reason then, the alternative (kinetically equivalent) reaction of the deprotonated enzyme reacting with **8** containing the uncharged sulfonamide group (2.2×10^7M^{-1}s^{-1}) is preferred.[49] A special problem arises when one of the acidic partners is water. Does the activated complex now arise from an rds between AH and OH$^-$, or between A$^-$ and H$_2$O, or even from the reaction of A$^-$ alone? This ambiguity has been particularly vexing in the study of the base hydrolysis of cobalt(III) complexes, and was a point of discussion for many years (see Sec. 4.3.4).[50]

8

Finally, when the rate law indicates that there is more than one activated complex of impor- tance, the composition but not the order of appearance of the activated complexes in the reac- tion scheme is defined by the rate law. Haim[6] has drawn attention to this in considering the reduction of V(III) by Cr(II) in acid solution.[51] The second-order rate constant k in the rate law

$$d[Cr^{III}]/dt = k[Cr^{II}][V^{III}] \tag{2.80}$$

is dependent on $[H^+]$:

$$k = \frac{a}{b + [H^+]} \tag{2.81}$$

The limiting forms of the rate law yield the compositions of the activated complexes. These will be, at low $[H^+]$, $[VCr^{5+}]^{\ddagger}$, and at high $[H^+]$, $[VCr(OH)^{4+}]^{\ddagger}$. Thus two mechanisms are possible. In one of these, $[VCr^{5+}]^{\ddagger}$ precedes $[VCr(OH)^{4+}]^{\ddagger}$ (mechanism 1):

$$Cr^{2+} + V^{3+} + H_2O \rightleftharpoons Cr(OH)V^{4+} + H^+ \qquad k_1, k_{-1} \tag{2.82}$$

$$Cr(OH)V^{4+} \rightleftharpoons CrOH^{2+} + V^{2+} \qquad k_2, k_{-2} \tag{2.83}$$

$$CrOH^{2+} + H^+ \rightarrow Cr^{3+} \qquad \text{rapid} \tag{2.84}$$

for which

$$\frac{d[Cr^{3+}]}{dt} = \frac{k_1 k_2 [V^{3+}][Cr^{2+}]}{k_2 + k_{-1}[H^+]} \tag{2.85}$$

which is of the required form with $a = k_1 k_2/k_{-1}$ and $b = k_2/k_{-1}$ (with k_{-2} ignored).

In the other scheme, $[VCr^{5+}]^{\ddagger}$ occurs after $[VCr(OH)^{4+}]^{\ddagger}$ in the reaction sequence (mechanism 2)

$$V^{3+} + H_2O \rightleftharpoons VOH^{2+} + H^+ \qquad \text{rapid, } K \tag{2.86}$$

$$Cr^{2+} + VOH^{2+} \rightleftharpoons Cr(OH)V^{4+} \qquad k_1, k_{-1} \tag{2.87}$$

$$Cr(OH)V^{4+} + H^+ \rightarrow Cr^{3+} + V^{2+} + H_2O \qquad k_2, k_{-2} \tag{2.88}$$

$$\frac{d[Cr^{3+}]}{dt} = \frac{k_1 k_2 K [V^{3+}][Cr^{2+}]}{k_{-1} + k_2[H^+]} \tag{2.89}$$

Equation (2.89) is of the form (2.81) with $a = k_1 K$ and $b = k_{-1}/k_2$ (with k_{-2} ignored). Indirect arguments for the validity of the second mechanism have been presented.[6] At high acidity, $[H^+] > 0.5$ M, the concentration and thus contribution of the very reactive VOH^{2+} is so reduced that an outer sphere

$$V^{3+} + Cr^{2+} \rightarrow V^{2+} + Cr^{3+} \tag{2.90}$$

reaction becomes detectable.[52]

An alternative presentation of the mechanisms in (2.82)–(2.84) and (2.86)–(2.88) is shown in (2.91).[6] This depiction is popular in the complex mechanisms encountered in metallo-enzyme chemistry.

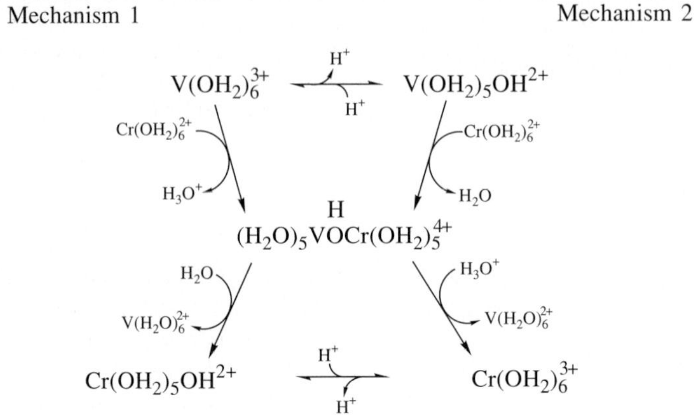

Mechanism 1   Mechanism 2

$$
\begin{array}{ccc}
V(OH_2)_6^{3+} & \xrightarrow[H^+]{H^+} & V(OH_2)_5OH^{2+} \\
Cr(OH_2)_6^{2+} \searrow & & \swarrow Cr(OH_2)_6^{2+} \\
H_3O^+ \searrow & \quad H \quad & \nwarrow H_2O \\
& (H_2O)_5VOCr(OH_2)_5^{4+} & \\
H_2O \searrow & & \nwarrow H_3O^+ \\
V(H_2O)_6^{2+} \searrow & & \swarrow V(H_2O)_6^{2+} \\
Cr(OH_2)_5OH^{2+} & \xrightarrow[H^+]{H^+} & Cr(OH_2)_6^{3+}
\end{array}
$$

(2.91)

2.2 Further Checks of Mechanism

So far we have assigned mechanisms mainly on the basis of the rate law. This can give only a somewhat crude picture, detailing at the most the number of the steps involved. Some evidence for the correctness of the mechanism can be obtained by consideration of the rate constants for these steps. In the reaction of $Co(edta)^{2-}$ with $Fe(CN)_6^{3-}$, for example, the formation of a "dead-end" complex is preferred (Sec. 1.6.4 (d)). The outer sphere redox reaction which results from this interpretation, (direct reaction of $Co(edta)^{2-}$ with $Fe(CN)_6^{3-}$) is also supported by agreement of the rate constant with that calculated using Marcus theory (Sec. 2.5).[53,54] Distinguishing an "active" from a "dead-end"complex, i.e. scheme (1.101) from scheme (1.107) and (1.108) is a vexing problem, particularly in the interaction of small inorganic reactants with proteins.[55] Solutions to these problems and finer mechanistic detail can often be produced by subsidiary experiments, usually chemical in nature, as will now be detailed.

2.2.1 The Detection and Study of Intermediates[56]

(a) *Direct*. It may happen that the form of the rate law can be accomodated only by a mechanism where intermediates are postulated. Therefore, strong evidence for such a mechanism is the detection of these intermediates. In some cases this may present little difficulty since the intermediate may accumulate in relatively large amounts and therefore be easily detected during the course of the reaction. V^{2+} mixed with VO^{2+} produces a more rapid loss of V^{2+} than a gain of V^{3+}, the ultimate product. In concentrated solution, an intermediate brown color (ascribed to VOV^{4+}) can actually be discerned.[57] Much more difficult is the support for intermediates of fleeting existence. Then special means must often be used, involving sophisticated equipment and techniques. The mode and power of the approach is well illustrated by the work on detecting the $HO_2^{\bullet}$ radical in aqueous solution.

The kinetics of the Ce(III)-Ce(IV) exchange reaction catalyzed by H_2O_2[58] and a later study of the kinetics of the Ce(IV) reaction with H_2O_2 in H_2SO_4 by stopped-flow methods,[59] argues for the following mechanism for the Ce(IV)-H_2O_2 reaction,

$$Ce^{IV} + H_2O_2 \rightleftharpoons Ce^{III} + HO_2^{\bullet} + H^+ \qquad k_1, k_{-1} \qquad (2.92)$$

$$Ce^{IV} + HO_2^{\bullet} \rightarrow Ce^{III} + O_2 + H^+ \qquad k_2 \qquad (2.93)$$

in which $k_1 = 1.0 \times 10^6 M^{-1}s^{-1}$ and $k_2/k_{-1} = 13$. It is apparent from these results that a sizable amount of $HO_2^{\bullet}$ should be produced, at least for a short while, by mixing large amounts of H_2O_2 (0.1 M) with small amounts of Ce(IV) (10^{-3}M), which are used up in step (2.92) and thus cannot remove $HO_2^{\bullet}$ in step (2.93). If this is carried out in an efficient mixer, and the mixed solutions examined within 10 ms by electron spin resonance (esr), then $HO_2^{\bullet}$ is detected in the flow tube.[60]

The trapping of reactive intermediates at low temperatures in a rigid medium prevents them from reacting, and allows a leisurely examination. This, combined with esr examination, has been important in studies of certain metalloproteins, e. g. nitrogenase.[61]

(b) *Indirect.* By far the most usual approach for sensing the presence of intermediates is to add reagents ("scavengers") which will react rapidly and effectively with an intermediate, but not the reactants. Either the rate law may be modified, perhaps only slightly, or new products may result to the exclusion of, or in addition to, the normal product.

The species $O_2^{\bullet-}$ or the form in acid, $HO_2^{\bullet}$, is often considered a logical immediate product of reactions of O_2. It has been indirectly detected by scavenging additives modifying the reaction.[62]

Co(sep)$^{2+}$ (9) reacts with O_2 in acid solution with an assumed mechanism

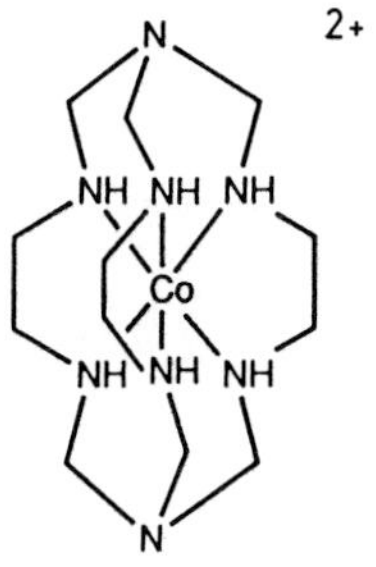

9

$$Co(sep)^{2+} + O_2 \xrightarrow{H^+} Co(sep)^{3+} + HO_2^{\bullet} \qquad k_1 \qquad (2.94)$$

$$Co(sep)^{2+} + HO_2^{\bullet} \xrightarrow{H^+} Co(sep)^{3+} + H_2O_2 \qquad \text{fast} \qquad (2.95)$$

based on a rate law

$$-d[Co(sep)^{2+}]/dt = 2k_1[Co(sep)^{2+}][O_2] \qquad (2.96)$$

If the reaction is carried out in the presence of Cu^{2+} the rate is reduced by a factor of approximately two. This is excellent evidence for the intermediacy of $HO_2^{\bullet}$ since this would react with Cu^{2+} at a close to diffusion-controlled rate

$$Cu^{2+} + HO_2^{\bullet} \rightarrow Cu^+ + O_2 + H^+ \tag{2.97}$$

This reaction would replace (2.95) and a modified rate law would operate[63]

$$-d\,[Co(sep)^{2+}]/dt = k_1\,[Co(sep)^{2+}]\,[O_2] \tag{2.98}$$

Transient radicals $R^{\bullet}$ can be visualized by reacting with diamagnetic spin traps T to form a persistent spin adduct RT which can be analyzed leisurely by esr,[64,65] (Sec. 1.11).

Competition experiments, in which the intermediate is scavenged by two reactants and the amount and nature of products examined, have played an important role in establishing the mechanism for base hydrolysis of cobalt(III) complexes. The intermediate produced in the base hydrolysis of $Co(NH_3)_5X^{n+}$, $n = 3, 2$ or 1 is $Co(NH_3)_4(NH_2)^{2+}$ if the D_{cb} mechanism is correct (Sec. 4.3.4). Normally this intermediate is considered to react with H_2O to produce $Co(NH_3)_5OH^{2+}$ but if another nucleophile $(Y-Z)^{m-}$ is also present and this can attack the intermediate also, then $Co(NH_3)_5(Y-Z)^{(3-m)+}$ and $Co(NH_3)_5(Z-Y)^{(3-m)+}$ will also result (Scheme (2.99)). These are linkage isomers (Sec. 7.4) and are spectrally distinguishable

$$Co(NH_3)_5X^{n+} + OH^- \rightleftharpoons Co(NH_3)_4(NH_2)X^{(n-1)+} + H_2O \rightleftharpoons Co(NH_3)_4NH_2^{2+} + X^{(3-n)-}$$

$$Co(NH_3)_4(NH_2)(H_2O)^{2+} \xrightarrow{\text{fast}} Co(NH_3)_5OH^{2+}$$

$$\overset{H_2O}{\diagup}$$

$$Co(NH_3)_4(NH_2)^{2+}$$

$$\underset{(Y-Z)^{m-}}{\overset{(Y-Z)^{m-}}{}}$$

$$Co(NH_3)_4(NH_2)(Z-Y)^{(2-m)+} \xrightarrow[\text{fast}]{H_2O} Co(NH_3)_5(Z-Y)^{(3-m)+} + OH^-$$

$$Co(NH_3)_4(NH_2)(Y-Z)^{(2-m)+} \xrightarrow[\text{fast}]{H_2O} Co(NH_3)_5(Y-Z)^{(3-m)+} + OH^-$$

$$\tag{2.99}$$

If a series of complexes of different charges and X groups are hydrolyzed in the presence of a constant concentration of $(Y-Z)^{m-}$ then a constant competition ratio

$$\frac{[Co(NH_3)_5(Y-Z)^{(3-m)+}] + [Co(NH_3)_5(Z-Y)^{(3-m)+}]}{[Co(NH_3)_5OH^{2+}]} \tag{2.100}$$

should result because a common intermediate is involved, independent of X and n in $Co(NH_3)_5X^{n+}$. A selection of some recent results are contained in Table 2.2[66-69]

The ratio of linkage isomers is remarkably constant and independent of the charge, steric bulk or reactivity of leaving group. The ratios are 2.0 ± 0.1 S/N (SCN^-); 2.0 ± 0.2 O/N (NO_2^-) and 2.3 ± 0.3 S/O $(S_2O_3^{2-})$. For a particular entering ligand $(Y-Z)^{m-}$ the percentage total anion capture is very constant within each charge group $(n = 1-3)$ although it increases

Table 2.2. Percentage of $(Y-Z)^{m-}$ Capture in the Base Hydrolysis of $Co(NH_3)_5X^{n+}$ in the Presence of $(Y-Z)^{m-}$ at 25°C (Usually in 1.0 M $(Y-Z)^{m-}$ and 0.1 M OH^-)[66-69]

n	X	$(Y-Z)^{m-}$			
		SCN^-[ab]	N_3^-	NO_2^- [ac]	$S_2O_3^{2-}$ [ab]
3	$OP(OCH_3)_3$	17.3 (11.9)	12.5	7.9 (5.0)	12.2 (8.7)
3	$OS(CH_3)_2$	18.0 (12.0)	12.3	8.8 (5.6)	12.6 (9.1)
2	I^-	13.6 (8.9)	10.0	4.5	11.3 (8.0)
2	Cl^-	–	8.5	–	–
2	$OS(O)_2CF_3^-$	13.6 (9.0)	9.7	7.0 (4.6)	10.8 (7.7)
1	OSO_3^{2-}	6.8 (3.7)	5.8	–	–

[a] Corrected value for subsequent reaction of $Co(NH_3)_5(Y-Z)^{(3-m)+}$.
[b] Value in parenthesis is % S bound linkage isomer.
[c] Value in parenthesis is % O bound linkage isomer.

systematically with the charge of substrate $3+>2+>1+$. The conclusion is that there is a 5-coordinated intermediate which is so short-lived that it retains the original ion-atmosphere of $Co(NH_3)_5X^{n+}$ but not the X group (Sec. 4.3.5). These experiments are very difficult to set up, carry out and analyze, but are very telling support for the D_{cb} mechanism.

In some cases the identification of the product, even without use of scavengers, can be important in the elucidation of mechanism. Thus the quenching of $*Ru(bpy)_3^{2+}$ by cobalt(III) complexes (Eqn. 1.32, B = Co(III)) may go by electron transfer

$$*Ru(bpy)_3^{2+} + Co(III) \rightarrow Ru(bpy)_3^{3+} + Co(II) \tag{2.101}$$

or energy transfer

$$*Ru(bpy)_3^{2+} + Co(III) \rightarrow Ru(bpy)_3^{2+} + *Co(III) \tag{2.102}$$

Product analysis has been useful in deciding the preferred path. Quantum yields for Co(II) production (number of molecules produced/number of quanta absorbed) measured by steady-state photolysis of $Co(NH_3)_6^{3+}$ and $Co(en)_3^{3+}$ are 0.45 and 0.11 respectively. These show that exclusive electron transfer (when the quantum yield would be 1.0) does not occur.[70]

With specially designed reactants, the determination of the product structure can be very informative. Mercury(II)-catalyzed aquations of Co(III) complexes are believed to proceed via a 5-coordinated intermediate (Sec. 4.3.2). The shape of this intermediate is of interest. The Hg^{2+}-catalyzed aquation of $Co(NH_3)_4 (ND_3)X^{2+}$ in which the ND_3 and X groups are *trans* to one another gives substantially *trans*-$Co(NH_3)_4(ND_3)(H_2O)^{3+}$. This is excellent evidence for a square pyramidal intermediate in the reaction. A trigonal-bipyramidal intermediate would be expected to lead to substantial scrambling of the ND_3 and NH_3 groups (Fig. 2.1).[71] The *definite* but very small amount ($2.8 \pm 0.4\%$) of *cis* product recently reported using $^{15}NH_3$ instead of ND_3 attests to the sensitivity of current nmr machines.[72]

The use of highly enriched ^{13}CO and $C^{18}O$ and ir or nmr monitoring has been a powerful combination for studying the photochemical and thermal substitution processes of metal carbonyls and derivatives. The site of substitution and the nature of the reactive intermediates (their geometry and flexibility) have been elucidated.[73]

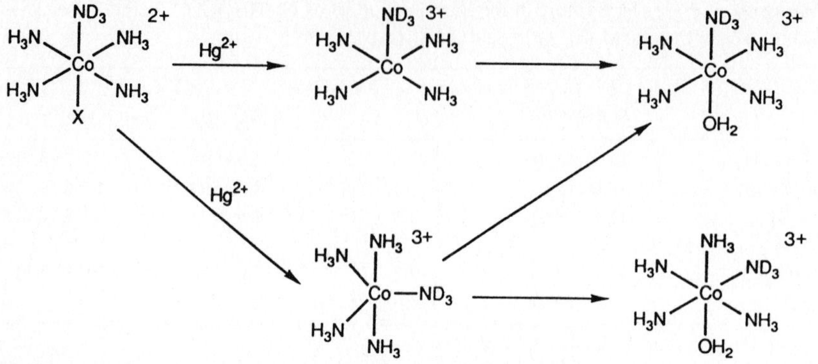

Fig. 2.1 Expected products from the Hg(II)-catalyzed aquation of *trans*-$Co(NH_3)_4(ND_3)X^{2+}$ on the basis of a square-pyramid or trigonal-bipyramid intermediate. [71]

2.2.2 The Determination of Bond Cleavage

Obtaining definite evidence regarding the occurrence of bond cleavage is helped considerably by using isotopes. The following two examples illustrate different approaches.

Reactions of $Co(NH_3)_5OH^{2+}$ or $Co(NH_3)_5H_2O^{3+}$ with certain ligands are unusually rapid (Chap. 8. Co(III)). This suggests that Co-O bond breakage does not occur during the substitution, else it likely would be very much slower. This can be verified by using isotopes e. g. [74,75]

$$Co(NH_3)_5{}^{18}OH_2^{3+} + NO_2^- \rightarrow Co(NH_3)_5{}^{18}ONO + H_2O \tag{2.103}$$

$$Co(NH_3)_5{}^{17}OH^{2+} + SO_2 \rightarrow Co(NH_3)_5{}^{17}OSO_2H^{2+} \rightarrow Co(NH_3)_5{}^{17}OH_2^{3+} + SO_2 \tag{2.104}$$

The nmr signal due to bound $^{17}OH_2$ is unchanged after the cycle (2.104) is completed (Fig. 2.2). [75] This observation also rules out linkage rotation (2.105)

$$(NH_3)_5Co{-}^{17}O{-}S\overset{O^+}{\underset{O}{\diagdown}} \rightleftharpoons (NH_3)_5Co{-}O{-}S\overset{O^+}{\underset{^{17}O}{\diagdown}} \tag{2.105}$$

Such a rearrangement, although seemingly unlikely, does occur with the nitrito complex as demonstrated by ^{17}O-nmr experiments [76]

$$(NH_3)_5Co{-}^{17}ONO^{2+} \rightleftharpoons (NH_3)_5Co{-}ON^{17}O^{2+}$$

$$(NH_3)_5Co{-}N\overset{^{17}O^{2+}}{\underset{O}{\diagup}} \tag{2.106}$$

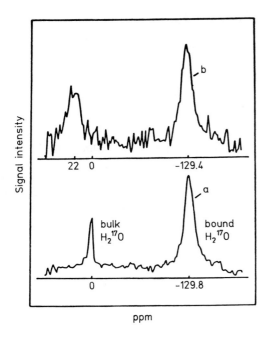

Fig. 2.2 ^{17}O nmr spectrum of $Co(NH_3)_5{}^{17}OH_2^{3+}$ dissolved in $H_2{}^{16}O$ (a) before and (b) after formation and acidification of sulfito complex, Eqn. (2.104). Resonance for the metal-bound water is upfield (-130 ppm) relative to bulk water (0 ppm) and is markedly broader. After acidification, little intensity changes occur for the bound and bulk water, proving that the $Co-{}^{17}OH_2$ bond remains substantially intact during the cycle. The broadening and shifting of the bulk water during the process results from paramagnetic Co^{2+} formed during a side redox reaction.[75] Reprinted with permission from R. van Eldik, J. von Jouanne, and H. Kelm, Inorg. Chem., **21**, 2818 (1982). © (1982) American Chemical Society.

The formation of a double-bridged Co(III) complex from a single peroxo-bridged complex can take place by either (2.107) or (2.108), $L = CH_3NH_2$,

$$(tren)LCoO_2CoL(tren)^{4+} \rightleftharpoons 2\,(tren)Co(H_2O)_2^{2+} + O_2 + 2\,L$$
$$\mathbf{10}$$
$$\rightleftharpoons (tren)Co(O_2,OH)Co(tren)^{3+} \qquad (2.107)$$
$$\mathbf{11}$$

$$(tren)LCoO_2CoL(tren)^{4+} \rightleftharpoons (tren)CoO_2Co(tren)^{4+} + 2\,L$$
$$\mathbf{10} \qquad\qquad\qquad | \quad |$$
$$OH_2\ OH_2 \qquad\qquad (2.108)$$
$$\rightleftharpoons (tren)Co(O_2,OH)Co(tren)^{3+}$$
$$\mathbf{11}$$

The first mechanism is preferred since the $^{18}O_2$-labelled bridged complex **10** exchanges with O_2 *at the same rate* as the formation of the bridged product **11**.[77,78]

How the position of bond cleavage can be determined is illustrated by consideration of thermally initiated ligand (L) substitution in $Re_2(CO)_{10}$. This can occur by a dissociative mechanism involving $Re-CO$ bond breakage,

$$Re_2(CO)_{10} \rightleftharpoons Re_2(CO)_9 + CO \qquad (2.109)$$

$$Re_2(CO)_9 + L \rightleftharpoons Re_2(CO)_9L$$

or by a radical mechanism involving metal-metal bond homolysis

$$Re_2(CO)_{10} \rightleftharpoons 2\,Re(CO)_5^\bullet$$

$$Re(CO)_5^\bullet + L \rightleftharpoons Re(CO)_4L^\bullet + CO \qquad (2.110)$$

$$2\,Re(CO)_4L^\bullet \rightleftharpoons Re_2(CO)_8L_2$$

The former mechanism is favored. When CO exchange (L = CO) is studied using a mixture of $^{185}Re_2(CO)_{10}$ and $^{187}Re_2(CO)_{10}$ no $^{185}Re\,^{187}Re(CO)_{10}$ appears as product (analyzed by mass spectrometry). The mixture of isotopic carbonyls do however "scramble" (slowly) in the absence of L or after photolysis, showing that Re – Re bond scission is at least feasible.[79] The *kinetic isotope effect* is extensively used to support mechanism.[80] A large deuterium kinetic isotope effect $(k_H/k_D > 5)$ is strong support for an X-H (or X-D) bond cleavage in the rate determining step. The 10.5 fold higher rate constant for the MnO_4^- oxidation of $Co(NH_3)_5OCHO^{2+}$, compared with the rate constant for $Co(NH_3)_5OCDO^{2+}$, supports a mechanism in which the rds is a one-electron oxidation via H atom abstraction,[81] for example

$$Co(NH_3)_5OCHO^{2+} + MnO_4^- \rightarrow Co(NH_3)_5(OCO^\bullet)^{2+} + HMnO_4^- \qquad (2.111)$$

$$
Co(NH_3)_5(OCO^\bullet)^{2+}
\begin{array}{l}
\xrightarrow{k_1} Co(NH_3)_5^{2+} + CO_2 \\[1em]
\xrightarrow[MnO_4^-]{k_2} Co(NH_3)_5H_2O^{3+}
\end{array}
\qquad (2.112)
$$

The first step accounts for the observed second-order kinetics and produces the large isotope effect. The second step leads to a Co(II), Co(III) mixture.[82,83] Even larger isotope effects have been noted (e. g. in C – H abstraction from 2-propanol by Ru(IV) compounds[84,85]).

It is difficult to correlate the k_H/k_D values with the operation of a hydride, proton or H atom transfer. The temperature dependence of the kinetic isotope effect, which is now easier to measure accurately, is more diagnostic of mechanism but has been applied mainly to organic systems.[86]

^{13}C kinetic isotope effects $(k\,(^{12}C)/k\,(^{13}C))$ are more difficult to measure accurately. The values for a variety of metal ion-catalyzed decarboxylations of oxaloacetate (2.113) are similar (1.04–1.05). This suggests that the transition state for decarboxylation (a) involves a marked breakage of the C – C bond and (b) is similar for the various metal ions, even though enhancement rates vary widely.[87] This apparent paradox is ascribed to an alteration of the distribution of oxaloacetate between the keto and enol forms.[88]

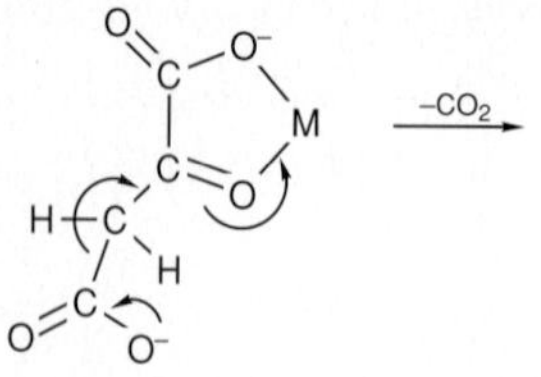

$$(2.113)$$

It is interesting that enzyme-catalyzed decarboxylations have a ^{13}C effect of ~ 1.00, thus invoking a different mechanism.

Finally in studying isomerization reactions, one must always consider whether these are intramolecular (no bond cleavage) or intermolecular (bond cleavage). The relation between the isomerization rate constants and the dissociation rate constants will resolve whether bond cleavage accounts for isomeric change. [89] Other examples of the use of isotopes to detect bond breakage or bond stretching in the activated complex will occur in Part II. Determination of the fractionation factor by using isotopes can also lead to similar information (Sec. 3.12.2).

2.3 Activation Parameters, Thermodynamic Functions and Mechanism

So far we have considered rate laws and chemical behavior for one temperature and at normal pressures only. Much information on the mechanism of a reaction may certainly be gleaned from this information alone. However, by carrying out measurements at a number of temperatures or pressures, or in different media, one may obtain useful information, even though the form of the rate law itself rarely changes.

2.3.1 The Effect of Temperature on the Rate of a Reaction

The rate of a chemical reaction may be affected by temperature in several ways, but the most common behavior by far is that observed by Arrhenius some 100 years ago. [90] The empirical expression

$$k = A \exp(-E_a/RT) \tag{2.114}$$

relates the rate constant k to the absolute temperature T (in K). It describes the behavior of a vast number of chemical reactions amazingly well, particularly over a fairly small temperature range, but in some instances over as large a range as 100 degrees. [91,92] Thus a plot of $\ln k$ vs T^{-1} is linear, with slope $-E_a/R$ and intercept $\ln A$. (Fig. 2.3) [91]

A similar relationship is also derived by the absolute reaction rate theory, which is used almost exclusively in considering, and understanding, the kinetics of reactions in solution. [93] The activated complex in the *transition state* is reached by reactants in the initial state as the highest point of the most favorable reaction path on the potential energy surface. The activated complex $X^{\ddagger}$ is in equilibrium with the reactants A and B, and the rate of the reaction V is the product of the equilibrium *concentration* of $X^{\ddagger}$ and the specific rate at which it decomposes. The latter can be shown to be equal to kT/h, where k is Boltzmann's constant and h is Planck's constant:

$$A + B \underset{}{\overset{K_c^{\ddagger}}{\rightleftharpoons}} X^{\ddagger} \rightarrow products \tag{2.115}$$

$$V = kT[X^{\ddagger}]/h = kTK_c^{\ddagger}[A][B]/h \tag{2.116}$$

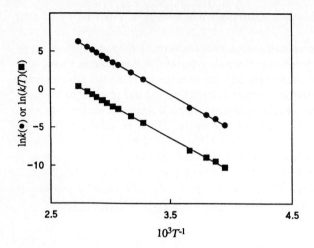

Fig. 2.3 Arrhenius ($\bullet$) and Eyring ($\blacksquare$) plots of data for the exchange of a single dmso molecule with Ga(dmso)$_6^{3+}$ in CD$_3$NO$_2$. Nmr line-broadening was used for the higher temperatures and stopped-flow nmr experiments for the lower temperatures.[91] For the Arrhenius plot, the slope and intercept (at $T^{-1} = 0$) are -9.2×10^3 and $+31.4$ respectively, leading to $E_a = 76$ kJ mol^{-1} and $\log A = 13.6$. For the Eyring plot, the slope and intercept are -8.77×10^3 and $+24.4$ respectively, leading to $\Delta H^{\ddagger} = 73$ kJ mol^{-1} and $\Delta S^{\ddagger} = 5$ J K^{-1}mol^{-1}.

so that the experimental second-order rate constant k is given by

$$k = kTK_c^{\ddagger}/h \tag{2.117}$$

The formation of the activated complex may be regarded as an equilibrium process involving an "almost" normal molecule (almost, since it is short one mode of vibrational energy). The free energy of activation $\Delta G^{\ddagger}$ can therefore be defined as in normal thermodynamics,

$$\Delta G^{\ddagger} = -RT \ln K_c^{\ddagger} = \Delta H^{\ddagger} - T\Delta S^{\ddagger} \tag{2.118}$$

leading to

$$k = \frac{kT}{h} \exp\left(\frac{-\Delta G^{\ddagger}}{RT}\right) = \frac{kT}{h} \exp\left(\frac{-\Delta H^{\ddagger}}{RT}\right) \exp\left(\frac{\Delta S^{\ddagger}}{R}\right) \tag{2.119}$$

where $\Delta H^{\ddagger}$ and $\Delta S^{\ddagger}$ are the enthalpy and entropy of activation. This equation strictly applies to nonelectrolytes in dilute solution and must be modified for ionic reactions in electrolyte solutions (see (2.179)).

Since $E_a = \Delta H^{\ddagger} + RT$, it is not difficult to show that

$$A = \frac{ekT}{h} \exp\left(\frac{\Delta S^{\ddagger}}{R}\right) \tag{2.120}$$

A plot of $\ln(k/T)$ against $1/T$ is linear from (2.119), with a slope, $-\Delta H^{\ddagger}/R$ and an intercept $(\ln k/h + \Delta S^{\ddagger}/R) = (23.8 + \Delta S^{\ddagger}/R)$, Fig. 2.3. This is sometimes referred to as the Eyring relationship. Both Arrhenius and Eyring plots are used and give very similar results. The quantities $\Delta H^{\ddagger}$ and $\Delta S^{\ddagger}$ are almost universally preferred for discussion of solution kinetics and will be used hereafter. The agreement in the values of $\Delta H^{\ddagger}$ and $\Delta S^{\ddagger}$ obtained by different workers investigating the same system is not always good. Nowhere has this been more evident than in the nmr studies of solvent exchange. For example, although the rate constant at 25 °C for exchange of $Ni(CH_3CN)_6^{2+}$ with solvent CH_3CN is generally reported in the relatively narrow range of 2.0×10^3 to $4.0 \times 10^3 s^{-1}$, $\Delta H^{\ddagger}$ and $\Delta S^{\ddagger}$ values vary enormously.[94] Both positive and negative values of $\Delta S^{\ddagger}$ are reported! This arises from the inaccuracy of obtaining $\Delta S^{\ddagger}$ by a long extrapolation of Eyring plots to $1/T = 0$. This problem can be partly circumvented by using as wide a temperature range as feasible.[95] $\Delta V^{\ddagger}$ values (see below) give similar and in some respects superior information to those of $\Delta S^{\ddagger}$, and are more accurately, but not as easily, measured.

2.3.2 The Variation of E_a or $\Delta H^{\ddagger}$ with Temperature

Most investigators are content to use the best linear plots of $\ln k$ or $\ln(k/T)$ vs T^{-1} in estimating E_a or $\Delta H^{\ddagger}$ values, and to assume that these are constants over a narrow temperature range.[96] It is worth examining this aspect, however, since deviations from linearity of such plots might be ascribable to complexity in the reaction mechanism.

Since the heat of a reaction ΔH is rarely temperature-independent and since for a single step

$$\Delta H = \Delta H_f^{\ddagger} - \Delta H_r^{\ddagger} \tag{2.121}$$

it would be expected that the enthalpy of activation in the forward direction, $\Delta H_f^{\ddagger}$, or in the reverse direction, $\Delta H_r^{\ddagger}$, or both, also would generally be temperature-variant even for a simple reaction. Using a simple approach, the relationship

$$dE_a/dT = \Delta C_p^{\ddagger} \tag{2.122}$$

holds where $\Delta C_p^{\ddagger}$ is the heat capacity of activation at constant pressure.[97] Values for $\Delta C_p^{\ddagger}$ within the range -40 to -120 J K^{-1} mol^{-1} (same units as $\Delta S^{\ddagger}$) have been found for the aquation of a number of $Co(NH_3)_5X^{2+}$ complexes.[98] *Marked* variation of E_a or $\Delta H^{\ddagger}$ values with temperature (that is, nonlinear Arrhenius or Eyring plots) for reactions of complexes is unusual. This is perhaps surprising, since apart from the $\Delta C_p^{\ddagger}$ effect many observed rate constants, and therefore associated activation parameters, are composite values. It is easily seen that even if the rate constants for the individual steps vary exponentially with respect to temperature, the composite rate constants may well not. Therefore deviations should arise in reactions involving equilibria, parallel or consecutive processes.[97,99] Examples are rare (Sec. 2.6 [100,101]).

2.3.3 The Effect of Pressure on the Rate of a Reaction

The effect of pressure P on the rate constant k for a reaction can be summarized in the simplified expression

$$\left(\frac{d\ln k}{dP}\right)_T = \frac{-\Delta V^{\ddagger}}{RT} \tag{2.123}$$

The volume of activation, $\Delta V^{\ddagger}$, is the partial molar volume change when reactants are converted to the activated complex. $\Delta V^{\ddagger}$ is the volume of activation extrapolated to $P = 0$ (if necessary). P is usually expressed in megaPascals (MPa) $= 10^6\text{Pa} = 10$ bars $= 9.87$ atmospheres or 1 J cm^{-3}. At 298.2 K, $RT = 2.48 \times 10^3$ J mol^{-1}. If the slope of the $\ln(k/k_{P=0})$ vs P (MPa) is s, then $\Delta V^{\ddagger} = -2.48 \times 10^3 \times s$ cm^{3}mol^{-1} at 298.2 K. Consider the effect of pressure on the rate of water exchange (k_{obs}) on aqueous Cr(III) perchlorate. For this reaction

$$k_{\text{obs}} = k_0 + k_1[\text{H}^+]^{-1} \tag{2.124}$$

and k_0 and k_1 can be determined at a number of pressures (Fig. 2.4). From the slopes, $\Delta V_0^{\ddagger} = -10\text{cm}^3\text{mol}^{-1}$ and $\Delta V_1^{\ddagger} = -0.9\text{cm}^3\text{mol}^{-1}$ for the k_0 and the k_1 paths. [102] We shall consider these values in Sec. 2.8.

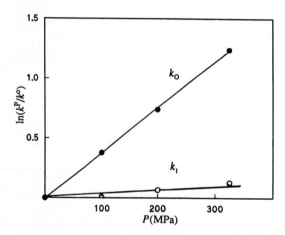

Fig. 2.4 Effect of pressure on the k_0 and k_1 terms in (2.124) for water exchange of aqueous Cr(III) perchlorate. [102] Since $\Delta V^{\ddagger} = -2.48 \times 10^3 \times$ slope (see text), $\Delta V_0^{\ddagger} = -2.48 \times 10^3 \times (3.8 \times 10^{-3}) = -10$ cm^{3}mol^{-1} and $\Delta V_1^{\ddagger} = -2.48 \times 10^3 \times (3.3 \times 10^{-4}) = -0.9$ cm^{3}mol^{-1}.

Although one of the earliest studies of pressure effects on a complex ion reaction was carried out over 30 years ago, [103] only the past decade has seen a concentrated number of studies of pressure effects on the rates of transition metal complex reactions. This is evidenced by the large number of recent reviews in this area. For a selection see Refs. 104–107. Instruments are now commercially available and flow, relaxation and nmr techniques have been adapted for use in high pressure kinetics (Ch. 3). [108] Fairly high pressures, 200–1000 MPa or 2.000–10.000 atmospheres, must be used to obtain sufficient effect on the rate. $\Delta V^{\ddagger}$ values usually lie in the range ± 25 cm^{3}mol^{-1} and can be obtained to ± 1 cm^{3}mol^{-1}. The values, resulting from the *slope* of $(d\ln k/dP)_T$ plots, are more likely to be accurate than those of $\Delta S^{\ddagger}$, arising from a long extrapolation to an intercept of a $(d\ln k/dT)_p$ plot, and with which they are sometimes compared. Although $\Delta V^{\ddagger}$ is conceptually easy to understand, there is one real problem in its interpretation. This arises because $\Delta V_{obs}^{\ddagger}$ can be considered as made up of two parts (a) $\Delta V_{intr}^{\ddagger}$, the intrinsic volume change when reactants are converted to the activated complex. It arises from changes in bond length, angles etc. and is diagnostic of the intimate mechanism. (b) $\Delta V_{solv}^{\ddagger}$ the volume change arising from solvation effects (electrostriction of solvent). Term (b) is unfortunately not easy to assess and this creates problems when it contributes substantially to the overall $\Delta V_{obs}^{\ddagger}$. Term (b) is less important when the charges of reactants and products are the same and for this reason, exchange reactions have been popular to study for mechanistic information.

2.3.4 The Variation of $\Delta V^{\ddagger}$ with Pressure

Sometimes $\Delta V^{\ddagger}$ is pressure-dependent because the compressibility of reactants and transition state is different. The most popular way of dealing with this is to use an empirical quadratic expression

$$\ln k = \ln k_{P=0} - \left(\frac{\Delta V^{\ddagger}}{RT}\right)_{P=0} P + \frac{\Delta\beta^{\ddagger}P^2}{2RT} \tag{2.125}$$

where $\Delta\beta^{\ddagger}$ is the compressibility coefficient of activation. The variation of $\Delta V^{\ddagger}$ with pressure appears to be, experimentally, a more important consideration than that of $\Delta H^{\ddagger}$ with temperature. The expression (2.125) is often used but the value of $\Delta\beta^{\ddagger}$ for interpretative purposes appears restricted at present and is often near zero.

2.3.5 Activation Parameters and Concentration Units

The use of accurate concentrations is vital in assessing correct rate constants and associated activation parameters. Concentrations are usually measured in terms of the number of moles of each reactant present in 1 liter of solution (M). Since liquids undergo thermal expansion or hydrostatic compression, the volume will change, leading to complications when using molarities. This can be avoided by expressing concentrations in units such as mole fractions or molalities (moles in 1 kg of solvent) which do not involve volumes. If these points are considered, an accurate value of $\Delta V^{\ddagger}$ (or $\Delta H^{\ddagger}$) can be obtained. Fortunately simple considerations [109] show that a true value of $\Delta V^{\ddagger}$ can be obtained using (2.125) and uncorrected rate constants at various pressures based on the *molarities at atmospheric pressure* . The dilemma and analysis have been carefully reviewed. [109]

2.3.6 Reaction Profiles

Activation parameters are often composite. Consider the value of $\Delta V_1^{\ddagger}$ in the exchange of Cr(III) with H_2O (Sect. 2.3.3). The values of k_1 and $\Delta V_1^{\ddagger}$ are related to the intrinsic rate and thermodynamic constants k_{OH} and $\Delta V_{OH}^{\ddagger}$ (parameters for water exchange on $Cr(H_2O)_5OH^{2+}$) by the equations

$$k_{OH} = k_1/K_a \tag{2.126}$$

and

$$\Delta V_{OH}^{\ddagger} = \Delta V_1^{\ddagger} - \Delta V_a \tag{2.127}$$

where K_a and ΔV_a refer to the ionization (2.128)

$$Cr(H_2O)_6^{3+} \rightleftharpoons Cr(H_2O)_5OH^{2+} + H^+ \qquad K_a, \Delta V_a \tag{2.128}$$

$$Cr(H_2O)_5OH^{2+} \rightleftharpoons \text{exchange} \qquad k_{OH}, \Delta V_{OH}^{\ddagger} \tag{2.129}$$

It is very important therefore to have information on the thermodynamic parameters, in this instance ΔV_a. These can be measured directly by dilatometry or from the relationship: $(d \ln K/dP)_T = -\Delta V/RT.$[108] Since $\Delta V_a = -3.8 \text{ cm}^3\text{mol}^{-1}$, $\Delta V_{OH}^{\ddagger} = -0.9 + 3.8 = 2.9 \text{ cm}^3\text{mol}^{-1}$, Ref. 102. We can represent the progress of this and any other reaction pictorially by a *reaction profile,* using the concept of the activated complex. The reaction profile shows, often in a qualitative but useful fashion, the change of any activation parameter ($\Delta G^{\ddagger}$, $\Delta H^{\ddagger}$, $\Delta S^{\ddagger}$ Ref. 110 or $\Delta V^{\ddagger}$ Ref. 111) as a function of the extent of the reaction (termed the reaction coordinate). Since each step in a reaction will have an associated transition state, and thus a separate reaction profile, we may have a continuous series of such profiles joining the reactants to the ultimate product.

Consider the reaction scheme

$$A \underset{k_{-1}}{\overset{k_1}{\rightleftharpoons}} B \xrightarrow{k_2} C \tag{2.130}$$

in which B is a reactive intermediate present in steady-state concentration

$$d[B]/dt = 0 = k_1[A] - (k_{-1} + k_2)[B] \tag{2.131}$$

$$d[C]/dt = k_2[B] = \frac{k_1 k_2 [A]}{k_{-1} + k_2} \tag{2.132}$$

The form of the reaction profile will depend on the relative values of the rate constants, Fig. 2.5. Several interesting points may be made.

(a) Although it is often stated that the rds is associated with the activated complex of highest energy, say $\ddagger_1$ in Fig. 2.5(A), this may be misleading because $\Delta G^{\ddagger}$ is calculated from the *rate constants.* Thus if in the first step of (2.130) another reactant D was involved, the forward rate associated with it would be $k_1[D]$ and it is consideration of rates which determine the rds. [112,113]

(b) Variations in the structures of the reactants may alter the energy level of A or $\ddagger$, or both, and these lead to changes in $\Delta G^{\ddagger}$. In the comparison of a reaction series, it is important to bear this in mind, and it is not always easy to diagnose which behavior is leading to the rate differences within the series. This has led to controversies on the meaning of $\Delta V^{\ddagger}$ values from water exchange with M^{2+} and $M(NH_3)_5H_2O^{3+}$ Refs. 114–117.

(c) If the reaction A → C is represented as in Fig. 2.5 moving from left to right, then the reverse reaction C → A is simply represented by the reverse process traversed from right to left in the same diagram. This embodies the *principle of microscopic reversibility* which states that the mechanism of a reversible reaction is the same in microscopic detail in one direction as in the other, under the same conditions. [118]

Using the activation parameters outlined in the previous sections, we probe their value in assessing mechanism.

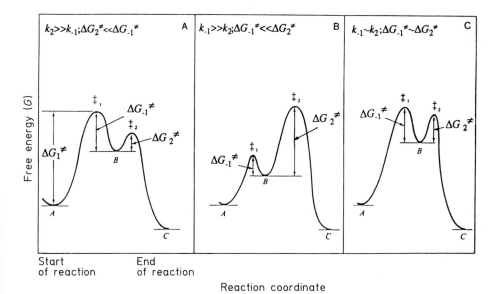

Fig. 2.5 Free energy reaction profile for (2.130) for various relative values of the associated rate constants.

2.4 Free Energy of Activation and Mechanism

The actual value of a rate constant for a reaction only infrequently gives a clue to its mechanism. Assessment of values within a reaction series may be more revealing, while comparisons of free energies of activation $\Delta G^{\ddagger}$ with free energies for the reactions ΔG, leading to the linear free-energy relationships (LFER), can be very useful in diagnosing mechanism.

The existence of similar rate constants for a series of reactions in which there is only a change in the central metal or in the ligand in one of the reactants, suggests that a common mechanism is operative for the whole reaction series. The rate constants for aquation of a series of complexes

$$M(NH_3)_5OCOCF_3^{2+} + H_2O \rightarrow M(NH_3)_5H_2O^{3+} + CF_3CO_2^- \qquad (2.133)$$

are similar for M = Co, Rh and Ir. This independence on metal is unusual and suggests that C−O rather than M−O bond cleavage is occurring, since the former process might be expected to be much less sensitive to the nature of M than would M−O breakage. [119]

Increasing steric hindrance in a reacting molecule will usually be relieved in the transition state if the mechanism is dissociative but aggravated if the reaction proceeds by an associative one. Thus the acid hydrolysis of $Co(NH_2CH_3)_5Cl^{2+}$ proceeds 23 times faster than that of $Co(NH_3)_5Cl^{2+}$. This supports the assigned dissociative (I_d) mechanism (Ch. 4). [120] With the Cr(III) analogs, the CH_3NH_2 substituted derivative reacts 33 times slower. This is used to suggest an associative (I_a) mechanism. These assignments have independent support, particularly from $\Delta V^{\ddagger}$ values (Sec. 4.2.6). The reversal of relative rates for the Cr(III) complexes has been rationalized in terms of small differences in Cr−Cl bond lengths in the two complexes. [121] This argument has been criticized because of the small variation of M−Cl bond distances anyway in solid metal complexes. [122] The suggestion does however emphasize the importance of examining the ground state as well as the activated complex in rationalizing activation free energy effects.

A subtle example of steric acceleration associated with a dissociative mechanism is shown in the reaction

$$\textit{cis-}Mo(CO)_4(R_3P)_2 + CO \xrightarrow{k} Mo(CO)_5(R_3P) + R_3P \tag{2.134}$$

As the "size" of R_3P increases so does the value of k. A popular method of assessing the "size" of R_3P is by using Tolman's cone angle. [123] This is the apex angle of the cone centered on P which just encloses the van der Waals radii of the outermost atoms of R_3P. The data are shown in Table 2.3 for (2.134) in CO-saturated C_2Cl_4 at 70°C (Ref. 124)

Table 2.3 Effect of Tolman's Cone Angle on (2.134)

PR_3	Cone Angle	k, s^{-1}
$PPhMe_2$	122°	$< 10^{-6}$
PPh_2Me	136°	1.3×10^{-5}
PPh_3	145°	3.2×10^{-3}

Study of the relative rate constants for reactions of a series of similar complexes A with reagents B and C may reveal deviations by a particular complex of the series A from the general pattern and therefore suggest it is reacting by an anomalous mechanism. This rarely occurs, but a general correlation suggests a common mechanism for the reactions.

The pattern for reduction of alkyl halides by Ni(tmc)$^+$ (**12**) closely resembles that by Cr(15[ane]N$_4$)$^{2+}$ (**13**). This suggests a similar mechanism for both reductants. There is strong

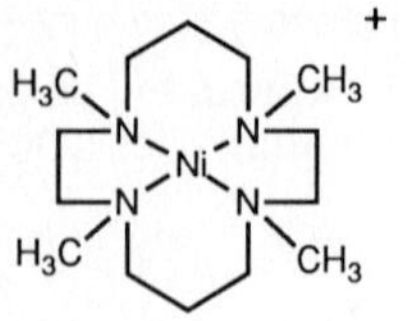

12

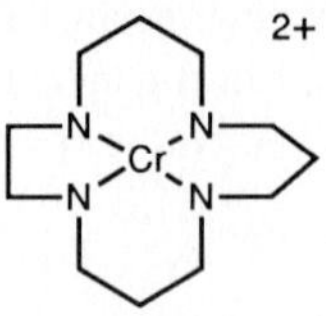

13

support for atom transfer for the Cr(II) reductions since inert halochromium(III) results (Sec. 5.3 (a)). Therefore reductions by Ni(tmc)$^+$ are also believed to occur by atom transfer, rather than by an outer sphere mechanism. [125]

$$Ni(tmc)^+ + RX \rightarrow XNi(tmc)^+ + R^\bullet \tag{2.135}$$

$$XNi(tmc)^+ \rightarrow Ni(tmc)^{2+} + X^- \tag{2.136}$$

Finally, there is a striking linear correlation of the rate constants for the reactions (N represents a ligand with an N-donor atom, e.g. NH_3):

$$Co(N)_5X^{2+} + Hg^{2+} + H_2O \xrightarrow{k_{Hg}} Co(N)_5H_2O^{3+} + HgX^+ \tag{2.137}$$

$$Co(N_5)X^{2+} + H_2O \xrightarrow{k_{H_2O}} Co(N_5)H_2O^{3+} + X^- \tag{2.138}$$

For the plot of log k_{Hg} vs log k_{H_2O} for 34 complexes the slope is 0.96 with little scatter (Fig. 2.6). A common mechanism with a 5-coordinated intermediate is indicated. [126]

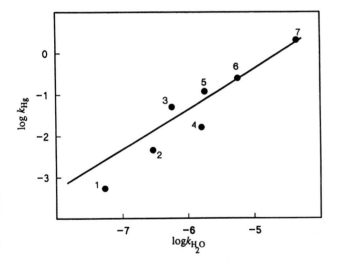

Fig. 2.6 Plot of log k_{Hg} vs log k_{H_2O} for a number of reactions of the type (2.137) and (2.138) respectively at μ = 1.0 M HClO$_4$ and 25°C. Only a few entries are selected from the 34 reactions tabulated in Ref. 126 but these illustrate the variety of complexes which conform to the correlation observed. Complexes: *cis*-Co(en)$_2$(NH$_2$CH$_2$CN)Cl^{2+} (1), *trans*-Co(en)$_2$(NH$_3$)Cl^{2+} (2), *cis*-Co(en)$_2$(imid)Cl^{2+} (3), *cis*-Co(en)$_2$(3 Mepy)Cl^{2+} (4), Co(NH$_3$)$_5$Cl^{2+} (5), *mer*-Co(dien)(tn)Cl^{2+} (6), *cis*-Co(tn)$_2$(C$_2$H$_5$NH$_2$)Cl^{2+} (7). The line is the best fit for the 34 reactions, slope = 0.96.

2.5 Linear Free-Energy Relationships $-\Delta G^{\ddagger}$ and ΔG

So far we have considered a limited series of rate relationships and their potential value in substantiating mechanism. We now examine more detailed linear free-energy relationships (LFER), a subject that has had full attention in organic chemistry but only recently has been exploited by the inorganic chemist. [126, 127]

In spite of the justifiable warnings not to confuse the kinetics and thermodynamics of a reaction, there are circumstances, for example in a closely related series of reactions, in which it might be expected that the free energies of activation, $\Delta G^{\ddagger}$ and reaction, ΔG would parallel one another. There is usually no problem in measuring or estimating the equilibrium constant, and hence ΔG, for many substitution and redox reactions by using formation constants or standard oxidation potentials. This information, together with the rate data, might then be used to test LFER. In turn, such LFER might be used to diagnose mechanism by determining the extent of bond formation or breakage in the transition state or by assessing the importance of electronic, polar, or steric effects on the rate.

Since

$$-\Delta G^{\ddagger} = RT \ln \frac{kh}{kT} = 2.3\,RT \log \frac{kh}{kT} \tag{2.139}$$

and

$$-\Delta G = RT \ln K = 2.3\,RT \log K \tag{2.140}$$

the linearity between the free energies of activation and reaction might be most easily expressed in the form

$$\log k = A \log K + B \tag{2.141}$$

Historically, the decadic log scale has been mainly used in LFER.

This idea can be tested by examining data for the aquation of a series of ions, $Co(NH_3)_5X^{(3-n)+}$,

$$Co(NH_3)_5X^{(3-n)+} + H_2O \rightleftharpoons Co(NH_3)_5H_2O^{3+} + X^{n-} \qquad k_1, k_{-1}, K_1 \tag{2.142}$$

The plot of $\log k_1$ vs $\log K_1$ is linear over a wide range of rate constants (Fig. 2.7). [128] Obviously, the faster the aquation, the more the reaction goes to completion! The slope A is 1.0 and this indicates that the activated complex and the products closely resemble one another, that is, that X^{n-} has substantially separated from the cobalt and that therefore the mechanism of these reactions is dissociative (I_d). Since $K_1 = k_1/k_{-1}$

$$\log k_1 = \log K_1 + \log k_{-1} \tag{2.143}$$

and (2.143) is of the form (2.141) with $A = 1.0$ and $B = \log k_{-1}$. The value of B is approximately constant if a dissociative mechanism applies to anation, since the $Co-H_2O$ bond

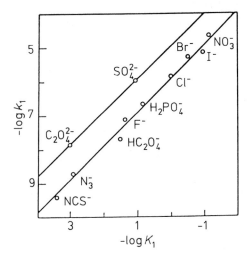

Fig. 2.7 Plot of $\log k_1$ vs $\log K_1$ for (2.142) at 25 °C. Ref. 128.

cleavage is common to all reactions and not much influenced by the nature of X^{n-}. Such linear plots for solvolysis rate constants vs equilibrium constants with slopes near unity are reported for reactions of some Cr(III) complexes [129] where L is a tetradentate Schiff base (2.144), and for Li and Na

$$Cr(L)(H_2O)X^+ + H_2O \rightleftharpoons Cr(L)(H_2O)_2^+ + X \qquad (2.144)$$

complexes of spherands and cryptands in a variety of solvents. [130, 131] These LFER afford good evidence for a strongly dissociative mode for substitution in these reactions. By contrast, an excellent linear correlation of $\log k_{-1}$ vs $\log K_1$ (slope 0.94) is observed for the hydrolysis of the blue and red forms of the square pyramidal Cu(II) complex, Cu(tet b)X (Sec. 7.9).

$$Cu(tet\,b)X + H_2O \rightleftharpoons Cu(tet\,b)H_2O + X \qquad k_1, k_{-1}, K_1 \qquad (2.145)$$

Now, it is the forward rate constant k_1 that is almost invariant (because there is a common nucleophile H_2O), whereas k_{-1} is very sensitive to the nucleophilic character of X, i.e. it is an associative type reaction. [132] Selectivity would be expected to lead to curved free energy plots but these have not yet been observed. [115]

Undoubtedly one of the most used LFER in transition metal chemistry involves electron transfer rate constants and associated equilibrium constants in *outer sphere redox reactions*. These are an unusual class of reactions in chemistry since bonds are only stretched or contracted in the formation of the activated complex. They therefore lend themselves well to theoretical treatment. We shall have more to say about these reactions in Chap. 5. It is sufficient here to state the simple form of the LFER [133] with an example (Fig. 2.8). [134] For the reaction

$$ox_1 + red_2 \rightleftharpoons red_1 + ox_2 \qquad k_{12}, k_{21}, K_{12} \qquad (2.146)$$

$$k_{12} = (k_{11} k_{22} K_{12} f)^{1/2} \qquad (2.147)$$

where

$$\log f = \frac{(\log K_{12})^2}{4 \log (k_{11} k_{22}/10^{22})} \qquad (2.148)$$

and k_{11} and k_{22} refer to the isotopic exchange reactions

$$\mathrm{ox}_1 + \mathrm{red}_1 \rightleftharpoons \mathrm{red}_1 + \mathrm{ox}_1 \qquad k_{11} \qquad (2.149)$$

$$\mathrm{ox}_2 + \mathrm{red}_2 \rightleftharpoons \mathrm{red}_2 + \mathrm{ox}_2 \qquad k_{22} \qquad (2.150)$$

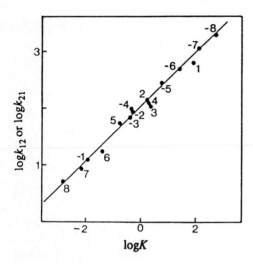

Fig. 2.8 Plot of $\log k_{12}$ or $\log k_{21}$ vs $\log K$ for the reaction (1.55).

$$\mathrm{Fe(III)} + \mathrm{PTZ} \rightleftharpoons \mathrm{Fe(II)} + \mathrm{PTZ}^{\ddagger} \qquad k_{12}, k_{21}, K_{12}$$

(PTZ = variety of N-alkylphenothiazines). Positive numbers use k_{12} and K_{12}. Negative numbers refer to k_{21} and $1/K_{12}$. X = OH, R = $(CH_2)_3N(CH_3)_2(1)$, X = H, R = $(CH_2)_3N\bigcirc NCH_3(2)$, X = OCH$_3$, R = $(CH_2)_3N(CH_3)_2(3)$, X = H, R = $(CH_2)_3N(CH_3)_2(4)$, X = Cl, R = $(CH_2)_3N(CH_3)_2(5)$, X = H, R = $(CH_2)_2N(C_2H_5)_2(6)$, X = H, R = $CH_2CH(CH_3)N(CH_3)_2(7)$, X = H, R = $CH(CH_3)CH_2N(CH_3)_2$ (8). [134] The slope of Fig. 2.8 is 0.5 as expected in (2.151). Reprinted with permission from E. Pelizetti and E. Mentasti, Inorg. Chem., **18**, 583 (1979). © (1979) American Chemical Society.

Although an extended form of (2.147) is sometimes used (Eqn. 5.37) the simple one can account for a surprisingly large number of redox reactions. When the oxidizing power of ox_1 and ox_2 are comparable, $f \sim 1$, and

$$\Delta G_{12}^{\ddagger} = 0.50 \, (\Delta G_{11}^{\ddagger} + \Delta G_{22}^{\ddagger} + \Delta G_{12}^0) \qquad (2.151)$$

The rate constant for one reaction may have to be correlated with the equilibrium constant not for that reaction but for a related one. Rate constants for reactions of metal complexes can be correlated with the proton affinity of one of the ligands, either in the coordinated or free state. Thus, the rate constant for CO_2 uptake by a variety of metal hydroxo complexes depends on the nucleophilicity of the "bound" hydroxide ion. It therefore correlates with the $O-H$ bond strength of the corresponding aqua complex, i.e. the pK_a of the coordinated water.[135] See also Figs. 6.5 and 8.4. As a second, more involved, example consider the internal electron transfer rate constant (k_{et}) within the ion pair of $Co(NH_3)_5O_2CR^{2+}$ with $Fe(CN)_6^{3-}$. It is observed that k_{et} varies inversely with the pK_a of RCO_2H. It is argued that the oxidation potential of Co(III) in the complex (and therefore the rate (see above)) will increase with decreasing σ-donor strength of the carboxylate ligand. Decreasing σ-donor strength is related to decreasing pK_a.[136,137] This indirect approach is the basis for most LFER discussed in the next sections.

2.5.1 Hammett Relationship

The Hammett equation,[138] one of the oldest and most useful of LFER, correlates the rates of reaction of a series of *meta-* and *para-*substituted aromatic compounds with a common substrate,

$$\log \frac{k}{k^0} = \rho \log \frac{K_a}{K_a^0} = \rho\sigma \tag{2.152}$$

where k and k^0 are the reaction rate constants for the X-substituted and unsubstituted aromatic compounds respectively, and K_a and K_a^0 are the dissociation constants for the X-substituted and unsubstituted benzoic acids. This is of the general form (2.141) with $A = \rho$ and $B = \log k^0 - \rho \log K_a^0$, which is a constant for a reaction series. The parameter σ depends on the substituent but is independent of the reaction series, whereas the value of ρ depends only on the actual reaction examined. Either $\log k$ or $\log (k/k^0)$ is plotted against σ. The slope of the line is ρ ($\Delta \log k/\Delta\sigma$) and is positive if k increases as the value of σ becomes more positive.

Second-order rate constants k for the replacement of a ring-substituted benzoate group by hydroxide ion in a number of complexes (base hydrolysis) have been carefully determined at 25 °C in 40% aqueous methanol:

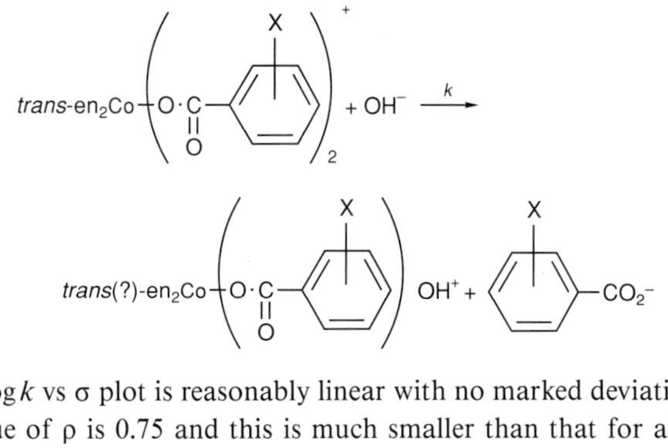

$$\tag{2.153}$$

The $\log k$ vs σ plot is reasonably linear with no marked deviations for seven X-substituents. The value of ρ is 0.75 and this is much smaller than that for alkaline hydrolysis of the free esters (1.8–2.5). Acyl-oxygen fission has been established for the esters. Since ρ is a measure of the sensitivity of the reaction series to ring substitution, the smaller value of ρ with the cobalt complexes could be rationalized by the reaction site being further removed from the aromatic ring than it is in the free esters i. e. $Co-O$ cleavage is involved for all seven reactions (2.153). [139]

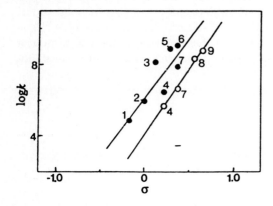

Fig. 2.9 Plots of $\log k$ for the reduction of $(H_2O)_5Cr(NC_5H_4X)^{3+}$ ($\bigcirc$) and $NC_5H_4XH^+$ ($\bullet$) by $^\bullet C(CH_3)_2OH$ vs the Hammett σ values for X, where X = $4-C(CH_3)_3(1)$, $H(2)$, $3-OH(3)$, $4-Cl(4)$, $3-CONH_2(5)$, $4-CONH_2(6)$, $3-Cl(7)$, $3-CN(8)$ and $4-CN(9)$. [140] Reprinted with permission of A. Bakač, V. Butkovic, J. H. Espenson, R. Marcec, and M. Orhanovic, Inorg. Chem., 25, 2562 (1986). © (1986) American Chemical Society.

A comparison of the reduction of coordinated and free pyridines by a variety of reducing agents is very informative (Fig. 2.9 [140]). Reduction of $(NH_3)_5M(NC_5H_4X)^{3+}$, M = Co and Ru, by $^\bullet C(CH_3)_2OH$, $Ru(NH_3)_6^{2+}$ and other strong reducing agents is known, or can be inferred, to occur by outer-sphere electron transfer to the metal. These reactions obey the Hammett LFER with ρ values of 1.1–1.9. A similar type of reduction of the free pyridines ($XC_5H_4NH^+$ in acid) has high values of ρ of 6.6–10.3. For the reductions (X = H, 4-CN, 3-CN, 4-Cl, 3-Cl, 4-t-Bu)

$$(H_2O)_5Cr(NC_5H_4X)^{3+} + {}^\bullet C(CH_3)_2OH + H_2O \rightarrow$$
$$Cr(H_2O)_6^{2+} + (CH_3)_2CO + HNC_5H_4X^+ \tag{2.154}$$

of the Cr analog by $^\bullet C(CH_3)_2OH$, $\rho = 6.6 \pm 1.9$. This value strongly suggests that the reduction is different from that of the other complexes (above) and occurs by the "chemical" mechanism in which reduction of the $Cr-NC_5H_4X$ moiety ($Cr^{III}(py)^{3+}$) is by electron transfer to the *pyridyl ring* (as with the free pyridines)

$$Cr^{III}(py)^{3+} + {}^\bullet C(CH_3)_2OH \rightarrow Cr^{III}(py^{\bar{\bullet}})^{2+} + (CH_3)_2CO + H^+ \tag{2.155}$$

$$Cr^{III}(py^{\bar{\bullet}})^{2+} \rightarrow Cr^{II}(py)^{2+} \xrightarrow{\ H^+\ } Cr^{2+} + pyH^+ \tag{2.156}$$

On the other hand, reduction of $(H_2O)_5Cr(NC_5H_4X)^{3+}$ by $Ru(bpy)_3^+$ has an associated value for ρ of 1.1, the same as for the Co and Ru ammine complexes and therefore a similar mechanism. [140] For other applications of Hammett LFER see Refs. 141–144.

2.5.2 Taft Relationship

The reactivity of aliphatic compounds, where steric hindrance near the substituent site is not important, is accommodated by the Taft modification of the Hammett equation[145]

$$\log \frac{k}{k^0} = \sigma^* \rho^* \tag{2.157}$$

where σ^* is the polar substituent constant and ρ^* is a reaction constant, analogous to the Hammett functions σ and ρ respectively. By definition $\sigma^* = 0.00$ for a CH_3 substituent. For the oxidation of the alkyl radicals $^{\bullet}CH_n(CH_3)_{3-n}$ by $IrCl_6^{2-}$ and $Fe(CN)_6^{3-}$ the Taft plot is shown in Fig. 2.10. The rate increase runs parallel with the increase in electron-donating power of the radicals. The oxidations by $IrCl_6^{2-}$ are all near diffusion-controlled. Those by $Fe(CN)_6^{3-}$ have a large $\rho^* = -13.2$, which indicates that the activated complex has ionic character, supporting an electron transfer mechanism.[146] See also Ref. 147.

Problems in the use of σ^* have led to a modification of (2.157)

$$\log \frac{k}{k^0} = \sigma_I \rho_I \tag{2.158}$$

where Taft's induction factor σ_I is based on σ_I being 0.0 for an H substituent.[148] Recent uses of the concept in complex ion chemistry are sparse.[149]

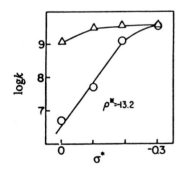

Fig. 2.10 Plots of $\log k$ for the reactions of $^{\bullet}CH_n(CH_3)_{3-n}$ with $IrCl_6^{2-}$ (△) and $Fe(CN)_6^{3-}$ (○) vs the Taft σ^* parameter.[146] Reprinted with permission from S. Steenken and P. Neta, J. Amer. Chem. Soc. **104**, 1244 (1982). © 1982 American Chemical Society.

2.5.3 Brønsted Relationship

The earliest LFER, advanced by Brønsted, correlates the acid dissociation constant (K_{AH}) and base strength $(1/K_{AH})$ of a species with its effectiveness as a catalyst in general acid (k_{AH}) and base (k_B)-catalyzed reactions respectively.[150] The relationships take the form

$$k_{AH} = A K_{AH}^{\alpha} \tag{2.159}$$

$$\log k_{AH} = \log A + \alpha \log K_{AH} \tag{2.160}$$

or

$$k_B = B \left(\frac{1}{K_{AH}} \right)^{\beta}$$ (2.161)

$$\log k_B = \log B - \beta \log K_{AH}$$ (2.162)

Equations (2.160) and (2.162) are both consistent with the generalized form (2.141). The values A and B and α and β are constants with $0 < \alpha, \beta < 1$.

Probably the most detailed study of the acid(HA)- and base(B)-catalyzed reactions of complex ions concerns the hydrolysis of the dichromate ion:

$$Cr_2O_7^{2-} + H_2O \rightleftharpoons 2\,HCrO_4^- \qquad k_1, k_{-1}$$ (2.163)

The hydrolysis rate constant k is given by

$$k = k_1[H_2O] + \sum k_B[B] + \sum k_{AH}[HA]$$ (2.164)

The value of k_B has been determined for a very large number of bases. There is a reasonable correlation of $\log k_B$ vs pK_{BH^+} in agreement with (2.162). The value of β is ~ 0.25, which indicates that general base catalysis holds. For specific base catalysis, β would equal 1, while for reactions in which solvent catalysis predominates β would be zero, since only the $k_1[H_2O]$ term would then be important. [151]

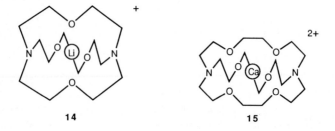

14   15

The dissociations of the metal cryptates $Li(2.1.1)^+$ **14** and $Ca(2.2.2)^{2+}$ **15**, abbreviated $M(cry)^{n+}$, are accelerated by H^+ and undissociated acid HA

$$-d[M(cry)^{n+}]/dt = (k_d + k_H[H^+] + k_{AH}[HA])[M(cry)^{n+}]$$ (2.165)

The plots of $\log k_{AH}$ vs pK_{AH} are linear with slopes (α) of -0.25 (Ca) and -0.64 (Li). These values require that a rate-determining proton transfer from HA to $M(cry)^{n+}$ occur. Alternatively, dissociation of M^{2+} or M^+ from an $AH \cdots (cry)M^{n+}$ adduct may be the rds. The point for k_H is well below the line, a not uncommon behavior. [152]

The Brønsted relationship can be strictly accurate only over a certain range of acid and base strengths. When k_{AH} has diffusion-controlled values, which of course cannot be exceeded, the linear plot of $\log k_{AH}$ vs $\log K_{AH}$ must level off to a zero slope, that is $\alpha = 0$. [153] As well as being reported, although rarely, in simple metal complexes, [154] the resultant curvature in the Brønsted plot is also shown by the zinc enzyme carbonic anhydrase (Chap. 8. Zn(II)). In one of the steps in the overall catalysis, buffers are believed to participate as proton-transfer agents (EZnH$_2$O = enzyme; A = buffer)

$$EZnH_2O + A \underset{k_{-1}}{\overset{k_1}{\rightleftharpoons}} EZnOH^- + AH^+ \tag{2.166}$$

The variation of log k_1 with pK_{AH} of the buffer (AH$^+$) is shown in Fig. 2.11.

A pK_{AH} = 7.6 for the donor group on the enzyme is deduced from the curve. When the pK_{AH} of the buffer is much greater than 7.6, a saturating $k \simeq 10^9 M^{-1}s^{-1}$ is observed. When $pK_{AH} \ll 7.6$, the log k_1 vs pK_{AH} is of unity slope and conforms to (2.160) with α = 1.0.[155] Similar types of plots accompany atom transfer reactions (Fig. 8.11).

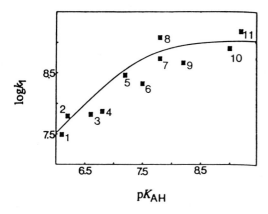

Fig. 2.11 The variation of log k_1 with pK_{AH} of the buffer for reaction (2.166). The buffers used are Mes(1), 3,5-lutidine(2), 3,4-lutidine(3), 2,4-lutidine(4), 1-Meimid(5), Hepes(6), triethanolamine(7), 4-Meimid(8), 1,2-diMeimid(9), Ted(10) and Ches(11). The curve drawn is calculated for k_1 = 1.1 $\times$ 10$^9 M^{-1}s^{-1}$ and a pK_{AH} for the enzyme donor group of 7.6.[155] Reprinted with permission from R. S. Rowlett, and D. N. Silverman, J. Amer. Chem. Soc. **104**, 6737 (1982). © (1982) American Chemical Society.

2.5.4 Swain-Scott Relationship

A completely empirical LFER can also be constructed with recourse only to kinetic data. This has been the case in the setting up of a scale of nucleophilic power for ligands substituting in square-planar complexes based on the Swain-Scott approach.[156] The second-order rate constants k_Y for reactions in MeOH of nucleophiles Y with *trans*-Pt(py)$_2$Cl$_2$, chosen as the standard substrate

$$\textit{trans-}Pt(py)_2Cl_2 + Y^- \rightarrow \textit{trans-}Pt(py)_2ClY + Cl^- \tag{2.167}$$

are compared with the rate constant for solvolysis (Y = CH$_3$OH)

$$k_s = \frac{k_{CH_3OH}}{[CH_3OH]} = \frac{k_{CH_3OH}}{26}$$

Then it is found that (s = 1),

$$\log \frac{k_Y}{k_s} = sn_{Pt} \tag{2.168}$$

On the basis of this equation, an index of nucleophilicity n_{Pt} can be assigned to each nucleophile Y (see Table 4.13). It is found, moreover, that a plot against n_{Pt} of log k_Y, for reaction of Y with another Pt(II) neutral substrate, is also often linear. Thus, Eq. (2.168) applies, and s is termed the nucleophilic discrimination factor (Sec. 4.7.1). Some of the departures from linearity of plots of k_Y vs n_{Pt} which have been observed, disappear if the Pt reference substrate chosen is of the same charge as the Pt reactants. [157] The value of n_{Pt} for a bulky nucleophile has also to be modified to allow for steric hindrance features. [158]

2.6 Enthalpy of Activation and Mechanism

As we have noted in Sec. 2.3.2, there are few authentic examples of changing $\Delta H^{\ddagger}$ values with temperature that are understandable on the basis of (and therefore evidence for) a proposed mechanism. A rds in a multistep mechanism which changes with temperature might account for such behavior.

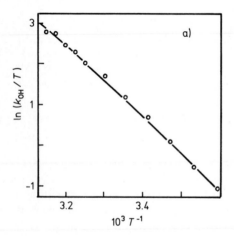

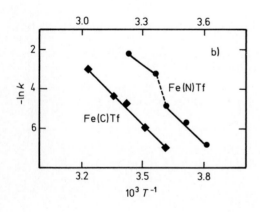

Fig. 2.12 Examples of non-linear Arrhenius (or Eyring) plots (a) $\ln(k_{OH})T^{-1}$) vs T^{-1} for the base hydrolysis of *trans*-Co(en)$_2$Cl$_2^+$. Curvature may result when $k_{-1} \sim k_2$ and $\Delta H^{\ddagger}_{-1}$ not equalling $\Delta H^{\ddagger}_2$ in the conjugate-base mechanism (Sec. 4.3.4). [100] Reprinted with permission from C. Blakeley and M. L. Tobe, J. Chem. Soc. Dalton Trans. 1775 (1987). (b) ln$k$ vs T^{-1} for iron removal from C- and N-terminal monoferric transferrin (lower and upper scales respectively). [101] Transferrin contains two iron binding sites ≈ 35 Å apart. Either of the two sites, designated C- and N-terminal, can be exclusively labelled by Fe(III) ions and these may be removed by a strong ligand such as a catechol (see Sec. 4.11). Reprinted with permission from S. A. Kretschmar and K. N. Raymond, J. Amer. Chem. Soc. **108**, 6212 (1986). © (1986) American Chemical Society.

One such example may arise in the base hydrolysis of *trans*-$Co(en)_2Cl_2^+$. Figure 2.12(a) shows curvature in the $\ln(k_{OH}/T)$ vs T^{-1} plot above 25°C. This arises from a change in the rate-determining step from formation of the 5-coordinated intermediate to deprotonation of *trans*-$Co(en)_2Cl_2^+$ see (Eqn. 2.99) and Sec. 4.3.4. [100] Since this mechanistic change had been anticipated, great care was taken to examine for non-Arrhenius behavior. It is a salutory lesson to consider that such behavior might be more prevalent than supposed. It is always tempting to draw a straight line through the extreme points shown in Figure 2.12(a).

In Figure 2.12(b) is shown the temperature dependence of the rate constant for iron removal from N-terminal monoferric transferrin. There is an obvious break between 12 and 20°C and this is ascribed to a temperature-induced conformational change. [101] The effect becomes less distinct when the ionic strength is increased from 0.13 to 2.0 M, [101]. See also Sec. 4.11.

Negative or very small values of $\Delta H^{\ddagger}$ are rare. They obviously cannot arise from a single step, and they give overwhelming evidence for a multistep process that includes a pre-equilibrium. Negative or near-zero values for $\Delta H^{\ddagger}$ for a few inner-sphere and outer-sphere redox reactions indicate the occurrence of intermediates, and rule out a single step, with a single activated complex (Sec. 5.5).

2.7 Entropy of Activation and Mechanism

Values for $\Delta S^{\ddagger}$ can be positive or negative. The same types of arguments and difficulties in understanding the magnitude and sign of ΔS values for a reaction will be encountered in interpreting the values of $\Delta S^{\ddagger}$ in the formation of the activated complex from the reactants. The interpretation of these values is easiest with "extreme" mechanisms. Thus from general considerations, one might expect large and negative values of $\Delta S^{\ddagger}$ for the associative mechanisms encountered in substitution in square-planar complexes (Sec. 4.6.2) or in reactions of octahedral carbonyls

$$V(CO)_6 + Ph_3P \rightarrow V(CO)_5Ph_3P + CO \tag{2.169}$$

$$-d[V(CO)_6]/dt = k[V(CO)_6][Ph_3P] \tag{2.169a}$$

where $k = 0.25\ M^{-1}s^{-1}$, $\Delta H^{\ddagger} = 41.8\ kJ\ mol^{-1}$ and $\Delta S^{\ddagger} = -116\ J\ K^{-1}mol^{-1}$ in hexane at 25°C Ref. 159. Similarly, one might expect that $\Delta S^{\ddagger}$ would be more positive for reactions accompanied by topological change than for similar ones that proceed with retention of configuration. This is generally observed, for example, in the aquation of *cis* and *trans* cobalt(III) complexes. [160] This also means that the steric course is determined in the rds and yields clues as to the structure of any five-coordinated intermediates (Sec. 4.3.5)

2.7.1 $\Delta S^{\ddagger}$ and the Charge of the Reactants

In an area of chemistry dominated by ionic reactions, it is not surprising that entropies of activation are largely charge-controlled. The reaction between unlike charged species is often attended by a positive $\Delta S^{\ddagger}$ because the solvent molecules are less restricted around an activated complex of reduced charge and thus are released in forming it. For the outer-sphere oxidations of *cis*-hydroxooxyvanadium(IV) chelates by negatively and positively charged oxidants e. g.

$$V^{IV}O(L)OH^{n-} + IrCl_6^{2-} \rightarrow \text{products} \tag{2.170}$$

there is a linear relationship between $\Delta S^{\ddagger}$ (and $\Delta H^{\ddagger}$) and the charge product ($z_R z_O$) of the redox couple ($2n$ in (2.170)). The values of $\Delta S^{\ddagger}$ (J K^{-1} mol^{-1}) span the range $+107$ ($z_R z_O = -8$) to -2 ($z_R z_O = +4$).[161] For a number of reactions this type of behavior does not even approximately hold. Newton and Rabideau have examined a large number of redox reactions of transition and actinide ions, involving both net chemical reactions and isotopic exchange. They showed that the molar entropy of the activated complex $S^{\ddagger}$ defined by

$$S^{\ddagger} = \Delta S^{\ddagger} + \sum S^0 \text{(reactants)} - \sum S^0 \text{(products in net activation process)} \tag{2.171}$$

is very much controlled by its charge.[5,162]

2.8 Volume of Activation and Mechanism

The volume of activation is probably the easiest parameter to understand conceptually. Consider again the water exchange of Cr(III), Sec. 2.3.3. The $\Delta V^{\ddagger}$ value of -10 cm^{3}mol^{-1} for $Cr(H_2O)_6^{3+}$ indicates that an associative process pertains (I_a) since $Cr(H_2O)_6^{3+}$ plus one H_2O will occupy more volume than the activated complex which has seven waters associated with the Cr. The volume of coordinated water has been estimated as anywhere between ~ 5 and 9 cm^{3}mol^{-1}, so that $\Delta V^{\ddagger} \leq -9$ cm^{3}mol^{-1} for an associative mechanism. Conversely, the value of $+2.9$ cm^{3}mol^{-1} for water exchange on $Cr(H_2O)_5OH^{2+}$ suggests a dissociative activation mode for the exchange.[102] More success in interpreting $\Delta V^{\ddagger}$ values is likely for reactions in which the $\Delta V_{solv}^{\ddagger}$ is small, or at least a small component of the overall $\Delta V_{obs}^{\ddagger}$. The study of the pressure effects on the rates of solvent exchange with M(III) and M(II) ions (Fig. 2.13) by nmr methods has been very rewarding,[94,104,105] and essential, for understanding the trend from associative to dissociative character as the d-orbitals are increasingly filled

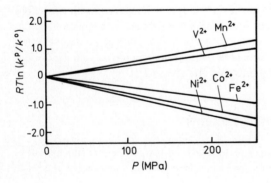

Fig. 2.13 Effect of pressure on solvent exchange with $M(H_2O)_6^{2+}$, $M = V$, Mn, Fe, Co and Ni.[105] Reprinted with permission from Y. Duccomun, A. E. Merbach, Inorganic High Pressure Chemistry. (R. van Eldik ed.). Elsevier, Amsterdam, 1986.

Table 2.4. Examples of Applications of $\Delta V^{\ddagger}$ Values to the Investigation of Mechanism

Reactants	$\Delta V^{\ddagger}$ $cm^3 mol^{-1}$	Conclusions	Technique with high pressure	Ref.
$Cr_2O_7^{2-}$ + base	Range from $-18(OH^-)$ to -26(2,6 lutidine)	Support I_a mechanism	Stopped flow	165
Ni(II) 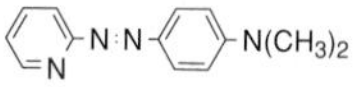 + (structure) $-N\cdot N-$ (ring) $-N(CH_3)_2$	8.2(H_2O), 11.5(DMF) and 11.3(DMSO)	Similar to values of $\Delta S^{\ddagger}$ for solvent exchange. Support I_d	Laser-flash	166
$M(tpps)(H_2O)_2^{3-}$ $+ SCN^- \rightarrow$ mono-complex	15.4(Co(III), 8.8 (Rh(III) and 7.4 (Cr(III))	Support D, I_d and I_d mechanisms respectively	High pressure spectral cell in stopped-flow or spectro-photometer	167
Racemization of Cr(III) complexes	Range from 3.3 $(Cr(phen)_3^{3+}$ to $-16.3(Cr(C_2O_4)_3^{3-}$	Consistent with intramolecular twist mechanisms for $Cr(phen)_3^{3+}$ and one-ended dissociation for $Cr(C_2O_4)_3^{3-}$ for racemization	Batch technique	168
Low Spin $\underset{k_{-1}}{\overset{k_1}{\rightleftharpoons}}$ High Spin 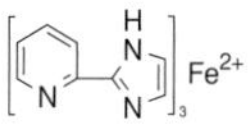 (structure) Fe^{2+}	$\Delta V_1^{\ddagger} = 4.9$; $\Delta V_{-1}^{\ddagger} = -5.4$ (acetone); $\Delta V_1^{\ddagger}$ solvent dependent	Longer Fe$-$N and larger partial molar volumes for high spin isomer. Transition state geometry intermediate between spin isomers	Pulsed laser	169

(Sec. 4.2.1). The interpretation of reactions involving charges is more difficult, because of the importance of ion solvation changes.[163,164] All types of reactions have been studied and will be encountered in Part II. Representative examples are shown in Table 2.4.[165-169] It is worth reemphasizing the value of determining both ΔV and $\Delta V^{\ddagger}$ for a reaction system. The construction then of volume profiles enables a detailed description of the mechanism. The reaction steps for interaction of CO_2 with $Co(NH_3)_5OH_2^{3+}$ are considered to be

$$Co(NH_3)_5OH_2^{3+} \rightleftharpoons Co(NH_3)_5OH^{2+} + H^+ \qquad K_{AH} \qquad (2.172)$$

$$Co(NH_3)_5OH^{2+} + CO_2 \rightleftharpoons Co(NH_3)_5OCO_2H^{2+} \qquad k_1, k_{-1} \qquad (2.173)$$

$$Co(NH_3)_5OCO_2H^{2+} \rightleftharpoons Co(NH_3)_5OCO_2^+ + H^+ \qquad K_B \qquad (2.174)$$

For bond formation $\Delta V_1^{\ddagger} = -10.1 \text{ cm}^3\text{mol}^{-1}$ and for bond cleavage $\Delta V_{-1}^{\ddagger} = +6.8 \text{ cm}^3\text{mol}^{-1}$ and these are considered $\Delta V_{intr}^{\ddagger}$. Comparison of these data with known ΔV values gives a volume profile shown in Fig. 2.14.[170]

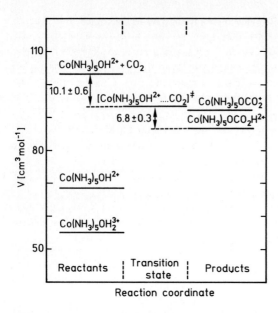

Fig. 2.14 The volume profile for reaction (2.173) at 25°C, Ref. 170. The partial molar volumes of CO_2, $Co(NH_3)_5H_2O^{3+}$ and $Co(NH_3)_5OH^{2+}$ were measured with a digital density apparatus. Reprinted with permission from U. Spitzer, R. van Eldik and H. Kelm, Inorganic Chemistry, **21**, 2821 (1982). © (1982) American Chemical Society.

The activated complex **16** is about halfway between reactants and products and there is considered to be about 50% formation or breakage in forming the activated complex in both directions[170]

16

For other examples of the use of volume profiles, see Refs. 171 (base hydrolysis of $Co(NH_3)_5X^{n+}$), 172 (nickel complexing with glycolate and lactate) and general reviews.[173,174] We now consider the use of more than one activation parameter and any relationships between them.

2.8.1 $\Delta H^{\ddagger}$ and $\Delta S^{\ddagger}$ Values — The Isokinetic Relationship

For a series of reactions of a similar type, a small range in rate constants may arise from parallel and much larger changes in $\Delta H^{\ddagger}$ and $\Delta S^{\ddagger}$. This is evident from a rearrangement of (2.118)

$$\Delta H^{\ddagger} = T\Delta S^{\ddagger} + \Delta G^{\ddagger} \tag{2.175}$$

The slope of any linear plot of $\Delta H^{\ddagger}$ vs $\Delta S^{\ddagger}$ is T. This is the value of the absolute temperature at which all the reactions represented on the line occur at the same rate and is termed the *isokinetic temperature, T.* Although there are a large number of such plots recorded in the literature, the mechanistic information extractable from them is limited. Linear plots for $\Delta H^{\ddagger}$ vs $\Delta S^{\ddagger}$ hold for a large number of redox reactions involving actinide ions, [175] for substitution at Pt(II) centers, Sec. 4.6.2, interaction of $CrOH^{2+}$ with a number of ligands, [176] exchange of oxoanions with H_2O, [177] formation and decarboxylation of metal carbonate complexes (seventeen reactions), [135] reactions of Fe(III) with hydroxamic acid derivatives, [144] and dissociation of Ni(II) macrocycles in 1 M HCl. [178] A common mechanism for each reaction series is supported. Enthusiasm for the mechanistic value of the concept should be tempered by considering the following: there is a striking correlation of $\Delta H^{\ddagger}$ and $\Delta S^{\ddagger}$ for second-order electron transfer reactions involving a number of metalloproteins with inorganic complexes. The linear isokinetic plot does *not* depend on whether the reaction is outer- or inner-sphere, on the protein charge, or on the magnitude of the rate constant or the driving force! The charge of the inorganic reactant only is important. [179]

2.8.2 $\Delta S^{\ddagger}$ and $\Delta V^{\ddagger}$ Values

These parameters often parallel one another since they are related to similar characteristic of the system (change in number of particles involved in the reaction etc.). The catalyzed hydrolysis of $Cr_2O_7^{2-}$ by a number of bases is interpreted in terms of a bimolecular (I_a) mechanism, and both $\Delta S^{\ddagger}$ and $\Delta V^{\ddagger}$ values are negative. [165] In contrast the aquation of $Co(NH_2CH_3)_5L^{3+}$ (L = neutral ligands) is attended by positive $\Delta S^{\ddagger}$ and $\Delta V^{\ddagger}$ values. The steric acceleration noted for these complexes (when compared with the rates for the ammonia analogs) is attributed to an I_d mechanism. [117] There is a remarkably linear $\Delta V^{\ddagger}$ vs $\Delta S^{\ddagger}$ plot for racemization and geometric isomerization of octahedral complexes when dissociative or associative mechanisms prevail, but not when twist mechanisms are operative (Fig. 2.15). [180] For other examples of parallel $\Delta S^{\ddagger}$ and $\Delta V^{\ddagger}$ values, see Refs. 103 and 181. In general $\Delta V^{\ddagger}$ is usually the more easily understandable, calculable and accurate parameter and $\Delta V^{\ddagger}$ is

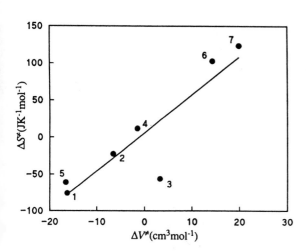

Fig. 2.15 Plot of $\Delta S^{\ddagger}$ ($J\ K^{-1}mol^{-1}$) vs $\Delta V^{\ddagger}$ ($cm^3 mol^{-1}$) for racemization and geometrical isomerization of a variety of octahedral metal complexes. Only a few entries are selected from the 27 reactions tabulated in Ref. 180. The deviation of (four) Cr(III) complexes represented by $Cr(phen)_3^{3+}$ (3) from the linear plot (best fit for 23 complexes) may indicate that these recemize by twist, and not dissociative, mechanisms. Racemization of $Cr(C_2O_4)_3^{3-}$(1), $Co(Ph_2dtc)_3$(2), $Cr(phen)_3^{3+}$(3), $Ni(phen)_3^{2+}$(4). Geometrical isomerization of *trans*-$Cr(C_2O_4)_2(H_2O)_2^-$ (5), *trans*-$Co(en)_2(H_2O)_2^{3+}$(6), β-$Co(edda)en^+$(7).

more likely than $\Delta S^{\ddagger}$ to give consistent data for a series of reactions of similar mechanism. Thus the reaction in dmf of $Ni(dmf)_6^{2+}$ with 5 different ligands gives positive $\Delta V^{\ddagger}$ values ranging from 8.8 to 12.4 $cm^3 mol^{-1}$, whereas $\Delta S^{\ddagger}$ values span the range $+63$ to -82 $J K^{-1} mol^{-1}$. The mechanism is considered dissociative (D). [182]

2.8.3 Use of All Parameters

A careful compilation of as many kinetic parameters as possible can lead to overwhelming support for a mechanism. Only occasionally are such comprehensive data available. The occurrence is nicely illustrated by the exchange reaction (M = Nb, Ta and Sb; X = Cl and Br; L = variety of neutral ligands):

$$MX_5L + {}^*L \rightleftharpoons MX_5{}^*L + L \qquad (2.176)$$

A few data (Table 2.5) are taken from a comprehensive compilation so as to illustrate the major points. [104,183]

Table 2.5. Exchange Data for (2.176) at 0°C in CH_2Cl_2

MX_5	L	Rate Law (order)	k	$\Delta H^{\ddagger}$ kJ mol^{-1}	$\Delta S^{\ddagger}$ J K^{-1} mol^{-1}	$\Delta V^{\ddagger}$ cm^3 mol^{-1}
$TaCl_5$	Me_2O	1st	$0.7 \ s^{-1}$	83	$+59$	$+28$
$TaBr_5$	Me_2O	1st	$54 \ s^{-1}$	74	$+62$	$+31$
$TaCl_5$	Me_2S	2nd	$10^3 M^{-1} s^{-1}$	22	-108	-20
$TaBr_5$	Me_2S	2nd	$90 \ M^{-1} s^{-1}$	29	-102	-13

With the first two entries the exchange is first-order (rate independent of [L]) and $\Delta H^{\ddagger}$, $\Delta S^{\ddagger}$ and $\Delta V^{\ddagger}$ are markedly larger or more positive than with the third and fourth entries for which the rate law is second-order. Consideration of the magnitudes of the values, particularly $\Delta V^{\ddagger}$, lead the authors to assign a D mechanism to the first two exchanges and an I_a (or possibly A) mechanism to the other two. Even steric acceleration and retardation for the two groups is observed. [104,183]

2.9 Medium Effects on the Rate

The effect of the medium on the rate of a reaction does not usually play an important role in the deduction of mechanism. However it is vital that its impact on rate is always assessed.

2.9.1 The Effect of Electrolytes

The rate constant for a reaction is obtained by working at constant ionic strength or alternatively by extrapolating data obtained at different ionic strengths to zero or occasionally "infinite" ionic strength. [184] This procedure is necessitated by the fact that ions (derived both from the reactants and from added electrolytes) often affect the rate of a reaction. This must be allowed for, or removed by "swamping" with electrolyte, in the derivation of a true rate law. The effect of ions on the rate constant for a reaction is easily derived from the absolute reaction rate theory. Now (see Eqn. 2.117),

$$k = \frac{kT}{h} K_c^{\ddagger} \tag{2.177}$$

and

$$K_a^{\ddagger} = K_c^{\ddagger} \frac{f_{X^{\ddagger}}}{f_A f_B} \tag{2.178}$$

The terms $K_a^{\ddagger}$ and $K_c^{\ddagger}$ are "activity" and "concentration" equilibrium constants and $f_{X^{\ddagger}}$, f_A, and f_B are activity coefficients of the activated complex and of the reactants. Thus,

$$k = \frac{kT}{h} K_a^{\ddagger} \frac{f_A f_B}{f_{X^{\ddagger}}} \tag{2.179}$$

This is equivalent to the earlier Brønsted-Bjerrum equation

$$k = k_0 \frac{f_A f_B}{f_X} \tag{2.180}$$

where k_0 is the rate constant at zero ionic strength, and X is a "critical" complex, the forerunner of the activated complex. Now, [185]

$$-\log f_i = \alpha z_i^2 F(\mu) \tag{2.181}$$

with $\alpha = 0.52$ at $25\,°C$ and $F(\mu)$ some function of the ionic strength μ. Thus since the charge on $AB^{\ddagger} = (z_A + z_B)$,

$$\log k = \log k_0 - (z_A^2 + z_B^2 - (z_A + z_B)^2)\, \alpha\, F(\mu) \tag{2.182}$$

leading to

$$\log k = \log k_0 + 2\alpha\, z_A z_B\, \mu^{1/2} \tag{2.183}$$

if the simple form $F(\mu) = \mu^{1/2}$ is used, and

$$\log k = \log k_0 + \frac{2\alpha\, z_A z_B\, \mu^{1/2}}{1 + \mu^{1/2}} \tag{2.184}$$

if the extended form

$$F(\mu) = \frac{\mu^{1/2}}{1 + \mu^{1/2}} \tag{2.185}$$

is substituted. Occasionally the expression

$$\log k = \log k_0 + \frac{2 \alpha z_A z_B \mu^{1/2}}{1 + \mu^{1/2}} - \beta \mu \tag{2.186}$$

is used where α and β are treated as free parameters but maintained at approximate values of 0.5 and 0.1 respectively. The additional term helps correct for ion pairs.[183]

These equations have been used quite successfully for correlating salt effects with the rate constants for reactions involving ions, even at relatively high ionic strengths, (with the extended equation). They are the basis of the well-known Livingston and LaMer diagrams, in which plots of $\log k$ vs $\mu^{1/2}$ or $\mu^{1/2} (1 + \mu^{1/2})^{-1}$ are linear, with slopes $\sim z_A z_B$ and intercept values $\log k_0$.[184] The equations predict a positive salt effect (that is, an increasing rate constant with increasing salt concentration) when reactants have charges of the same sign and a negative salt effect when the charges are of opposite sign. For reactions between ions of unlike sign, (2.183) or (2.184) is usually reasonably obeyed, especially if uni-univalent electrolytes (often $NaClO_4$) are used to adjust the ionic strength.[126] However if higher charged electrolytes are used[186] or reactions are carried out in nonaqueous solution, where ion-pairing is much more important,[187] then marked deviations from the expected behavior occur even for reactants of opposite charge. This is usually ascribed to the formation of ion pairs which will not only lead to reduction in the computed ionic strength but, more important, might very easily produce species (ion pairs) that will react with rate constants different from those for the constituent ions (for, at the least, Coulombic reasons). This effect can be accomodated by equating the observed rate constants to those for the free ions and for the ion pairs. The approach is detailed in studies of bivalent anion effects on the second-order reaction between $Co(NH_3)_5Br^{2+}$ and OH^-,[186,188] and between $FeBr_4^-$ and Cl^- in CH_2Cl_2.[187]

For reactions between ions of like sign, the dependence of rate constant on ionic strength is less straightforward. Now the rate is influenced by the concentration and nature of the supporting ion of opposite charge, *even at constant ionic strength*.[184] The ionic strength dependence of the $Ru_2(NH_3)_{10}pz^{5+} - Ru(NH_3)_5py^{2+}$ reaction does however conform very well to (2.184) when $CF_3SO_3^-$ is used to adjust the ionic strength.[189] The rate constant for the reaction

$$cis\text{-}Co(en)_2(imid)Cl^{2+} + Hg^{2+} \rightarrow cis\text{-}Co(en)_2(imid)H_2O^{3+} + HgCl^+ \tag{2.187}$$

is independent of $[H^+]$ from 0.3 to 1.0 M $HClO_4$ at constant ionic strength. It is however markedly dependent on the anion used to provide this medium.[190] The reactions of $Cr_2O_7^{2-}$, $V_{10}O_{28}^{6-}$ and $Mo_7O_{24}^{6-}$ with OH^- are influenced by the cation present. The second-order rate constant $(M^{-1}s^{-1})$ for the $Cr_2O_7^{2-}/OH^-$ reaction changes from 7.6×10^2 (Na^+) to 35 (Et_4N^+) at $\mu = 1.0$ M.[191] These effects can be treated by the use of empirical expressions. Extended ones[184] such as

$$k = \frac{k_a + k_b[B] + k_c[B]^2}{1 + K_b[B] + K_c[B]^2} \tag{2.188}$$

where [B] is the formal concentration of the added ion, and k_a, k_b, k_c, K_b and K_c are adjustable parameters, fit the extensive data on the effects of *cations* on the $Fe(CN)_6^{3-}$, $Fe(CN)_6^{4-}$ electron transfer.[192] A simpler form

$$k = k_a + k_b[B] \qquad (2.189)$$

takes care adequately of alkali metal cation effects on the manganate, permanganate exchange reaction. Activation parameters (including $\Delta V^{\ddagger}$) values associated with the two terms are carefully analyzed.[193]

For the reaction between an ion and a dipolar molecule, the rate is largely uninfluenced by ionic strength. A relation of the type

$$\log k = \log k_0 - c\mu \qquad (2.190)$$

sometimes holds. Thus the reversible hydrolysis

$$Rh(NH_3)_5Cl^{2+} + H_2O \rightleftharpoons Rh(NH_3)_5H_2O^{3+} + Cl^- \qquad (2.191)$$

is attended by small ionic strength effects on the rate constants for the forward and large effects on the reverse direction.[194] The effect of ionic strength on reactions whose kinetic order is greater than two or non-simple has been considered, but is rarely used.[184,195]

When some dissipation of charge occurs as with larger reactants the assignment of charge to Eqn. (2.184) becomes difficult. For example, the reactions of porphyrins with peripheral charges ($tppsH_2^{4-}$ and $tmpypH_2^{4+}$) give slopes for the Brønsted-Bjerrum plots of the expected sign, but with lower values than expected on the basis of (2.184). The marked effects of ionic strength on these reactions stress the importance of not using one set of conditions (e.g. $\mu = 0.5$ M) when comparing different and highly charged reactants. In addition specific salt effects of the type outlined above should be considered.[195]

The problem of the definition of charge and consideration of the sizes of reactants (the diameters of the reactant ions and the activated complex are assumed equal in the derivation of (2.184)) is most acute with reactions of metalloproteins. Probably the most used expression for the effect of ionic strength on such reactions is[196]

$$\ln k = \ln k_{\infty} - \frac{3.58 \, z_A \, z_B}{r_A + r_B} \left[\frac{\exp(-\kappa r_A)}{1 + \kappa r_B} + \frac{\exp(-\kappa r_B)}{1 + \kappa r_A} \right] \qquad (2.192)$$

where κ equals $0.33 \, \mu^{1/2} A^{-1}$ in H_2O at 25°C. The quantities r_A and r_B are the radii for the two reactants charges z_A and z_B. For proteins, molecular weight M, $r = 0.717 \, M^{1/3}$ is often used,[197] and $k_{\infty} = $ rate constant at "infinite" ionic strength. In this latter condition the reactants are fully shielded from each other by electrolyte ions and this is an alternative approach to that of extrapolating to zero ionic strength so as to negate reactant-reactant interactions. With (2.192) also, increasing ionic strength gives increased k_{obs} for a reaction between similarly charged reactants. Agreement of (2.192) with experimental data has been observed in a number of systems, including the reaction of blue copper proteins with $Fe(edta)^{2-}$ ref. 197, cytochrome b_5 (modified at the heme center) with $Fe(edta)^{2-}$ ref. 198 and a large number of cytochromes with oxidizing and reducing anions.[199,200] Equations other than (2.192) and (2.184) have also been employed with variable success.[199,200] In many cases the use of the overall protein charge calculated from the pI value and the aminoacid composition (which may range from $+8$ to -9[200]) fits the equation (2.192) and others well, particularly if low ionic strengths are used.[196,199,200]

Charge on the protein $-$ An estimate of the charge on the protein may be made from a knowledge of the amino acid composition. At neutral pH, where the majority of the studies are carried out, the glutamates and aspartates are negative ions, the lysines are protonated, the histidines are half-protonated and the arginines are in a monopositive form. The charges on the metal ion and other ligand groups are also taken into account but are relatively unimportant. As an example, azurin *(P. aeruginosa)* has 11 Lys $(11+)$, 1 Arg $(1+)$, 2 His $(1+)$, 4 Glu $(4-)$, and 11 Asp $(11-)$. If deprotonated amide and thiolate are ligands to the Cu(II) the metal site is zero-charged. The overall charge $(2-)$ at pH ~ 7 is consistent with the pI value of 5.2. S. Wherland and H. B. Gray, in Biological Aspects of Inorganic Chemistry, Wiley-Interscience, New York, 1977, Chap. 10.

Equation (2.184) has rarely been used to determine the charge on one of the reactants since this is invariably known for small reactants. The value close to -6.0 for the slope of the plot of $\log k$ against $\mu^{1/2} (1 + \mu^{1/2})^{-1}$ for the reaction of a nickel(III) macrocycle (L) with SO_4^{2-} strongly supports a $+3$ charge for the complex and therefore its formulation as $Ni(L)^{3+}$ rather than as $Ni(L)OH^{2+}$ in the pH 3.2–4.7 range.[201] Charges have been assigned to e_{aq}^-, Cu^+ and CrH^{2+} on the basis of ionic strength effects on the rates of their reactions.[202-204] The assignment of charge with proteins is a complex subject. The charges on cytochrome c_{552} (pI $= 5.5$) and horse cytochrome (pI $= 10.1$) are negative and positive respectively in neutral pH. The effect of ionic strength on rate constants for electron transfer involving these proteins are as predicted using these charges. This tends to rule out the common heme moiety as the site for electron transfer since in that case a similar charge (that of the heme moiety) and therefore a similar effect of ionic strength on the rate constants for the two proteins might have been anticipated.[200] A positive patch (Sec. 5.9) on cytochrome f for binding of $(CN)_5FeCNFe(CN)_5^{5-}$ has been implicated by ionic strength effects.[179]

Ionic strength will also influence other rate parameters. Its effect on $\Delta V^{\ddagger}$ is shown to be

$$\Delta V^{\ddagger} = -RT \left(\frac{d \ln k_0}{dP} \right) + X z_A z_B \tag{2.193}$$

X is a complex term containing $\mu^{1/2}$. At $\mu = 0.1$ M, X $= 0.62$. k_0 is the rate constant at $\mu = 0$. The increase of ionic strength produces a screening effect which decreases electrostriction and increases molal volumes. The effect is more important in reactions between like charges.[205]

2.9.2 The Effect of Electrolytes — Medium or Mechanistic?

The determination of the variation of the rate with the concentrations of ions in solution is the keystone to the rate law. It is apparent from the above discussion that one has to be cautioned that medium effects rather than definite reaction pathways might be responsible for this variation. For example the reaction of Hg^{2+} with *trans*-$CrCl_2^+$ studied at a constant ionic strength made up of $HClO_4$ and $LiClO_4$ shows decreasing values for the computed second-order rate constant (k) with increasing concentrations of Hg^{2+}. This trend might reasonably be ascribed to the formation of appreciable amounts of an adduct such as $CrCl_2Hg^{2+}$ (see Sec. 1.6.3 and 1.6.4). Addition of Ba^{2+} rather than Li^+ to maintain the concentration of *dipositive ions constant* produces a constant k however and indicates that the variation in k observed using Li^+ counter-ion, is a medium-based effect. [206] Distinguishing a mechanistic from a medium effect may be difficult and the problem has been most encountered when there is a relatively small influence of H^+ on the rate. The dilemma arises because of the breakdown of the principle of constant ionic strength, particularly when large and highly charged ions are reactants. [207] The examining media $LiClO_4 - HClO_4$ and also $LiCl - HCl$ are particularly useful since activity coefficients of ionic species are reasonably constant in such mixtures at constant ionic strength. [35] Any observed H^+ effects in these electrolyte media are more likely to arise from mechanistic pathways. [208-210] An allowance can be made for the effect of replacement by H^+ of another ion, say Na^+ or Li^+, in examining the effect of pH on the rate at a constant ionic strength.

The reduction of $Co(NH_3)_5H_2O^{3+}$ ions by $Cr(II)$ has been examined in a ClO_4^- medium at $\mu = 1.0$ M over an $[H^+]$ range of $0.096-0.79$ M, with the ionic medium held constant with $LiClO_4$. [211] The data (Table 2.6) could be interpreted in two ways. It could be as the result of a two-term rate law of the form

$$k = a + b[H^+]^{-1} \tag{2.194}$$

Least-square analysis yields a *negative* value for $a = -0.72 \pm 0.14$ $M^{-1}s^{-1}$ and for $b = 3.12 \pm 0.05$ s^{-1}. Alternatively a single-term rate law in which the Harned correction term takes care of changing activity coefficient [212] could explain the data,

$$k = c[H^+]^{-1} \exp(-\beta[H^+]) \tag{2.195}$$

with $c = 3.13 \pm 0.05$ s^{-1} and $\beta = 0.25 \pm 0.05$ M^{-1}. The fit of Eqn. (2.195) to the data is very good (Table 2.6). The values of b and c in (2.194) and (2.195) are almost identical, so that there is no ambiguity in the rate constant for the hydroxo form in either formulation. Reactivity ascribable to the aqua form, negative in (2.194) or zero in (2.195), must in either case be very small.

One can usually be confident that H^+-containing terms in a rate law relate to a reaction pathway in a number of circumstances: (a) all the terms in the rate expression for the aquation of $Cr(H_2O)_5N_3^{2+}$

$$-d\ln[CrN_3^{2+}]/dt = k_1[H^+] + k_0 + k_2[H^+]^{-1} + k_3[H^+]^{-2} \tag{2.196}$$

Table 2.6. Rate Constants for the Reduction of $Co(NH_3)_5H_2O^{3+}$ by Cr^{2+} at Various $[H^+]$ at 25.1°C[211]

$[H^+]$, M	k, $M^{-1}s^{-1}$	$k_{calcd.}$,[a] $M^{-1}s^{-1}$
0.794	3.39±0.03	3.22
0.654	4.06±0.04	4.12
0.560	4.83±0.03	4.85
0.494	5.43±0.02	5.59
0.438	6.53±0.04	6.39
0.411	7.51±0.16	6.86
0.386	6.72±0.21	7.35
0.271	10.9 ±0.1	10.8
0.202	14.9 ±0.1	14.8
0.131	23.3 ±0.2	23.0
0.114	27.1 ±0.3	26.1
0.096	31.6 ±2.5	31.7

[a] Calculated from $k = 3.13[H^+]^{-1}\exp(-0.25[H^+])$.

can be important, contributing $\geqslant 50\%$ towards the rate at different pH's, and are therefore unlikely to be due to medium effects[213] (b) a pK derived for a pH-dependent kinetic term (Sec. 1.10) agrees with a value determined spectrally[35] (c) the reduction of *cis*-$Co(en)_2(HCO_2)_2^+$ by Cr^{2+} is *inhibited* by acid and the hydrolysis of the Co(III) complex is *enhanced* thereby. Similar pK values are derived in the two cases; these almost certainly cannot be due to medium effects.[214]

For the reaction

$$Cr(NH_3)_5H_2O^{3+} + Br^- \rightleftharpoons Cr(NH_3)_5Br^{2+} + H_2O \qquad k_1, k_{-1}, K_1 \qquad (2.197)$$

the value of K_1 (estimated from the ratio k_1/k_{-1} (0.30 M^{-1})) is much larger than that determined by spectral analysis of equilibrated mixtures (0.02 M^{-1}) even though the ionic strengths are similar ($\mu = 0.8$–1.0 M at 25°C). The difference has been experimentally verified and arises from medium and specific ion effects. The value of K_1 estimated at equilibrium varies with $[Br^-]$ even if the ionic strength is maintained constant with Br^-/ClO_4^- mixtures. The discrepancy is only important because of the high reactant concentrations ($[Br^-] \approx [ClO_4^-]$) which are necessary with (2.197) to effect substantial change in a reaction with a relatively small K_1.[215]

2.9.3 The Solvent Effect and Mechanism

We can arbitrarily divide solvents into three categories: *protic,* including both proton donors and acceptors; *dipolar aprotic,* solvents with dielectric constants >15 but without hydrogen capable of forming hydrogen bonds; and *aprotic,* having neither acidic nor basic properties, for example, CCl_4.[216] These may be expected to interact in widely different ways with complex ions containing large internal charges. The effect of solvent on the rates of reactions has been extensively explored. As a means of interpreting mechanisms it has had variable success, but complex ions have proved to be valuable substrates for examining solvent structural features.[217] There are basically two ways in which the solvent may be regarded, although assessing their distinction and relative importance is very difficult.

(a) The solvent may be regarded as an "inert" medium. In this case, the dielectric constant of the solvent is the most important parameter although viscosity may play an important role.[218] The dielectric effect can be semiquantitatively evaluated for ion-ion or ion-dipolar reactant mixtures, where electrostatic considerations dominate.

For a reaction between two ions, of charge z_A and z_B, the rate constant k (reduced to zero ionic strength) is given by

$$\ln k = \ln k_0 - \frac{e^2}{2DkT} \left[\frac{(z_A + z_B)^2}{r_{\ddagger}} - \frac{z_A^2}{r_A} - \frac{z_B^2}{r_B} \right] \tag{2.198}$$

where k_0 is the hypothetical rate constant in a medium of infinite dielectric constant, D is the dielectric constant, and r_A, r_B, and $r_{\ddagger}$ are the radii of the reactant ions A and B and the activated complex respectively.

If $r_A = r_B = r_{\ddagger}$, (2.198) becomes

$$\ln k = \ln k_0 - \frac{z_A z_B e^2}{DkT r_{\ddagger}} \tag{2.199}$$

and in a reaction between an ion z_A and a polar molecule (that is, $z_B = 0$), (2.198) becomes

$$\ln k = \ln k_0 + \frac{z_A^2 e^2}{2DkT} \left[\frac{1}{r_A} - \frac{1}{r_{\ddagger}} \right] \tag{2.200}$$

There should be a linear plot of $\log k$ vs $1/D$ with a negative slope if the charges of the ions are of the same sign and with a positive slope if the ions are oppositely charged. For the reaction of an ion and a polar molecule (common with solvolysis reactions), the linear plot of $\log k$ vs $1/D$ will have a positive slope irrespective of the charge of the ion since $(1/r_A - 1/r_{\ddagger})$ is positive.

These expressions appear more applicable to nonpolar solvents or mixtures than to polar solvents. The nature of the solvation process (and the radii and so forth of the solvated reactants) may stay approximately constant in the first situation but almost certainly will not in the second. The function $(D_{op}^{-1} - D_s^{-1})$ features in the reorganisation term λ_0 which is used for estimating rate constants for redox reactions (Eqn. 5.23). D_{op} is the optical dielectric constant and D_s the static dielectric constant ($=$ refractive index^2).

(b) The solvent may act as a nucleophile and an active participator in the reaction. It is extremely difficult to assess the function of the solvent in solvolysis reactions. Some attempts to define the mechanism for the replacement of ligand by solvent in octahedral complexes have been made using mixed solvents (Sec. 4.2.1) or the solvating power concept.[219]

The rate of solvolysis (k_s) of tert-butyl chloride in a particular solvent S compared with the rate (k_0) in 80% v/v aqueous ethanol is used as a measure of that solvent's ionizing power, Y_s:

$$Y_s = \log \frac{k_s}{k_0} \tag{2.201}$$

For any other substrate acting by an S_N1 mechanism, it might be expected that

$$\log \frac{k_s}{k_0} = mY_s \tag{2.202}$$

where m depends on the substrate, and equals 1.0 for t-BuCl. It has not proved very useful as diagnostic of mechanism e. g. for Hg(II)-assisted reactions.[220] More data and patterns of behavior are required before the concept is likely to be more valuable.

References (B refers to Selected Bibliography)

1. (a) J. F. Bunnett in B5, Chap. IV, From Kinetic Data to Reaction Mechanism
   (b) J. H. Espenson in B5, Chap. VII, Homogeneous Inorganic Reactions.
2. R. van Eldik in B17, Chap. 1.
3. D. A. House and H. K. J. Powell, Inorg. Chem. **10**, 1583 (1971).
4. J. L. Kurz, Acc. Chem. Res. **5**, 1 (1972).
5. T. W. Newton and S. W. Rabideau, J. Phys. Chem. **63**, 365 (1959).
6. A. Haim, Prog. Inorg. Chem. **30**, 299 (1983).
7. P. Moore, S. F. A. Kettle and R. G. Wilkins, Inorg. Chem. **5**, 466 (1966).
8. S. Funahashi, F. Uchida and M. Tanaka, Inorg. Chem. **17**, 2784 (1978).
9. One useful definition of the rds is that for which a change in rate constant produces the largest effect on the overall rate, W. J. Ray, Biochemistry, **22**, 4625 (1983).
10. E. J. Billo, Inorg. Chem. **23**, 2223 (1984).
11. R. K. Steinhaus and B. I. Lee, Inorg. Chem. **21**, 1829 (1982).
12. D. V. Stynes, S. Liu and H. Marcus, Inorg. Chem. **24**, 4335 (1985).
13. Y. Sasaki, K. Z. Suzuki, A. Matsumoto and K. Saito, Inorg. Chem. **21**, 1825 (1982) and previous work cited.
14. S. Fallab, Adv. Inorg. Bioinorg. Mechs. **3**, 311 (1984).
15. S. C. F. Au-Yeung and D. R. Eaton, Inorg. Chem. **23**, 1517 (1984).
16. H. J. Price and H. Taube, J. Amer. Soc. **89**, 269 (1967).
17. Q. H. Gibson, Biochem. Soc. Trans. **18**, 1 (1990); R. G. Wilkins, in Oxygen Complexes and Oxygen Activation by Transition Metals, A. E. Martell, ed. Plenum, 1988, p. 49.
18. K. A. Jongeward, D. Magde, D. J. Taube, J. C. Marsters, T. G. Traylor and V. S. Sharma, J. Amer. Chem. Soc. **110**, 380 (1988) also use a four-state model, with higher rate constants for some steps.
19. F. Basolo, P. H. Wilks, R. G. Pearson and R. G. Wilkins, J. Inorg. Nucl. Chem. **6**, 161 (1958).
20. W. R. Mason, Coordn. Chem. Revs. **7**, 241 (1972).
21. L. Drougge and L. I. Elding, Inorg. Chim. Acta **121**, 175 (1986).
22. J. Halpern and M. Pribanić, Inorg. Chem. **9**, 2616 (1970).
23. H. U. Blaser and J. Halpern, J. Amer. Chem. Soc. **102**, 1684 (1980).
24. J. H. Baxendale and P. George, Trans. Faraday Soc. **46**, 736 (1950).
25. G. Kramer, J. Patterson, A. Poë and L. Ng, Inorg. Chem. **19**, 1161 (1980).
26. W. C. E. Higginson and A. G. Sykes, J. Chem. Soc. 2841 (1962).
27. N. A. Dougherty and T. W. Newton, J. Phys. Chem. **68**, 612 (1964).
28. G. A. Tondreau and R. G. Wilkins, Inorg. Chem. **25**, 2745 (1986).
29. Derived in B8, p. 79–80.
30. For the derivation in a similar scheme $Mn_2(CO)_{10} \rightleftharpoons 2\,Mn(CO)_5^\bullet$; $Mn(CO)_5^\bullet + O_2 \rightarrow$ products; see J. P. Fawcett, A. Poë and K. R. Sharma, J. Amer. Chem. Soc. **98**, 1410 (1976).

31. L. W. Wilson and R. D. Cannon, Inorg. Chem. **24,** 4366 (1985) and earlier references.

32. For the derivation for a related scheme see B8, p. 135.

33. A. S. Goldman and J. Halpern, J. Amer. Chem. Soc. **109,** 7537 (1987) for rate laws containing $+1/2$ and $-1/2$ exponents.

34. M. A. -Halabi, J. D. Atwood, N. P. Forbus and T. L. Brown, J. Amer. Chem. Soc. **102,** 6248 (1980).

35. E. Deutsch and H. Taube, Inorg. Chem. **7,** 1532 (1968).

36. K. M. Jones and J. Bjerrum, Acta Chem. Scand. **19,** 974 (1965).

37. S. F. Lincoln and D. R. Stranks, Aust. J. Chem. **21,** 67 (1968).

38. The second term could also be expressed as (k_2/K_w) [Fe(III)] [Cl$^-$] [OH$^-$]. In keeping with the tendency of expressing the rate law in terms of the predominant species, the form shown is preferred for data in acid medium.

39. S. L. Bruhn, A. Bakač and J. H. Espenson, Inorg. Chem. **25,** 535 (1986).

40. M. Orhanović and R. G. Wilkins, J. Amer. Chem. Soc. **89,** 278 (1967).

41. D. Seewald and N. Sutin, Inorg. Chem. **2,** 643 (1963).

42. J. H. Espenson and S. R. Helzer, Inorg. Chem. **8,** 1051 (1969).

43. R. B. Wilhelmy, R. C. Patel, and E. Matijević, Inorg. Chem. **24,** 3290 (1985).

44. K. Ishihara, S. Funahashi and M. Tanaka, Inorg. Chem. **22,** 194 (1983); with Fe(III) complexing with 4-isopropyltropolone, the assessment is made on the basis of $\Delta V^{\ddagger}$ values.

45. C. P. Brink and A. L. Crumbliss, Inorg. Chem. **23,** 4708 (1984).

46. J. H. Espenson, Inorg. Chem. **8,** 1554 (1969).

47. B. Perlmutter-Hayman and E. Tapuhi, Inorg. Chem. **16,** 2742 (1977).

48. A. Campisi and P. A. Tregloan, Inorg. Chim. Acta **100,** 251 (1985).

49. P. W. Taylor, R. W. King, and A. S. V. Bergen, Biochemistry **9,** 3894 (1970).

50. R. G. Pearson, J. Chem. Educ. **55,** 720 (1978).

51. J. H. Espenson, Inorg. Chem. **4,** 1025 (1965); A. Adin and A. G. Sykes, J. Chem. Soc. A 351 (1968).

52. F. P. Rotzinger, Inorg. Chem. **25,** 4870 (1986).

53. D. H. Huchital and J. Lepore, Inorg. Chem. **17,** 1134 (1978).

54. J. Phillips and A. Haim, Inorg. Chem. **19,** 1616 (1980).

55. A. G. Sykes, Chem. Soc. Revs. **14,** 283 (1985).

56. M. L. Bender in Investigation of Rates and Mechanisms of Reactions, Part I, S. L. Friess, E. S. Lewis and A. Weissberger, eds. Interscience, NY, 1961, Chap. 25.

57. T. W. Newton and F. B. Baker, Inorg. Chem. **3,** 569 (1964).

58. P. B. Sigler and B. J. Masters, J. Amer. Chem. Soc. **79,** 6353 (1957).

59. G. Czapski, B. H. J. Bielski and N. Sutin, J. Phys. Chem. **67,** 201 (1963).

60. E. Saito and B. H. J. Bielski, J. Amer. Chem. Soc. **83,** 4467 (1961).

61. R. N. F. Thorneley and D. J. Lowe, in Molybdenum Enzymes, T. G. Spiro, ed. Wiley-Interscience, NY, 1985, Chap. 5.

62. D. S. Sigman, J. Amer. Chem. Soc. **102,** 5421 (1980).

63. A. Bakač, J. H. Espenson, I. I. Creaser and A. M. Sargeson, J. Amer. Chem. Soc. **105,** 7624 (1983).

64. E. Janssen, in Free Radicals in Biology 4, W. A. Pryor, ed. Academic, NY, 1980, p. 115.

65. C. R. Johnson, T. K. Myser and R. E. Shepherd, Inorg. Chem. **27,** 1089 (1988) for studying OH$^{\bullet}$ production from complexes and H_2O_2 or O_2.

66. N. E. Dixon, W. G. Jackson, W. Marty and A. M. Sargeson, Inorg. Chem. **21,** 688 (1982). (N$_3^-$)

67. W. G. Jackson, M. L. Randall, A. M. Sargeson and W. Marty, Inorg. Chem. **22,** 1013 (1983). (NO$_2^-$)

68. W. G. Jackson, D. P. Fairlie and M. L. Randall, Inorg. Chim. Acta **70,** 197 (1983). (S$_2$O$_3^{2-}$)

69. W. G. Jackson and C. N. Hookey, Inorg. Chem. **23,** 668 (1984). (SCN$^-$)

70. Z. Khurram, W. Böttcher and A. Haim, Inorg. Chem. **24,** 1966 (1985).

71. D. A. Buckingham, I. I. Olsen and A. M. Sargeson, Aust. J. Chem. **20,** 597 (1967).

72. N. E. Brasch, D. A. Buckingham, C. R. Clark and K. S. Finnie, Inorg. Chem. **28,** 3386 (1989).

73. D. J. Darensbourg, Adv. Organomet. Chem. **21,** 113 (1982).

74. R. K. Murmann and H. Taube, J. Amer. Chem. Soc. **78**, 4886 (1956).
75. R. van Eldik, J. von Jouanne and H. Kelm, Inorg. Chem. **21**, 2818 (1982).
76. W. G. Jackson, G. A. Lawrance, P. A. Lay and A. M. Sargeson, J. Chem. Soc. Chem. Communs. 70 (1982).
77. S. Fallab, H. P. Hunold, M. Maeder and P. R. Mitchell, J. Chem. Soc. Chem. Communs. 469 (1981).
78. S. Fallab and P. R. Mitchell, Adv. Inorg. Bioinorg. Mechs. **3**, 311 (1984).
79. N. J. Coville, A. M. Stolzenberg and E. L. Muetterties, J. Amer. Chem. Soc. **105**, 2499 (1983).
80. W. H. Saunders, Jr., in B5, Chap. VIII.
81. J. P. Candlin and J. Halpern, J. Amer. Chem. Soc. **85**, 2518 (1963).
82. H. Taube, Electron Transfer Reactions of Complex Ions in Solution, Academic, NY, 1970, p. 92.
83. N. V. Brezniak and M. Z. Hoffman, Inorg. Chem. **18**, 2935 (1979).
84. M. S. Thompson and T. J. Meyer, J. Amer. Chem. Soc. **104**, 4106 (1982).
85. J. T. Chen and J. H. Espenson, Inorg. Chem. **22**, 1651 (1983).
86. H. Kwart, Acc. Chem. Res. **15**, 401 (1982).
87. C. B. Grissom and W. W. Cleland, J. Amer. Chem. Soc. **108**, 5582 (1986).
88. D. L. Leussing and M. J. Emly, J. Amer. Chem. Soc. **106**, 443 (1984).
89. J. N. Armor and H. Taube, J. Amer. Chem. Soc. **92**, 2561 (1970).
90. S. Arrhenius, Z. Phys. Chem. **4**, 226 (1889).
91. A. E. Merbach, P. Moore, O. W. Howarth and C. H. McAteer, Inorg. Chim. Acta **39**, 129 (1980). The wide temperature range used to study the $Ga(dmso)_6^{3+}$-dmso exchange improves markedly the accuracy of the derived activation parameters.
92. J. V. Beitz, J. R. Miller, H. Cohen, K. Wieghardt, and D. Meyerstein, Inorg. Chem. **19**, 966 (1980). The formation and breakdown of the radical anion $Co(NH_3)_5(p\text{-}NO_2C_6H_4CO_2^-)^{2+}$ obeys Arrhenius from 200 to 300 K in 50% H_2O/ethyleneglycol.
93. For a full account see B11, Chap. 5, and B20, pp. 62–73, 109; also S. Glasstone, K. J. Laidler and H. Eyring, The Theory of Rate Processes, McGraw-Hill, NY, 1941; M. M. Kreevoy and D. G. Truhlar, in B5, Chap. 1.
94. K. E. Newman, F. K. Meyer and A. E. Merbach, J. Amer. Chem. Soc. **101**, 1470 (1979); A. E. Merbach, Pure Appl. Chem. **59**, 161 (1987).
95. C. H. McAteer and P. Moore, J. Chem. Soc. Dalton Trans. 353 (1983).
96. G. Kohnstam, Adv. Phys. Org. Chem. **5**, 121 (1967), for a discussion of heat capacities of activation and their uses in mechanistic studies.
97. J. R. Hulett, Quart. Rev. **18**, 227 (1964).
98. W. E. Jones, R. B. Jordan and T. W. Swaddle, Inorg. Chem. **8**, 2504 (1969).
99. J. F. Bunnett, in B5, Chap. 4.
100. C. Blakeley and M. L. Tobe, J. Chem. Soc. Dalton Trans. 1775 (1987). There is however no variation of $\Delta V^{\ddagger}$ with temperature, which is puzzling; V. Kitamura, G. A. Lawrance and R. van Eldik, Inorg. Chem. **28**, 333 (1989).
101. S. A. Kretchmar and K. N. Raymond, J. Amer. Chem. Soc. **108**, 6212 (1986); Inorg. Chem. **27**, 1436 (1988).
102. F. C. Xu, H. R. Krouse, and T. W. Swaddle, Inorg. Chem. **24**, 267 (1985).
103. H. R. Hunt and H. Taube, J. Amer. Chem. Soc. **80**, 2642 (1958).
104. A. E. Merbach, Pure Appl. Chem. **54**, 1479 (1982).
105. Y. Ducommun and A. E. Merbach in B17, Chap. 2.
106. R. van Eldik, T. Asano and W. J. Le Noble, Chem.Revs. **89**, 549 (1989).
107. P. Moore, Pure Appl. Chem. **57**, 347 (1985).
108. M. Kotowski and R. van Eldik, Coordn. Chem. Revs. **93**, 19 (1989) for a succinct account of high pressure equipment used in various techniques.
109. S. D. Hamann and W. J. Le Noble, J. Chem. Educ. **61**, 658 (1984).
110. J. K. Beattie, M. T. Kelso, W. E. Moody and P. A. Tregloan, Inorg. Chem. **24**, 415 (1985).

111. Y. Ducommun, P. J. Nichols and A. E. Merbach, Inorg. Chem. **28**, 2643 (1989).
112. D. Kost and A. Pross, Educ. in Chem. **14**, 87 (1977).
113. J. R. Murdoch, J. Chem. Educ. **58**, 32 (1981).
114. T. W. Swaddle, J. Chem. Soc. Chem. Communs. 832 (1982).
115. T. W. Swaddle, Adv. Inorg. Bioinorg. Mechs. **2**, 95 (1983).
116. P. A. Lay, Inorg. Chem. **26**, 2144 (1987).
117. N. J. Curtis and G. A. Lawrance, Inorg. Chem. **25**, 1033 (1986).
118. R. M. Krupka, H. Kaplan and K. J. Laidler, Trans. Faraday Soc. **62**, 2754 (1966). R. L. Burwell, Jr. and R. G. Pearson, J. Phys. Chem. **70**, 300 (1966).
119. F. Monacelli, F. Basolo and R. G. Pearson, J. Inorg. Nucl. Chem. **24**, 1241 (1962).
120. N. J. Curtis, Inorg. Chem. **25**, 1033 (1986).
121. P. A. Lay, Inorg. Chem. **26**, 2144 (1987).
122. L. Mønsted and O. Mønsted, Coordn. Chem. Revs. **94**, 109 (1989).
123. C. A. Tolman, Chem. Revs. **77**, 313 (1977). Ligand intermeshing must also be considered; H. C. Clark, Isr. J. Chem. **15**, 210 (1977).
124. D. J. Darensbourg and A. H. Graves, Inorg. Chem. **18**, 1257 (1979).
125. A. Bakac and J. H. Espenson, J. Amer. Chem. Soc. **108**, 713 (1986).
126. D. A. House, Inorg. Chim. Acta, **51**, 273 (1981).
127. E. S. Lewis in B5, Chap. 13.
128. C. H. Langford, Inorg. Chem. **4**, 265 (1965); A. Haim, Inorg. Chem. **9**, 426 (1970).
129. D. R. Prasad, T. Ramasami, D. Ramasamy and M. Santappa, Inorg. Chem. **21**, 850 (1982) − These unusually rapid reactions of Cr(III) are promoted by lengthening (distortion) of the axial water and thus enhancing the dissociative character. They are *not* typical of Cr(III) behavior.
130. B. G. Cox, J. Garcia-Ross and H. Schneider, J. Amer. Chem. Soc. **103**, 1054 (1981).
131. D. J. Cram and G. M. Lein, J. Amer. Chem. Soc. **107**, 3657 (1985).
132. D.-T. Wu and C.-S. Chung, Inorg. Chem. **25**, 4841 (1986).
133. R. Marcus, Ann. Rev. Phys. Chem. **15**, 155 (1964).
134. E. Pelizzetti and E. Mentasti, Inorg. Chem. **18**, 583 (1979).
135. R. van Eldik, D. A. Palmer, H. Kelm and G. M. Harris, Inorg. Chem. **19**, 3679 (1980).
136. E. Kremer, G. Cha, M. Morkevicius, M. Seaman, and A. Haim, Inorg. Chem. **23**, 3028 (1984).
137. S. Goldstein and G. Czapski, Inorg. Chem. **24**, 1087 (1985).
138. C. Hansch, A. Leo and R. W. Taft, Chem. Revs. **91**, 165 (1991).
139. F. Aprile, V. Cagliotti and G. Illuminati, J. Inorg. Nucl. Chem. **21**, 325 (1961).
140. A. Bakač, V. Butković, J. H. Espenson, R. Marcec and M. Orhanovic, Inorg. Chem. **25**, 2562 (1986).
141. J. P. Leslie and J. H. Espenson, J. Amer. Chem. Soc. **98**, 4839 (1976).
142. R. J. Mureinik, M. Weitzberg and J. Blum, Inorg. Chem. **18**, 915 (1979).
143. S. S. Eaton and G. R. Eaton, Inorg. Chem. **16**, 72 (1977).
144. C. P. Brink and A. L. Crumbliss, Inorg. Chem. **23**, 4708 (1984). A number of LFER are included with a short discussion of the strict statistical criteria for isokinetic relationships.
145. R. W. Taft, J. Amer. Chem. Soc. **75**, 4231 (1953).
146. S. Steenken and P. Neta, J. Amer. Chem. Soc. **104**, 1244 (1982).
147. R. S. Berman and J. K. Kochi, Inorg. Chem. **19**, 248 (1980).
148. D. F. DeTar, J. Amer. Chem. Soc. **102**, 7988 (1980).
149. H. Cohen, S. Efrima, D. Meyerstein, M. Nutkovich and K. Wieghardt, Inorg. Chem. **22**, 688 (1983).
150. J. N. Brønsted and K. J. Pederson, Z. Phys. Chem. **108**, 185 (1924).
151. R. Bahwad, B. Perlmutter-Hayman and M. A. Wolff, J. Phys. Chem. **73**, 4391 (1969), and references therein.
152. B. G. Cox and H. Schneider, J. Amer. Chem. Soc. **102**, 3628 (1980).
153. M. Eigen, Angew. Chem. Int. Ed. Engl. **3**, 1 (1964).
154. D. A. Buckingham, C. R. Clark and T. W. Lewis, Inorg. Chem. **18**, 2041 (1979).

155. R. S. Rowlett and D. N. Silverman, J. Amer. Chem. Soc. **104**, 6737 (1982); D. N. Silverman and S. Lindskog, Accs. Chem. Res. **21**, 30 (1988).

156. C. G. Swain and C. B. Scott, J. Amer. Chem. Soc. **75**, 141 (1953).

157. G. Annibale, L. Canovese, L. Cattalini, G. Marangoni, G. Michelon and M. L. Tobe, Inorg. Chem. **20**, 2428 (1981).

158. M. Becker and H. Elias, Inorg. Chim. Acta **116**, 47 (1986).

159. Q.-Z. Shi, T. G. Richmond, W. C. Trogler and F. Basolo, J. Amer. Chem. Soc. **106**, 71 (1984).

160. M. L. Tobe, Inorg. Chem. **7**, 1260 (1968).

161. M. Nishizawa, Y. Sasaki and K. Saito, Inorg. Chem. **24**, 767 (1985).

162. N. A. Daugherty and J. K. Erbacher, Inorg. Chem. **14**, 683 (1975).

163. J. Burgess, A. J. Duffield and R. Sherry, J. Chem. Soc. Chem. Communs. 350 (1980). It is difficult to rationalize $\Delta V^{\ddagger}_{obs}$ values of ~ 20 cm^{3}mol^{-1} with an associative mechanism for nucleophiles reacting with low spin Fe(II) diimine complexes. This value however represents a balance between contributions from ΔV_{intr} (negative) and anion desolvation in forming activated complex (positive $\Delta V^{\ddagger}$).

164. I. Krack and R. van Eldik, Inorg. Chem. **25**, 1743 (1986).

165. P. Moore, Y. Ducommun, P. J. Nichols and A. E. Merbach, Helv. Chim. Acta **66**, 2445 (1983).

166. E. F. Caldin and R. C. Greenwood, J. Chem. Soc. Faraday Trans. I, **77**, 773 (1981).

167. J. G. Leipoldt, R. van Eldik and H. Kelm, Inorg. Chem. **22**, 4146 (1983).

168. G. A. Lawrance and D. R. Stranks, Inorg. Chem. **16**, 929 (1977).

169. J. J. McGarvey, I. Lawthers, K. Heremans and H. Tofflund, J. Chem. Soc. Chem. Communs. 1575 (1984). See also J. DiBenedetto, V. Arkle, H. A. Goodwin and P. C. Ford, Inorg. Chem. **24**, 455 (1985).

170. U. Spitzer, R. van Eldik and H. Kelm, Inorg. Chem. **21**, 2821 (1982).

171. Y. Kitamura, R. van Eldik and H. Kelm, Inorg. Chem. **23**, 2038 (1984).

172. T. Inoue, K. Sugahara, K. Kojima and R. Shimozawa, Inorg. Chem. **22**, 3977 (1983).

173. A comprehensive collection of volume profiles and their value is contained in Ref. 106.

174. R. van Eldik, Angew. Chem. Int. Ed. Engl. **25**, 673 (1986).

175. T. W. Newton and F. B. Baker, Adv. Chem. **71**, 268 (1967).

176. D. Thusius, Inorg. Chem. **10**, 1106 (1971).

177. H. von Felten, B. Wernli, H. Gamsjäger and P. Baertschi, J. Chem. Soc. Dalton Trans. 496 (1978).

178. A. Ekstrom, L. F. Lindoy and R. J. Smith, Inorg. Chem. **19**, 724 (1980).

179. D. B.-Betts and A. G. Sykes, Inorg. Chem. **24**, 1142 (1985).

180. G. A. Lawrance and S. Suvachittanont, Inorg. Chim. Acta **32**, L13 (1979).

181. F. K. Meyer, W. L. Earl and A. E. Merbach, Inorg. Chem. **18**, 888 (1979).

182. P. J. Nichols, Y. Fresard, Y. Ducommun and A. E. Merbach, Inorg. Chem. **23**, 4341 (1984).

183. H. Vanni and A. E. Merbach, Inorg. Chem. **18**, 2758 (1979).

184. A. D. Pethybridge and J. E. Prue, Inorganic Reaction Mechanisms, Part 2, p. 327, for a comprehensive account.

185. Again we use decadic logarithms because of traditional and current usage.

186. P. B. Abdullah and C. B. Monk, J. Chem. Soc. Dalton Trans. 1175 (1983) and previous studies.

187. G. P. Algra and S. Balt, Inorg. Chem. **20**, 1102 (1981).

188. B. Perlmutter-Hayman and Y. Weissman, J. Phys. Chem. **68**, 3307 (1964); M. R. Wendt and C. B. Monk, J. Chem. Soc. A, 1624 (1969).

189. U. Fürholz and A. Haim, J. Phys. Chem. **90**, 3686 (1986).

190. Despite this problem, the literature data for (2.187) and many related reactions can be converted to 1.0 M HClO$_4$ as a standard state. The rate constants obtained (k_{Hg}) give a more consistent LFER when $\log k_{Hg}/\log k_{H_2O}$ for these reactions are plotted. [126]

191. B. W. Clare, D. M. Druskovich, D. L. Kepert and J. H. Kyle, Aust. J. Chem. **30**, 211 (1977).

192. R. J. Campion, C. F. Deck, P. King, Jr. and A. C. Wahl, Inorg. Chem. **6**, 672 (1967).

193. L. Spiccia and T. W. Swaddle, Inorg. Chem. **26**, 2265 (1987).

194. G. C. Lalor and G. W. Bushnell, J. Chem. Soc. A, 2520 (1968).

195. J. Nwaeme and P. Hambright, Inorg. Chem. **23**, 1990 (1984).

196. S. Wherland and H. B. Gray, Proc. Natl. Acad. Sci. USA **79**, 2950 (1976).

197. R. C. Rosenberg, S. Wherland, R. A. Holwerda and H. B. Gray, J. Amer. Chem. Soc. **98**, 6364 (1976).

198. L. S. Reid, M. R. Mauk and A. G. Mauk, J. Amer. Chem. Soc. **106**, 2182 (1984).

199. T. Goldkorn and A. Schejter, J. Biol. Chem. **254**, 12562 (1979). Figures representing four equations are presented and compared.

200. R. A. Marcus and N. Sutin, Biochim. Biophys. Acta **811**, 265 (1985), Table V.

201. H. Cohen, L. J. Kirschenbaum, E. Zeigerson, M. Jacobi, E. Fuchs, G. Ginzburg and D. Meyerstein, Inorg. Chem. **18**, 2763 (1979).

202. G. Czapski and H. A. Schwarz, J. Phys. Chem. **66**, 471 (1962).

203. O. J. Parker and J. H. Espenson, J. Amer. Chem. Soc. **91**, 1968 (1969).

204. D. A. Ryan and J. H. Espenson, Inorg. Chem. **20**, 4401 (1981).

205. S. Wherland, Inorg. Chem. **22**, 2349 (1983).

206. J. P. Birk, Inorg. Chem. **9**, 735 (1970).

207. In other words, there may be a relation of the form $\log f_H = \log f_O + a[H^+]$, where f_H and f_0 are the activity coefficients of reactants and activated complex at $[H^+] = H$ and O respectively at constant μ. From (2.179) this will lead to a rate law, $\log k \propto [H^+]$ from a purely medium effect (A. Pidcock and W. C. E. Higginson, J. Chem. Soc. 2798 (1963).

208. J. Doyle and A. G. Sykes, J. Chem. Soc. A, 795 (1967); R. Davies and A. G. Sykes, J. Chem. Soc. A, 2831 (1968).

209. A. Ekstrom, Inorg. Chem. **16**, 845 (1977).

210. W. D. Drury and J. M. DeKorte, Inorg. Chem. **22**, 121 (1983).

211. D. L. Toppen and R. G. Linck, Inorg. Chem. **10**, 2635 (1971).

212. R. A. Robinson and R. H. Stokes, Electrolyte Solutions, Butterworths, London, 1955, Chap. 15.

213. T. W. Swaddle and E. L. King, Inorg. Chem. **3**, 234 (1964).

214. J. R. Ward and A. Haim, J. Amer. Chem. Soc. **92**, 475 (1970).

215. L. Mønsted, T. Ramasami and A. G. Sykes, Acta Chem. Scand. A **39**, 437 (1985).

216. M. Chastrette, M. Rajzmann, M. Chanon and K. F. Purcell, J. Amer. Chem. Soc. **107**, 1 (1985) increase this to nine classes of (83) solvents, based on eight different variables, dipole moments etc.

217. M. J. Blandamer and J. Burgess, Pure Appl. Chem. **62**, 9 (1990).

218. Consider the striking effect of glycerol in reducing the solvent exchange rate constant (Sec. 4.2(a)) of all the first row transition metal ions.

219. E. Grunwald and S. Winstein, J. Amer. Chem. Soc. **70**, 846 (1948).

220. R. Banerjee, Coordn. Chem. Revs. **68**, 145 (1985).

Selected Bibliography

B19. R. P. Bell, The Proton in Chemistry, 2nd edit. Chapman and Hall, London, 1973. Full discussion of Brønsted relationships.

B20. J. E. Leffler and E. Grunwald, Rates and Equilibria of Organic Reactions, Wiley, NY, 1963.

Problems

1. Suggest mechanisms for the following reactions. In each case specify the units for the rate constants (Sec. 1.1) and the composition of the activated complex(es) (Sec. 2.1). Indicate

which mechanism is likely when there is more than one and suggest additional experiments to substantiate the choice.

(a) $6 \, Fe(CN)_5H_2O^{3-} \rightarrow 5 \, Fe(CN)_6^{4-} + Fe^{2+} + \cdot 6 \, H_2O$

$$-d \, [Fe(CN)_5H_2O^{3-}]/dt = k \, [Fe(CN)_5H_2O^{3-}]$$

The reaction is carried out in dilute solution with exclusion of oxygen. J. A. Olabe and H. O. Zerga, Inorg. Chem. **22**, 4156 (1983).

(b) $(H_2O)_5CrCH_2C_6H_5^{2+} + 2 \, Fe^{3+} + 2 \, H_2O \rightarrow Cr(H_2O)_6^{3+} + 2 \, Fe^{2+} + C_6H_5CH_2OH + H^+$

$$-d \, [(H_2O)_5CrCH_2C_6H_5^{2+}]/dt = k \, [(H_2O)_5CrCH_2C_6H_5^{2+}]$$

A similar rate law is obtained with O_2, Cu(II) and $Co(NH_3)_5Cl^{2+}$ with similar values for k, but with different organic products.
R. S. Nohr and J. H. Espenson, J. Amer. Chem. Soc. **97**, 3392 (1975).

(c) $(tpps)Fe^{III} - O - Fe^{III}(tpps)^{8-} + H_2O + S_2O_4^{2-} \rightarrow 2 \, (tpps)FeOH^{5-} \, (+ \, 2 \, SO_2)$

$$-d \, [(tpps)Fe - O - Fe(tpps)^{8-}]/dt = k \, [(tpps)Fe - O - Fe(tpps)^{8-}]$$

A. A. El-Awady, P. C. Wilkins and R. G. Wilkins, Inorg. Chem. **24**, 2053 (1985).

(d) $N_2H_4 + 2 \, H_2O_2 \xrightarrow{Cu^{2+}} N_2 + 4 \, H_2O$

$$\frac{d \, [N_2]}{dt} = \frac{a \, [N_2H_4] \, [H_2O_2] \, [Cu(II)]}{1 + b \, [N_2H_4]}$$

See Problem 2, Chap. 1.
C. R. Wellman, J. R. Ward and L. P. Kuhn, J. Amer. Chem. Soc. **98**, 1683 (1976).

(e) $2 \, Cu(II)azurin + S_2O_4^{2-} \rightarrow 2 \, Cu(I)azurin \, (+ \, 2 \, SO_2)$

$$-d \, [Cu(II)azurin]/dt = k_1 \, [Cu(II)azurin] \, [S_2O_4^{2-}] + k_2 \, [Cu(II)azurin] \, [S_2O_4^{2-}]^{1/2}$$

G. D. Jones, M. G. Jones, M. T. Wilson, M. Brunori, A. Colosimo and P. Sarti, Biochem. J. **209**, 175 (1983); Z. Bradić and R. G. Wilkins, J. Amer. Chem. Soc. **106**, 2236 (1984).

(f) $2 \, Cu^{2+} + 6 \, CN^- \rightarrow 2 \, Cu(CN)_2^- + (CN)_2$

$$d \, [Cu(CN)_2^-]/dt = k \, [Cu^{2+}]^2 [CN^-]^6$$

J. H. Baxendale and D. T. Westcott, J. Chem. Soc. 2347 (1959); R. Patterson and J. Bjerrum, Acta Chem. Scand. **19**, 729 (1965); A. Katagiri and S. Yoshizawa, Inorg. Chem. **20**, 4143 (1981).

(g) $Fe(phen)_3^{2+} + ox \rightarrow Fe(III) + red$

$(ox = P_2O_8^{4-}, \, ClO_2^-, \, BrO_4^-)$

$$-d \, [Fe(phen)_3^{2+}]/dt = \frac{a \, [Fe(phen)_3^{2+}] \, [ox]}{b \, [phen] + c \, [ox]}$$

A. M. Kjaer and J. Ulstrup, Inorg. Chem. **21**, 3490 (1982) and references cited.

(h) $HCrO_4^- + 2 H_2O_2 \rightarrow CrO_5(OH)^- + 2 H_2O$

$$-d[HCrO_4^-]/dt = k[HCrO_4^-][H^+][H_2O_2]$$

The rate law and value of k for the formation of the violet diperoxochromium(VI) are the same as for the formation of the blue CrO_5 species (Sec. 2.1)
S. N. Witt and D. M. Hayes, Inorg. Chem. **21**, 4014 (1982).

(i) $Cr(H_2O)_6^{3+} + H_3N^+CH_2CO_2^- \rightarrow Cr(NH_2CH_2CO_2)(H_2O)_4^{2+} + 2 H_2O + H^+$

$$d[Cr(NH_2CH_2CO_2)(H_2O)_4^{2+}]/dt = (k_2 + k_2'[H^+]^{-1})[Cr(III)]$$

M. Abdullah, J. Barrett and P. O'Brien, J. Chem. Soc. Dalton Trans. 1647 (1984); Inorg. Chim. Acta **96**, L35 (1985).

(j) $Fe^{III} + Cu^{I} \rightarrow Fe^{II} + Cu^{II}$

$$-d[Fe(III)]/dt = k[Fe(III)][Cu(I)][H^+]^{-1}$$

This is a common type of rate law for reactions between two acidic metal ions.
O. J. Parker and J. H. Espenson, Inorg. Chem. **8**, 1523 (1969).

(k) $Cr(NH_3)_5H_2O^{3+} + HN_3 \rightarrow Cr(NH_3)_5N_3^{2+} + H_3O^+$

$$-d[Cr(NH_3)_5H_2O^{3+}]/dt = k[Cr(NH_3)_5H_2O^{3+}][HN_3][H^+]^{-1}$$

S. Castillo-Blum and A. G. Sykes, Inorg. Chem. **23**, 1049 (1984).

(l) $*Cr(III) + Cr(VI) \rightarrow Cr(III) + *Cr(VI)$

$$V_{exch} = (k_1 + k_2[H^+]^{-2})[Cr^{(III)}]^{4/3}[H_2CrO_4]^{2/3}$$

C. Altman and E. L. King, J. Amer. Chem. Soc. **83**, 2825 (1961).

(m) $7 MoO_4^{2-} + 8 H^+ \rightarrow Mo_7O_{24}^{6-} + 4 H_2O$

$$d[Mo_7O_{24}^{6-}]/dt = k[MoO_4^{2-}]^7[H^+]^8$$

D. S. Honig and K. Kustin, Inorg. Chem. **11**, 65 (1972).

2. The dissociation of metal complexes of macrocycles is often speeded up by acid. For the nickel(II) complexes A, B and C (indicated NiL) the rate laws are

$$-d[NiL]/dt = \{k_1 + k_2[H^+]^n\}[NiL]$$

where for A, $n = 1$; B, $k_1 = 0$, $n = 2$; and C, $k_1 = 0$, $n = 3$. Suggest mechanisms for these reactions.

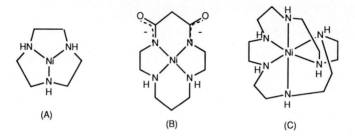

(A)

(B)

(C)

(A) L. J. Murphy, Jr., and L. J. Zompa, Inorg. Chem. **18**, 3278 (1979).
(B) R. W. Hay, M. P. Pujari and F. McLaren, Inorg. Chem. **23**, 3033 (1984).
(C) R. W. Hay, R. Bembi, W. T. Moodie and P. R. Norman, J. Chem. Soc. Dalton Trans. 2131 (1982).

3. The oxidation of Fe^{2+} by two-equivalent oxidants produces unstable oxidation states either of iron or of the oxidant. The immediate products of oxidation by (1) H_2O_2, (2) HOCl, and (3) O_3 in 0.1 M to 1.0 M $HClO_4$ are (1) $>99\%$ Fe^{3+} (and $FeOH^{2+}$), (2) $\approx 80\%$ Fe^{3+} and $\approx 15\%$ $Fe_2(OH)_2^{4+}$, and (3) $\approx 60\%$ Fe^{3+} and $\approx 40\%$ $Fe_2(OH)_2^{4+}$. Suggest reasons for this difference in behavior, and for the decreasing yield of $Fe_2(OH)_2^{4+}$ with increasing $[H^+]$ in reaction (2).
T. J. Conocchioli, E. J. Hamilton and N. Sutin, J. Amer. Chem. Soc. **87**, 926 (1965).

4. Neither Cr(VI) nor Fe(III) can oxidize iodide ions rapidly. However a mixture of Cr(VI) and Fe(II) forms iodine rapidly from iodide ions. The oxidation of I^- is said to be induced by the Fe(II)$-$Cr(VI) reaction. At high $[I^-]/[Fe^{2+}]$ ratios, the induction factor (ratio of I^- oxidized per Fe^{2+} oxidized) is 2.0. Interpret this behavior, detailing the intermediates involved.
J. H. Espenson, Accts. Chem. Res. **3**, 347 (1970).

5. The reaction between $Fe(CN)_5NO^{2-}$ and NH_2OH in basic solution gives N_2O quantitatively

$$(NC)_5FeNO^{2-} + NH_2OH + OH^- \rightarrow (NC)_5FeOH_2^{3-} + N_2O + H_2O$$

with a rate law

$$-d\,[(NC)_5FeNO^{2-}]/dt = k\,[(NC)_5FeNO^{2-}]\,[NH_2OH]\,[OH^-]$$

Using ^{15}N and ^{18}O labelled reactants show how the source of N_2O might be deduced, and suggest a mechanism for the reaction.
S. K. Wolfe, C. Andrade and J. H. Swinehart, Inorg. Chem. **13**, 2567 (1974).

6. Suggest a reason for the small but definite value for k_H/k_D of 2.1 for reaction of $Co(sep)^{2+}$ (**9**) with O_2^-, but not with O_2. The NH groups are deuterated.
A. Bakač, J. H. Espenson, I. I. Creaser and A. M. Sargeson, J. Amer. Chem. Soc. **105**, 7624 (1983).

7. In general,

$$k = CT^n \exp(-U/RT)$$

so that a plot of $\ln kT^{-n}$ vs $1/T$ is linear with slope $-U/R$. Show that with various values of n: 0, 1/2 and 1 one obtains the Arrhenius expression, the Eyring expression and the expression used in the collision theory for bimolecular reactions in the gas phase.

8. The plot of $\ln k_{OH}/I$ vs T^{-1} for the base hydrolysis of *trans*-Co $(RSSR[14]$aneN$_4)$Cl$_2^+$ is curved slightly (concave down). Bearing in mind the conjugate mechanism for base hydrolysis give a plausible explanation for this behavior.

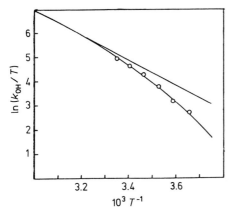

Problem 8. Plot of $\ln(k_{OH}T^{-1})$ against T^{-1} for the base hydrolysis of *trans*-Co(*RSSR*-cyclam)Cl$_2^+$. Reproduced with permission from J. Lichtig, M. E. Sosa, and M. L. Tobe, J. Chem. Soc. Dalton Trans. 581 (1984).

J. Lichtig, M. E. Sosa and M. L. Tobe, J. Chem. Soc. Dalton Trans. 581 (1984).

9. Cryokinetic studies of the plastocyanin-ferricyanide redox reactions in 50:50 v/v MeOH + H$_2$O, pH = 7.0, μ = 0.1 M reveal an Eyring plot shown for the second-order rate constant k from 25°C to -35°C. The reaction is irreversible over the whole temperature range and there is no evidence for a change in the Cu(I) active site. Recalling that these reactions may involve consecutive steps, explain the deviation from a linear Eyring plot. F. A. Armstrong, P. C. Driscoll, H. G. Ellul, S. E. Jackson and A. M. Lannon, J. Chem. Soc. Chem. Communs. 234 (1988).

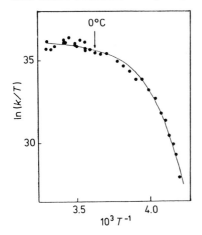

Problem 9. Eyring plot for the oxidation of PCu^I by Fe(CN)$_6^{3-}$ in 50:50 v/v CH$_3$OH$-$H$_2$O, pH = 7.0, μ = 0.1 M(NaCl).

10. For the exchange reaction

$$NbCl_5 \cdot RCN + *RCN \xrightleftharpoons{k_{exch}} NbCl_5 \cdot *RCN + RCN$$

there is a linear relationship between k_{exch} and K the stability constant of the adduct for a variety of R groups. What does this suggest for the mechanism of the exchange?
R. Good and A. E. Merbach, Inorg. Chem. **14**, 1030 (1975).

11. Why is there likely to be a parallelism between the dissociation rate constant for nickel(II) complexes with L and the $pK_a(LH^+)$?
P. Moore and R. G. Wilkins, J. Chem. Soc. 3454 (1964).
H. Hoffmann, Ber. Bunsenges. Phys. Chem. **73**, 432 (1969).

12. Comment on the ρ-values for the following reactions

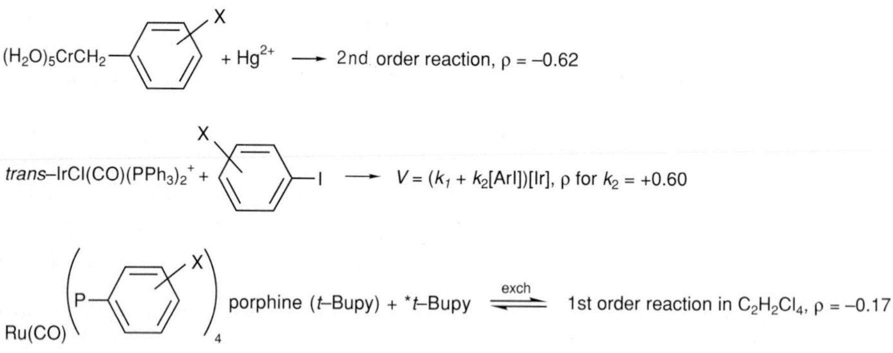

$(H_2O)_5CrCH_2$—⟨ring⟩—X $+ Hg^{2+}$ ⟶ 2nd order reaction, $\rho = -0.62$

$trans$–$IrCl(CO)(PPh_3)_2^+$ + X–⟨ring⟩–I ⟶ $V = (k_1 + k_2[ArI])[Ir]$, ρ for $k_2 = +0.60$

$Ru(CO)\left(P\text{—}⟨ring⟩\text{—}X \right)_4$ porphine $(t\text{–Bupy}) + *t\text{–Bupy} \xrightleftharpoons{exch}$ 1st order reaction in $C_2H_2Cl_4$, $\rho = -0.17$

J. P. Leslie and J. H. Espenson, J. Amer. Chem. Soc. **98**, 4839 (1976).
R. J. Mureinik, M. Weitzberg and J. Blum, Inorg. Chem. **18**, 915 (1979).
S. S. Eaton and G. R. Eaton, Inorg. Chem. **16**, 72 (1977).

13. Draw a reaction profile for a reaction in which a) the activated complex most resembles the reactants and b) the activated complex most resembles the products.
B17, p. 54.

14. Give a plausible explanation for
(a) a positive value for $\Delta V^{\ddagger}$ (+ 13 $cm^3 mol^{-1}$ at 298 K) for the second-order reaction of OH^- ion with an iron(II) chelate, $Fe(hxsb)^{2+}$ which would be expected to be negative intrinsically.

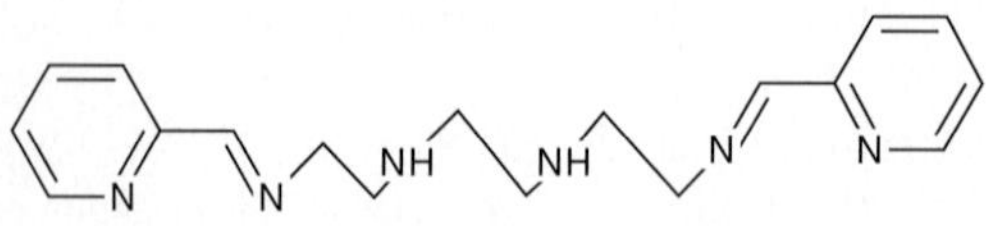

J. Burgess and C. D. Hubbard, J. Chem. Soc. Chem. Communs. 1482 (1983).

(b) negative and positive $\Delta V^{\ddagger}$ values for dmf exchange of solvent with $Co(Me_6tren)dmf^{2+}$ and $Cu(Me_6tren)dmf^{2+}$ respectively.

S. F. Lincoln, A. M. Hounslow, D. L. Pisaniello, B. G. Doddridge, J. H. Coates, A. E. Merbach and D. Zbinden, Inorg. Chem. **23**, 1090 (1984).

15. The values of $\Delta V_{expl}^{\ddagger}$ and ΔV_0 for the reaction

$$Co(NH_3)_5X^{(3-n)+} + OH^- \rightarrow Co(NH_3)_5OH^{2+} + X^{n-}$$

are shown in the Table. Also included are partial molar volumes $V(Co(NH_3)_5X^{(3-n)+})$ and $V(X^{n-})$, all $cm^3 mol^{-1}$

X^{n-}	$\Delta V_{expl}^{\ddagger}$ [a]	ΔV_0	$V(Co(NH_3)_5X^{(3-n)+}$ [b]	$V(X^{n-})$ [b]
Me_2SO	40	21	112	69
Cl^-	33	10	84	22
Br^-	33	11	89	29
SO_4^{2-}	22	-4	95	23

[a] $\mu = 10$–16 mM, value little dependent on μ.
[b] By dilatometry, data extrapolated to infinite dilution.

Consider the D_{cb} mechanism for base hydrolysis.
(a) Comment on the constancy of

$$\Delta V_{expl}^{\ddagger} - \Delta V_0 \text{ and its value.}$$

(b) satisfy yourself that

$$\Delta V_{expl}^{\ddagger} \approx V(Co(NH_3)_4NH_2^{2+}) + V(X^{n-}) + V(H_2O)$$
$$- V(OH^-) - V(Co(NH_3)_5X^{(3-n)+})$$

(c) Plot $\Delta V_{expl}^{\ddagger} + V(Co(NH_3)_5X^{(3-n)+})$ vs $V(X^{n-})$ and comment on the values of the slope and intercept. $(V(H_2O) - V(OH^-) = 17.6 \ cm^3 mol^{-1}$

(d) What does $\Delta V_{expl}^{\ddagger}$ comprise on the basis of a D_{cb} mechanism?

Y. Kitamura, R. van Eldik and H. Kelm, Inorg. Chem. **23**, 2038 (1984); R. van Eldik, Angew. Chem. Int. Ed. Engl. **25**, 673 (1986).

16. The plot of $\log k$ vs $\mu^{1/2}/1 + \mu^{1/2}$ for the second-order reaction between $Co(edta)^-$ and $Fe(CN)_6^{4-}$ (Sec. 1.6.3) showed an initial linear slope at low μ of 3.8, reached a maximum at $\mu \approx 0.1$ M and then decreased. The ionic strength was supplied by $NaClO_4$. Give a reasonable explanation for this behavior.

D. H. Huchital and J. Lepore, Inorg. Chim. Acta **38**, 131 (1980).

17. Determine the dependence of the first-order rate constant k for the reaction

$$Cr(H_2O)_5ONO^{2+} + H_3O^+ \rightarrow Cr(H_2O)_6^{3+} + HNO_2$$

on the proton concentration, $T = 10°C$, $\mu = 1.0$. (See accompanying table.)

$[H^+]$, M	$10^2 k$, s^{-1}	$[H^+]$, M	$10^2 k$, s^{-1}
0.0114	0.345	0.150	7.9
0.0216	0.706	0.200	11
0.030	1.16	0.300	20
0.040	1.59	0.400	31
0.050	2.07	0.500	45
0.060	2.53	0.600	59
0.080	3.50	0.700	81
0.100	4.66	0.800	104
		0.900	129
		0.993	158

Suggest a mechanism, and analyze whether medium effects are likely to be a cause of the acidity dependence.

T. C. Matts and P. Moore, J. Chem. Soc. A, 1997 (1969).

Chapter 3

The Experimental Determination of the Rate of Reaction

3.1 Essential Preliminaries

In determining experimentally the rate of a reaction, it is imperative to define the reaction completely, with respect to the reactants, the stoichiometry and even the products. This is preferably carried out before any detailed rate measurements are made, otherwise difficulties in understanding the rate data (particularly if these are complex) are likely to arise.

3.1.1 Reactant Species in Solution

It is essential to characterize the reactant species in solution. One of the problems, for example, in interpreting the rate law for oxidation by Ce(IV) or Co(III) arises from the difficulties in characterizing these species in aqueous solution, particularly the extent of formation of hydroxy or polymeric species.[1] We used the catalyzed decomposition of H_2O_2 by an Fe(III) macrocycle as an example of the initial rate approach (Sec. 1.2.1). With certain conditions, the iron complex dimerizes and this would have to be allowed for, since it transpires that the dimer is catalytically inactive.[2] In a different approach, the problems of limited solubility, dimerization and aging of iron(III) and (II)-hemin in aqueous solution can be avoided by intercalating the porphyrin in a micelle. Kinetic study is then eased.[3,4]

Inability or failure to characterize the reactant species may lead to problems in kinetic interpretation. In M(II)-edta complexes the degree of coordination (5- or 6-) by the ligand is uncertain. Since the mode of coordination appears sensitive to temperature and ionic strength, this may explain contradictions in the literature in the rate behavior of M(II)-edta complexes.[5] The ability of ligands to induce changes in protein structure has been probed by examining the reactivity of thiol groups in the protein towards organomercurials. However, ligands react with organomercurials and rate

$$ArHgOH + L^- \rightleftharpoons ArHgL + OH^- \qquad (3.1)$$

differences may arise simply because of different reactivities of ArHgOH and ArHgL.[6] See Prob. 1.

It must always be considered that a reactant is in fact a mixture of species. If these species are in labile equilibrium, or have similar reactivities, their presence will not show up as multiphasic kinetics.

Recent ^{1}H nmr studies of an equilibrated solution of Ni(cyclam)$^{2+}$ at 25 °C shows that it contains in addition to Ni(cyclam) (H$_2$O)$_2^{2+}$ (1) a mixture of about 15% of isomer *RSRS* **2**, and mainly *RRSS*, **3**, ((+ + + +) and (+ − − +) respectively where + indicates the H of the NH group is above the macrocycle plane). The results mean that previous studies of the Ni(II) macrocycle made with the assumption of only a single planar isomer in solution have to be reassessed.[7] For a pair of esoteric examples of the importance and difficulties of characterization consider the following:

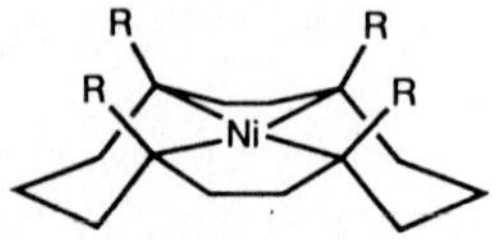

Trans I
RSRS
2

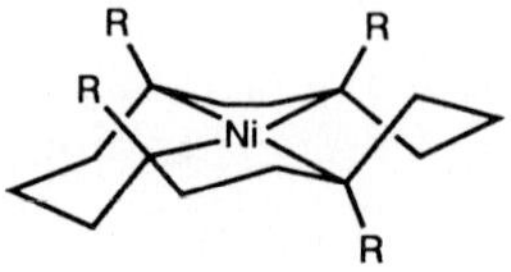

Trans II
RSRR

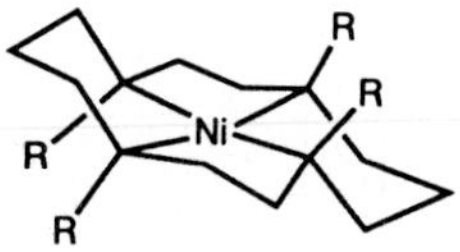

Trans III
RRSS
3

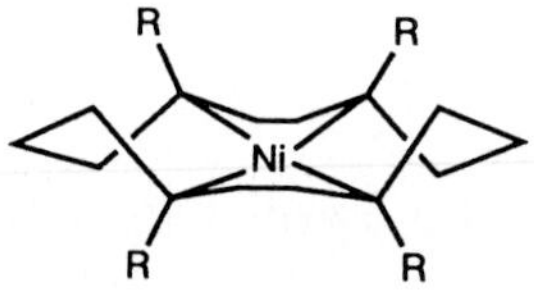

Trans IV
RSSR

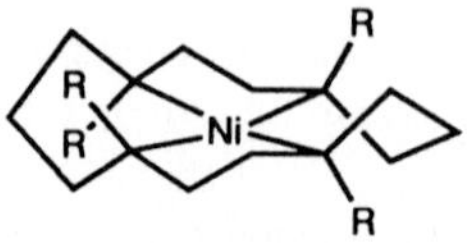

Trans V
RRRR

Configurational isomers of planar Ni(cyclam),$^{2+}$
R = H and Ni(Me$_4$cyclam), R = CH$_3$.

(a) Native myoglobin exists with the heme in two different orientations (one 10%). These have the same optical spectrum but different nmr, O$_2$ affinities and associated rate constants. Reconstituted myoglobin from apomyoglobin and heme contains equimolar mixtures of the two forms.[8]

(b) Some metal ions may exist in aqueous solution as dinuclear species bridged by hydrogen oxide ligands (H$_3$O$_2^-$), **4**. These may be reactive species not easily detected but precursors to olation.[9]

$$L_5M(H_2O)^{n+} + (HO)ML_5^{(n-1)+} \rightleftharpoons L_5M(H_3O_2)ML_5^{(2n-1)+} \tag{3.2}$$

$$L_5M(H_3O_2)ML_5^{(2n-1)+} \longrightarrow L_5M(OH)ML_5^{(2n-1)+} + H_2O \tag{3.3}$$

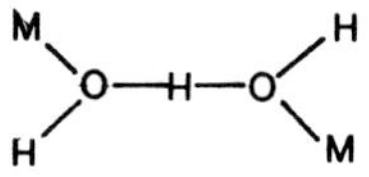

4

3.1.2 Stoichiometry of Reaction

The importance of knowing the stoichiometry of a reaction can be simply illustrated by considering the aquation of cis-$Cr(en)_2(NCS)Cl^+$ ion.[10]. Is Cl^- or NCS^- replaced in the initial step and is the product cis or $trans$ or both? Does the product of this first step aquate further, and if so what groups are then replaced? Chemical and spectral analysis answers these questions. The results reveal the surprising fact that the bidentate ligand en is lost at one stage, a behavior that appears more common with ammines of $Cr(III)$[11] than with $Co(III)$, where its occurrence is only occasionally[12] noted.

$$cis\text{--}Cr(en)_2(NCS)Cl^+ + H_2O \begin{array}{l} \xrightarrow{\leq 4\%} trans\text{--}Cr(en)_2(H_2O)NCS^{2+} + Cl^- \\ \xrightarrow{\leq 4\%} cis\text{--} \text{ and } trans\text{--}Cr(en)_2(H_2O)Cl^{2+} + SCN^- \\ \xrightarrow{\geq 92\%} cis\text{--}Cr(en)_2(H_2O)NCS^{2+} + Cl^- \end{array} \qquad (3.4)$$

$$cis\text{--}Cr(en)_2(H_2O)NCS^{2+} + H_2O \begin{array}{l} \xrightarrow{33 \pm 10\%} Cr(en)(H_2O)_3NCS^{2+} + en \\ \xrightarrow{67 \pm 10\%} cis\text{--} \text{ and } trans\text{--}Cr(en)_2(H_2O)_2^{3+} + SCN^- \end{array} \qquad (3.5)$$

Inconsistencies in the values of equilibrium constants obtained from measurements on systems at equilibrium with those derived from rate measurements may also reveal unexpected reaction paths.[13] See Chap. 8, Pd(II).

3.1.3 The Nature of the Products

If the rate constants for parallel reactions are to be resolved, then analysis of the products is essential (Sec. 1.4.2). This is vital for understanding, for example, the various modes of deactivation of the excited state (Sec. 1.4.2). Only careful analysis of the products of the reactions of $Co(NH_3)_5H_2O^{3+}$ with SCN^-, at various times after initiation, has allowed the full characterization of the reaction (1.95) and the detection of linkage isomers. Kinetic analysis by a number of groups failed to show other than a single second-order reaction.[14] As a third instance, the oxidation of 8-Fe ferredoxin with $Fe(CN)_6^{3-}$ produces a 3 Fe-cluster, thus casting some doubt on the reaction being a simple electron transfer.[15]

It is easily observed when a product precipitates from solution. Some reductions of Co(III) complexes yield insoluble Co(II) products. Addition of powerful ligands such as edta will complex with Co(II), maintain homogeneity and force irreversibility (Sec. 1.5.1). [16] It should be always checked whether the added ligand has an influence on the rate.

3.1.4 The Influence of Impurities

The materials used, including the solvent, should be as pure as possible. There are several instances recorded (and doubtless a number unrecognized) in which traces of impurities introduced inadvertently into a system have had catastrophic consequences.

The pseudo first-order rate constants (k) for reaction of $Co(CN)_5H_2O^{2-}$ with N_3^- concentration in excess were reported as curved (Sec. 1.6.3) and have been interpreted for some 20 years as evidence for a 5-coordinate intermediate and a D mechanism. If however $Co(CN)_5H_2O^{2-}$ is generated in solution by aquation of $Co(CN)_5Cl^{3-}$ (rather than from $Co_2(CN)_{10}O_2^{6-}$ as in the original studies) the $k/[N_3^-]$ plot does *not* deviate from linearity. Reasons are suggested why the latter behavior is correct and importantly, the original material could be shown to contain a slower reacting component which gives false data at higher N_3^- concentrations. The D mechanism remains a possibility however. [17, 18] Traces of impurities in a solution of $Fe(CN)_5H_2O^{3-}$ have led to false conclusions as to the pK_a of the coordinated water. [19, 20] The photolytic behavior of aqueous $Fe(CN)_6^{3-}$ depends on its history. [21] Copper(II) ion is a potent catalyst even in micromolar concentrations for a number of reactions of $Fe(CN)_6^{3-}$ ion. [22]

These catalytic effects are usually signaled by irreproducible behavior. If it is suspected that traces of metal ions may be causing peculiar rate effects, a strong ligand may be added to sequester the metal ion. The spontaneous decomposition of H_2O_2 has been reported as 4.7×10^{-7} $M^{-1}s^{-1}$ at pH 11.6 and 35 °C. This is the lowest recorded value and is obtained in the presence of strong chelators. [23] In a similar way the decomposition of permanganate in alkaline solution (3.6) is markedly slowed when the reactants are extensively purified

$$4MnO_4^- + 4OH^- \rightarrow 4MnO_4^{2-} + O_2 + 2H_2O \qquad (3.6)$$

and metal ion concentrations are reduced below 10^{-9} M. [24]

3.1.5 The Control of Experimental Conditions

Some general considerations applicable to all rate studies can be outlined. [25] Since the rate constants for many reactions are affected by the ionic strength of the medium (Sec. 2.9.1) it is necessary either to maintain a constant ionic strength with added electrolyte, or to carry out a series of measurements at different ionic strengths and extrapolate to infinite dilution. The former practice is usually followed and $NaNO_3$ or $NaClO_4$ have been popular as added salts. There are some advantages however in using the weakly-coordinating [26] p-toluenesulfonate or trifluoromethylsulfonate ions in place of perchlorate. A potential explosive hazard is avoided and problems of oxidation, e. g. with V(III), [27] do not arise.

Many complex-ion reactions are accompanied by a pH-change, and since the rate of these reactions is often pH-dependent, it is necessary to use buffers. A list of useful buffers for the pH region 5.5–11.0 is contained in Table 3.1.[28]

Table 3.1 Some Useful Buffers for Studying Complex-Ion Reactions.[28]

Buffer (Acronym)	pK_a (25°C)
2-[N-Morpholino]ethanesulfonic acid (MES)	6.1
Bis[2-hydroxyethyl]iminotris(hydroxymethyl)methane (Bis-Tris)	6.5
Piperazine-N,N'-bis[2-ethanesulfonic acid] (PIPES)	6.8
3-[N-Morpholino]propanesulfonic acid (MOPS)	7.2
N-[2-Hydroxyethyl]piperazine-N'-[2-ethanesulfonic acid] (HEPES)	7.5
N,N-bis[2-Hydroxyethyl]glycine (BICINE)	8.3
2-[N-Cyclohexylamino]-ethanesulfonic acid (CHES)	9.3
3-[Cyclohexylamino]-1-propanesulfonic acid] (CAPS)	10.4

Many of these are substantially non-nucleophilic and unlikely to effect the rate or course of the reaction, although this should always be checked. References 29 to 31 relate some problems in the use of some of these buffers. Occasionally, one of the reactants being used in excess may possess buffer capacity and this obviates the necessity for added buffer. The situation will often arise in the study of complex ion-ligand interactions when either reactant may be involved in an acid-base equilibrium.

Concentrations of reactants reported usually refer to room temperature and pressure. Any changes in concentrations caused by carrying out the runs at other temperatures or pressures may be either ignored or are not relevant (Sec. 2.3.5). It is often unnecessary to maintain the reaction temperature more constant than ± 0.05 degrees. Variations in rate constants due to such a temperature fluctuation are generally well within the experimental error. It is obviously wise to exclude light or air, if it is suspected that these might interfere with the reaction. Darkened reaction vessels and apparatus for anaerobic manipulation are described.[32] Special equipment must be used if the reaction is carried out at elevated pressure.[33] A simple device for working at high temperatures (say >100°C), when continual opening of the apparatus is to be avoided, has been described.[34] Finally, when all the results have been gathered in, analysis of the errors, precision and accuracy associated with the study has to be made.[35]

3.2 The Methods of Initiating Reaction and their Time Ranges

Obviously the speed with which it is necessary to initiate a reaction will depend on its rate. If it is a slow process with half-lives longer than about 20 s, then the reaction can be initiated simply by mixing the reactants. Even reactions with shorter reaction times may be studied without recourse to sophisticated equipment. Simple mixing devices fitting into a spectrophotometric cuvette[32, 36] or used to initiate and quench enzymatic reactions (down to 0.5 s and using 20 μl volumes)[37] can effectively reduce accessible reaction times. It is even possible to commence absorbance/time traces within 5 s of dissolution of a solid.[38] Flow and rapid mixing methods allow observation times to be as short as, and occasionally even much shorter than, milliseconds and have been considerably exploited. Relaxation methods circumvent the mixing limitations. These are extremely important for the measurement of rapid reactions but the relative simplicity of the associated kinetics (see Chap. 1) makes them also an attractive prospect for the study of slow reactions.[36] Any perturbation e.g. ionic strength or pH changes which alters concentrations of species at equilibrium can in principle be used, imposed by flow methods if necessary.[36,39-43] A special method for the rapid initiation of a reaction employs a large perturbation by a light pulse (usually with a laser) or an electron beam. Primary and secondary reducing and oxidizing radicals or excited states can be produced in very short times (even picoseconds, or less, but usually nano- or microseconds) and their physical properties and chemical reactivity (towards added substrates) can be studied.

The time coverage for the various rapid-reaction techniques is shown in Figure 3.1.[44]

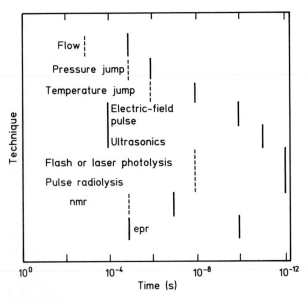

Fig. 3.1 Time coverage for various rapid-reaction techniques. Broken lines indicate the usual shorter time limits. Full lines indicate the shorter time limits attainable in some laboratories. Unless indicated, the longer time limit is usually unlimited.[44]

3.3 Flow Methods

Comprehensive discussions of flow methods are available in the literature.[45,46] There are basically three ways in which the reaction solution may be treated after mixing (Fig. 3.2 and Table 3.2)

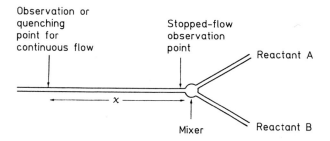

Fig. 3.2 The operation of flow methods. The distance x and the combined flow rate govern the time that elapses between mixing and when the combined solutions reach the observation, or quenching, point. In the stopped flow method, observation is made as near to the mixer as is feasible, and monitoring occurs after the solutions are stopped. In the pulsed accelerated flow method, observation is within the mixer.

Table 3.2 Rapid Mixing and Flow Methods

Flow-Mode	General Procedure	Characteristics
Continuous	*In situ* monitoring at a fixed point on the observation tube with various flow rates. Alternatively, the mixing chamber is incorporated into the observation tube with early monitoring	Tedious but leisurely analysis. Useful with sluggish monitoring probes. A 1–0.01 ms resolution. Large volumes of reactants used ($\geqslant 5$ ml). Not commercially available.
Quenched	Mixed solutions quenched after a predetermined time controlled by the distance between the mixer and quencher and the flow rate.	Tedious but leisurely analysis. Essential for the batch method used in rapid isotopic exchange and low temperature epr monitoring. A 10–20 ms resolution. Large volumes of reactants used ($\geqslant 5$ ml). Commercially available.
Stopped	Mixed solutions abruptly stopped and analyzed near mixer.	Most popular method. Easy analysis requiring rapidly responding monitor. A 1 ms resolution. Uses small (≈ 0.2 ml) volumes. A number of commercial apparatus in wide variety of modes and monitoring methods.

3.3.1 Continuous Flow

The continuous-flow method avoids the stopping features of the stopped-flow technique which are time limiting, but suffers from the disadvantage of consuming relatively large amounts of material, even in the latest developments. There has however been a resurgence of interest in the method, using integrating observation with a very fast jet mixer which is incorporated into the observation tube. The resolution time is short ($\leqslant 10$ μs) but relatively large volumes of solution are necessary.[47,48] The integrating observation feature has been combined with pulsed continuous flow (pulsed continuous-flow spectrometer), resulting in the use of smaller volumes of material (5 ml), enhanced optical sensitivity and short resolution times (4 μs).[49,50] It now represents a method for determining high rate constants ($\approx 10^5$ s^{-1} or 10^{10} M^{-1}s^{-1}) for irreversible reactions (which are not ameniable to relaxation methods). Unfortunately, the treatment of kinetic data is more involved than with the stopped-flow method. [47-50]

The continuous flow method is still necessary when one must use probe methods which respond only relatively slowly to concentration changes. These include pH,[51] O_2-sensitive electrodes,[52] metal-ion selective electrodes,[53] thermistors and thermocouples,[54] epr[55] and nmr detection. Resonance Raman and absorption spectra have been recorded in a flowing sample a few seconds after mixing horseradish peroxidase and oxidants. In this way spectra of transients (compounds I and II) can be recorded, and the effect of any photoreduction by the laser minimized. [56]

3.3.2 Quenched Flow

A number of simple pieces of apparatus for using the quenched-flow method have been described. [57-59] Commercial stopped-flow apparatus are available in which a double mixer arrangement allows quenching of a transient produced in a first mixer with quencher from a second mixer. [60-62] In one apparatus the aging interval for the transient (time between first and second mixer) can be adjusted from 20 ms to 11 s. This is accomplished by altering the volume of tubing between the mixers and/or altering the flow rate. A series of quenched solutions with different "ages" are thus obtained and can be analyzed (for details, see Ref. 63). See Prob. 2. Some varied examples of the applications of quenched flow are shown in Table 3.3. Quenching can be effected by several means including rapid cooling,[62] precipitation of one reactant[57,58,61,64] or chemical destruction by adding a complexing agent or acid. The method has been particulary useful for studying fairly rapid isotopic exchange reactions (Table 3.3). Some of the reactions have also been studied without the necessity of a separation procedure, and therefore more conveniently, by nmr line-broadening methods (Sec. 3.9.6). The latter still require fairly high reactant concentrations, and with these conditions subtle medium

Table 3.3 Some Rapid Isotopic Exchange Reactions Studied by Quenched Flow

Exchanging Pair	Rate Constant $M^{-1}s^{-1}$	Temp. °C	Quenching Method	Reference
$^{54}MnO_4^{2-}$, MnO_4^- (0.16 M OH^-)	7.1×10^2	0.0	$(C_6H_5)_4As^+$ coprecipitates MnO_4^- (in presence of ReO_4^-)	57
$IrCl_6^{3-}$, $^{192}IrCl_6^{2-}$	$\approx 2.3 \times 10^5$	25.0	2-Butanone extracts $IrCl_6^{2-}$ (stopped flow mixing)	58
$^{63}Cu^{II}$ stellacyanin, $^{65}Cu^{I}$ stellacyanin	1.2×10^5	20.0	Cool to $-120°C$ and epr analysis using slight differences in epr of $^{63}Cu^{II}$ and $^{65}Cu^{II}$ stellacyanin. Rapid mixing, reaction half times ≈ 10 msec at 0.5 mM concentrations.	62
VO_4^{3-}, $H_2^{18}O$ (in 0.5 M base)	$2.4 \times 10^2 (s^{-1})$	0.0	$Co(NH_3)_6^{3+}$ precipitates VO_4^{3-}. An injection device allows short $t_{1/2}$'s 8–44 s.	64
IO_3^-, $H_2^{18}O$	4 term rate law	5.0	Ag^+ precipitates IO_3^- which pyrolyzes cleanly to O_2 for isotopic analysis. Dionex double mixer, $t_{1/2}$ 20 ms–2 s.	61

effects are less easily recognized. The quenched-flow method has been particularly effective also for the study of the transients produced in rapid reactions of certain Mo [65,66] and Fe [67,68] proteins. Epr monitoring of rapidly quenched solutions fingerprints transients. The MoFe protein of nitrogenase in the reduced form is oxidized rapidly by $Fe(CN)_6^{3-}$ and frozen at $-140°C$ after various times have elapsed between oxidation (complete within milliseconds) and freezing. The epr examination of the frozen oxidized material of various ages allows a) the assignment of the product to a $[4Fe-4S]^+$ containing center and b) an indication of an epr signal diminishing with time after oxidation. The transient signal fades by an intramolecular process ($k = 4.1 \pm 0.8$ s^{-1}) at the same rate as that of enzyme turnover. Probably both processes are controlled by a conformational change in the protein. [65]

The disadvantage of the quenched-flow technique is the tedium associated with the batch method of assay. Additionally there is a relatively long reaction time limit, often >10 ms, necessitated by the extended quenching times. Offsetting these limitations are the simple equipment and the leisurely assay that are integral features of the method.

3.3.3 Stopped Flow

The stopped-flow technique is by far the most popular of the flow methods. [69] A block diagram is shown in Fig. 3.3. Stopped-flow systems have employed nearly all the usual monitoring methods and these will be discussed later (Table 3.7). The linking of the method with spectral monitoring is by far the most popular combination. A useful test reaction is the reduction of 2,6-dichlorophenol indophenol by *L*-ascorbic acid both for performance of the stopped-flow apparatus [70] and the double mixing arrangement. [71] The pseudo first-order rate constant varies linearly with the concentration of *L*-ascorbic acid, in excess, over 3 orders of magnitude. The time for mixing and moving the solution from the mixer to the observation chamber is referred to as the deadtime. [72] If the absorbance change accompanying a reaction is large, there may still be sufficient absorbance change after the absorbance loss due to the deadtime. The associated first-order rate constants will be large ($>10^2$ s^{-1}) but can be corrected for mixing effects. [73] A commercial stopped-flow apparatus can be modified so that unequal volumes of reactants (e.g. 50:1) can be mixed. [74] Details for mixing small amounts of organic solutions (e.g. containing dmso) with large amounts of aqueous solution have been described. Such an approach allows the examination of the reactions of O_2^- (stable in organic solution) with reactants (in H_2O in larger syringe) in a mainly aqueous solution. [75] Turbidity and cavitational problems can be largely circumvented by simultaneous observation of a reaction with two detector systems set at two wavelengths and computer-subtracting the two absorbances. [76] *Spurious* traces can arise from mixing effects. [77,78] Although the majority of reported studies relate to aqueous solution in ambient conditions, the stopped-flow method has been used in nonaqueous solution, [79] at subzero temperatures, [80] and at high pressure. [81-84] The stopped-flow apparatus can be easily converted into a double-mixing arrangement to examine a transient intermediate, [63] such as $VO_3^+ HO_2^-$ [85] and O_2^- Refs. 86, 87. The double-mixer has played an important role in studying the transients in the unfolding and folding of proteins. [71,88]

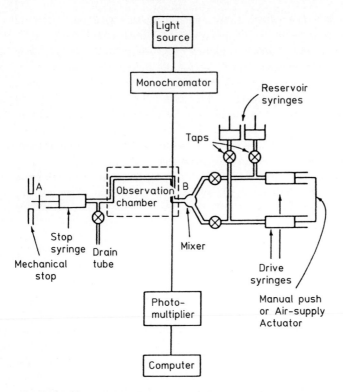

Fig. 3.3 Block diagram of stopped-flow apparatus. Reagents from the reservoir syringes are transferred to the 2-ml drive syringes via taps. A small portion (about 0.25 ml) of the reactants in each syringe is pushed through the mixer and observation chamber into a syringe, where the flow is abruptly stopped when A hits a mechanical stop. The progress of the reaction in the portion of stopped solution in B is monitored spectrally. The spent solution in the stop syringe is ejected through a drain tube and the process repeated. Ultraviolet, or visible, light through a monochromator passes through the observation chamber to a photomultiplier and oscilloscope or (more usually) an interfaced computer. Changing light intensity, arising from the absorbance changes in B as the reaction proceeds, is converted directly into absorbances. The design due to Q. H. Gibson and L. Milnes, Biochem. J. **91**, 161 (1964) has been the basis of the most successful stopped-flow apparatus, sold by Dionex.

3.4 Relaxation Methods

Even using the flow methods outlined in the previous sections, the reaction may still be too rapid to measure (i. e. it is complete within mixing). We must resort then to methods in which a reaction change is initiated by means other than mixing, the so-called relaxation methods. [89]

The amounts of species present in a chemical equilibrium may be changed by a variety of means. The rate of change of the system from the old to the new equilibrium, the *relaxation*, is dictated by (and is therefore a measure of) the rate constants linking the species at equilibrium (Sec. 1.8). Any perturbation which changes concentrations can in principle be used. It is in the area of rapid reactions that relaxation methods are most powerful and have been most applied. The perturbation must of course be applied more rapidly than the relaxation time of the system under study. However perturbations can be imposed in much shorter times than are involved in the mixing process. Since also monitoring of very rapid processes rarely presents insuperable difficulties, reaction times in the micro-to pico second range can be measured (Fig. 3.1).

The relaxation technique does not have the wide applicability associated with the flow method since one does not usually have, nor is it easy to induce, a reasonable degree of reversibility in a system. This is essential in order that measurable changes of concentration may be induced by the perturbation, although sensitive methods for detecting very small changes are now available. Relaxation methods cover the time ranges not generally attainable by flow techniques and have permitted the measurement of large first-order rate constants associated with many fundamental elementary reactions. In addition, relatively small volumes of solution are required for most relaxation techniques and these can often be used repeatedly. The various types of relaxation techniques associated with specific perturbation modes will now be considered. These perturbations can be of two types: stepwise, meaning one abrupt change, and continuous, usually imposed as an oscillating perturbation. Regardless of the complexity of the reaction, a set of first-oder kinetic equations is always obtained (Sec. 1.8.1).

3.4.1 Temperature Jump

The temperature jump is undoubtedly the most versatile and useful of the relaxation methods.[91] Since the vast majority of reactions have nonzero values for the associated ΔH, a variation of equilibrium constant K with temperature is to be expected:

$$d \ln K/dT = \Delta H/RT^2 \tag{3.7}$$

Figure 3.4 shows a block diagram of a temperature-jump apparatus. In the original[92] and still most popular form of the apparatus, electric heating is supplied by discharging a capacitor through the solution.[93, 94] Heating times are about 1 μs but if a coaxial cable is used as the capacitor, a heating time of 10 ns − 100 ns is possible.[95, 96] An electrically conducting solution is required, usually by adding 0.1 M KNO_3 or KCl, but this is avoided when microwave heating (1 μs) is used, although polar solvents must be employed.[97] Neither electrically-conducting nor polar solvents are necessary if laser heating is employed and a 1−10 ns heating time can be reached.[98, 99] Temperature jumps of 1°−10° are commonly used, but even if a small temperature change is used, repetitive jumps on flowing solutions with computer collection can be employed effectively (e. g. with microwave heating).[97] The most common monitoring method is absorption spectrometry but a few others have been employed occasionally (Table 3.7). Fluorescence monitoring has the advantage of high sensitivity which means that low concentrations of reactants can be used, thus better ensuring equilibrium conditions, even with reactions which have high equilibrium constants. The most effective ways to maximize the signal-to-noise, acquire and analyze data have been fully discussed.[91] Temperature jump can be used for reactions involving dissolved gases, but

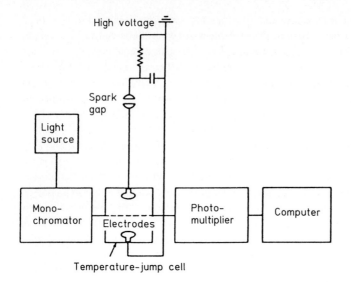

Fig. 3.4 Block diagram of temperature-jump apparatus. The condenser is charged to 30–50 kilovolts and then discharged via a spark gap through the metal electrodes of the cell. This heats up a small portion of the solution some 3–10 degrees. This portion is monitored spectrally in the same manner as with the flow apparatus. An apparatus based on this arrangement was first manufactured by Messanlagen Studiengesellschaft, Göttingen, Germany.

cavitational effects may last as long as 50 μs and thus limit its value. [100] The combination of the stopped-flow and temperature-jump methods for the study of the relaxation behavior of transients has not yet been extensively applied. [101-104] A commercial piece of equipment combines stopped flow with temperature jump and although the heating times are relatively long (>50 μs), the apparatus has been used effectively to examine the primary reaction of sodium nitroprusside with thiols before further redox reactions occur within seconds. [104] The temperature-jump method has been effectively combined with high pressure. [91]

3.4.2 Pressure Jump

Most chemical reactions occur with a change in volume. The equilibrium position will be therefore changed by an applied pressure, which can therefore be used as a perturbation. [105] Nearly always the progress of the reaction is observed at ambient pressures *after* the applied pressure has been terminated.

The expression

$$\left(\frac{d \ln K}{dP}\right)_s = \frac{-\Delta V}{RT} + \frac{\Delta H}{RT^2} \frac{\alpha T}{\rho C_p} \tag{3.8}$$

relates the variation of the equilibrium constant K with pressure P under adiabatic conditions. ΔV = standard volume of reaction; ΔH = enthalpy of reaction; α = coefficient of thermal expansion of the solution; ρ = density and C_p = specific heat. Normally in aqueous solution, the first term is the major contributor (>90% of total).[105] Usually the pressure change imposed is of the order of 50–150 atmospheres and since the response of the equilibrium constant to pressure change is much less than to temperature changes, sensitive conductivity monitoring is usually employed. Occasionally, other monitoring methods have been reported (Table 3.7). The method has the advantage that a wide range of solvents and a medium of low ionic strangth can be used. The time range can be extended to long times since the pressure (unlike the temperature) remains constant after the perturbation. The method lacks the power of the temperature jump in reaching very short times (usually ≥ 20 μs) although attempts to shorten the working time[106, 107] and improve the accuracy[106] have been made. The method has been used to study, for example, metal complex formation in solution[105, 108] and protein self-association, using light scattering monitoring.[109] Pressure-jump relaxations at various (high) applied pressures allow the determination of volume of activation.[110] (Prob. 3.) No commercial set-up is offered but the component parts are readily available and there have been good descriptions of the system.[105]

3.4.3 Electric-Field Jump

Any reacting system occuring with a change in electric moment, ΔM, will show a dependence of the associated equilibrium constant K on the electric-field strength E[111]

$$\left(\frac{d \ln K}{dE}\right)_{P,T} = \frac{\Delta M}{RT} \tag{3.9}$$

There is a modest increase in the electrical conductance with an increase in the electric-field gradient, an effect that operates with both strong and weak electrolytes (the first Wien effect). More important in the present context is the *marked* increase in electrical conductance of weak electrolytes when a high-intensity electric field is applied (second Wien effect). The high field promotes an increase in the concentration of ion pairs and free ions in the equilibrium

$$AB \rightleftharpoons A^+ \| B^- \rightleftharpoons A^+ + B^- \tag{3.10}$$

Commonly, a 10^5 volt/cm field will produce a 1% change in conductance of weak electrolytes. The measurement of very short relaxation times (≈ 50 ns) is possible by the electric-field jump method but the technique is generally complicated and mainly restricted to ionic equilibria.[111]

The five-coordinate $Ni(L)Cl_2$ ($L = Et_2N(CH_2)_2NH(CH_2)_2NEt_2$) acts as a moderately weak electrolyte in acetonitrile where equilibrium with a planar form is assumed

$$Ni(L)Cl_2 \underset{\approx 7 \times 10^5 s^{-1}}{\overset{6.7 \times 10^5 s^{-1}}{\rightleftharpoons}} Ni(L)Cl^+ \| Cl^- \underset{\approx 2 \times 10^9 M^{-1} s^{-1}}{\overset{1.6 \times 10^5 s^{-1}}{\rightleftharpoons}} Ni(L)Cl^+ + Cl^- \tag{3.11}$$

Perturbation by an electric field jump (conductivity monitoring) produces a single relaxation. These data and other considerations give the rate constants shown in (3.11). Perturbation by a laser pulse of (3.11) using spectral monitoring (Sec. 3.5.1) gives reasonably concordant kinetic data. Irradiation at 1.06 μm, where $Ni(L)Cl_2$ absorbs, increases the concentration of ion pairs and ions.[112]

3.4.4 Ultrasonic Absorption

The methods so far discussed involve a single discrete perturbation of the chemical system with direct observation of the attendant relaxation. An oscillating perturbation of a chemical equilibrium can also lead to a hysteresis in the equilibrium shift of the system. This effect can lead to the determination of a relaxation time. The process will obviously be more complex than with discrete perturbations, and there will be problems in the monitoring.[113]

Sound waves provide a periodic oscillation of pressure and temperature.[114] In water, the pressure perturbation is most important; in non-aqueous solution, the temperature effect is paramount. If ω (= $2\pi f$, where f is the sound frequency in cps) is very much larger than τ^{-1} (τ, relaxation time of the chemical system), then the chemical system will have no opportunity to respond to the very high frequency of the sound waves, and will remain sensibly unaffected. If $\omega \ll \tau^{-1}$, then the changing concentrations of chemical species demanded by the oscillating perturbation can easily follow the low frequency of the sound waves. In both cases there will be net absorption of sound. Of greatest interest is the situation in which the relaxation time is of the same order of magnitude as the periodic time of the sound wave, that is, $\tau \approx \omega^{-1}$. An amplitude and phase difference between the perturbation and the responding system develops and this leads to an absorption of power from the wave. It can be shown that the sound absorption is proportional to $\omega\tau (1 + \omega^2\tau^2)^{-1}$ and that this value passes through a maximum at $\omega\tau = 1$. Experimentally, one has then to measure the maximum attenuation of the wave as the ultrasonic frequency is changed.

In quantitative terms, for a single equilibrium

$$\frac{\alpha}{f^2} = \frac{A}{1 + \omega^2\tau^2} + B \tag{3.12}$$

α = absorption coefficient in Np cm^{-1}, the experimentally determined value; f = frequency (Hz); ω = angular frequency = $2\pi f$; τ = relaxation time; A = constant, the relaxation amplitude; B = background absorption which includes all relaxations other than due to chemical system ("classical" absorption — due to viscous and thermal losses). The plot of (3.12) is shown in Fig. 3.5. When $\omega^2\tau^2 \gg 1$, $\alpha/f^2 = B$. When $\omega^2\tau^2 \ll 1$, $\alpha/f^2 = A + B$. At $\omega\tau = 1$, there is an inflection and from this the relaxation time τ can be obtained. As a bonus, from the value of A, ΔV for the reaction can be determined. There are a number of variations in the presentation of the data.[113-117] Computer fitting to equations such as (3.12) is increasingly popular.

A number of techniques are necessary to cover the five decades in frequency from 10^4 to 10^9 Hz corresponding to relaxation times of 10^{-4} to 10^{-9} s. A wide frequency range can be spanned with only two or three methods, requiring only a few ml of sample and relatively straightforward equipment. In the 10–100 kHz range considerable volumes of sample are still required (although the material is recoverable!).

Ultrasonic absorption played a major historical role in an understanding of the mechanisms of metal complex formation.[118] It has also been used to study stereochemical change in metal

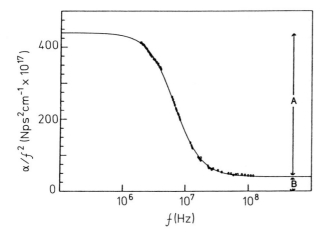

Fig. 3.5 One representation of ultrasonic data in terms of (3.12) for a single relaxation (J. K. Beattie, Adv. Inorg. Chem. **32**, 1 (1988)). Reproduced with permission from J. K. Beattie, Adv. Inorg. Chem. **32**, 1 (1988).

complexes (Sec. 7.2.3),[115,116] spin equilibria (Sec. 7.3),[119] proton transfer, aggregation phenomena, and helix-coil transitions of biopolymers.[113]

3.5 Large Perturbations

Photolysis and pulse radiolysis are powerful methods for producing sizeable amounts of reactive transients whose physical and chemical properties may be examined. Structural information on the transient and the characterization of early (rapid) steps in an overall reaction can be very helpful for understanding the overall mechanism in a complex reaction. Chemical equilibria may be disturbed by photolysis or radiolysis since one of the components may be most affected by the beam and its concentration thereby changed. The original equilibrium will be reestablished on removing the disturbance and the associated change can be examined just as in the relaxation methods. The approach has been more effectively used in laser photolysis and since very short perturbations are possible the rates associated with very labile equilibria may be measured.

3.5.1 Flash or Laser Photolysis

This involves the application of a pulse of high intensity light of short duration to a solution containing one or more species. In the original Nobel prize winning studies a flash lamp of a few microseconds duration was used.[120] Now a laser pulse is more often utilized and times as short as picoseconds or less may be attained. Several set-ups have been described.[121] Their complexity and cost are related to the time resolution desired. An inexpensive system using a photographic flash lamp ($\sim$ 5 ms) and two laser systems with time resolutions of 1–5 µs and 30 ns have been described in detail.[122] (See also Refs. 123–124.) One of these is shown in Fig. 3.6.[122] Laser photolysis in the ns-ps region is obviously very sophisticated but quite extensively used.[125] In Ref. 126 an apparatus for picosecond resonance Raman spectroscopy of carboxyhemoglobin is described.[126]

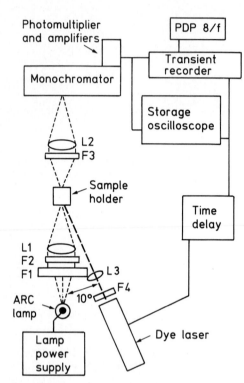

Fig. 3.6 Schematic diagram of a dye laser photolysis set up for relaxation times $\geq$ µs. The photolyzing light pulse is produced by a dye laser and enters the sample at about 10° to the axis of the sample beam. The observation beam originates from a 75-W xenon arc lamp. The apparatus is supplied by OLIS, Athens, Georgia USA.[122] Reproduced with permission from C. A. Sawicki and R. J. Morris, Flash Photolysis of Hemoglobin, in Methods in Enzymology (E. Antonini, L. R. Bernardi, E. Chiancone eds.), **76**, 667 (1981).

The intense pulse of light will likely produce a highly reactive excited state or states in a matter of femtoseconds. The subsequent steps that are possible can be illustrated by examining the case of $Ru(bpy)_3^{2+}$, undoubtedly the most photochemically studied complex ion.[127] Irradiation at 503 nm produces the excited species $*Ru(bpy)_3^{2+}$ which is generally believed to have the transferred electron on a single ligand. The subsequent behavior of $*Ru(bpy)_3^{2+}$ is shown as *A in (1.32). Of most interest to us is the strong reducing ($E^0 = -0.86$ V) and moderate oxidizing ($E^0 = 0.84$ V) properties of the excited species, compared with the ground state. Thus if Fe^{3+} is also present in the irradiated solution (B in scheme 1.32) the following sequence of events occurs:[128]

$$Ru(bpy)_3^{2+} \quad\quad\quad \rightarrow \; *Ru(bpy)_3^{2+} \tag{3.13}$$

$$*Ru(bpy)_3^{2+} + Fe^{3+} \; \rightarrow \; Ru(bpy)_3^{3+} + Fe^{2+} \tag{3.14}$$

$$Ru(bpy)_3^{3+} + Fe^{2+} \; \rightarrow \; Ru(bpy)_3^{2+} + Fe^{3+} \tag{3.15}$$

The very rapid reaction (3.15) with a large $-\Delta G$ can thus be measured.[129] We therefore have an effective method for generating very rapidly *in situ* a powerful reducing or oxidizing agent. One of the most impressive applications of these properties is to the study of internal electron transfer in proteins.[130, 131]

The $Ru(NH_3)_5^{2+}$ moiety can be attached to histidine-83 on the azurin surface. It can then be oxidized to Ru(III) without altering the conformation of the protein. This ruthenated protein is mixed with $Ru(bpy)_3^{2+}$ and laser irradiated. The sequence of events which occurs is shown in the scheme

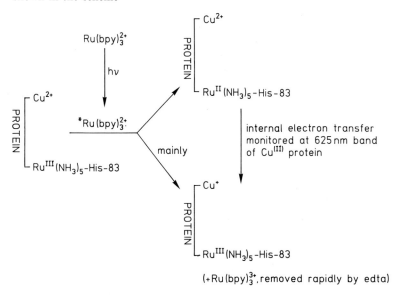

(+Ru(bpy)$_3^{3+}$, removed rapidly by edta)

In the first step, considerable amounts of the final product are produced as well as smaller amounts of a transient in which the oxidation states are "incorrect". Internal electron transfer redresses this imbalance. The species $Ru(bpy)_3^{3+}$ produced must be removed rapidly (by scavenging with edta) so that it cannot oxidize the Ru(II) protein and interfere with the final step.[130] See Sec. 5.9. Some other examples of the application of the photolytic method to a variety of systems are shown in Table 3.4.[132-139] (See Probs. 4 and 5)

Table 3.4 Some Transients Generated by Photolytic Methods

Irradiated System	Conditions	Results	Applications	Ref
Naphthols	pH 4–7; subnanosecond irradiation	Photoexcited naphthols with much lower pK_a than in ground state dissociate[133]	Rapid (50 ns) pH (4 units) drop. Probe macromolecules[134, 135]	132
FeXsal$_2$trien$^+$, X = H or OCH$_3$. Low spin, high spin mixture	530 nm, nano-second laser; nonaqueous solvents	Transient bleaching of low spin species followed by thermal relaxation τ's 46–192 ns at 255–205 K	Kinetics of low spin $\rightleftharpoons$ high spin Fe(III) Sec. 7.3	137

Table 3.4 (continued)

Irradiated System	Conditions	Results	Applications	Ref
Ni(pad) (H$_2$O)$_4^{2+}$ [See **6**]	3 μs laser pulse; aqueous solution	Deactivation of excited state partly by photo-substitution and producing sizeable amounts of Ni(pada)(H$_2$O)$_5^{2+}$	Direct analysis of ring closure, not possible by temperature or pressure jump	138
Oxy and carboxy-hemo proteins	Varying laser pulse times	Photodeligation to varying degrees	Measurement of combination rates of nonequilibrated forms of hemopro-teins (see Eqn. 2.20)	139

3.5.2 Pulse Radiolysis

This technique has similarities to photolysis in that a large perturbation is involved and reactive transients can be produced and examined.[140-142] Van der Graff accelerators or, more popular, microwave linear electron accelerators are used to produce a high energy electron pulse typically within ns to μs. The set-up is illustrated in Figure 3.7. Reducing and oxidizing radicals result:[143]

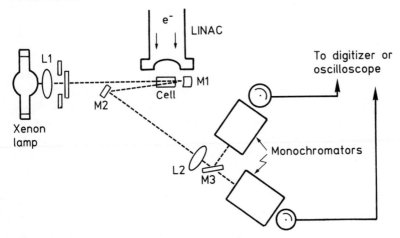

Fig. 3.7 Schematic set-up for pulse radiolysis. Electron beam and analyzing light beams are at right angles. The light beam is split for simultaneous recording at two wavelengths.[142]

$$4\,H_2O \rightarrow 2.6\,e^-_{aq} + 2.6\,OH^\bullet + 0.6\,H^\bullet + 2.6\,H^+ + 0.4\,H_2 + 0.7\,H_2O_2 \qquad (3.16)$$

The mixture of radicals would be difficult to examine or use. Fortunately, by the careful choice of an added substrate, some of the radicals produced in (3.16) are rapidly removed so as to leave the desired radical for examination. For example, to examine e_{aq}^-, pulse radiolysis of solutions containing CH_3OH are used which rapidly removes $H^{\bullet}$ and $OH^{\bullet}$ radicals. Table 3.5 lists the more important radicals and gives references to the extensive compilation of properties and kinetic data for reactions of those radicals.[144-150] We thus have a means for the rapid production of highly reducing or oxidizing species. These can, in turn, be used to generate, for example, less usual oxidation states of metal ions and complexes (Chap. 8). Metastable oxidation states in polypeptides and proteins can also be produced and intramolecular electron transfer examined, just as by the photolytic method (Sect. 3.5.1).[151] Other examples are contained in Table 3.6. It is essential to have a concentration of substrate (S) sufficient to produce a higher rate of reaction with the radical (R), k [S] [R], than that of the spontaneous decay of R, $k_1[R]^2$, which occurs usually by disproportionation (Figure 3.8).[149]

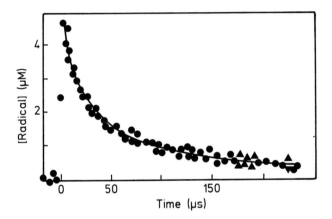

Fig. 3.8 Second-order decay of $N_3^{\bullet}$ ($2 N_3^{\bullet} \xrightarrow{2k} 3 N_2$) measured at 274 nm. $2 k = 8.8 \times 10^9 M^{-1} s^{-1}$. Loss of 1–5 μM (initial $N_3^{\bullet}$ concentrations) all conform to the solid curve. There is no evidence of any significant first-order component even at $[N_3^{\bullet}] = 0.5$ μM.[149] Reproduced with permission from Z. B. Alfassi and R. H. Schuler, J. Phys. Chem. **89**, 3359 (1985). © (1985) American Chemical Society.

Table 3.5 Some Radicals Generated by Radiolysis

Radical	Production (added substances underlined)[a]	Remarks
e_{aq}^-	$\underline{CH_3OH}^{(b)} + OH^\bullet \rightarrow {}^\bullet CH_2OH + H_2O$	Powerful reducing agent ($E^0 = -2.9\,V$)
	$\underline{CH_3OH}^{(b)} + H^\bullet \rightarrow {}^\bullet CH_2OH + H_2$	Reacts diffusion controlled with many substrates. [144, 145]
$H^\bullet$	$e_{aq}^- + H^+ \rightarrow H^\bullet$; $\underline{CH_3OH}, \underline{N_2O}$	Powerful reducing agent ($E^0 = -2.3\,V$). [145, 146]
$CO_2^{\overline{\bullet}}$	$\underline{N_2O} + e_{aq}^- \rightarrow N_2 + OH^- + OH^\bullet$	$E^0 = -2.0\,V$; reactivity often intermediate
	$\underline{HCO_2^-} + OH^\bullet \rightarrow CO_2^{\overline{\bullet}} + H_2O$	between e_{aq}^- and $SO_2^{\overline{\bullet}}$ Refs. 147–149.
	$\underline{HCO_2^-} + H^\bullet \rightarrow CO_2^{\overline{\bullet}} + H_2$	
$O_2^{\overline{\bullet}}$	$\underline{O_2} + H \rightarrow O_2^{\overline{\bullet}} + H^+$	Often acts as one-electron slow reductant
	$\underline{O_2} + e_{aq}^- \rightarrow O_2^{\overline{\bullet}}$	($E^0 = -0.33\,V$). [150]
	$\underline{HCO_2^-} + OH^\bullet \rightarrow CO_2^{\overline{\bullet}} + H_2O$	
	$\underline{O_2} + CO_2^{\overline{\bullet}} \rightarrow O_2^{\overline{\bullet}} + CO_2$	
$OH^\bullet$	$\underline{N_2O} + e_{aq}^- \rightarrow N_2 + OH^- + OH^\bullet$	Strong oxidant ($E^0 = 1.9\,V$). [145]
	$\underline{N_2O} + H^\bullet \rightarrow N_2 + OH^\bullet$	
$(SCN)_2^{\overline{\bullet}}$	$\underline{N_2O} + e_{aq}^- \rightarrow N_2 + OH^- + OH^\bullet$	Ref. 148, 149. At maxm, 472 nm, $\epsilon = 7.6$
	$2\,\underline{SCN^-} + OH^\bullet \rightarrow (SCN)_2^{\overline{\bullet}} + OH^-$	$\times 10^3\,M^{-1}cm^{-1}$ and useful for dosimetry
$Br_2^{\overline{\bullet}}$	$\underline{N_2O} + e_{aq}^- \rightarrow N_2 + OH^- + OH^\bullet$	Ref. 148, 149. Strong oxidant ($E^0 = 1.63\,V$)
	$2\,\underline{Br^-} + OH^\bullet \rightarrow Br_2^{\overline{\bullet}} + OH^-$	
$N_3^\bullet$	$\underline{N_2O} + e_{aq}^- \rightarrow N_2 + OH^- + OH^\bullet$	Ref. 148. Rapid, selective [149] strong
	$\underline{N_3^-} + OH^\bullet \rightarrow N_3^\bullet + OH^-$	oxidant ($E^0 = 1.30\,V$)

(a) Usually $1-10\ \mu M$ radical generated in $0.2-2.0\ \mu s$. 1% methanol or saturated N_2O or $\approx 0.1\,M$ HCO_2^- or halide used. Phosphate buffers.
(b) $(CH_3)_3COH$ also used to scavenge. Radicals produced are relatively unreactive.

Table 3.6 Examples of Use of Radicals Generated by Pulse Radiolysis

Irradiated System	Pulse Radical	Result	Ref.
(HIPIP) reduced	e_{aq}^- (but not $CO_2^{\overline{\bullet}}$)	First production in aqueous solution of (HIPIP) superreduced	152
Co(III) ammine complexes e.g. $Co(NH_3)_6^{3+}$ in H^+	e_{aq}^-	Production of $Co(NH_3)_6^{2+}$ Can then observe subsequent acid hydrolysis $Co(NH_3)_6^{2+} \xrightarrow{H^+} Co^{2+} + 6\,NH_4^+$ using conductivity monitoring	153
$Co^{III}(NH_3)_5OCO\!-\!\langle\ \rangle\!-\!NO_2^{2+}$	$CO_2^{\overline{\bullet}}$ or ${}^\bullet C(CH_3)_2OH$	Formation of $Co^{III}(NH_3)_5OCO\!-\!\langle\ \rangle\!-\!NO_2^{\overset{+}{\bullet}}$ ($\epsilon = 2 \times 10^4\,M^{-1}cm^{-1}$ at 330 nm). Followed by intramolecular electron transfer ($k_{et} = 2.6 \times 10^3 s^{-1}$), k_{et} independent of [Co(III)]. (Sec. 5.8.2)	154 see also 155, 156

3.5.3 Comparison of Large Perturbation Methods

In general, photolysis induces substitutional and redox-related changes, whereas pulse radiolysis primarily promotes redox chemistry. Indeed one of the unique features of the latter method is to induce *unambiguous* one electron reduction of multi-reducible centers.[157] Metalloproteins can be rapidly reduced to metastable conformational states and subsequent changes monitored.[158]

Monitoring of events following perturbations can be achieved in much shorter times by photolysis. A variety of monitoring techniques have been linked to both methods (Table 3.7). It is valuable to obtain kinetic data by more than one method, when possible. The measurement of spin-change rates have, for example, been carried out by a variety of rapid-reaction techniques, including temperature-jump, ultrasonics and laser photolysis with consistent results (Sec. 7.3).

3.6 Competition Methods

In this method, the reaction under study competes with another fast process, which may be spin relaxation (nmr and epr), fluorescence or diffusion towards an electrode. Monitoring of the competition is generally internal, making use of the characteristics of the fast process itself. This approach will be treated in some of the next sections.

3.7 Accessible Rate Constants Using Rapid Reaction Methods

It is the half-life of a reaction that will govern the choice of the initiation method (Fig. 3.1) and it is the character of the reaction that will dictate the monitoring procedure (Sec. 3.8 on).

The half-life of a reaction with a kinetic order higher than one is lengthened as the concentrations of the reactants are decreased (Sec. 1.3). Provided that there is still a sufficient change of concentration during the reaction to be accurately monitored, quite large rate constants may be measured if low concentration of reactants are used, even without recourse to the specialized techniques described in the previous section.

The second-order redox reaction

$$Fe^{2+} + Co(C_2O_4)_3^{3-} \xrightarrow{6\,H^+} Fe^{3+} + Co^{2+} + 3\,H_2C_2O_4 \qquad (3.17)$$

can be followed using even micromolar reactant concentrations, because of the high molar absorptivity of $Co(C_2O_4)_3^{3-}$ ($2.6 \times 10^4 M^{-1}cm^{-1}$ at 245 nm). The observed second-order rate constant ($1.2 \times 10^3 M^{-1}s^{-1}$ at 25 °C, μ very low) corresponds to reaction times of minutes with μM concentrations.[159] However if flow techniques and higher concentrations of reactants are used additional features of the reaction show up. Two steps are noted at 310 nm, a fast increase and slower decrease in absorbance corresponding to the production and loss of an intermediate $Fe(C_2O_4)^+$, which has a higher molar absorptivity at 310 nm than reactants or products.[160]

$$Fe^{2+} + Co(C_2O_4)_3^{3-} \xrightarrow{4H^+} Fe(C_2O_4)^+ + Co^{2+} + 2H_2C_2O_4 \tag{3.18}$$

$$Fe(C_2O_4)^+ \xrightarrow{2H^+} Fe^{3+} + H_2C_2O_4 \tag{3.19}$$

Large second-order rate constants, even diffusion controlled, can be measured by flow methods using low concentrations of reactants.[161] Use of pulsed accelerated flow (monitoring reaction times as low as 10 μs) and the large perturbation methods ($\leqslant 1$ μs) can allow the measurement of rate constants $> 10^9$ $M^{-1}s^{-1}$ for *irreversible* reactions which are not amenable to relaxation methods. The concentration of a reactant and therefore the rate of a reaction may be drastically reduced by adjustment of the pH or by the addition of chelating agents.[162] It is worth recalling that forward and reverse rate constants are related by an equilibrium constant for the process. Relatively high rate constants for the formation of metal complexes (k_f) can be estimated from the relationship $k_f = k_d K$ in cases where K (the formation constant) is large and k_d is a small and easily measured dissociation rate constant. Many early data for Ni(II) were obtained in this way.[163]

The power of the relaxation and large perturbation methods lie in their ability to allow the measurement of large *first-order* rate constants. The half-lives of first-order reactions are concentration-independent and values $\leqslant 1$ ms ($k \gtrsim 10^3$ s^{-1}) are outside the ability of most flow methods. First-order rate constants ranging from 10^4 s^{-1} to as large as $10^{11}-10^{12}$ s^{-1} can be measured by the methods shown in Fig. 3.1. This encompasses such disparate processes as conformational change, intersystem crossing, excited state deactivation and small displacements of ligands from heme centers. The rate of any reaction with a finite heat of activation will be reduced by lowering the temperature. Some reactions that are fast at normal temperatures were studied nearly forty years ago by working in methanol at $-75°C$, when quite long reaction times were observed.[164] A combination of lowered temperatures (to 252 K) with fast-injection techniques has allowed the measurement of the electron self-exchange by ^{17}O nmr:[165]

$$Ru(H_2O)_6^{2+} + Ru(H_2{}^{17}O)_6^{3+} \rightleftharpoons Ru(H_2O)_6^{3+} + Ru(H_2{}^{17}O)_6^{2+} \tag{3.20}$$

The lowered temperature approach has been linked to flow,[80, 166] temperature jump,[167] photolysis,[168] and nmr[165] methods. Cryoenzymology allows the characterization of enzyme intermediates which have life-times of only milliseconds at normal temperatures, but are stable for hours at low temperatures. Mixed aqueous/organic solvents or even concentrated salt solutions are employed and must always be tested for any adverse effects on the catalytic or structural properties of the enzyme.[167, 169]

3.8 The Methods of Monitoring the Progress of a Reaction

The rate of a reaction is usually measured in terms of the change of concentration, with time, of one of the reactants or products, $- d$ [reactant]$/dt$ or $+ d$ [products]$/dt$, and is usually expressed as moles per liter per second, or $M s^{-1}$. We have already seen how this information might be used to derive the rate law and mechanism of the reaction. Now we are concerned, as kineticists, with measuring experimentally the concentration change as a function of the time that has elapsed since the initiation of the reaction. In principle, any property of the reactants or products that is related to its concentration can be used. A large number of properties have been tried.

It is obviously advantageous from a point of view of working up the data if the reactant property and concentration are linearly related. It is much easier and more likely to be accurate to monitor the reaction continuously *in situ* without disturbing the solution than to take samples periodically from the reaction mixture and analyze these separately, in the so-called batch method. The batch method cannot, however, be avoided when an assay involves a chemical method (which "destroys" the reaction). Separation of reactants or products or both also is necessary when an assay involving radioisotopes is employed. Separation prior to analysis is sometimes helpful when the system is complicated by a number of equilibria, or when a variety of species is involved.

Methods that have been used for monitoring reactions will be discussed in the next sections. Those applicable to rapid reactions are shown in Table 3.7. Of the hundreds of possible references to the literature, only a few key ones are given. Several considerations will dictate the method chosen. If it is suspected that the reaction may be complex, then more than one method of analysis ought to be tried, so as to show up possible intermediates and characterize the reaction paths in more detail.

Table 3.7 Monitoring Methods for Rapid Reactions

Monitoring Method	Flow	T-jump	P-jump	E-jump	Photolysis	Radiolysis
UV and visible spectrophotometry	46	170	171	172	122	173
Infrared	174	–	–	–	175	–
Raman	176. 177	–	–	–	178	179
Light scattering	180	181	109	–	–	–
Fluorescence	182	183	184	–	–	–
Polarimetry (cd)	185, 186	187	188	189	190	–
nmr	191, 192	–	–	–	193	194
epr	195	–	–	–	196	197
Conductivity	198, 199	200	199	201	202, 203	204
Thermal	205–207					

3.9 Spectrophotometry

Nearly all the spectral region has been used in one kinetic study or another to follow the progress of a chemical reaction. *In toto* it represents by far the most powerful and utilized method of monitoring.

3.9.1 Ultraviolet and Visible Regions

These regions are particularly useful since few, if any, reactions of transition metal complexes are unaccompanied by spectral absorption changes in these regions. We first show how optical absorbances may be substitued for the concentration changes required in deriving the rate law.

The optical absorbance D shown by a single chemical species A in solution is related to its concentration by the Beer-Lambert law:

$$D = \log \frac{I_0}{I_t} = \varepsilon_\lambda \cdot l \cdot [A] \tag{3.21}$$

where I_0 and I_t = incident and transmitted light intensities at wavelength λ.

$\quad\quad\quad\quad \varepsilon_\lambda$ = molar absorptivity at a wavelength λ.

$\quad\quad\quad\quad l$ = light path, in cm.

$\quad\quad\quad [A]$ = molar concentration of the species A.

Mixtures of species A, B, ... usually give additive absorbances:

$$\left(\frac{D}{l}\right) = \varepsilon_A [A] + \varepsilon_B [B] + \cdots \tag{3.22}$$

and so it is possible to analyze changes in the concentrations of specific reactants or products from absorbance changes in the reaction mixture.

It is not difficult to show that the optical absorbance, or other properties for that matter, can be used directly to measure first-order rate constants, without converting to concentrations. Consider

$$A \rightarrow B \tag{3.23}$$

Omitting brackets to denote concentrations, we find:

At zero time, $D_0 = \varepsilon_A A_0 + \varepsilon_B B_0$ $\tag{3.24}$

At time t, $D_t = \varepsilon_A A_t + \varepsilon_B B_t$ $\tag{3.25}$

At equilibrium $D_e = \varepsilon_B B_e = \varepsilon_B (A_0 + B_0) = \varepsilon_B (A_t + B_t)$ $\tag{3.26}$

Therefore,

$$A_0 = \frac{(D_0 - D_e)}{\varepsilon_A - \varepsilon_B} \quad A_t = \frac{(D_t - D_e)}{\varepsilon_A - \varepsilon_B} \tag{3.27}$$

$$\ln \frac{A_0}{A_t} = \ln \frac{(D_0 - D_e)}{(D_t - D_e)} = kt \tag{3.28}$$

For orders of reaction other than one, a knowledge of the molar absorptivities is necessary.[208] See Fig. 1.3.

Commercial spectrophotometers, both in the conventional and stopped-flow mode display, usually on a computer screen, the optical absorbance changes directly. There are distinct advantages in using absorbance in the visible region since there is less likelihood of interference from buffers and added electrolytes, which are usually transparent in this region. However, the molar absorptivities of complex ions are often much higher in the ultraviolet, since they are based on charge transfer rather than on *d-d* transitions. Consequently, lower concentrations of reactants with ultraviolet monitoring will, as the reaction proceeds, give absorbance changes that are comparable with the absorbance changes when the visible region is used. A lower concentration means a greater economy in materials, but sometimes more important, a longer $t_{1/2}$ for all but first-order reactions. We have seen the value of this already (Sec. 3.7). Even small absorbance changes can be measured accurately using the sensitive detectors and associated data acquisition equipment now available. Many runs can be accumulated and processed to obtain high signal/noise ratios even with absorbance changes as low as 0.001 units.[209] Small absorbance changes inherent in the system may be amplified by the introduction of a chromophoric group into one of the reactants. The modification of carboxypeptidase A by attachment of the intensely colored arsanilazo group to the tyrosine-248 residue 5 allows

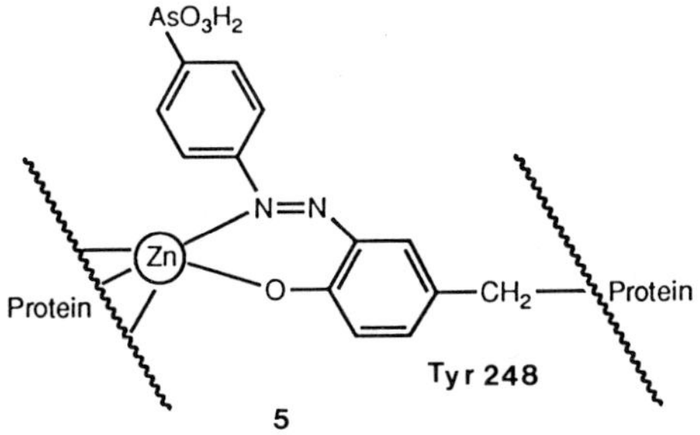

5

the ready visualization of interactions of the enzymes with substrates and inhibitors (Sec. 1.8.2)[210] or of the apoenzyme with metal complexes.[211] Care has to be taken that the modification does not alter radically the characteristics of the unmodified reactant.[212] Other methods of enhancing spectral changes are to add indicators to monitor pM[213, 214] or pH[215] changes.[216] Colored complexes may be used to probe reaction sites e.g. Zn(pad)$^{2+}$ 6 is a probe for two adjacent carboxylate groups of the catalytic residues of acid proteinases.[217]

Reaction complexity can often show up (and be resolved) by studying the reaction at more than one wavelength, when different rate patterns may be observed (Sec. 3.7).[218] However the collection of a family of spectra directly as the reaction proceeds can be of enormous value.

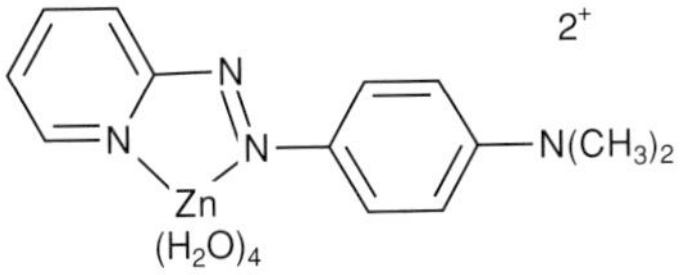

6

As might be expected, the problem of obtaining spectra of a reacting system increases as the time resolution involved decreases. The spectral changes associated with a reaction may be constructed by wavelength point-by-point measurements. The method, although tedious and costly on materials, is still used. However rapid-scan spectrophotometry, linked to stopped-flow, is now more readily available and reliable. Two systems are used, shown schematically in (3.29)[219-221] and (3.30).[222, 223] An example of its use is shown in Fig. 3.9.[222] Rapid scan

$$\text{white light} \rightarrow \text{mixing chamber} \rightarrow \text{diffraction grating} \rightarrow \sim 30 \text{ photodiodes} \quad (3.29)$$

$$\text{white light} \rightarrow \text{moving diffraction grating} \rightarrow \text{mixing chamber} \rightarrow \text{photomultiplier} \quad (3.30)$$

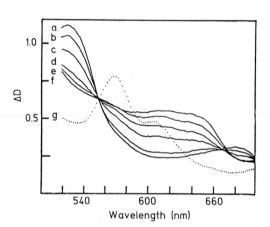

Fig. 3.9 Spectra of $Fe(tpps)H_2O^{3-}$ (a), $Fe(tpps)OH^{4-}$ (f) and mixtures (curve b, pH 6.5; curve c, pH 7.0; curve d, pH 7.46, curve e, pH 7.9). Curves b–f were obtained within 3.8 ms after mixing $Fe(tpps)H_2O^{3-}$ at pH $\approx$ 5 with buffer at the designated pH. The spectrum of $Fe(tpps)OH^{4-}$ was obtained by plunging into pH 9.1. The spectrum of $(tpps)Fe-O-Fe(tpps)^{8-}$ is shown in curve g. In all cases, the final concentration of iron(III) is 103 μM. Path length in the Dionex stopped-flow instrument is 1.72 cm (fluorescence observation chamber).[222] Reproduced with permission from A. A. El-Awady, P. C. Wilkins and R. G. Wilkins, Inorg. Chem. **24**, 2053 (1985). © (1985) American Chemical Society.

spectrophotometry is valuable for obtaining spectra following radiolysis[224] or photolysis.[225] The acquiring of spectra after very rapid initiation (10^{-9} to 10^{-14} s) presents large technical problems[125] but has been used effectively to shed light on the early events following photodeligation of oxy- and carboxyhemoproteins[139] (Sec. 2.1.2). A stopped-flow rapid scanning spectrophotometer for use at low temperatures has been described.[226] It has been used to obtain the absorption (and epr) spectra of peptide and ester intermediates formed within 0.5 s in 4.5 M NaCl at $-20\,°C$ by the hydrolysis of dansyl oligopeptides and esters catalyzed by cobalt carboxypeptidase. The results indicate that the metal plays a vital but different role in the two hydrolyses.[226]

The occurrence or not of *isosbestic points* during a reaction is very informative. Isosbestic points are wavelengths at which the absorbance remains constant as the reactant and product composition changes. One (or preferably more) isosbestic points during a reaction strongly suggest that the original reactant is being replaced by *one* product, or if more than one product, that these are always in a strictly constant ratio. The occurrence of isosbestic points implies the absence of appreciable amounts of reaction intermediates, thus, for example, supporting a scheme

$$A \underset{\searrow}{\overset{\nearrow}{}} \begin{matrix} B \\ C \end{matrix} \qquad (3.31)$$

rather than

$$A \rightarrow B \rightarrow C \tag{3.32}$$

Seven isosbestic points in the relatively fast reaction of Hg(tpp) (7) with Zn(II) in pyridine (Fig. 3.10) show that Hg(tpp) converts to Zn(tpp) without formation of appreciable amounts of free tpp base, which has a different spectrum from either complex.[227] The occurrence of

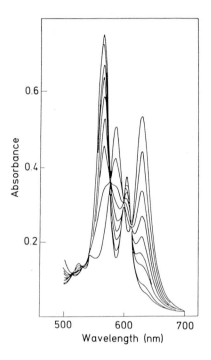

Fig. 3.10 Seven isosbestic points observed during the reaction of Zn(II) with Hg(tpp).[227]

7

an isosbestic point at 435 nm strongly suggests that Compound I is the *sole* product of the reaction between catalase and methyl hydroperoxide or peracetic acid. Compound I is only stable for a few hundred milliseconds and rapid-scan stopped-flow must be used.[228]

3.9.2 Infrared Region

Direct monitoring by infrared absorption is not commonly used for the study of complex-ion reactions in water, because the solvent and dissolved electrolytes often absorb in this region. Infrared analysis had found some use in studying H-D exchange processes in both substitution and redox reactions, using D_2O as the solvent since this does not absorb extensively in the near infrared. The nmr method (Sec. 3.9.5) has largely replaced infrared for monitoring these exchanges. Infrared monitoring sometimes provides structural information not obtainable using uv-visible spectroscopy. The CO, NO and NC stretches are very sensitive to the metal environment. Therefore substitution,[229] including CO exchange,[230,231] geometrical isomeriza-tion (3.33)[232] and redox reactions[233] of organometallics have been studied in

$$cis\text{-}Mo(CO)_4(Bu_3P)_2 \;\rightleftharpoons\; trans\text{-}Mo(CO)_4(Bu_3P)_2 \tag{3.33}$$

nonaqueous solvents using infrared detection. The potential for the examination of metal oxo compounds is illustrated by the demonstration of oxo-group transfer in a Mo(III) $\rightarrow$ Mo(V) transformation

$$\tag{3.34}$$

$(L \;=\; 1,4,7\text{-triazacyclononane})\ \mathbf{8}.[234]$ The rate law

$$d\,[Mo_2^V]/dt \;=\; k\,[Mo_2^{III}]\,[NO_3^-] \tag{3.35}$$

8

indicates that substitution of H_2O by NO_3^- is the rds. Using 90% enriched $N^{18}O_3^-$ the Mo(V) product shows a predominant $v\,(Mo{=}O_{term}) \;=\; 868\ cm^{-1}$ and $v_{as}\,(Mo{-}O_{bridge}) \;=\; 730\ cm^{-1}$. Comparison with the unlabelled Mo(V) product which shows $v\,(Mo{=}O_{term}) \;=\; 918\ cm^{-1}$ and $v_{as}\,(Mo{-}O_{bridge}) \;=\; 730\ cm^{-1}$ and with one terminal $M \;=\; O_{term}$ ($v \;=\; 887\ cm^{-1}$ and $730\ cm^{-1}$) indicates that the terminal oxo groups in the product of (3.34) must have originated from the NO_3^- group.[234]

Fast time resolved infrared attached to flow[174] and to uv-vis flash photolysis[175] has been an important development for the study of rapid substitution, e. g. in $Co_2(CO)_8$ in hexane[174] and for the detection of transient organometallic species, with μs resolution.[175] (Sec. 3.13.1) A unidentate intermediate is observed in the reaction of $M(CO)_6$, M = Cr, Mo and W, with 4,4'-dialkyl-2,2'-bipyridine (L-L) using rapid-scan Fourier transform ir.[235]

$$M(CO)_6 \underset{\Delta}{\overset{h\nu}{\rightleftharpoons}} M(CO)_5 + CO \tag{3.36}$$

$$M(CO)_5 + \text{L-L} \xrightarrow[\text{fast}]{\Delta} M(CO)_5\text{L-L} \xrightarrow{\Delta} M(CO)_4(\text{L-L}) + CO \tag{3.37}$$

Raman, like infrared, gives sharp detailed spectra and structural information. The sensitivity is however poorer than uv-vis and the equipment more complex. By choosing an excitation frequency matching that of a strong absorption band, ordinary Raman intensities may be boosted some five orders of magnitude (resonance Raman).[236] Small amounts of a transient in a high solute background may be characterized. For examining some short time transients, two laser pulses are used. One photolyses the sample. The second laser acts as a resonance Raman probe. This allows picosecond imaging in very fast photochemically induced reactions.[126] Resonance Raman is a powerful structural tool for characterizing metalloproteins[236] and has been successfully linked to flow and photolytic, but not (so far) to the relaxation methods.

3.9.3 Fluorescence

The attenuation of fluorescence can be used as a sensitive monitor for the progress of a reaction. Even reactant concentrations as low as 0.5-2.5 μM can be detected, thus bringing the reaction rates into the stopped-flow region even although second-order rate constants as high as $10^7 \text{ M}^{-1} \text{ s}^{-1}$ may be involved.[237] Fluorescence has been little used to monitor complex-ion reactions, although its enhanced sensitivity, compared with that of uv-visible should be an important consideration, not only for reactions involving aromatic-type ligands e.g. porphyrins[238] but also others.[239]

The luminescence of Tb^{3+} increases when coordinated water is replaced by a ligand such as edta. Thus the reaction

$$\text{Ca(edta)}^{2-} + Tb^{3+} \rightarrow Ca^{2+} + \text{Tb(edta)}^- \tag{3.38}$$

can be monitored at 545 nm at right angles to the exciting beam (360-400 nm).[239] Flow- and relaxation apparatus with spectral monitoring can be easily adapted to follow fluorescence changes.[182, 183]

It is in the study of reactions involving proteins that fluorescence comes into its own.[240] Thus, the reaction of carbonic anhydrase with a wide variety of sulfonamides (important inhibitors) results in a substantial quenching of the fluorescence of the native protein (Sec. 1.10.2 (b)) The binding of concanavalin A to certain sugars is most conveniently monitored by attaching a 4-methylumbelliferyl fluorescent label **9** to the sugar and examining by stopped-flow the fluorescence quenching of the sugar on binding.[241, 242] See Structure **10,** Chap. 1. The kinetics of binding of "colorless" sugars can also be studied in competition with the fluorescent one.[243] Alternatively, the protein may be chromophorically modified. Since Ca^{2+}, an important component of many metalloproteins, can be replaced by La^{3+} without

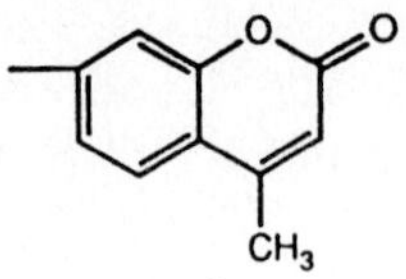

9

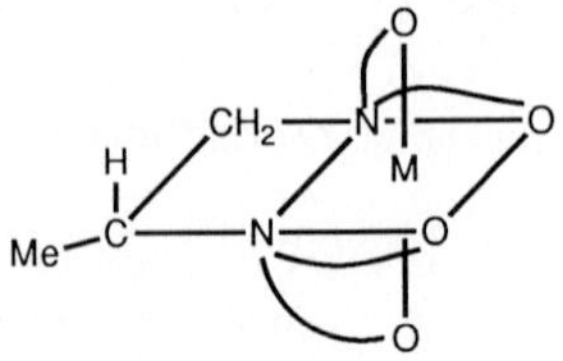

Δ-[M(R(-)pdta)]$^{n-}$

10a

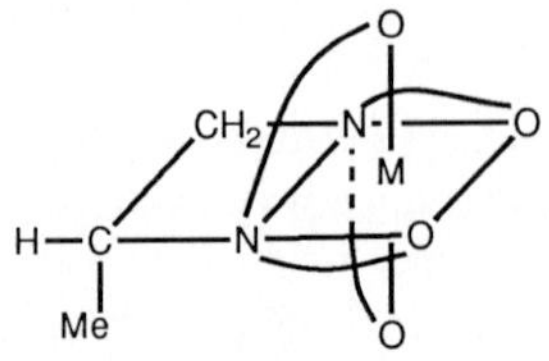

Λ- [M(R(-)pdta)]$^{n-}$

10b

gross changes in protein structure, the La(III) derivative can be monitored by fluorescence methods, and give, indirectly, information on the native protein.[244]

The measurement of *fluorescence life-times* has great value for probing structural features of proteins. It requires expensive equipment since very rapid extinction of the exciting nanosecond pulse is necessary and the rapid decay of the emission must then be measured.[245] The decay of the tryptophan fluorescence of LADH is biphasic with $\tau = 3.9$ and 7.2 ns and these are assigned to buried Trp-314 and exposed Trp-15, respectively.[246]

3.9.4 Polarimetry

The polarimetric method naturally must be used when stereochemical change involving chiral compounds is being investigated. The optical rotatory power of an optically active ligand is dependent on the environment. This fact can be used to follow the interaction of resolved ligands with metal ions.[247] Optical rotation changes accompanying reactions of proteins (folding, conformational changes, helix-coil transformations) are a useful monitor especially now that stopped-flow and relaxation techniques have been successfully linked with circular dichroism.[185]

The polarimetric method is sometimes useful for the study of exchange reactions in which large rotation changes but no net chemical change is involved:[249,250]

$$\Delta\text{-Co}^{III}(-)(\text{pdta})^- + \Lambda\text{-Co}^{II}(+)(\text{pdta})^{2-} \rightleftharpoons \Delta\text{-Co}^{II}(-)(\text{pdta})^{2-} + \Lambda\text{-Co}^{III}(+)(\text{pdta})^- \tag{3.39}$$

$$\Lambda\text{-[Co(cage)]}^{3+} + \Delta\text{-[Co(cage)]}^{2+} \rightleftharpoons \Lambda\text{-[Co(cage)]}^{2+} + \Delta\text{-[Co(cage)]}^{3+} \tag{3.40}$$

(cage = a variety of cage ligands,[250] See Sect. 2.2.1(a)). Even when a very slight absorbance change is involved as in (3.41)

$$(\pm)\text{-Pb(edta)}^{2-} + R(-)\text{pdta}^{4-} \rightleftharpoons \Delta\text{-}(+)\text{-Pb-}R(-)(\text{pdta})^{2-} + \text{edta}^{4-} \tag{3.41}$$

the large rotational differences between the Pb-pdta complex and the pdta ligand **(10)** at 365 nm is more sensitive and may be utilized.[247] Although polarimetry is rarely superior to the other spectral methods as an analytical tool, probing by more than one monitoring device can be informative. In the unfolding of apomyoglobin, monitoring by cd indicates a more rapid change than when fluorescence is examined. This is ascribed to changes in α-helical structure (responding to cd) being faster than those occurring in the local environment of tryptophans which show up in fluorescence.[251] See also Refs. 88 and 252.

3.9.5 Nmr Region

Nmr can be used simply as an analytical tool in which the strength of the signal is a measure of the concentration of a particular species. With the advent of reliable machines, more sensitive to small changes in concentration, and with the availability of multinuclear probes, the method is finding increasing use for studying a variety of reactions. One disadvantage still is the relatively high concentrations (typically $0.05-0.25$ M) of solute that must be used, this posing problems of availability, solubility or stability of material. Although we saw in Chapter 1 an excellent example of the nmr analysis of the three participants in an $A \rightarrow B \rightarrow C$ sequence,[253] its real advantage lies in the study of exchange processes, using both the analytical and the line broadening (next section) aspects of nmr. A selection of exchange reactions

Table 3.8 Analytical Applications of Nmr

Type of Reaction	Example	Nucleus Used	Ref. & Notes
Water exchange in inert aqua complexes	*cis*-Co(en)$_2$(H$_2^{17}$O)$_2^{3+}$ $-$ H$_2$O exchange	^{17}O	254. Signal for coordinated water at 126 ppm decreases; signal for free water at 0 ppm increases.
	Pt(H$_2$O)$_4^{2+}$ $-$ H$_2^{17}$O	^{17}O	255. High pressure probe allows determination of $\Delta V^{\ddagger}$. Could also use ^{195}Pt,[256] but not as accurate.
Proton exchange in ammines	Co(tren)(NH$_3$)$_2^{3+}$ in D$_2$O	^{1}H	257. Shows H exchange of various ammine centers
	Co(NH$_3$)$_6^{3+}$/D$_2$O, H$_2$O	^{59}Co	258. Figure 3.11
Electron Transfer	Co$_o^{II}$ttp, Co$_o^{III}$tpp$^+$ in CDCl$_3$ (ttp = 5, 10, 15, 20 tetra-p-tolylporphine and tpp = 5, 10, 15, 20 tetraphenylporphine)	^{1}H p-methyl peak Co^{III}(ttp) δ = 2.63 ppm Co^{II}(ttp) δ = 4.13 ppm (shifts from Me$_4$Si)	259. Anions accelerate, then must use line broadening (coalescence temperature, next section)
	Ru(H$_2$O)$_6^{3+/2+}$ electron transfer	^{17}O and ^{99}Ru	260. Injection technique. Also by line broadening with consistent results.
Substitution	VS$_4^{3-}$ + H$_2$O $\rightarrow$ VO$_4^{3-}$ + S^{2-}	^{51}V	261. Can detect VOS$_3^{3-}$ during slow reaction

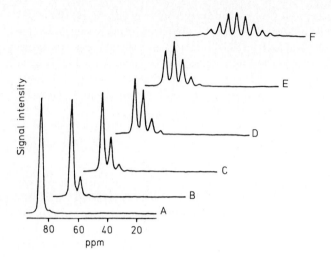

Fig. 3.11 ^{59}Co nmr spectra obtained as a function of time at 71.7 MHz and 25°C for 0.05 M Co(NH$_3$)$_6$Cl$_3$ in D$_2$O/H$_2$O (3:1) at pH 4.0. 1.7 min (A); 18 min (B); 38 min (C); 68 min (D); 107 min (E) and 140 min (F). There is a 5.6 ppm isotope shift experienced at the Co nucleus for every proton replaced by D on the N. Each H/D isotopomer can be observed. The final distribution will depend on the H/D ratio in the solvent. A plot of the H$_{18}$ isotopomer vs. time is first order, $k = 3.5 \times 10^{-5}\,\text{s}^{-1}$, Ref. 258. Reproduced with permission from J. G. Russell, R. G. Bryant and M. M. Kreevoy, Inorg. Chem. **23**, 4565 (1984). © (1984), American Chemical Society.

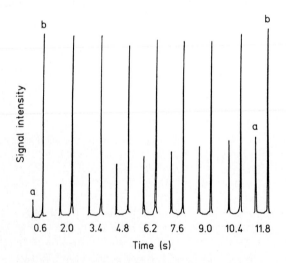

Fig. 3.12 Parts of successively recorded ^{1}H nmr spectra obtained by stopped-flow nmr after mixing Fe(dmso)$_6^{3+}$ with [^{2}H$_6$]dmso in CD$_3$NO$_2$ solution at 243.4 K. Fe(III) = 10 mM [^{2}H$_6$]dmso = 1.0 M. Resonances: a = dmso; b = SiMe$_4$.[270] Reproduced with permission from C. H. McAteer, P. Moore, J. Chem. Soc. Dalton Trans. 353 (1983).

studied by nmr monitoring is shown in Table 3.8[254-261] and illustrated in Fig. 3.11.[258] ^{17}O can replace ^{18}O in tracer applications, and continuous nmr monitoring then replaces the tedious analysis associated with the batch technique. Nowhere is this more evident than in the study of the exchange of the 7 types of O in $V_{10}O_{28}^{6-}$ **11** with H_2O. Using ^{18}O, excellent first-order traces indicating *identical* exchange rates for all O's were obtained.[262] The individual O's (7 types) are identified by nmr and the exchange data indicate some *very slight* differences in their exchangeability (but close to the values by ^{18}O).[263] For other comparative examples see Ref. 264–266. One can also examine ^{18}O–^{16}O exchange by using the fact that ^{13}C,[267,268] as well as ^{195}Pt[256] and other nuclei show different nmr signals when adjacent to ^{18}O compared with ^{16}O. ^1H Nmr has been extensively used for diamagnetic complexes such as those of Co(III) but there are problems with Cr(III) because of the long electron spin relaxation times associated with this paramagnetic ion. Using ^2H nmr may be advantageous here since 40 fold narrower lines result.[269] Relatively fast processes can be examined by nmr by using fast injection techniques[260] or flow methods,[191,192,270] Fig. 3.12.[270]

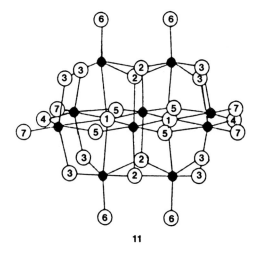

11

3.9.6 Nmr Line Broadening

The determination of the rates of fast exchange processes by nmr linebroadening experiments has been one of the most significant factors in the success in understanding the mechanisms of complex-ion reactions of all types. Among the nuclei which have now been utilized, sometimes routinely, one finds ^1H, ^{13}C, ^{14}N, ^{17}O, ^{19}F, ^{23}Na, ^{31}P, ^{39}K and ^{55}Mn. For exchange, the proton is most used because of the favorable 100% abundance and 1/2 spin quantum number. However the proton is invariably far from the influence of the metal site and chemical shifts between bound and free ligand are small. For this reason, ^{13}C, ^{14}N, ^{17}O or ^{31}P which are nearer the central atom are also used even though they are less sensitive nuclei. There have been several comprehensive accounts of the application of nmr to the measurement of exchange rates.[271,272]

The same nucleus (say methyl protons) in different chemical environments A and B will generally have nuclear magnetic resonances at different frequencies. If the exchange of protons between A and B is sufficiently slow, sharp lines corresponding to A and B will be recorded. As the exchange rate increases however, it is observed that at first there is an initial broadening of the signals; this is followed by their coalescing, and finally, at high exchange rates, narrowing of the single signal occurs.

This behavior is well typified in Fig. 3.13 involving the exchange[259]

$$Co^{II}(ttp) + Co^{III}(ttp)Cl \rightleftharpoons Co^{III}(ttp)Cl + Co^{II}(ttp) \tag{3.42}$$

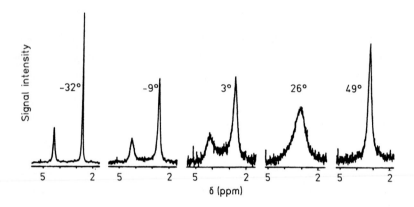

Fig. 3.13 Temperature dependence of the p-Me line shapes in 90-MHz ^{1}H nmr spectra of a mixture of Co^{II}(ttp) and Co^{III}(ttp) Cl in CDCl$_3$. ttp = 5, 10, 15, 20-tetra-p-tolylporphine.[259] (Temperatures in °C). Reproduced with permission from R. D. Chapman and E. B. Fleischer, J. Amer. Chem. Soc. **104**, 1582 (1982). © (1982) American Chemical Society.

in CDCl$_3$. At temperatures around $-30°C$, the spectra of mixtures are additive and exchange is too slow to affect the signals. Near 0°C, as exchange becomes important, the lines broaden until they coalesce at 26°C. Above this temperature, the methyl line appears at the average position and continually narrows as the temperature is raised, and the exchange is very fast. The region between $-10°C$ and near to 26°C is termed the *slow-exchange region*. That around the coalescence temperature is the *intermediate-exchange region,* and the region above about 40°C is the *fast-exchange region*. Only the temperature region between $-10°C$ and about 50°C can be used to assess exchange rates (see following discussion). For another example of this sequence of events see Ref. 273.

(a) Slow-Exchange Region. If the exchange rate is very slow, the lines due to A and B in the mixture correspond to the lines of the single components in both position and linewidth.

The broadening of the signal, say due to A, as the exchange rate increases, is the difference in full linewidth at half height between the exchange-broadened signal W_A^E, and the signal in the absence of exchange W_A^O. This difference is often denoted $\Delta v_{1/2}^E - \Delta v_{1/2}^O$. If the broadening of the lines is still much smaller than their separation and the widths are expressed in cps, or Hertz, then in this, *the slow-exchange region.*[274]

$$W_A^E - W_A^O = (\pi \tau_A)^{-1} \tag{3.43}$$

The width W is related to the transverse relaxation time T_2 by the expression

$$W = (\pi T_2)^{-1} \tag{3.44}$$

Now τ_A is the required kinetic information, since it represents the mean lifetime of the nucleus (for example, a proton) in the environment A.

$$\tau_A = \frac{[A]}{d[A]/dt} \tag{3.45}$$

τ_A^{-1} is the first-order rate constant k_A for transfer of the nucleus out of site A. Similarly for the signal due to B,

$$W_B^E - W_B^O = (\pi \tau_B)^{-1} \tag{3.46}$$

$$\tau_B = \frac{[B]}{d[B]/dt} \tag{3.47}$$

and

$$\tau_A/\tau_B = [A]/[B] = k_B/k_A \tag{3.48}$$

The values of τ_A and τ_B and the manner in which they vary with the concentrations of A and B yield information on the rate law and rate constants for the exchange. (Prob. 11.)

The linewidth of the diamagnetic $^{55}MnO_4^-$ resonance, W_A^O, is broadened on the addition of the paramagnetic MnO_4^{2-}, but there is no shift in the signal position, which is proof that we are in the slow-exchange region. The broadened line width W_A^E is a linear function of added MnO_4^{2-}

$$\pi W_A^E - \pi W_A^O = k_1[MnO_4^{2-}] = k_A \text{(from 3.43)} \tag{3.49}$$

$$\pm d[MnO_4^-]/dt = k_A[MnO_4^-] \tag{3.50}$$

$$= k_1[MnO_4^{2-}][MnO_4^-] \tag{3.51}$$

indicating that the exchange is second-order with a rate constant, k_1.[275] This exchange has also been examined by quenched flow using radioactive manganese (see Table 3.3); there is good agreement in the results. (See Problem 13.) The slow-exchange region has been very useful for studying the rates of complex ion reactions. It has been used to study fast electron transfer processes involving the couples Mn(VII) − Mn(VI), (above), Cu(II) − Cu(I),[276] V(V) − V(IV),[277] Ru(III) − Ru(II),[278] Mn(II) − Mn(I),[279] Cu(III) − Cu(II),[280] Cu(taab)$^{2+/+}$ (**14** in Chap. 8).[281] The slow-exchange region has also been used in studies of the exchange of ligands between the free and complexed states,[282,283] ligand exchange in tetrahedral complexes,[284] optical inversion rates in Co(III) chelates,[285,286] and a most impressive series of studies of exchange of solvent molecules between solvated cations and the bulk solvent (Tables 4.1–4.6) which merit a separate section (d).

(b) Intermediate-Exchange Region. In the intermediate-exchange region, the separation of the peaks is still discernible. This separation Δv_E (in cps) is compared with the separation in the *absence* of exchange Δv_0. With the conditions of equal populations of A and B, that is, $P_A = P_B$ and $P_A + P_B = 1$, $\tau_A = \tau_B$, and no spin coupling between sites:

$$\tau^{-1} = 2^{1/2}\pi(\Delta v_O^2 - \Delta v_E^2)^{1/2} = \tau_A^{-1} + \tau_B^{-1} \tag{3.52}$$

Coalescence of lines occurs at

$$\tau^{-1} = 2^{1/2}\pi\Delta v_0 \tag{3.53}$$

and

$$k_A = k_B = 1/2\tau \tag{3.54}$$

The use of the intermediate region to determine rate constants is less straightforward, although it is relatively simple to obtain an approximate rate constant at the coalescence temperature. [287-289]

(c) Fast-Exchange Region. The two lines have now coalesced to a single line. Exchange is still slow enough to contribute to the width, however. Eventually, when the exchange is very fast, a limiting single-line width is reached. Under certain conditions in the fast-exchange region [290-292]

$$\tau^{-1} = 4\pi P_A P_B(\Delta v_0)^2(W^E - W^0)^{-1} \tag{3.55}$$

where W^E and W^0 are the widths of the single broadened and final lines respectively.

A number of studies have used the slow-, intermediate-, and fast-exchange regions with consistent results. [282] Both the slow- and the fast-exchange regions were used to analyze the exchange between $Pt[P(OEt)_3]_4$ and $P(OEt)_3$ [273] with very good agreement between the resultant activation parameters. The intermediate region was not useful because of serious overlap of lines. [273] The most reliable method involves matching observed spectra with a series of computer-calculated spectra with a given set of input parameters including τ. [293] Fig. 3.14.

(d) Exchange Involving a Paramagnetic Ion. When the nuclei examined can exist in two environments, one of which is close to a paramagnetic ion, then the paramagnetic contribution to relaxation is extremely useful for determining the exchange rate of the nuclei between the two environments. This is undoubtedly the area on which nmr line broadening technique has had its greatest impact in transition metal chemistry, particularly in studying the exchange of solvent between metal coordinated and free (bulk) solvent. A variety of paramagnetic metal ions in aqueous and nonaqueous solvents have been studied (Tables in Ch. 4).

When a paramagnetic ion is dissolved in a solvent there will be, in principle, two resonance lines (due perhaps to ^{17}O or ^1H) resulting from the two types of solvent, coordinated and free. The paramagnetic species broadens the signal due to the diamagnetic one (free solvent) via the chemical exchange. Often only the diamagnetic species will have a signal because the paramagnetic one is broadened completely by the interaction of an unpaired electron or electrons and the nucleus. The sole signal due to solvent can still be used for rate analysis, however. The broadening of the line will follow the sequence outlined above as the temperature of the

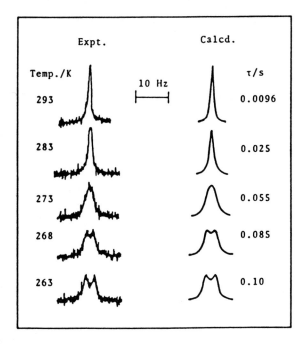

Fig. 3.14 Experimental (left) and best fit calculated ^1H nmr line shapes of $UO_2(acac)_2Me_2SO$ in $o\text{-}C_6H_4Cl_2$. Temperature and best fit τ values are shown at left and right sides of the figure.[293] Reproduced with permission from Y. Ikeda, H. Tomiyasu and H. Fukutomi, Inorg. Chem. **23**, 1356 (1984). © (1984) American Chemical Society.

solution is raised. A rigorous treatment of two-site exchange, considering the effects of a relaxation time and a chemical shift, has been made by Swift and Connick,[294] with some later modifications.[295,296]

The relaxation time is given by

$$\pi\,(W_A^E - W_A^0) = \frac{1}{T_2} - \frac{1}{T_{2A}} = \frac{P_M}{\tau_M}\,\frac{\dfrac{1}{T_{2M}^2} + \dfrac{1}{\tau_M T_{2M}} + \Delta\omega_M^2}{\left[\dfrac{1}{T_{2M}} + \dfrac{1}{\tau_M}\right]^2 + \Delta\omega_M^2} + \frac{P_M}{T_{20}} \tag{3.56}$$

where

T_{2A}, T_2 = transverse relaxation times for bulk solvent nuclei alone, and with solute (concentration [M]), respectively.

T_{2M} = relaxation time in the environment of the metal.

T_{20} = nuclear relaxation times of solvent molecules outside the first coordination shell.

τ_M = average residence time of the solvent molecule in the metal coordination sphere (coordination number n).

P_M = mole fraction of solvent that is coordinated to the metal $\sim n[M]/[solvent]$ for dilute solutions; $P_M/\tau_M = 1/\tau_A$.

$\Delta\omega_M$ = chemical shift between the two environments in the absence of exchange, (in radians s^{-1}).

The subscripts A and M refer to bulk and coordinated solvent. The term P_M/T_{20} is included in (3.56) when solvent exchange using ^{1}H nmr is studied. The value $P_M^{-1}(T_2^{-1} - T_{2A}^{-1})$ is sometimes termed T_{2r}^{-1}, the reduced line width of the free solvent. [296]

We can obtain the exchange regions outlined previously by considering various terms in (3.56) dominant. These might be induced by temperature changes. There are four exchange regions I–IV shown in Figure 3.15. [297] Only rarely is the complete behaviour displayed by a single system. [297,287] Although caution must be exercised in the use of approximate forms of (3.56), regions II and III can provide us with exchange rate constants.

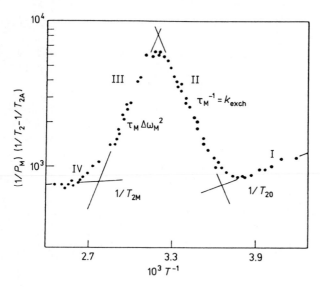

Fig. 3.15 Temperature dependence of $(1/P_M)(1/T_2 - 1/T_{2A})$ for protons in CH_3CN solutions of $Ni(CH_3CN)_6^{2+}$ at 56.4 MHz. [297]

Region II. If the chemical exchange is slow compared with the relaxation mechanism (which is incorporated in the terms $\Delta\omega_M$ or T_{2M}), that is,

$$\tau_M^{-2}, \quad T_{2M}^{-2} \ll \Delta\omega_M^2 \tag{3.57}$$

or

$$\tau_M^{-2}, \quad \Delta\omega_M^{-2} \ll T_{2M}^{-2} \tag{3.58}$$

then in either case

$$\frac{1}{T_2} - \frac{1}{T_{2A}} = \frac{P_M}{\tau_M} \tag{3.59}$$

Relaxation is thus controlled by ligand exchange between bulk and coordinated ligand. This, the slow-exchange region (II in Fig. 3.15), is most useful for obtaining kinetic data and for studying effects of pressure. [298] Since $\tau_M^{-1} = k_1$, the pseudo first-order exchange rate constant, then a semilog plot of $1/P_M(1/T_2 - 1/T_{2A})$ vs $1/T$ will give an Arrhenius-type plot, from which k_1 at any temperature, and the activation parameters for exchange, may be directly determined.

$$\tau_M^{-1} = \frac{kT}{h} \exp\left(\frac{-\Delta H^{\ddagger}}{RT} + \frac{\Delta S^{\ddagger}}{R}\right) \tag{3.60}$$

It is clearly shown, for example, in Fig. 3.15, that at $T = 298.2$ K $(1/T = 3.36 \times 10^{-3}$ $K^{-1})k_1 \simeq 3 \times 10^3 s^{-1}$.

Region III. If relaxation is controlled by the difference in precessional frequency between the free and coordinated states, that is, if

$$\tau_M^{-2} \gg \Delta\omega_M^2 \gg (\tau_M T_{2M})^{-1} \tag{3.61}$$

then

$$\frac{1}{T_2} - \frac{1}{T_{2A}} = P_M \tau_M \Delta\omega_M^2 \tag{3.62}$$

Now the line widths decrease rapidly with decreasing τ_M (increasing temperature) and anti-Arrhenius behaviour is shown (Region III). Nevertheless the exchange rate constant can be determined if $\Delta\omega_M$ is known. It is useful if data from Regions II *and* III can be used. The $Ni(dmso)_6^{2+}$ exchange with solvent using ^{13}C was studied in the fast-exchange region.[299(a)] If ^{17}O nmr is used, because $\Delta\omega_M$ is much larger, all of the data relate to the slow exchange region, since (3.57) and (3.59) apply.[299(b)] There is good agreement in the data from the two regions.

Region IV. If the T_{2M} process controls relaxation, that is,

$$(T_{2M}\tau_M)^{-1} \gg T_{2M}^{-2}, \Delta\omega_M^2 \tag{3.63}$$

then

$$\frac{1}{T_2} - \frac{1}{T_{2A}} = \frac{P_M}{T_{2M}} \tag{3.64}$$

Now the linewidths no longer depend on the exchange rate and only vary slightly with temperature. The chemical shift as well as the linewidth changes as chemical exchange becomes important. Chemical shift is used less frequently to estimate exchange rates with paramagnetic systems but is important in the study of diamagnetic ones.[294, 296]

3.9.7 Epr Region

Electron paramagnetic resonance can be valuable as a monitor when paramagnetic species are being consumed or formed, as in the acid-catalyzed reaction[300] (see also Ref. 301)

$$Cr(CN)_2(NO)(H_2O)_3 \xrightarrow{H_3O^+} Cr(CN)(NO)(H_2O)_4 + HCN \tag{3.65}$$

Since there are small differences in the epr of isotopes of a particular nucleus, these may be used to study electron transfer reactions (Table 3.3)

$$^{61}\text{Ni(cyclam)}^{2+} + {}^{58}\text{Ni(cyclam)}^{3+} \rightleftharpoons {}^{61}\text{Ni(cyclam)}^{3+} + {}^{58}\text{Ni(cyclam)}^{2+} \qquad (3.66)$$

Only the Ni(III) complexes display epr signals. Those or ^{61}Ni show a splitting of the $g_{\parallel}$ feature owing to the nuclear spin ($I = 3/2$) of ^{61}Ni. This is missing in the corresponding ^{58}Ni complexes.[302]

Epr is most effective for detecting free radicals that may occur as intermediates in oxidation and reduction reactions involving transition metal ions. Since these transients are invariably quite labile, epr is combined with continuous flow,[55,303] (more conveniently) stopped-flow,[304-306] flash photolysis,[196] and pulse radiolysis.[197]

A sharp epr signal observed 0.1 s after mixing Cr(III) (which has a broad epr spectrum) and H_2O_2 in base can be assigned to a transient Cr(V) intermediate. The subsequent loss of the epr signal parallels the absorbance loss at 500 nm.[305]

Radicals such as OH$^{\bullet}$ and O$_2^{\bullet-}$ can be stabilized, and detected by epr, by spin-trapping with dmpo, **12**[307,308]

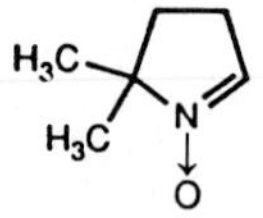

12

3.9.8 Epr Line Broadening

Another example of the use of spectral line broadening for rate measurements is by epr, although the method has been much less used than nmr. It covers a range of very short lifetimes, $10^{-4} - 10^{-10}$ s. The very rapid interaction of ligands with square-planer complexes at the axial positions are suitable for treatment by epr line broadening.[309] The fast self-exchange reactions involving Cr(I) and Cr(O) complexes of bpy[310] and η^6-arenes[311] have also been studied by epr line broadening and full details and equations are provided.[310,311]

Anionic radical ligand derived from alloxan **13** shows epr which are markedly affected by coordination to diamagnetic Mg^{2+}, Ca^{2+} and Zn^{2+} ions.

The splitting patterns due to ligand protons and ^{14}N nuclei differ in the free ligand and the M^{2+} complexes. There is rapid intramolecular exchange of M^{2+} between two possible sites (shown in **13**) and this leads to line broadening which can be used to measure the exchange rate constants (5×10^4 s^{-1} to 2×10^7 s^{-1}).[312] See Table 7.4.

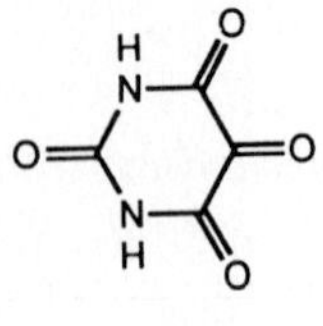

13

3.10 Non-Spectrophotometric Methods

Non-spectral methods are generally less useful although in specific cases, they may represent the only monitoring method.

3.10.1 [H⁺] Changes

Many complex-ion and metalloprotein reactions are accompanied by a change in the hydrogen ion concentration. This may be as a direct result of the reaction under study or because of fast concomitant secondary reactions that monitor the primary reaction. The change of [H⁺] with time may therefore be used *qualitatively,* to give insight into the number of changes occurring during a reaction, or as is more usual, *quantitatively,* to measure the rate of the reaction or reactions. There are basically two ways in which [H⁺] changes are measured in kinetic studies.

(a) Glass Electrode. The relationship

$$pH = -\log a_{H^+} \tag{3.67}$$

means that pH values read from a meter must usually be converted from activities into concentrations of H⁺, [H⁺], by using activity coefficients, calculated, for example, from Davies' equation

$$\log \gamma_{\pm} = \frac{-[A]z_1 z_2 \mu^{1/2}}{1 + \mu^{1/2}} + B\mu \tag{3.68}$$

At $\mu = 0.1$ M, $A = 0.507$ and $B = 0.1$, and therefore

$$-\log [H^+] = pH - 0.11 \tag{3.69}$$

For experiments in D_2O, $pD = pH + 0.40,$[313] while for work in mixed aqueous solvents, an operational pH scale has been used.[314] A small correction must therefore be made in estimating [H⁺] or [OH⁻] from the pH, when these concentrations have to be used for calculating the rate constants in [H⁺]-dependent rate laws. However, $d[pH]/dt$ can be used directly as a measure of $d[H^+]/dt$ provided that only small pH changes are involved in the reaction. Since the rates of many reactions are pH-sensitive, it is obviously sensible in any case to avoid a large pH change. Specialized apparatus has been developed that will measure a change as small as 0.001 pH unit in an overall change of 0.05 pH unit during the reaction. Such equipment has been successfully employed in the study of certain enzyme reactions.[315] Obviously, the reaction must be sufficiently slow so that the meter or recorder response does not become rate-limiting. In certain cases, it may be necessary to add a small amount of buffer to keep the pH change reasonably small.

In a clever variation of the use of a pH change to monitor rate, the pH change is minimized by the controlled and registered addition of acid or base to maintain a constant pH during the reaction. The rate of addition of reagent is thus a measure of the pH change and the reaction rate. The so-called pH-*stat method* has the decided advantage that reactions can be studied at a constant pH without recourse to buffers. It can only be used for reaction half-lives in the range 10 sec to a few hours, because of the slow response time, and possible electrode drift, over longer times.

(b) Indicators. It may be more convenient, or even essential in certain cases, to avoid the glass electrode and register $d[H^+]/dt$ using an appropriate acid-base indicator. This is usually necessary in the study of rapid reactions, although glass electrodes have been incorporated into continuous-flow apparatus (Sec. 3.3.1). The acid-base equilibrium involving the indicator is usually established rapidly and will not be rate-limiting with flow measurements, but may have to be considered with temperature-jump experiments, carried out at the shortest times. Some useful indicators that have been used are bromochlorophenol blue ($pK = 4.0$), bromocresol green (4.7), chlorophenol red (6.0), bromothymol blue (7.1), and phenol red (7.5). The figure in parenthesis is the pK, and therefore the optimum pH value at which the indicators may be used. For a fuller list see Refs. 316 and 317. As always, one must take care that the indicator does not interfere with the system under study. Thymol blue and phenol blue, but not bromocresol green, interact with certain metals ions at pH 4.5.[318] The acid-base properties of certain napthols are modified in the excited state.[133] Large pH changes can thereby be initiated by electronic means within nanoseconds and monitored by indicators,[132] (Table 3.4).

The types of reaction for which pH monitoring is useful are as follows. The mode of monitoring, (a) or (b) above, will be determined by the rate.

(1) Spontaneous and metal-ion catalyzed base hydrolysis of esters, amides, etc.[319]

$$(3.70)$$

pH changes linked to indicators have been effectively used to probe the carbonic anhydrase catalysis of the reaction[317]

$$CO_2 + H_2O \rightleftharpoons H^+ + HCO_3^-$$ (3.71)

and (indirectly) enzyme catalyzed hydrolysis of fructose biphosphate.[320]

Reduction of *D*-proline by *D*-amino acid oxidase at pH 8 shows two steps when monitored at 640 nm. These are interpreted as the build-up and breakdown of a reduced enzyme-imino acid charge transfer complex. If the reaction is monitored using phenol red the same two rates are observed but additionally the release of ≈ 1 proton for each step can be assessed and interpreted. The indicator changes are followed at 505 nm and 385 nm, which are isosbestic wavelengths for the two steps (without indicator),[321]

(2) Formation and dissociation of complexes containing basic ligands.[322-324] The polymerization of Cr(III) in basic solution is a complex reaction the kinetics of which only recently have been probed. Monitoring the pH changes using a pH-stat aids considerably in the understanding of the species involved.[324]

The acid-base properties of a transient can be assessed directly by placing a glass electrode in a streaming fluid that is generating that transient a few ms after the mixer. The pH is measured after the electrode has equilibrated. A series of measurements of the extent of protonation vs pH yields a pK_a value in the usual way. The pK_a's of H_3VO_4 (and lower protonated species) and $Ni(H_2O)_6^{2+}$ could be determined, without interference by slightly slower polymerization.[325]

3.10.2 Cationic and Anionic Probes

Electrodes are now available for the selective determination of the concentrations of a large number of cations and anions.[53] Halide-sensitive electrodes have been used to monitor reactions,[326] but their relatively slow response has restricted their use. They may have particular utility in the study of reactions with low spectral absorbance changes and also in an ancillary role to the kinetics determination.

Although the rate of hydrolysis of nitroammine cobalt(III) complexes can be easily measured spectrally, it proved important to monitor for loss of NH_3 or NO_2^- groups in the early stages by using selective electrodes. The dominant loss of NH_3 or NO_2^- in the first stage depended on the complex.[327]

3.10.3 Conductivity

The conductivity method of monitoring has occasionally been valuable for studying specific reactions. It is a colligative property and rarely shows up the fine reaction detail possible with spectral measurements. It can cope with a wide variety of rates and is convenient to use with flow, relaxation, laser photolysis and pulse radiolysis techniques (Table 3.7).[198-204] It has decided value in relaxation techniques since in these the changes in concentration of reactants are usually small and conductivity is a very sensitive and rapidly registered property. High concentrations of nonreacting electrolyte are to be avoided however, since otherwise the conductivity changes would be relatively small, superimposed on a large background conductivity.

Conductivity monitoring is most valuable for studying reactions which have very small spectral changes but which are accompanied by pH changes. The interaction of group 1 and 2 metal ions with cryptands and diaza-crown ethers has been studied by flow/conductivity methods.[198] Conductivity monitoring has been linked to reactions which may follow pulse radiolysis, for example, in examining the

$$VO_{aq}^{2+} \xrightarrow[8 \times 10^{10} M^{-1}s^{-1}]{e_{aq}^-} VO^+ \xrightarrow[1.5 \times 10^{10} M^{-1}s^{-1}]{(H^+)} VOH^+ \qquad (3.72)$$

rapid conductivity decrease corresponding to loss of 1 H^+ in the second step of (3.72)[204] or following laser photolysis of Cr(III) complexes.[202, 203] (Probs. 5 and 6.) Of course, pH-indicators might also be used (Sec. 3.10.1 (b)) but might be effected by the perturbation. A conductivity cell for use at high pressures has been employed to measure $\Delta V^{\ddagger}$ for substitution reactions of *cis*- and *trans*-$Pt(L)_2ClX$ with pyridine in various solvents.[328]

3.10.4 Thermal Changes

If a reaction is accompanied by a change in ΔH, then the temperature of that reaction sensed with time is a measure of the rate. Although the method has found little use generally, it can be linked to a stopped-flow apparatus and this allows the determination of the thermal properties of a transient.[205-207] The heat of formation of an intermediate, which decomposes with $k = 0.27$ s^{-1}, in the complicated luciferase-FMNH$_2$ reaction with O$_2$ can be measured by stopped-flow calorimetry.[205]

3.10.5 Pressure Changes

Reactions accompanied by gas evolution or absorption may be followed by measuring the change in pressure in a sealed vessel equipped with a manometer. The method has been used to study decarboxylation reactions, the decomposition of hydrogen peroxide[329] and the homogeneous reduction of transition metal complexes by H$_2$.[330] It is essential to check that the rate of equilibration between dissolved gas in and above the solution, or that the manipulation times in using the manometer, or both, are not being measured and thus mistaken for a (faster) reaction rate. Again a combination of monitoring methods can be helpful, as in the study of the Ag(I) catalyzed decomposition of S$_2$O$_8^{2-}$. Both the loss of S$_2$O$_8^{2-}$ (polarographically) and the gain of O$_2$ (Warburg meter) were monitored.[331]

3.10.6 Electrochemical Methods

Polarography can function, like nmr and epr, in a dual manner, both as an analytical and a competitive probe. For polarography to be potentially useful analytically, one of the species in the reaction must give a polarographic wave. The limiting current at a given potential is a measure of the concentration of the species involved and so a change of waveheight with time is recorded. The polarographic method rarely has any obvious advantages over other methods, although it is straightforward to use. It is increasingly being used in laboratories where electrochemical measurements are perhaps the main thrust, and the equipment and expertise are readily available.

A graphite rotating disk electrode maintained at 0.5 V is used to monitor the reaction of Ru(NH$_3$)$_6^{2+}$ as it is being oxidized by O$_2$ to Ru(NH$_3$)$_6^{3+}$. The limiting current is proportional to [Ru(NH$_3$)$_6^{2+}$] and there is no interference by O$_2$ or the product. The electrode is rotated at 3600 rpm to ensure rapid mixing of reactants within seconds, since reaction times are 20–30 s.[332] See Ref. 333. Square-wave amperometry has been linked to stopped-flow to measure reaction half-lives as short as 5 ms.[334]

Polarographic probes that respond specifically to concentrations of O$_2$, CO$_2$ or SO$_2$ are very useful. They have decided advantages over the more clumsy manometric monitoring. Their use is limited to slow reactions or the continuous-flow approach, because of the relatively long response time of the probe.[335] An O$_2$-electrode system for incorporation in a spectrophotometer cuvette, for simultaneous monitoring of [O$_2$] and spectral changes, has been described.[336]

The polarographic technique can be used to measure the rates of rapid reactions.[337] Because an "internal" process is examined the problem of mixing is avoided, as it is in the relaxation and other non-flow methods. The rate of diffusion of a species (which can be oxidized or reduced) to an electrode surface competes with the rate of a chemical reaction of that species, for example

$$\text{Cd} + 4\,\text{CN}^- \xleftarrow[\text{electrode}]{2e^-} \text{Cd(CN)}_4^{2-} \underset{k_{-1}}{\overset{k_1}{\rightleftharpoons}} \text{Cd(CN)}_3^- + \text{CN}^- \tag{3.73}$$

Values of k_1 and k_{-1} may be extracted from the polarographic data, although the treatment is complex.[338] Examples of its use to measure the rate constants for certain redox reactions are given in Refs. 339 and 340 which should be consulted for full experimental details. The values obtained are in reasonable agreement with those from stopped-flow and other methods. The technique has still not been used much to collect rate constants for homogenous reactions.[341] The availability of ultramicroelectrodes has enabled cyclic voltammograms to be recorded at speeds as high as $10^6\ \text{V}\,\text{s}^{-1}$. Transients with very short lifetimes ($\ll$ μs) and their reaction rates may be characterised.[342]

3.11 Batch Methods

All the methods described above have been amenable to continuous monitoring as the reaction proceeded. In this section the batch procedure is described, in which aliquots of the reaction mixture are removed at various times and analyzed. Although the batch method is tedious it must be used to study certain exchange reactions and when the quenched-flow technique is used (Sec. 3.3.2). Recent events have suggested that batch analysis of a reacting system may give vital information not easily obtained by routine spectral analysis, see the next section.

(a) Chemical Methods

In years past, the hydrolysis of ions such as $\text{Co(NH}_3)_5\text{Cl}^{2+}$ or $cis\text{-Co(en)}_2\text{F}_2^+$ has been studied by volumetric analysis of liberated chloride or fluoride ions respectively. Such purely chemical methods of assay have been largely superseded by continuous monitoring of the hydrolyzed substrate, based on some physical property. Chemical methods do have the advantage, however, of analyzing *specific* reactants, usually after a separation, and should always be used (at least in one run) to check the stoichiometry of the reaction if this is in doubt and to monitor reactions that may have irritating side reactions.[343] Only recently has it been realized that the reaction of $\text{Co(NH}_3)_5\text{H}_2\text{O}^{3+}$ with SCN^- leads initially to both S-(26%) and N-bound thiocyanate cobalt(III) complexes. This can be shown by early sampling of the reaction mixture combined with ion-exchange separation. Previous kinetic studies with spectral monitoring failed to observe the S-bound isomer. Since the S-bound isomerizes completely to the N-bound form in the same time frame as the anation reaction, accurate anation rate constants are best obtained by observation at the isosbestic point for $\text{Co(NH}_3)_5\text{SCN}^{2+}$ and $\text{Co(NH}_3)_5\text{NCS}^{2+}$, Ref. 344.

Another classical reaction which has been reinvestigated by the batch method (with quick HPLC separation of samples withdrawn at various times) is the acid-catalyzed substitution of $Co(CN)_5N_3^{2-}$ by SCN^-. The products are $Co(CN)_5H_2O^{2-}$, $Co(CN)_5SCN^{3-}$ and $Co(CN)_5NCS^{3-}$. The kinetics of these reactions are difficult to disentangle by *in situ* spectral examination alone, and indeed this procedure has led to erroneous conclusions.[345] Care must be observed that the separation procedure used in the batch method does not alter the conditions of the system.[346]

(b) Isotopic Analysis

The use of radioisotopes as an analytical tool, although somewhat laborious, has decided value. The reaction

$$Co(NH_3)_5PO_4 + H_2O \rightarrow Co(NH_3)_5H_2O^{3+} + PO_4^{3-} \tag{3.74}$$

has been studied using ^{32}P-labelled complexes. The spectral changes are insufficient for easy analysis.[347] Because of the high sensitivity of radioisotope assay, μM or less concentrated solutions of complexes can be examined. In this way the effect on the rate of low ionic strength (for which the Debye-Hückel treatment is most applicable) or low concentration of ions can be assessed. In using radioisotopes to study exchange reactions, we must necessarily use the batch method.

Full experimental details for assaying ^{18}O, commonly used for studying exchange reactions, have been given by leading practitioners.[348] Usually the oxygen-containing compound is converted into a gas, CO_2 or O_2 for example, which is analyzed usually by mass spectrometry but also by densimetry.[348] The exchange reaction between $Au([15]NH_3)_4^{2+}$ and NH_3 has been studied[349] by precipitating at various times the gold complex (as phosphate) to quench the reaction, carrying out Dumas destruction of the precipitate at 400 °C and analyzing the evolved gas by optical emission catalysis. The latter method is effective for ^{15}N and ^{13}C analysis.[350] ^{18}O-sampling techniques lead to the sum of all exchanging oxygens and occasionally it is difficult to distinguish between oxygens of very similar exchangability. In this respect the nmr method is superior, as well as representing an *in situ* technique. Nmr can be used to determine the [18]O environment of a species (and an exchange rate) by examining the nmr signal of adjacent atoms e.g. [13]C, ^{31}P etc. (Sec. 3.9.5).

3.12 Competition Methods[351]

Occasionally it is as useful to obtain *relative* constants for a series of reactants acting on a common substrate, as it is to have actual rate values. Relative rate constants are obtained by competition methods, which avoid the kinetic approach entirely. The method is well illustrated by considering the second-order reactions of two Co(III) complexes Co_A^{III} and Co_B^{III} (which might, for example, be $Co(NH_3)_5Cl^{2+}$ and $Co(NH_3)_5Br^{2+}$), with a common reductant (Cr(II)) (leading in this case to $CrCl^{2+}$ and $CrBr^{2+}$ respectively):

$$\text{Co}_A^{III} + \text{Cr}^{II} \rightarrow \text{Co}^{II} + \text{Cr}_A^{III} \quad k_A \tag{3.75}$$

$$\text{Co}_B^{III} + \text{Cr}^{II} \rightarrow \text{Co}^{II} + \text{Cr}_B^{III} \quad k_B \tag{3.76}$$

If Cr(II) is used in deficiency, that is, if the starting conditions are $[\text{Cr}^{II}]_0 < [\text{Co}_A^{III}] + [\text{Co}_B^{III}]$, then when the reaction is complete it is clear that the ratio of Co(III) complexes remaining, $[\text{Co}_A^{III}]_e/[\text{Co}_B^{III}]_e$, will be related to their relative reactivities. Great care is necessary to ensure thorough mixing of all reagents particularly when, as in this example, rapid reactions are involved, otherwise fallacious results might result.[352] The problem of efficient mixing is circumvented by generating the scavenged material *in situ* in the presence of the two competing species. The generating process may be much slower than its scavenging and yet still be used successfully (see Prob. 17).[353] The approach has been particularly useful for studying certain radicals with low optical absorbancies.

The reaction of µM concentrations of OH• radicals (produced by pulse radiolysis) with substrates such as Co(en)_3^{3+} produces very small absorbance changes. The reaction of OH• with SCN⁻ on the other hand ($k = 6.6 \times 10^9 \text{ M}^{-1}\text{s}^{-1}$ at 25°C) yields the highly absorbing $(\text{SCN})_2^{\bullet-}$. Competition of Co(en)_3^{3+} with SCN⁻ for OH• can be used to measure the relative rate constants and hence the value for OH• with Co(en)_3^{3+} Ref. 354, 355. This approach is useful for the study of reactions of $\text{CO}_2^{\bullet-}$[356] and $\text{O}_2^{\bullet-}$[357] with pairs of reactants.

3.12.1 Stern-Volmer Relationship

The decay of the excited state *S is through a number of processes shown in Eqn (1.32) with a combined first-order rate constant k_0

$$\text{S} \xrightarrow{h\nu} \text{*S} \xrightarrow{k_0} (\text{products})_0 \tag{3.77}$$

When a quencher Q (B in (1.32)) is added there is an additional path for deactivation

$$\text{S} \rightarrow \text{*S} \xrightarrow{(+Q)k_q} (\text{products})_q \tag{3.78}$$

It is easy to see that if τ_0 and τ are the excited state lifetimes in the absence and in the presence of quencher, respectively, then

$$\frac{\tau_0}{\tau} = \frac{k_0 + k_q[Q]}{k_0} = 1 + \left(\frac{k_q}{k_0}\right)[Q] \tag{3.79}$$

This is the Stern-Volmer relationship with $K_{SV} = k_q/k_0$, and is an important basis for determining quenching rate constants after *pulsed* excitation. The quantum yield of (product)$_0$ can be measured without (ϕ_0) and with (ϕ) quencher under *continuous* excitation (ϕ = moles of product/einsteins of light absorbed by system). Assuming that a steady state concentration of *S exists in both cases,

$$\frac{\phi_0}{\phi} = 1 + K_{SV}[Q] \tag{3.80}$$

This represents a competitive, non-kinetic, method for determining relative rate constants for the photochemical system. The value of k_q may be obtained after one direct determination of k_0 has been made. For an extensive compilation of quenching rate constants for excited states of metal complexes see Ref. 358.

3.12.2 Isotope Fractionation

Another powerful application of the competition method is in *isotope fractionation experiments*. These allow determination of the relative rates of reactants with different isotopic composition. Co_A^{III} in (3.75) might be $Co(NH_3)_5H_2{}^{16}O^{3+}$ and Co_B^{III} in (3.76) might be $Co(NH_3)_5H_2{}^{18}O^{3+}$. By examining the $^{16}O/^{18}O$ contents of the Co(III) complex remaining after all the other reactant (Cr(II) or V(II), for example) has disappeared, one can determine k_{16}/k_{18} ($=f$), their relative rate constants, obtaining thereby the isotopic fractionation factor f. Since the rate constants are so close, separate studies of the two reactions could not possibly yield the ratio with the desired accuracy. The $^{16}O/^{18}O$ ratio in the complex, converted to H_2O by heating and thence equilibrating with CO_2 for assay, can however be accurately determined by mass spectrometry. An accurate value for f can thence be obtained, although the treatment and procedure are complicated.

In the reaction of a cobalt(III) aqua complex containing ^{16}O and ^{18}O with another reagent,[359]

$$f = \frac{k_{16}}{k_{18}} = \frac{d \ln [^{16}O]}{d \ln [^{18}O]} = \frac{[^{18}O]}{d [^{18}O]} \cdot \frac{d [^{16}O]}{[^{16}O]} \tag{3.81}$$

It can be shown that[360]

$$f = \frac{\ln \alpha (1 - N_t)/(1 - N_0)}{\ln \alpha (N_t /N_0)} \sim \frac{\log \alpha}{\log \alpha (N_t /N_0)} \tag{3.82}$$

where $\alpha = $ $[Co(III)]_t /[Co(III)]_0$, the fraction of complex sample left unreduced at time t, which may conveniently be designated as that when all the other reagent is consumed.

$N_t,\ N_0$ = mole fractions of ^{18}O in complex after partial reduction and initially.

^{13}C-kinetic isotope effects have been used to determine the degree of $C-C$ breakage in metal and non-metal catalyzed decarboxylation of oxalacetic acid.[361] (Sec. 2.2.2).

3.13 The Study of Transients

A knowledge of the properties of transients is an integral feature in understanding the rate and mechanism of reactions. We concentrate here on labile species, because the properties of transients arising in slow reactions can be usually determined by conventional methods on equilibrated solutions. Flow and the massive perturbation techniques (but not the relaxation methods) can, as we have seen, generate sizeable amounts of reactive transients. The technique used to examine their properties depends on their reactivity. The lifetime of transients is controlled by self-reactions, reaction with solvent, buffer, etc.

3.13.1 Spectral Properties of Transients

The presence of intermediates suggested by a mechanism has been supported in a number of cases by rapid scan spectrophotometry. Examples include Ni(II) polyamine complexes during acid dissociation,[362,363] protonated and ring-opened cobalt(III) carbonate complexes during acid dissociation,[364] copper(III) intermediates in the oxidation of Cu(II) by OCl^-,[220] and Mn(V) generated in the course of the MnO_4^-, SO_3^{2-} reaction.[221] Streak camera recordings of transient spectra in 200 μs intervals demonstrates two distinct transients in the reaction of $Cr(en)_2(SCH_2CO_2)^+$ with $OH^•$ radicals, generated by pulse radiolysis.[224]

An important development for detection of transient organometallic species is fast time-resolved infrared.[175] The transient is generated rapidly by uv-visible flash photolysis and monitored by ir with μs resolution. Spectra are obtained from a series of kinetic traces recorded at about 4 cm^{-1} interval.

Uv-vis flash photolysis of $[CpFe(CO)_2]_2$ (14) and MeCN in cyclohexane at 25°C gives the ir/time display shown in Figure 3.16. This shows that 14 is destroyed in the flash and that there are two intermediates $CpFe(\mu CO)_3FeCp$ (16) and in much smaller amounts, $CpFe(CO)_2^•$, (15). The principal intermediate reacts relatively slowly with MeCN to give $Cp_2Fe_2(CO)_3(MeCN)$ (17) and the decay of 16 mirrors exactly the increase in 17 ($k = 7.6 \times 10^5$ M^{-1}s^{-1} at 24°C

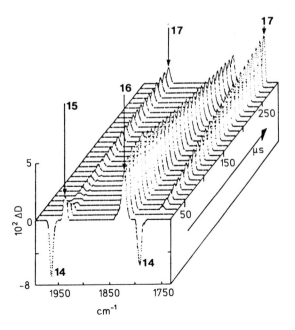

Fig. 3.16 Time resolved ir spectra obtained by uv flash photolysis of $[CpFe(CO)_2]_2$(14) (0.6 mM) and MeCN(6mM) in cyclohexane solution at 25°. Only ~5% of 14 is destroyed by the flash so that the concentration of 16 ≪ 14. The spectra have been reconstituted from ≈70 kinetic traces recorded at intervals of ~4 cm^{-1} from 1750 cm^{-1} to 1950 cm^{-1}. The first three spectra correspond to the duration of the firing of the flash lamp and subsequent spectra are shown at intervals of 10 μs. The negative peaks in the first spectrum (subsequently omitted) are due to material destroyed by the flash.[365] Reproduced with permission from A. J. Dixon, M. A. Healy, M. Poliakoff and J. J. Turner, J. Chem. Soc. Chem. Comm. 994 (1986)

by separate kinetic experiments). The radical **15** disappears rapidly (probably by recombination) but does not produce much **17**.[365] The impact of these observations on the understanding of mechanism is self-evident.

The three stages in the replacement of three dmso molecules in $Al(dmso)_6^{3+}$ by one terdentate ligand, trpy, (**18**) in nitromethane solution is resolved beautifully by flow/nmr. The
method is effective because nmr can monitor specifically for free dmso.[366] See also Fig. 3.12.

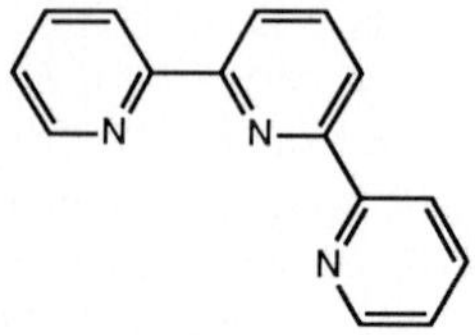

18

3.13.2 Thermodynamic Properties of Transients

In the equilibria

$$Zn^{2+} + SO_4^{2-} \xrightleftharpoons{K_1} Zn^{2+}\|SO_4^{2-} \xrightleftharpoons{K_2} ZnSO_4 \tag{3.83}$$

electric field jump (Sec. 3.4.3) allows the measurement of K_1 ($1.3 \times 10^3 M^{-1}$). Low-field conductance measurements give an overall equilibrium constant $K_1 K_2 = 7.5 \times 10^6 M^{-1}$, from
which it follows that the ratio of contact to solvent separated pairs (K_2) is 5.8×10^3, Ref.
367.

If a transient is formed rapidly within mixing time, but undergoes further reactions slowly,
then the thermal properties of the transient may be determined using thermal monitoring.
When Cr(VI) containing $HCrO_4^-$ and $Cr_2O_7^{2-}$ is treated with base, reaction (3.84) occurs
within mixing and the associated heat change can be measured, isolated from the much slower
(3.85)[368]

$$HCrO_4^- + OH^- \rightleftharpoons CrO_4^{2-} + H_2O \tag{3.84}$$

$$Cr_2O_7^{2-} + 2\,OH^- \rightarrow 2\,CrO_4^{2-} + H_2O \tag{3.85}$$

Similarly, the spectrum of a mixture of $Fe(tpps)H_2O^-$ and $Fe(tpps)(OH)^{2-}$ can be
measured by rapid scan/stopped-flow at various pH's within a few milliseconds after generation (Fig. 3.9). In this short time, dimerization is unimportant so that the spectrum of
$Fe(tpps)OH^{2-}$ can be measured and the pK_a of $Fe(tpps)H_2O^-$ estimated.[222]

3.13.3 Chemical Reactivity of Transients

Different chemical reactivity may be used to diagnose structurally different groups. The use
of flow simply widens the dimension of this use. The reactivity of the enol but not the keto

form of aldehydes as a substrate towards horseradish peroxidase I (by stopped-flow) can be used to estimate (as 0.01–0.001%) the enol form in the equilibrium mixture.[369]

The reactions of an unstable intermediate can be studied by the use of multiple mixers. The transient is generated in one mixer and then mixed with a reagent in a second mixer. Monitoring after the second mixer is usually by spectra.[63,85] The protonated superoxotitanium(IV) species $TiO(HO_2^+)$ is formed by mixing a Ce(IV) solution with one containing H_2O_2 and TiO^{2+} (which gives $Ti(O_2)^{2+}$).

$$Ti(O_2)^{2+} + Ce(IV) \xrightarrow[-H^+]{H_2O} TiO(HO_2^+)^{2+} + Ce(III) \quad k = 1.1 \times 10^5 M^{-1}s^{-1} \quad (3.86)$$

The mixture is flowed for 0.23–0.31 s^{-1} before being allowed to react in a second mixer with the examining reductant. In this time, there is a maximum production of $TiO(HO_2^+)^{2+}$, and insufficient time has elapsed for its decomposition to be marked.[63]

$$TiO(HO_2^+)^{2+} \rightarrow TiO^{2+} + HO_2^+ \quad k = 0.11 s^{-1} \quad (3.87)$$

Transients such as O_2^- or HO_2^+ can be generated in solution by pulse radiolysis of O_2. If such solutions are contained in one syringe of a stopped-flow apparatus they may be mixed with substrate and the final mixture examined spectrally. For flow experiments these transients must, of course, have lifetimes longer than a few millisecond. For the examination of more labile transients, production may be by laser photolyses or pulse radiolysis, and the substrate under examination must be then incorporated in the pulsed solution. Care has now to be taken that substantial amounts of the substrate are not lost (by reaction) as a result of the pulse.

References

1. A. Samuni and G. Czapski, J. Chem. Soc. A, 487 (1973); G. Davies and B. Warnquist, Coordn. Chem. Revs. **5**, 349 (1970); M. Knoblowitz, L. Miller, J. I. Morrow, S. Rich, and T. Scheinbart, Inorg. Chem. **15**, 2847 (1976).
2. A. C. Melnyk, N. K. Kildahl, A. R. Rendina and D. H. Busch, J. Amer. Chem. Soc. **101**, 3232 (1979).
3. J. Simplicio, K. Schwenzer and F. Maenpa, J. Amer. Chem. Soc. **97**, 7319 (1975).
4. E. L. Evers, G. G. Jayson and A. J. Swallow, J. Chem. Soc. Faraday Trans. I **74**, 418 (1978).
5. R. F. Evilia, Inorg. Chem. **24**, 2076 (1985) and references therein.
6. R. G. Khalifah, G. Sanyal, D. J. Strader and W. McI. Sutherland, J. Biol. Chem. **254**, 602 (1979); G. Sanyal and R. G. Khalifah, Arch. Biochem. Biophys. **196**, 157 (1979).
7. P. J. Connolly and E. J. Billo, Inorg. Chem. **26**, 3224 (1987).
8. D. J. Livingston, N. L. Davis, G. N. LaMar and W. D. Brown, J. Amer. Chem. Soc. **106**, 3025 (1984).
9. M. Ardon and B. Magyar, J. Amer. Chem. Soc. **106**, 3359 (1984); M. Ardon and A. Bino, Structure and Bonding, **65**, I (1987); P. Andersen, Coordn. Chem. Revs. **94**, 47 (1989). This structure persists in concentrated solution.
10. J. M. Veigel and C. S. Garner, Inorg. Chem. **4**, 1569 (1965).
11. C. Narayanaswamy, T. Ramasami and D. Ramasamy, Inorg. Chem. **25**, 4052 (1986); H. Bruce, D. Reinhard, M. T. Saliby and P. S. Sheridan, Inorg. Chem. **26**, 4024 (1987).
12 S. Balt and C. Dekker, Inorg. Chem.**15**, 1025 (1976).
13. R. Koren and B. Permutter-Hayman, Israel J. Chem. **8**, 1 (1970).
14. W. G. Jackson, S. S. Jurisson and B. C. McGregor, Inorg. Chem. **24**, 1788 (1985).

15. A. J. Thomson, A. E. Robinson, M. K. Johnson, R. Cammack, K. K. Rao and D. O. Hall, Biochim. Biophys. Acta **637**, 423 (1981).
16. A. J. Miralles, R. E. Armstrong and A. Haim, J. Amer. Chem. Soc. **99**, 1416 (1977); D. Pinnell and R. B. Jordan, Inorg. Chem. **18**, 3191 (1979).
17. M. G. Burnett and M. Gilfallen, J. Chem. Soc. Dalton Trans. 1578 (1981).
18. A. Haim, Inorg. Chem. **21**, 2887 (1982).
19. G. Davies and A. R. Garafalo, Inorg. Chem. **19**, 3543 (1980); D. H. Macartney and A. McAuley, Inorg. Chem. **20**, 749 (1981).
20. H. E. Toma, A. A. Batista and H. B. Gray, J. Amer. Chem. Soc. **104**, 7509 (1982).
21. M. W. Fuller, K.-M. F. LeBrocq, E. Leslie and I. R. Wilson, Aust. J. Chem. **39**, 1411 (1986).
22. K. Madlo, A. Hasnedl and S. Veprek-Siska, Coll. Czech Chem. **41**, 7 (1976); F. R. Duke, Inorg. Nucl. Letters **12**, 107 (1976).
23. D. F. Evans and M. W. Upton, J. Chem. Soc. Dalton Trans. 2525 (1985).
24. J. Veprek-Siska and V. Ettel, J. Inorg. Nucl. Chem. **31**, 789 (1969).
25. J. F. Bunnett in B5, Chap. III, directs attention to the various points that should be considered before and after a kinetic study.
26. G. A. Lawrance, Chem. Revs. **86**, 17 (1986).
27. A. D. Hugi, L. Helm and A. E. Merbach, Helv. Chim. Acta **68**, 508 (1985).
28. W. J. Ferguson, K. I. Braunschweiger, W. R. Braunschweiger, J. R. Smith, J. J. McCormick, C. C. Wasmann, N. P. Jarvis, D. H. Bell and N. E. Good, Anal. Biochem. **104**, 300 (1980).
29. K. Hegetschweiler and P. Saltman, Inorg. Chem. **25**, 107 (1986).
30. D. Masi, L. Mealli, M. Sabat, A. Sabatini, A. Vacca, F. Zanobini, Helv. Chim. Acta **67**, 1818 (1984).
31. F. Wang and L. M. Sayre, Inorg. Chem. **28**, 169 (1989).
32. R. N. Smith, Anal. Biochem. **93**, 380 (1979).
33. M. Kotowski and R. van Eldik, Coordn. Chem. Revs. **93**, 19 (1989).
34. J. R. Graham and R. J. Angelici, Inorg. Chem. **6**, 2082 (1967).
35. J. O. Edwards, F. Monacelli and G. Ortaggi, Inorg. Chim. Acta **11**, 47 (1974).
36. J. H. Swinehart and G. W. Castellan, Inorg. Chem. **3**, 278 (1964); J. H. Swinehart, J. Chem. Educ. **44**, 524 (1967), gives a full description of the study of the equilibrium $Cr_2O_7^{2-} + H_2O \rightarrow 2\,HCrO_4^-$ by changing the total concentration of Cr(VI) and watching the concomitant spectral changes on a recording spectrophotometer.
37. J. F. Eccleston, R. G. Messerschmidt and D. W. Yates, Anal. Biochem. **106**, 73 (1980).
38. D. A. Buckingham, P. J. Cresswell, A. M. Sargeson and W. G. Jackson, Inorg. Chem. **20**, 1647 (1981).
39. B. F. Peterman and C.-W. Wu, Biochemistry **17**, 3889 (1978).
40. Z. A. Schelly in B21, p. 35–39.
41. H. Kihara, E. Takahashi, K. Yamamura and I. Tabushi, Biochem. Biophys. Res. Communs. **95**, 1687 (1980).
42. S. Saigo, J. Biochem. Japan **89**, 1977 (1981).
43. L. E. Erickson, H. L. Erickson and T. Y. Meyer, Inorg. Chem. **26**, 997 (1987).
44. From R. G. Wilkins, Adv. Inorg. Bioinorg. Mechs. **2**, 139 (1983), slightly amended.
45. B. H. Robinson in B6, Chap. 1.
46. B13, Chap. 2.
47. H. Gerischer, J. Holzworth, D. Seifert, and L. Strohmaier, Ber. Bunsenges. Gesellschaft **73**, 952 (1969).
48. J. F. Holzwarth in B21, p. 13.
49. S. A. Jacobs, M. T. Nemeth, G. W. Kramer, T. Y. Ridley, and D. W. Margerum, Anal. Chem. **56**, 1058 (1984).
50. M. T. Nemeth, K. D. Fogelman, T. Y. Ridley and D. W. Margerum, Anal. Chem. **59**, 283 (1987).
51. J. A. Sirs, Trans. Faraday Soc. **54**, 207 (1958).

52. M. R. Luzzana and J. T. Penniston, Biochim. Biophys. Acta **396**, 157 (1975) monitor the hemoglobin-O_2 reaction by continuous flow combined with a Clark O_2 electrode and obtain a rate profile similar to that with the spectral method.

53. G. A. Rechnitz, Anal. Chim. Acta **180**, 289 (1986); R. L. Solsky, Anal. Chem. **60**, 106R (1988).

54. P. Bowen, B. Balko, K. Blevins, R. L. Berger and H. P. Hopkins Jr., Anal. Biochem. **102**, 434 (1980).

55. D. C. Borg, Nature, London **201**, 1087 (1964).

56. T. Ogura and T. Kitagawa, J. Amer. Chem. Soc. **109**, 2177 (1987).

57. J. C. Sheppard and A. C. Wahl, J. Amer. Chem. Soc. **79**, 1020 (1957).

58. P. Hurwitz and K. Kustin, Trans. Faraday Soc. **62**, 427 (1966).

59. W. J. Ray, Jr., and J. W. Long, Biochemistry **15**, 3990 (1976).

60. P. A. Benkovic, W. P. Bullard, M. M. de Maine, R. Fishbein, K. J. Schray, J. J. Steffens and S. J. Benkovic, J. Biol. Chem. **249**, 930 (1974).

61. H. von Felten, H. Gamsjäger and P. Baertschi, J. Chem. Soc. Dalton Trans. 1683 (1976).

62. S. Dahlin, B. Reinhammar and M. T. Wilson, Biochem. J. **218**, 609 (1984).

63. G. C. M. Bourke and R. C. Thompson, Inorg. Chem. **26**, 903 (1987).

64. R. K. Murmann, Inorg. Chem. **16**, 46 (1977).

65. B. E. Smith, D. J. Lowe, C. G. Xiong, M. J. O'Donnell and T. R. Hawkes, Biochem. J. **209**, 207 (1983); R. N. F. Thorneley and D. J. Lowe, Kinetics and Mechanism of the Nitrogenase Enzyme System in Molybdenum Enzymes, T. G. Spiro, ed. Wiley Interscience NY, 1985, Chap. 5 includes a description of the rapid freeze technique.

66. J. P. G. Malthouse, J. W. Williams and R. C. Bray, Biochem. J. **197**, 421 (1981).

67. P. Reisberg, J. S. Olson and G. Palmer, J. Biol. Chem. **251** , 4379 (1976).

68. R. Hille, G. Palmer and J. S. Olson, J. Biol. Chem. **252**, 403 (1977).

69. A number of stopped-flow systems are commercially available.[46] Three of the most used are manufactured by Atago Bussan (formerly Union Giken), Japan; Dionex (formerly Durrum), USA and Hi-Tech Scientific, UK. These also manufacture rapid scan spectrophotometers, multimixer, temperature-jump and flash photolysis equipment.

70. B. Tonomura, H. Nakatani, M. Ohnishi, J. Yamaguchi-Ito and K. Hiromi, Anal. Biochem. **84**, 370 (1978).

71. P. J. Hagerman, F. X. Schmid and R. L. Baldwin, Biochemistry **18**, 293 (1979).

72. C. Paul, K. Kirschner and G. Haenisch, Anal. Biochem. **101**, 442 (1980) determine the deadtime for a stopped-flow apparatus using a disulfide exchange reaction.

73. P. N. Dickson and D. W. Margerum, Anal. Chem. **58**, 3153 (1986), estimate that $k_{corrected)} = k_{obs}(1 - k_{obs}/k_{mix})$ and $k_{mix} \approx 1700 s^{-1}$ and $\approx 1000 s^{-1}$ for Dionex and Hi-Tech stopped-flow instruments respectively.

74. M. A. Abdallah, J.-F. Biellmann and P. Lagrange, Biochemistry **18**, 836 (1979).

75. G. J. McClune and J. A. Fee, FEBS Lett. **67**, 294 (1976); G. H. McClune and J. A. Fee, Biophys. J. **24**, 65 (1978).

76. J. T. Coin and J. S. Olson, J. Biol. Chem. **254**, 1178 (1979).

77. J. D. Ellis, K. L. Scott, R. K. Wharton, and A. G. Sykes, Inorg. Chem. **11**, 2565 (1972), who observe optical density changes when 1 M acid solutions are mixed with water in a Durrum-Gibson stopped-flow apparatus. Such traces could be incorrectly assigned to chemical reactions.

78. J. H. Sutter, K. Colquitt and J. R. Sutter, Inorg. Chem. **13**, 1444 (1974).

79. S. S. Hupp and G. Dahlgren, Inorg. Chem. **15**, 2349 (1976).

80. D. S. Auld, K. Geoghegan, A. Galdes and B. L. Vallee, Biochemistry **25**, 5156 (1986); C. Balny, T.-L. Saldana and N. Dahan, Anal. Biochem. **163**, 309 (1987) describe a high-pressure, low temperature, stopped-flow apparatus.

81. P. J. Nichols, Y. Ducommun and A. E. Merbach, Inorg. Chem. **22**, 3993 (1983) and references therein.

82. S. Funahashi, Y. Yamaguchi and M. Tanaka, Inorg. Chem. **23**, 2249 (1984).

83. R. van Eldik, D. A. Palmer, R. Schmidt and H. Kelm, Inorg. Chim. Acta **50**, 131 (1981) for a detailed description.
84. K. Ishihara, H. Miura, S. Funahashi and M. Tanaka, Inorg. Chem. **27**, 1706 (1988), describe a high-pressure, stopped-flow arrangement with conductivity monitoring.
85. J. D. Rush and B. H. J. Bielski, J. Phys. Chem. **89**, 1524 (1985).
86. C. Bull, G. J. McClune and J. A. Fee, J. Amer. Chem. Soc. **105**, 5290 (1983).
87. Z. Bradić and R. G. Wilkins, J. Amer. Chem. Soc. **106**, 2236 (1984).
88. P. S. Kim and R. L. Baldwin, Ann. Rev. Biochem. **59**, 631 (1990).
89. Selected Bibliography.
90. T.-E. Dubois, Pure Appl. Chem. **50**, 801 (1978); H. Kruger, Chem. Soc. Revs. **11**, 227 (1982).
91. D. H. Turner in B6, Chap. III.
92. G. H. Czerlinski and M. Eigen, Z. Electrochem. **63**, 652 (1959).
93. List of commercially available equipment (not cheap!) in Hiromi p. 159 and Bernasconi Ed. 4, p. 171.
94. A. S. Verkman, A. A. Pandiscio, M. Jennings and A. K. Solomon, Anal. Biochem. **102**, 189 (1980).
95. D. Porschke, Rev. Sci. Instrum. **47**, 1363 (1976)
96. M. F. Perutz, T. K. M. Sanders, D. H. Chenery, R. W. Noble, R. R. Pennelly, L. W.-M. Fung C. Ho, I. Giannini, D. Pörschke and H. Winkler, Biochemistry **17**, 3640 (1978) show relaxation of azidomethemoglobin ($\Delta T = 4°$, T = 260 ns for human and < 100 ns for carp).
97. J. Aubard, J. M. Nozeran, P. Levoir, J. J. Meyer and J. E. Dubois, Rev. Sci. Instrum. **50**, 52 (1979).
98. J. V. Beitz, G. W. Flynn, D. H. Turner and N. Sutin, J. Amer. Chem. Soc. **92**, 4130 (1970), **94**, 1554 (1972).
99. K. A. Reeder, E. V. Dose and L. J. Wilson, Inorg. Chem. **17**, 1071 (1978).
100. K. Kustin, I. A. Taub and E. Weinstock, Inorg. Chem. **5**, 1079 (1966).
101. A. J. Miralles, R. E. Armstrong and A. Haim, J. Amer. Chem. Soc. **99**, 1416 (1977).
102. C. F. Bernasconi and M. C. Muller, J. Amer. Chem. Soc. **100**, 5530 (1978).
103. A. S. Verkman, J. A. Dix and A. A. Pandiscio, Anal. Biochem. **117**, 164 (1981).
104. M. D. Johnson and R. G. Wilkins, Inorg. Chem. **23**, 231 (1984).
105. W. Knoche in B6, Chap. IV.
106. H. R. Halvorson, Biochemistry **18**, 2480 (1979).
107. R. M. Clegg and B. W. Maxfield, Rev. Sci. Instrum. **47**, 1383 (1976).
108. M. A. Lopez Quintela, W. Knoche and I. Veith, T. Chem, Soc. Faraday Trans. I **80**, 2313 (1984).
109. G. Kegeles, Methods in Enzym. **48**, 308 (1978) P-jump/light scattering arrangments are reviewed.
110. T. Inoue, K. Sugahara, K. Kojima and R. Shimozawa, Inorg Chem. **22**, 3977 (1983).
111. E. M. Eyring and P. Hemmes in B6, Chap. V.
112. H. Hirohara, K. J. Ivin, J. J. McGarvey and J. Wilson, J. Amer. Chem. Soc. **96**, 4435 (1974).
113. J. E. Stuehr in B6, Ed. 4, Chap. VI.
114. S. Harada, Y. Uchida, M. Hiraishi, H. L. Kuo and T. Yasunaga Inorg. Chem. **17**, 3371 (1978); B. Perlmutter-Hayman, J. Chem. Educ. **47**, 201 (1970); N. Purdie and M. F. Farrow, Coordn. Chem. Revs. **11**, 189 (1973) for excellent presentations.
115. Y. Funaki, S. Harada, K. Okumiya and T. Yasunaga, J. Amer. Chem. Soc. **104**, 5325 (1982).
116. J. K. Beattie, M. T. Kelso, W. E. Moody and P. A. Tregloan, Inorg. Chem. **24**, 415 (1985).
117. B22, p. 45.
118. M. Eigen and K. Tamm, Ber. Bunsenges. Gesellschaft **66**, 107 (1962); L. G. Jackopin and E. Yeager, J. Phys. Chem. **74**, 3766 (1970).
119. R. A. Binstead, J. K. Beattie, E. V. Dose, M. F. Tweedle and L. J. Wilson, J. Amer. Chem. Soc. **100**, 5609 (1978).
120. R. G. W. Norrish and G. Porter, Nature London **164**, 658 (1949).
121. M. A. West, In B6, Chap. VIII.
122. C. A. Sawicki and R. J. Morris, Flash Photolysis of Hemoglobin, Methods in Enzym. **76**, 667 (1981).

123. D. G. Nocera, J. R. Winkler, K. M. Yocum, E. Bordignon and H. B. Gray, J. Amer. Chem. Soc. **106**, 5145 (1984).

124. P. Connolly, J. H. Espenson and A. Bakac, Inorg. Chem. **25**, 2169 (1986); M. A. Hoselton, C.-T. Lin, H. A. Schwarz and N. Sutin, J. Amer. Chem. Soc. **100**, 2383 (1978).

125. G. R. Fleming, "Chemical Applications of Ultrafast Spectroscopy", Oxford Univ. Press, NY, 1986. The book covers the range 10^{-9} to 10^{-14} s, and full details on generating, characterizing and probing ultrashort light pulses are given.

126. J. Terner and M. A. El-Sayed, Acc. Chem. Res. **18**, 331 (1985).

127. A. Juris, V. Balzani, F. Barigelletti, S. Campagna, P. Belser and A. von Zelewsky, Coordn. Chem. Revs. **84**, 85 (1988).

128. C. R. Bock, T. J. Meyer and D. G. Whitten, J. Amer. Chem. Soc. **96**, 4710 (1974).

129. Since Ru(bpy)$_3^{3+}$ and Fe^{2+} must of necessity be produced in equal concentrations, (3.15) is set up for second-order conditions (Sec. 1.4.3).

130. H. B. Gray, Chem. Soc. Revs. **15**, 17 (1986).

131. A. G. Sykes, Chem. in Britain **24**, 551 (1988) and references therein.

132. S. P. Webb, S. W. Yeh, L. A. Philips, M. A. Tolbert and J. H. Clark, J. Amer. Chem. Soc. **106**, 7286 (1984) and references therein.

133. J. F. Ireland and P. A. H. Wyatt, Adv. Phys. Org. Chem. **12**, 131 (1976).

134. R. Yam, E. Nachliel and M. Gutman, J. Amer. Chem. Soc. **110**, 2636 (1988)

135. M. Gutman and E. Nachliel, Biochem. Biophys. Acta **1015**, 391 (1990).

136. G. Ferraudi, Inorg. Chem. **17**, 2506 (1978) generates methyl radicals photolytically.

137. I. Lawthers and J. J. McGarvey, J. Amer. Chem. Soc. **106**, 4280 (1984).

138. B. H. Robinson and N. C. White, J. Chem. Soc. Faraday Trans. I **74**, 2625 (1978).

139. Q. H. Gibson, J. Biol. Chem. **264**, 20155 (1989); Biochem. Soc. Trans. **18**, 1 (1990).

140. J. H. Baxendale and M. A. J. Rodgers, Chem. Soc. Revs. **7**, 235 (1978).

141. M. S. Matheson and L. M. Dorfman, Pulse Radiolysis, MIT Press, Cambridge, Mass. 1969.

142. L. M. Dorfman and M. C. Sauer in B6, Chap. IX.

143. G. V. Buxton and R. M. Sellers, Coordn. Chem. Revs. **22**, 195 (1977).

144. M. Anbar, M. Bambenek, A. B. Ross, National Bureau of Standards Publication NSRDS-NBS43 1973 and Supplement, 1975.

145. G. V. Buxton, C. L. Greenstock, W. P. Helman and A. B. Ross, J. Phys. Chem. Ref. Data **17**, 513 (1988).

146. M. Anbar, F. Ross and A. B. Ross, NSRDS-NBS51, 1975.

147. A. J. Swallow, Prog. React. Kinetics **9**, 195 (1978) − an account of reactions of free radicals (including CO$_2^-$) produced by irradiation of organic compounds.

148. A. B. Ross and P. Neta NSRDS-NBS65, 1979; P. Neta, R. E. Huie and A. B. Ross, J. Phys. Chem. Ref. Data **17**, 1027 (1988).

149. Z. B. Alfassi and R. H. Schuler, J. Phys. Chem. **89**, 3359 (1985).

150. B. H. J. Bielski, D. E. Cabelli and R. L. Arudi, J. Phys. Chem. Ref. Data **14** 1041 (1985).

151. S. S. Isied, Prog. Inorg. Chem. **32**, 443 (1984).

152. J. Butler, A. G. Sykes, G. V. Buxton, P. C. Harrington and R. G. Wilkins, Biochem. J. **189**, 641 (1980).

153. J. Lilie, N. Shinohara and M. G. Simic, J. Amer. Chem. Soc. **98**, 6516 (1976).

154. M. Z. Hoffman and M. Simic, J. Amer. Chem. Soc. **94**, 1757 (1972).

155. K. D. Whitburn, M. Z. Hoffman, N. V. Brezniak, and M. G. Simic, Inorg. Chem. **25**, 3037 (1986) and references therein.

156. J. V. Beitz, J. R. Miller, H. Cohen, K. Wieghardt and D. Meyerstein, Inorg. Chem. **27**, 966 (1988).

157. R. G. Wilkins, Adv. Inorg. Bioinorg. Mechs. **2**, 139 (1983).

158. L. A. Blumenfeld and R. M. Davidov, Biochim. Biophys. Acta **549**, 255 (1979).

159. H. Barrett and J. H. Baxendale, Trans. Faraday Soc. **52**, 210 (1956).

160. A. Haim and N. Sutin, J. Amer. Chem. Soc. **88**, 5343 (1966); R. D. Cannon and T. S. Stillman J. Chem. Soc. Dalton Trans. 428 (1976).

161. J. P. Candlin and J. Halpern, Inorg. Chem. **4**, 766 (1965).

162. M. J. Carter and J. K. Beattie, Inorg. Chem. **9**, 1233 (1970).

163. R. G. Wilkins, Acc. Chem. Res. **3**, 408 (1970); Comments Inorg. Chem. **2**, 187 (1983).

164. J. Bjerrum and K.G. Poulsen, Nature London **169**, 463 (1952).

165. P. Bernhard, L. Helm, A. Ludi and A. E. Merbach, J. Amer. Chem. Soc. **107**, 312 (1985).

166. D. S. Auld, Methods in Enzym. **61**, 318 (1979); A. Galdes, D. S. Auld and B. L. Vallee, Biochemistry **22**, 1888 (1983).

167. P. Douzou and G. A. Petsko, Adv. Protein Chem. **36**, 245 (1984); S. T. Cartwright and S. G. Waley, Biochemistry **26**, 5329 (1987) for a critical analysis of the technique.

168. R. H. Austin, K. W. Beeson, L. Eisenstein, H. Frauenfelder, and I. C. Gunsalus, Biochemistry **14**, 5355 (1975).

169. A. L. Fink, Acc. Chem. Res. **10**, 233 (1977).

170. A. S. Verkman, A. A. Pandiscio, M. Jennings, and A. K. Solomon, Anal. Biochem. **102**, 189 (1980).

171. K. Murakami, T. Sano and T. Yasunaga, Bull. Chem. Soc. Japan **54**, 862 (1981). This is unusual case where there are no temperature-jump relaxations. The interaction of bovine serum albumin with bromophenol blue is accompanied by four relaxations which are attributed to a fast second-order interaction followed by three first-order steps.

172. S. L. Olsen, L. P. Holmes and E. M. Eyring, Rev. Sci. Instrum. **45**, 859 (1974).

173. A. Raap, J. W. Van Leeuwen, H. S. Rollema and S. H. de Bruin, Eur. J. Biochem. **88**, 555 (1978).

174. M. Absi-Halabi, J. D. Atwood, N. P. Forbus and T. L. Brown, J. Amer. Chem. Soc. **102**, 6248 (1980); M. S. Corraine and J. D. Atwood, Inorg. Chem. **28**, 3781 (1989).

175. M. Poliakoff and E. Weitz, Adv. Organomet. Chem. **25**, 277 (1986).

176. T. Ogura and T. Kitagawa, J. Amer. Chem. Soc. **109**, 2177 (1987).

177. M. Foster, R. E. Hester, B. Cartling and R. Wilbrandt, Biophys. J. **38**, 111 (1982).

178. W. K. Smothers and M. S. Wrighton, J. Amer. Chem. Soc. **105**, 1067 (1983); B. Cartling and R. Wilbrandt, Biochim. Biophys. Acta **637**, 61 (1981).

179. K. B. Lyons, J. M. Freidman, P. A. Fleury, Nature London **275**, 565 (1978); R. F. Dallinger, W. H. Woodruff and M. A. J. Rodgers, J. Appl. Spectros. **33**, 522 (1979).

180. M. Brouwer, C. Bonaventura and J. Bonaventura, Biochemistry **20**, 1842 (1981).

181. D. Thusius in B23, p. 339.

182. S. M. J. Dunn, J. G. Batchelor and R. W. King, Biochemistry **17**, 2356 (1978).

183. R. Rigler, C.-R. Rabl and T. M. Jovin, Rev. Sci. Instrum. **45**, 580 (1974).

184. M. J. Hardman, Biochem. J. **197**, 773 (1981) Pressure-jump combined with protein fluorescence changes are used to study LADH catalyzed reduction of acetaldehyde. The results show that the rate determining step is isomerization.

185. P. M. Bayley, Prog. Biophys. Molec. Biol. **37**, 149 (1981).

186. H. Nielsen and P. E. Sørensen, Acta Chem. Scand. **37**, 105 (1983), describe the modification of a commercial stopped-flow for polarimety.

187. M. Anson, S. R. Martin and P. M. Bayley, Rev. Sci. Instrum. **48**, 953 (1977).

188. B. Gruenwald and W. Knoche, Rev. Sci. Instrum. **49**, 797 (1978).

189. A. L. Cummings and E. M. Eyring, Biopolymers. **14**, 2107 (1975).

190. X. Xie and J. D. Simons, J. Amer. Chem. Soc. **112**, 7802 (1990).

191. A. E. Merbach, P. Morre, O. W. Howarth, and C. H. McAteer, Inorg. Chim. Acta **39**, 129 (1980).

192. S. Lanza, D. Minniti, P. Moore, J. Sachinidis, R. Romeo and M. L. Tobe, Inorg. Chem. **23**, 4428 (1984), measure exchange of $Me_2SO[^2H_6]$ with cis-$Pt(C_6H_5)_2(Me_2SO)_2$ in $CDCl_3$ by stopped-flow ^1H NMR. There is a steady decrease in the signal at δ 2.82 due to coordinated Me_2SO and increase at $\delta 2.61$ (uncoordinated Me_2SO).

193. R. J. Miller and G. L. Closs, Rev. Sci. Instr. **52**, 1876 (1981).

194. A. D. Trifunac, K. W. Johnson and R. H. Lowers, J. Amer. Chem. Soc. **98**, 6067 (1976).

195. R. C. Bray, Adv. Enzym. **51**, 107 (1980).

196. K.A. McLauchlan and D. G. Stevens, Acc. Chem. Res. **21**, 54 (1988).

197. V. Jagannadham and S. Steenken, J. Amer. Chem. Soc. **106**, 6542 (1984). The reaction of RĊHOH, generated by pulse radiolysis, was studied with p-substituted nitrobenzenes using time-resolved optical and conductance detection. The radical anion of the nitrobenzene is produced directly and indirectly.

198. B. G. Cox. P. Firman, I. Schneider and H. Schneider, Inorg. Chem. **27**, 4018 (1988).

199. T. Okubo and A. Enokida, J. Chem. Soc. Faraday Trans. I, 1639, (1983) flow and pressure-jump with conductance.

200. H. Hoffman, E. Yaeger and J. Stuehr, Rev. Sci. Instrum. **39**, 649 (1968).

201. T. Sano and T. Yasunaga, Biophys. Chem. **11**, 377 (1980).

202. J. Lilie, W. L. Waltz, S. H. Lee and L. L. Gregor, Inorg. Chem. **25**, 4487 (1986).

203. W. L. Waltz, J. Lilie and S. H. Lee, Inorg. Chem. **23**, 1768 (1984).

204. B. Fourest, K. H. Schmidt and J. C. Sullivan, Inorg. Chem. **25**, 2096 (1986). For description of pulse radiolysis/conductivity see K. H. Schmidt. S. Gordon, M. Thompson, J. C. Sullivan and W. A. Mulac, Radiat. Phys. Chem. **21**, 321 (1983).

205 T. Nakamura, J. Biochem. Japan **83**, 1077 (1978); J. V. Howarth, N. C. Millar and H. Gutfreund, Biochem. J. **248**, 677, 683, (1987) describe the construction and testing of a thermal/stopped-flow apparatus.

206. P. Bowen, B. Balko, K. Blevens, R. L. Berger and H. P. Hopkins, Jr., Anal. Biochem. **102**, 434 (1980).

207. G. W. Liesegang, J. Amer. Chem. Soc. **103**, 953 (1981).

208. B 16, Chap. 3; B 12, p. 45.

209. S. Yamada, K. Ohsumi and M. Tanaka, Inorg. Chem. **17**, 2790 (1978).

210. L. W. Harrison and B. L. Vallee, Biochemistry **17**, 4359 (1978).

211. J. Hirose and R. G. Wilkins, Biochemistry **23**, 3149 (1984); J. Hirose, M. Noji, Y. Kidani and R. G. Wilkins, Biochemistry **24**, 3495 (1985).

212. D. A. Malencik, S. R. Anderson, Y. Shalitin and M. I. Schimerlik, Biochem. Biophys. Res. Communs. **101**, 390 (1981).

213. J. R. Blinks, W. G. Wier, P. Hess and F. G. Prendergast, Prog. Biophys. Mol. Biol. **40**, 1 (1982) − measurement of Ca^{2+} in living cells.

214. P. L. Dorogi, C.-R. Rabl and E. Neumann, Biochem. Biophys. Res. Communs. **111**, 1027 (1983).

215. R. Koren and G. G. Hammes, Biochemistry **15**, 1165 (1976).

216. D. H. Devia and A. G. Sykes, Inorg. Chem. **20**, 910 (1981).

217. K. Nakatani, K. Kitagishiam and K. Hiromi, J. Biochem. Japan, **87**, 563 (1980).

218. D. Barber, S. R. Parr and C. Greenwood, Biochem. J. **173**, 681 (1978).

219. S. C. Koerber, A. K. H. MacGibbon, H. Dietrich, M. Zeppezauer and M. F. Dunn, Biochemistry **22**, 3424 (1983).

220. E.T. Gray Jr., R. W. Taylor and D. W. Margerum, Inorg. Chem. **16**, 3047 (1977).

221. L. I. Simándi, M. Jáky, C. R. Savage and Z. A. Schelly, J. Amer. Chem. Soc. **107**, 4220 (1985), which gives a detailed description of the arrangement.

222. A. A. El-Awady, P. C. Wilkins and R. G. Wilkins, Inorg. Chem. **24**, 2053 (1985).

223. A system which has been used successfully by a number of groups for some years is put together by On Line Instrument Systems, Jefferson, Gerogia 30549, USA.

224. J. C. Sullivan, E. Deutsch, G. E. Adams, S. Gordon, W. A. Mulac and K. H. Schmidt, Inorg. Chem. **15**, 2864 (1976).

225. M. Tsuda, Biochim. Biophys. Acta **545**, 537 (1979).

226. K. F. Geoghegan, A. Galdes, R. A. Martinelli, B. Holmquist, D. S. Auld and B. L. Vallee, Biochemistry **22**, 2255 (1983), give a schematic diagram of a low-temperature, stopped-flow, rapid scanning spectrometer and its use for recording spectra of intermediates in cobalt carboxypeptidase A catalyzed hydrolysis of very reactive dansyl oligo-peptides and -esters.

227. C. Grant, Jr., and P. Hambright, J. Amer. Chem. Soc. **91**, 4195 (1969).

228. M. M. Paleic and H. B. Dunford, J. Biol. Chem. **255**, 6128 (1980).

229. F. Zingales, A. Trovati and P. Uguagliati, Inorg. Chem. **10**, 510 (1971).

230. W. G. Jackson, Inorg. Chem. **26**, 3004 (1987), indicates the problems in interpreting the kinetics and stereochemistry of compounds of the type $Mn(CO)_5X$.

231. T. L. Brown, Inorg. Chem. **28**, 3229 (1989).

232. D. J. Darensbourg, Inorg. Chem. **18**, 14 (1979).

233. R. W. Callahan and T. J. Meyer, Inorg. Chem. **16**, 574 (1977).

234. K. Wieghardt, K. Woeste, P. S. Roy and P. Chaudhuri, J. Amer. Chem. Soc. **107**, 8276 (1985).

235. R. J. Kazlauskas and M. S. Wrighton, J. Amer. Chem. Soc. **104**, 5784 (1982).

236. I. D. Campbell and R. A. Dwek, Biological Spectroscopy, Benjamin/Cummings, Menlo Park, 1984, Chap. 9.

237. N. Capelle, J. Barbet, P. Dessen, S. Blanquet, B. P. Roques and J.-B. Le Pecq, Biochemistry **18**, 3354 (1979).

238. V. H. Rao and V. Krishnan, Inorg. Chem. **24**, 3538 (1985).

239. P.J. Breen, W. DeW. Horrocks, Jr. and K. A. Johnson, Inorg. Chem. **25**, 1968 (1986).

240. M. R. Eftink and C. A. Ghiron, Anal. Biochem. **114**, 199 (1981), review solute fluorescence quenching of proteins in the study of structure and dynamics.

241. P. C. Harrington and R. G. Wilkins, Biochemistry **17**, 4245 (1978).

242. R. D. Farina and R. G. Wilkins, Biochim. Biophys. Acta **631**, 428 (1980).

243. R. M. Clegg, F. G. Loontiens, A. V. Landschoot and T. M. Jovin, Biochemistry **20**, 4687 (1981).

244. P. J. Breen, K. A. Johnson and W. D. Horrocks, Jr., Biochemistry **24**, 4997 (1985) and references therein.

245. M. G. Badea and L. Brand. Methods in Enzym. **61**, 378 (1979).

246. J. B. A. Ross, C. J. Schmidt and L. Brand, Biochemistry **20**, 4369 (1981).

247. S. J. Simon, J. A. Boslett Jr. and K. H. Pearson, Inorg. Chem. **16**, 1232 (1977).

248. M. Abdullah, J. Barrett and P. O'Brien, Inorg. Chim. Acta **96**, L35 (1985).

249. Y. Ae Im and D. H. Busch, J. Amer. Chem. Soc. **83**, 3362 (1961). Models show that when the Me group of the (-)pdta ligand is equatorial (**10a**) there is less steric interaction with other hydrogens than in the axial Me group arrangement in (**10b**). The Δ-configuration about the metal (Sec. 7.6.1) will therefore be retained regardless of the lability of the complex. A beginning net-rotation thus becomes zero at equilibrium in (3.39).

250. I. I. Creaser, A. M. Sargeson and A. W. Zanella, Inorg. Chem. **22**, 4022 (1983).

251. H. Kihara, E. Takahashi, K. Yamamura and I. Tabushi, Biochem. Biophys. Res. Communs. **95**, 1687 (1980).

252. L. F. McCoy, Jr., E. S. Rowe and K.-P. Wong, Biochemistry **19**, 4738 (1980).

253. P. Hendry and A. M. Sargeson, Inorg. Chem. **25**, 865 (1986).

254. S. Aygen, H. Hanssum and R. van Eldik, Inorg. Chem. **24**, 2853 (1985). The results are very similar to older data obtained by the ^{18}O batch method (Sec. 3.11(b)), W. Kruse and H. Taube, J. Amer. Chem. Soc. **83**, 1280 (1961).

255. L. Helm, L. I. Elding and A. E. Merbach, Inorg. Chem. **24**, 1719 (1985).

256. O. Gröning, T. Drakenberg and L. I. Elding, Inorg. Chem. **21**, 1820 (1982).

257. D. A. Buckingham, C. R. Clark and T. W. Lewis, Inorg. Chem. **18**, 2041 (1979).

258. J. G. Russell, R. G. Bryant and M. M. Kreevoy, Inorg. Chem. **23**, 4565 (1984); see also R. K. Harris and R. J. Morrow, J. Chem. Soc. Faraday Trans. I, **80**, 3071 (1984).

259. R. D. Chapman and E. B. Fleischer, J. Amer. Chem. Soc. **104**, 1575; 1582 (1982).

260. P. Bernhard, L. Helm, A. Ludi and A. E. Merbach, J. Amer. Chem. Soc. **107**, 312 (1985) also I. Rapaport, L. Helm, A. E. Merbach, P. Bernhard and A. Ludi, Inorg. Chem. **27**, 873 (1988).

261. Y. T. Hayden and J. O. Edwards, Inorg. Chim. Acta **114**, 63 (1986).

262. R. K. Murmann, J. Amer. Chem. Soc. **96**, 7836 (1974); R. K. Murmann and K. C. Giese, Inorg. Chem. **17**, 1160 (1978).

263. P. Comba and L. Helm, Helv. Chim. Acta **71**, 1406 (1988).

264. R. L. Kump and L. J. Todd, Inorg. Chem. **20**, 3715 (1981).

265. D. T. Richens, L. Helm, P.-A. Pittet, A. E. Merbach, F. Nicolò and G. Chapuis, Inorg. Chem. **28**, 1394 (1989); G. D. Hinch, D. E. Wycoff and R. K. Murmann, Polyhedron, **5**, 487 (1986).

266. R. van Eldik, J. von Jouanne and H. Kelm, Inorg. Chem. **21**, 2818 (1982).

267. J. M Risley and R. L. Van Etten, J. Amer. Chem. Soc. **101**, 252 (1979).

268. T. G. Wood, O. A. Weisz and J. W. Korarich, J. Amer. Chem. Soc. **106**, 2222 (1984).

269. W. D. Wheeler, S. Kaizaki and J. I. Legg, Inorg. Chem. **21**, 3250 (1982).

270. P. Moore, Pure Appl. Chem. **57**, 347 (1985); C. H. McAteer and P. Moore, J. Chem. Soc. Dalton Trans. 353 (1983).

271. E. D. Becker, High Resolution NMR. Theory and Chemical Applications, Academic Press, NY, 1980.

272. G. Fraenkel in B6, Chap. X.

273. M. Meier, F. Basolo and R. G. Pearson, Inorg. Chem. **8**, 795 (1969).

274. These and subsequent equations are treated in: F. A. Bovey, Nuclear Magnetic Resonance Spectroscopy, 2nd. Edit. Academic Press, San Diego, CA, 1988, p. 118, 119, 191–203.

275. L. Spiecia and T. W. Swaddle, Inorg. Chem. **26**, 2265 (1987) is the more recent of a series of studies of this electron transfer.

276. H. M. McConnell and H. E. Weaver, J. Chem. Phys. **25**, 307 (1956).

277. C. R. Giuliau and H. M. McConnell, J. Inorg. Nucl. Chem. **9**, 171 (1959); K. Okamoto, W.-S. Jung, H. Tomiyasu and H. Fukutomi, Inorg. Chim. Acta **143**, 217 (1988) – see Prob. 5(b) Chap. 8.

278. P. J. Smolenaers and J. K. Beattie, Inorg. Chem. **25**, 2259 (1986).

279. R. M. Nielson, J. P. Hunt, H. W. Dodgen and S. Wherland, Inorg. Chem. **25**, 1964 (1986).

280. C. A. Koval and D. W. Margerum, Inorg. Chem. **20**, 2311 (1981).

281. E. J. Pullian and D. R. McMillan, Inorg. Chem. **23**, 1172 (1984).

282. D. L. Rabenstein and R. J. Kula, J. Amer. Chem. Soc. **91**, 2492 (1969); G. E. Glass, W. B. Schwabacher and R. S. Tobias, Inorg. Chem. **7**, 2471 (1968).

283. P. W. Taylor, J. Feeney and A. S. V. Burgen, Biochemistry **10**, 3866 (1971), employ nmr to study the binding of acetate (using ^{1}H) and fluoroacetate (^{19}F) to carbonic anhydrase.

284. L. H. Pignolet and W. D. Horrocks, Jr., J. Amer. Chem. Soc. **90**, 922 (1968).

285. G. N. La Mar, J. Amer. Chem. Soc. **92**, 1806 (1970).

286. P. R. Rubini, Z. Poaty, J.-C. Boubel, L. Rodenhüser and J.-J. Delpuech, Inorg. Chem. **22**, 1295 (1983).

287. P. R. Rubini, L. Rodenhüser and J. J. Delpuech, Inorg. Chem. **18**, 2962 (1979).

288. J. M. Lehn and M. E. Stubbs, J. Amer. Chem. Soc. **96**, 4011 (1974).

289. G. Binsch in Dynamic Nuclear Magnetic Resonance Spectroscopy, L. M. Jackman and F. A. Cotton, eds. Academic Press, NY, 1975, Chap. 3.

290. A. Allerhand, H. S. Gutowsky, J. Jones and R. A. Meinzer, J. Amer. Chem. Soc. **88**, 3185 (1966).

291. C. B. Storm, A. H. Turner and M. B. Swann, Inorg. Chem. **23**, 2743 (1984).

292. M.-S. Chan and A. C. Wahl, J. Phys. Chem. **82**, 2542 (1978).

293. G. Binsch, Top. Stereochem. **3**, 97 (1968); Y. Ikeda, H. Tomiyasu and H. Fukutomi, Inorg. Chem. **23**, 1356 (1984) study intramolecular exchange in $UO_2(acac)_2Me_2SO$ in $o\text{-}C_6H_4Cl_2$.

294. T. J. Swift and R. E. Connick, J. Chem. Phys. **37**, 307 (1962).

295. Z. Luz and S. Meriboom, J. Chem. Phys. **40**, 1058, 2686 (1964); R. Murray, H. W. Dodgen and J. P. Hunt, Inorg. Chem. **3**, 1576 (1964); H. H. Glaeser, H. W. Dodgen and J. P. Hunt, Inorg. Chem. **4**, 1061 (1965); S. Funahashi and R. B. Jordan, Inorg. Chem. **16**, 1301 (1977); S. F. Lincoln, A. M. Hounslow and A. N. Boffa, Inorg. Chem. **25**, 1038 (1986); A. Kioki, S. Funahashi, M. Ishii and M. Tanaka, Inorg. Chem. **25**, 1360 (1986).

296. T. R. Stengle and C. H. Langford, Coordn. Chem. Revs. **2**, 349 (1967) and S. F. Lincoln, Prog. React. Kinetics **9**, 1 (1977) for comprehensive discussions of complete line shape analyses.

297. D. K. Ravage, T. R. Stengle and C. H. Langford, Inorg. Chem. **6**, 1252 (1967).

298. A. E. Merbach, Pure Appl. Chem. **59**, 161 (1987).

299. (a) P. J. Nichols and M. W. Grant, Aust. J. Chem. **31**, 258 (1978). (b) C. H. McAteer and P. Moore, J. Chem. Soc. Dalton Trans. 353 (1983).

300. J. Burgess, B. A. Goodman and J. B. Raynor, J. Chem. Soc. A, 501 (1968).

301. J. N. Marov, M. N. Vargaftik, V. K. Belyaeva, G. A. Evtikova, E. Hoyer, R. Kirmse and W. Dietzsch, Russ. J. Inorg. Chem. **25**, 101 (1980).

302. A. McAuley, D. H. Macartney and T. Oswald, J. Chem. Soc. Chem. Communs. 274 (1982).

303. G. Czapski, J. Phys. Chem. **75**, 2957 (1971).

304. J. Stach, R. Kirmse, W. Dietzsch, G. Lassmann, V. K. Belyaeva and I. N. Marov, Inorg. Chim. Acta **96**, 55 (1985).

305. M. Knoblowitz and J. I. Morrow, Inorg. Chem. **15**, 1674 (1976).

306. S. A. Jacobs and D. W. Margerum, Inorg. Chem. **23**, 1195 (1984).

307. E. Finkelstein, G. M. Rosen and E. J. Rauckman, Arch. Biochem. Biophys. **200**, 1 (1980).

308. A. J. F. Searle and A. Tomasi, J. Inorg. Biochem. **17**, 161 (1982).

309. B. J. Corden and P. H. Rieger, Inorg. Chem. **10**, 263 (1971); J. B. Farmer, F. G. Herring and R. L. Tapping, Can. J. Chem. **50**, 2079 (1972).

310. T. Saji and S. Aoyagui, Bull. Chem. Soc. Japan **46**, 2101 (1973).

311. T. T.-T. Li and C. H. Brubaker, Jr., J. Organomet. Chem. **216**, 223 (1981).

312. C. Daul, J.-N. Gex, D. Perret, D. Schaller and A. von Zelewsky, J. Amer. Chem. Soc. **105**, 7556 (1983).

313. P. K. Glasoe and F. A. Long, J. Phys. Chem. **64**, 188 (1960).

314. R. G. Bates, M. Paabo and R. A. Robinson, J. Phys. Chem. **67**, 1833 (1963); B. B. Hasinoff, H. B. Dunford and D. G. Horne, Can. J. Chem **47**, 3225 (1969).

315. F. Millar, J. M. Wrigglesworth and P. Nicholls, Eur. J. Biochem. **117**, 13 (1981) − using glass microelectrode and pH meter followed changes of ±0.005 pH at neutral pH.

316. B 13, p. 178.

317. R. S. Rowlett and D. N. Silverman, J. Amer. Chem. Soc. **104**, 6737 (1982); D. N. Silverman and S. Lindskog, Acc. Chem. Res. **21**, 30 (1988).

318. L. S. W. L. Sokol, T. D. Fink and D. B. Rorabacher, Inorg. Chem. **19**, 1263 (1980).

319. R. W. Hay and P. Banerjee, J. Chem. Soc. Dalton Trans. 362 (1981).

320. P. A. Benkovic, M. Hegazi, B. A. Cunningham and S. J. Benkovic, Biochemistry **18**, 830 (1979) use continuous monitoring of inorganic phosphate product of enzyme catalyzed hydrolysis of fructose biphosphate. The pH change monitored by phenol red results from an acid, base adjustment of the liberated phosphate. This is a detailed valuable, if complicated, account.

321. P. F. Fitzpatrick and V. Massey, J. Biol. Chem. **257**, 9958 (1982).

322. A. K. Shamsuddin Ahmed and R. G. Wilkins, J. Chem. Soc. 3700 (1959).

323. R. Gresser, D. W. Boyd, A. M. Albrecht-Gary and J. P. Schwing, J. Amer. Chem. Soc. **102**, 651 (1980).

324. F. P. Rotzinger, H. Stünzi and W. Marty, Inorg. Chem. **25**, 489 (1986), use a weakish base such as ethanolamine to avoid local excesses of OH⁻.

325. G. Schwarzenbach and G. Geier, Helv. Chim. Acta **46**, 906 (1963); G. Geier, Chimia, **25**, 401 (1971).

326. C. Marques and R. A. Hasty, J. Chem. Soc. Dalton Trans. 1269 (1980).

327. S. Balt and C. Dekker, Inorg. Chem. **15**, 2370 (1976).

328. M. Kotowski, D. A. Palmer and H. Kelm, Inorg. Chem. **18**, 2555 (1979).

329. L. J. Csányi, Z. M. Galbács and L. Nagy, J. Chem. Soc. Dalton Trans. 237 (1982).

330. R. G. Dakers and J. Halpern, Can. J. Chem. **32**, 969 (1954).

331. M. Kimura, T. Kawajiri and M. Tanida, J. Chem. Soc. Dalton Trans. 726 (1980).

332. F. C. Anson, C.-L. Ni and J. M. Saveant, J. Amer. Chem. Soc. **107**, 3442 (1985).

333. E. L. Yee, O. A. Gansow and M. J. Weaver, J. Amer. Chem. Soc. **102**, 2278 (1980).

334. B. G. Cox and W. Jedral, J. Chem. Soc. Faraday Trans. I **80**, 781 (1984).

335. R. F. Jameson and N. J. Blackburn, J. Chem. Soc. Dalton Trans. 9, (1982).

336. R. Hamilton, D. Maguire and M. McCabe, Anal. Biochem. **93**, 386 (1979).

337. B 18, p. 42.

338. A. J. Bard and L. R. Faulkner, Electrochemical Methods, Wiley, NY 1980 Ch 11.

339. T. Matusinovic and D. E. Smith, Inorg. Chem. **20**, 3121 (1981).

340. K. Shigehara, N. Oyama and F. C. Anson, Inorg. Chem. **20**, 518 (1981).

341. H. Krüger, Chem. Soc. Revs. **11**, 227 (1982) for a short account of electrochemical (as well as other) methods for studying fast reactions.

342. R. M. Wightman and D. O. Wipf, Acc. Chem. Res. **23**, 64 (1990). C. P. Andrieux, P. Hapiot and J.-M. Saveant, Chem. Revs. **90**, 723 (1990).

343. W. E. Jones and J. D. R. Thomas, J. Chem. Soc. A, 1481 (1966).

344. W. G. Jackson, S. S. Jurisson and B. C. McGregor, Inorg. Chem. **24**, 1788 (1985).

345. M. H. M. Abou-El-Wafa, M. G. Burnett and J. F. McCullagh, J. Chem. Soc. Dalton Trans. 2083 (1986).

346. M. F. Perutz, Biochem. J. **195**, 519 (1981).

347. S. F. Lincoln and D. R. Stranks, Aust. J. Chem. **21**, 37; 67 (1968).

348 H. Gamsjäger and R. K. Murmann, Adv. Inorg. Bioinorg. Chem. **2**, 317 (1983).

349. B. Bronnum, H. S. Johansen and L. H. Skibsted, Inorg. Chem. **27**, 1859 (1988).

350. H. S. Johansen and V. Middelboe, Appl. Spectrosc. **36**, 221 (1982).

351. R. M. Noyes in B 5, Chap. V (Sec. 4).

352. H. Ogino, E. Kikkawa, M. Shimura and N. Tanaka, J. Chem. Soc. Dalton Trans. 894 (1981).

353. A. Bakač and J. H. Espenson, Inorg. Chem. **20**, 953 (1981) generate Cr^{2+} slowly by the reaction $CrCH_2OH^{2+} + VO^{2+} + H^+ \rightarrow Cr^{2+} + V^{3+} + CH_2O + H_2O$ and then the relative reaction rates with $(NH_3)_5CoX^{2+}$ and VO^{2+} both in large excess are assessed by the Co^{2+} produced (color with SCN^-).

354. N. Shinohara and J. Lilie, Inorg. Chem. **18**, 434 (1979).

355. H. Boucher, A. M. Sargeson, D. F. Sangster and J. C. Sullivan, Inorg. Chem. **20**, 3719 (1981).

356. M. Z. Hoffman and M. Simic, Inorg. Chem. **12**, 2471 (1973).

357. R. F. Pasternack and B. Halliwell, J. Amer. Chem. Soc. **101**, 1026 (1979).

358. M. Z. Hoffman, F. Bolleta, L. Moggi and G. L. Hug, Rate Constants for the Quenching of Excited States of Metal Complexes in Fluid Solution, J. Phys. Chem. Ref. Data, **18**, 219 (1989).

359. F. A. Posey and H. Taube, J. Amer. Chem. Soc. **79**, 255 (1957).

360. I. Dostrovsky and F. S. Klein, Anal. Chem. **24**, 414 (1952).

361. M. H. O'Leary, Methods in Enzym. **64**, 83 (1980); C. B. Grissom and W. W. Cleland, J. Amer. Chem. Soc. **108**, 5582 (1986).

362. T. J. Kemp, P. Moore and G. R. Quick, J. Chem. Soc. Dalton Trans. 1377 (1979).

363. K. J. Wannowius, K. Krimm and H. Elias, Inorg. Chim. Acta **127**, L 43 (1987).

364. R. van Eldik, U. Spitzer and H. Kelm, Inorg. Chim. Acta **74**, 149 (1983).

365. A. J. Dixon, M. A. Healy, M. Poliakoff and J. J. Turner, J. Chem. Soc. Chem. Commun. 994 (1986).

366. A. J. Brown, O. W. Howarth, P. Moore and G. E. Morris, J. Chem. Soc. Dalton Trans. 1776 (1979).

367. P. Hemmes and J. J. McGarvey, Inorg. Chem. **18**, 1812 (1979).

368. A. Lifshitz and B. Perlmutter-Hayman, J. Phys. Chem. **65**, 2098 (1961).

369. C. Bohne, I. D. MacDonald and H. B. Dunford, J. Amer. Chem. Soc. **108**, 7867 (1986).

Selected Bibliography

B21. W. J. Gettins and E.Wyn-Jones (eds.) Techniques and Applications of Fast Reactions in Solution, D. Reidel, Boston, 1979.

B22. D. N. Hague, Fast Reactions, Wiley-Interscience, New York, 1971.

B23. I. Pecht and R. Rigler (eds.) Chemical Relaxation in Molecular Biology, Springer-Verlag, New York, 1977.

Problems

1. The buried Cys-212 of human carbonic anhydrase B ($3\,\mu M$) is virtually unreactive towards 2-chloromercuric-4-nitrophenol ($60\,\mu M$) at pH 9.2, but upon the addition of only $40\,\mu M$ CN^-, the half-life drops to 10 minutes which is an, at least, 75-fold rate enhancement. On first analysis, this would suggest that inhibitor binding to the enzyme has produced a conformational change or altered the $-SH$ environment of the $Cys-212$. This is unexpected. How would you prove by kinetic experiments that the CN^- is binding to the mercury compound and not the enzyme and that this is changing the reactivity. The rate reaches a constant value at high $[CN^-]$.
   R. G. Khalifah, G. Sanyai, D. J. Straker and W. McI. Sutherland, J. Biol. Chem. **254**, 602 (1979).

2. A number of Co(III) complexes, such as $Co(edta)^-$ and $Co(phen)_3^{3+}$, can be resolved into optical isomers and are extremely stable towards racemization. The Co(II) analogs are configurationally labile and resolution has proved impossible. Suggest how with a double mixing apparatus it might be possible to measure half-lives in the $10^{-3}-1$ s range for the first-order racemization of the Co(II) complexes.
   E. L. Blinn and R. G. Wilkins, Inorg. Chem. **15**, 2952 (1976).

3. Figure 1 shows pressure-jump relaxation traces with conductivity monitoring at pressures, P, of 1 kg cm^{-2} (lower curve) and 1000 kg cm^{-2} for the relaxation of a 6.06 mM nickel (II) glycolate solution at 20°C. Time scale: 2 ms/division. Only the $1:1$ complex ($K = 210\ M^{-1}$ at $P = 1$ kg cm^{-2} and $K = 100\ M^{-1}$ at $P = 1000$ kg cm^{-2}) need be considered. Estimate the $\Delta V^{\ddagger}$ value for the formation and dissociation of Ni(II) glycolate from these data, and after reading Chap. 4 account for the data.
   T. Inoue, K. Sugahara, K. Kojima and R. Shimozawa, Inorg. Chem. **22**, 3977 (1983).

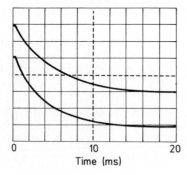

0   10   20
Time (ms)

Problem 3. Reproduced with permission from T. Inone, K. Sugahara, K. Kojima and R. Shimozawa, Inorg. Chem. **22**, 3977 (1983). © (1983) American Chemical Society.

4. Geminate recombination of iron(II) porphyrin with a number of isocyanides and 1-methylimidazole has been observed by T. G. Traylor, D. Magde, D. Taube and K. Jongeward, J. Amer. Chem. Soc. **109**, 5864 (1987).

(a) What is geminate recombination and what is its significance to the measured quantum yield?

(b) The suggested mechanism (with rate constants) for the binding of MeCN(L) to the 1-methylimidazole (B) complex of protoheme dimethyl ester in toluene/B is

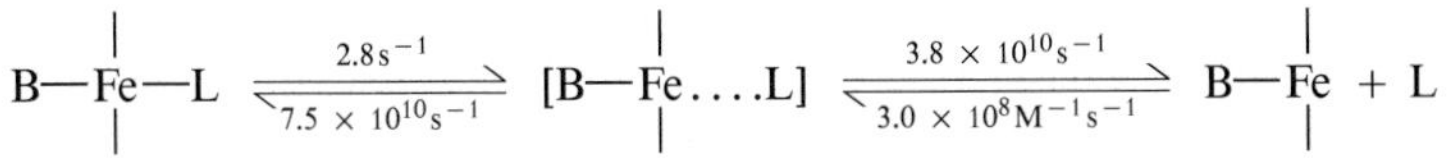

$$
\text{B}-\underset{|}{\overset{|}{\text{Fe}}}-\text{L} \underset{7.5 \times 10^{10}\text{s}^{-1}}{\overset{2.8\,\text{s}^{-1}}{\rightleftharpoons}} [\text{B}-\underset{|}{\overset{|}{\text{Fe}}}\ldots.\text{L}] \underset{3.0 \times 10^{8}\text{M}^{-1}\text{s}^{-1}}{\overset{3.8 \times 10^{10}\text{s}^{-1}}{\rightleftharpoons}} \text{B}-\underset{|}{\overset{|}{\text{Fe}}} + \text{L}
$$

How are these individual rate constants determined?

5. Laser pulses of 265 nm and 20 ns duration were delivered into a quartz cell through which a solution of dearated 0.2 mM $[\text{Co(NH}_3)_5\text{Cl}]\text{Cl}_2$ in 5 mM HCl was flowing. Both spectral and conductivity changes could be monitored. The very fast absorbance change at 340 nm (only a slight conductivity increase) and the slower conductivity changes (no absorbance changes) are shown in Fig. 2. The absorbance at the end of (a) corresponds to Cl_2^-. Interpret the results.

J. Lilie, J. Amer. Chem. Soc. **101**, 4417 (1979).

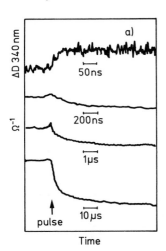

Problem 5. Reproduced with permission from I. Lilie, J. Amer. Chem. Soc. **101**, 4417 (1979), © (1979) American Chemical Society.

6. Nitroprusside ion $\text{Fe(CN)}_5\text{NO}^{2-}$ is an important drug for the alleviation of severe hypertension. A. R. Butler and C. Glidewell, Chem. Soc. Revs. **16**, 361 (1987).

Radiolysis of $\text{Fe(CN)}_5\text{NO}^{2-}$ with a number of reducing radicals e.g. e_{aq}^-, CO_2^-, $(\text{CH}_3)_2\dot{\text{C}}\text{OH}$, and $\text{H}^{\bullet}$ gives a common transient, **A** with maxima at 345 nm and 440 nm. What will be the approximate magnitude of rate constants for production of **A** from the radicals and what is likely to be the structure of **A**?

A undergoes spectral changes in the ms range by a first-order process (k) to give a stable product (in the absence of O_2 and light). k is $2.8 \times 10^2\text{s}^{-1}$ from pH 4.6–8.5. Its value is increased if free CN^- is added to the pulse radiolyzed solution (at pH 6.7, mainly HCN). Now,

$$k_{obs} = 2.8 \times 10^2 + 4 \times 10^6 [CN^-]$$

Monitoring by conductivity shows a very rapid increase (during the reduction) and a slower decrease $(2.6 \times 10^2 s^{-1})$. Explain. R. P. Cheney, M. G. Simic, M. Z. Hoffman, I. A. Taub and K.-D. Asmus, Inorg. Chem. **16**, 2187 (1977).

7. The conductivity changes following pulse radiolysis of a mixture of 0.5 mM $Co(acac)_3$, 0.04 mM $HClO_4$ and 0.1 M tert-butyl alcohol are shown in Figure 3 (the units of G $\times$ $\Delta\Lambda$ are (molecules/100 eV) $\Omega^{-1} cm^2 M^{-1}$). The very first increase in conductivity also appears in solutions containing no $Co(acac)_3$. The decreases in conductivity are speeded up in acid. Account for this behavior.
D. Meisel, K. H. Schmidt and D. Meyerstein, Inorg. Chem. **18**, 971 (1979).

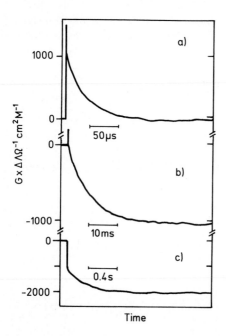

Problem 7. Reproduced with permission from D. Meisel, K.-H. Schmidt and D. Meyerstein, Inorg. Chem. **18**, 971 (1979), © (1979) American Chemical Society.

8. Vitamin B_{12r} a cobalt(II) complex designated Co(II) is oxidized by Br_2:

$$2 Co(II) + Br_2 \rightarrow 2 Co(III) + 2 Br^-$$

Since the reaction is second-order, the rate limiting step is considered to be:

$$Co(II) + Br_2 \rightarrow Co(III) + Br_2^- (or\ Co(III)-Br + Br^·) \qquad (1)$$

This would be followed by the faster reactions (2) or (3)

$$Br_2^- + Co(II) \rightarrow Co(III) + 2 Br^- (or\ Co(III)-Br + Br^-) \qquad (2)$$

or

$$Br_2^- + Br_2^- \rightarrow Br_2 + 2\,Br^-\tag{3}$$

Show how, using pulse-radiolytically generated Br_2^-, you might distinguish between (2) and (3) and how you might decide whether the reaction is inner sphere.

D. Meyerstein, J. H. Espenson, D. A. Ryan and W. A. Mulac, Inorg. Chem. **18**, 863 (1979).

9. How would you verify that the intermediate in the Fe^{2+}, $Co(C_2O_4)_3^{3-}$ reaction (3.17) is the $Fe(C_2O_4)^+$ (3.18) ion?

At high $Co(C_2O_4)_3^{3-}$ concentrations, this reaction obeys the rate law

$$-d[Fe^{2+}]/dt = k_1[Fe^{2+}][Co(C_2O_4)_3^{3-}] + k_2[Fe^{2+}][Co(C_2O_4)_3^{3-}]^2$$

Suggest a reason for the second term.

R. D. Cannon and J. S. Stillman, J. Chem. Soc. Dalton Trans. 428 (1976).

10. The second-order rate constants k for the base hydrolysis of a number of cobalt(III) complexes were measured with a simple flow apparatus using conductivity as a monitoring device. Equal concentrations (A_0) of reactants were used. Show that a plot of $R_t/R_e - R_t$ vs time is linear, having slope s, and that

$$k = \frac{(R_e - R_0)s}{R_0 A_0}$$

where R_0, R_t and R_e are the resistance of the solution at times zero, t and at equilibrium, respectively.

R. G. Pearson, R. E. Meeker and F. Basolo, J. Amer. Chem. Soc. **78**, 709 (1956).

11. Suppose that the mechanism for exchange between A and B is a dissociative one:

$$A \rightleftharpoons B + C \qquad k_1, k_{-1}$$

Deduce the dependence of the broadening of the lines A and B on the concentrations of the reactants and the rate constant k_1.

T. L. Brown, Acc. Chem. Res. **1**, 25 (1968).

12. For the $1:1$ Ca^{2+} complex of the ligand **A**, at $4\,°C$ in D_2O, two sets of four ^{13}C resonances are observed arising from C_R^N, C_R^O, C_B^N and C_B^O. The signals for C_R^O and C_R^N have twice the intensity of the other two. When the temperature is raised, the signals of the same intensity within each set coalesce at $40\,°C$. The separation of ^{13}C signals for C_R^N, Δv, is 48 Hz at $4\,°C$. Excess ligand or Ca^{2+} do not affect the result. The averaged four ^{13}C lines of the Ca^{2+} complex and the four ^{13}C signals of free ligand (in about equal amounts) remain sharp as the temperature is raised but finally broaden and coalesce at $\approx 105\,°C$. The separation of ^{13}C signals for C_R^N is 21 Hz at $32\,°C$. Calculate the exchange rate constants and free energies of activation at the two coalescent temperatures and account for the behavior.

J. M. Lehn and M. E. Stubbs, J. Amer.Chem. Soc. **96**, 4011 (1974).

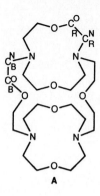

A

13. The CN^- exchange with $Pd(CN)_4^{2-}$ (0.117 M) was studied by ^{13}C nmr complex line broadening at 24°C in D_2O with the following results:

CN^-, M	$W_A^E - W_A^O$, Hz
0.00	1.0
0.036	1.5
0.104	3.5
0.214	7.5
0.321	12.5
0.447	18

Estimate k_A for each CN^- concentration and hence the order of the exchange and the value of the rate constant.

J. J. Pesek and W. R. Mason, Inorg. Chem, **22**, 2958 (1983).

14. The mechanism for the reaction

$$HOCl + SO_3^{2-} \rightarrow Cl^- + SO_4^{2-} + H^+$$

may involve either direct oxygen or Cl^+ transfer. Show how by using indicators e. g. phenolphthalein ($pK = 9.55$) or thymol blue ($pK = 9.20$), it should be possible to differentiate between the two mechanisms.

B. S. Yiin and D. W. Margerum, Inorg. Chem. **27**, 1670 (1988).

15. Two plausible mechanisms for the reaction of $Co(CN)_5^{3-}$ with H_2 invoke either homolytic splitting of H_2:

$$H_2 + 2\,Co(CN)_5^{3-} \rightarrow 2\,Co(CN)_5H^{3-}$$

or heterolytic cleavage:

$$H_2 + 2\,Co(CN)_5^{3-} \rightarrow Co_2(CN)_{10}H^{7-} + H^+ \rightarrow 2\,Co(CN)_5H^{3-}$$

An attempt was made to distinguish between these by carrying out the reaction in D_2O and examining the ratio $[Co(CN)_5H^{3-}]/[Co(CN)_5D^{3-}]$. The $Co-D/Co-H$ ir stretching intensity ratio $(I_{1340\,cm^{-1}}/I_{1840\,cm^{-1}})$ R is 0.53 for an $Co-D/Co-H$ ratio of 1.0. The following results were obtained:

Time (mins)	R	Time (mins)	R
5	0.26	30	0.71
10	0.58	50	0.75
15	0.36	90	1.04
20	0.68 (0.43)[a]		

[a] Repeat experiment.

Which mechanism do these data support?

J. Halpern, Inorg. Chim. Acta **77**, L 105 (1983).

16. The $-SH$ group in proteins $(P-SH)$ can be estimated by the addition of Ellmans reagent (ESSE)

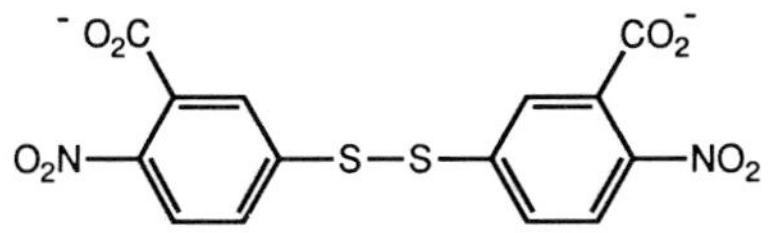

ESSE

which gives the colored ES^- group $(pK_{ESH} = 4.50)$

$$P-SH + ESSE \rightarrow P-SSE + ES^-$$

Devise a competition method which allows the determination of the rate of reaction of $P-SH$ with another disulfide RSSR which does *not* lead to a colored species

$$P-SH + RSSR \rightarrow P-SSR + RSH$$

(Hint: RSH reacts rapidly with ESSE to give ES^-)

J. M. Wilson, D. Wu, R. M. DeGrood and D. J. Hupe, J. Amer. Chem. Soc. **102**, 359 (1980).

17. The reaction of VO^{2+} with Cr^{2+}:

$$VO^{2+} + Cr^{2+} + 2H^+ \xrightarrow{k_1} V^{3+} + Cr^{3+} + H_2O$$

is too fast to be measured by stopped-flow. The reaction of $Co(NH_3)_5F^{2+}$ with Cr^{2+}:

$$Co(NH_3)_5F^{2+} + Cr^{2+} + 5H^+ \xrightarrow{k_2} Co^{2+} + CrF^{2+} + 5NH_4^+$$

has been measured by stopped-flow, since low concentrations of reactants can be used. The relative rate constants k_1/k_2 can be measured by a competition method in which a Cr^{2+} solution is added (in fact generated *in situ*) to a well-stirred solution of a mixture of VO^{2+} and $Co(NH_3)_5F^{2+}$ both well in excess of the Cr^{2+} concentration. The initial concentrations are $[VO^{2+}]_0$ and $[Co(NH_3)_5F^{2+}]_0$. After measuring, the solution was analyzed for Co^{2+} $([Co^{2+}]_e)$ with the following results:

$[VO^{2+}]_0$ mM	$[Co(NH_3)_5F^{2+}]_0$ mM	$[Cr^{2+}]_0$ mM	$[Co^{2+}]_e$ mM
10.0	18.0	0.90	0.48
10.0	10.0	0.90	0.35
20.0	10.0	0.95	0.25
25.0	10.0	0.96	0.19

Estimate (preferably using a graphical method) the value of k_1/k_2. A. Bakač and J. H. Espenson, Inorg. Chem. **20**, 953 (1981).

18. Suggest a suitable method (other than uv-vis spectral) for monitoring the following reactions and give details:

a. 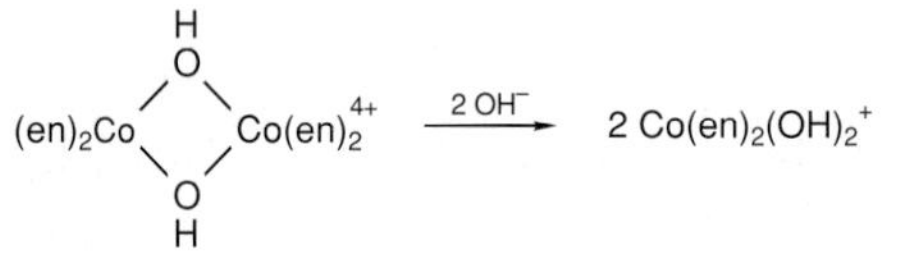
$$\text{(en)}_2\text{Co}\underset{\substack{O\\H}}{\overset{\substack{H\\O}}{\diamond}}\text{Co(en)}_2^{4+} \xrightarrow{\text{2 OH}^-} 2\,\text{Co(en)}_2(\text{OH})_2^{+}$$

A. A. El-Awady and Z. Z. Hugus, Jr., Inorg. Chem. **10**, 1415 (1971).

b. $Pd(PR_3)_2(CH_3)Cl + py \xrightarrow{\text{MeOH}} Pd(PR_3)_2(CH_3)py^+ + Cl^-$

F. Basolo, J. Chatt, H. B. Gray, R. G. Pearson and B. L. Shaw, J. Chem. Soc. 2207 (1961).

c. $Cd(pdta)^{2-} + H_2pdta^{2-} \rightarrow Cd(pdta)^{2-} + H_2pdta^{2-}$

B. Bosnich, F. P. Dwyer and A. M. Sargeson, Aust. J. Chem. **19**, 2213 (1966).

d. $Mo(CO)_6 + Ph_3As \xrightarrow{\text{decalin}} Mo(CO)_5Ph_3As + CO$

J. R. Graham and R. J. Angelici, Inorg. Chem. **6**, 2082 (1967).

e. $Zn^{2+} +$ apocarbonic anhydrase $\rightarrow$ carbonic anhydrase

(this is the regeneration of the enzyme from the demetallated form and zinc ion).
R. W. Henkens and J. M. Sturtevant, J. Amer. Chem. Soc. **90**, 2669 (1968).

f. $MA_2 + MB_2 \rightleftharpoons 2\,MAB$

A and B are different N,N-disubstituted dithiocarbamates (see Sec. 4.7.6(a)) M. Moriyasu and Y. Hashimoto, Bull. Chem. Soc. Japan **54**, 3374 (1981). J. Stach, R. Kirmse, W. Dietzsch, G. Lassmann, V. K. Belyaeva and I. N. Marov, Inorg. Chim. Acta **96**, 55 (1985).

g. $Co(en)_3^{3+} + OH^\bullet \rightarrow$ products

N. Shinohara and J. Lilie, Inorg. Chem. **18**, 434 (1979).

h. $Cu(II)$ azurin $+ Cu(I)$ azurin (self-exchange rate constant $\approx 10^6 M^{-1}s^{-1}$)

C. M. Groeneveld, S. Dahlin, B. Reinhammer and G. W. Canters, J. Amer. Chem. Soc. **109**, 3247 (1987) and previous references.

Chapter 4
Substitution Reactions

4.1 The Characteristics of Substitution Reactions

Substitution involves the replacement of a ligand coordinated to a metal by a free ligand in solution or the replacement of a coordinated metal ion by a free metal ion. No change of oxidation state of the metal occurs during the substitution, but a change may take place as a result of the substitution. The kinetics of the process have been studied for all the important stereochemistries but most intensely investigated for octahedral and square planar complexes. A very wide span of rates, almost 18 orders of magnitude, is found as indicated in Figure 4.1 which shows the water exchange rate constants for metal ions.[1] Thus the whole armory of

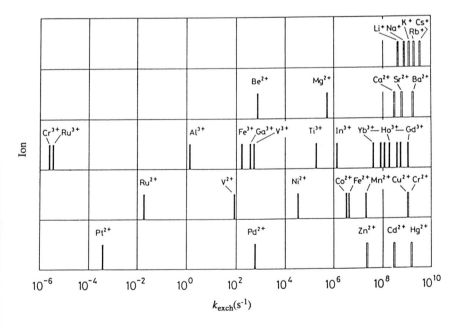

Fig. 4.1 Rate constants (s^{-1}) for water exchange of metal cations, measured directly by nmr or estimated from the rate constants for complex formation.[1] Reproduced with permission from Y. Ducommun and A. E. Merbach in Inorganic High Pressure Chemistry (R. von Eldik Ed), Elsevier, Amsterdam, 1986. ▬ nmr ▭ complex formation

techniques must be used to measure substitution rate constants. Metal ions or complexes that generally react rapidly (within a matter of seconds) are termed *labile,* whereas if they substitute slowly, taking minutes or longer for completion, they are considered *inert.*[2]

Ligand interchange in metal complexes can occur in two ways, either (a) by a combination of solvolysis and ligation e. g.[3]

$$Co(CN)_5Cl^{3-} + H_2O \rightarrow Co(CN)_5H_2O^{2-} + Cl^- \tag{4.1}$$

$$Co(CN)_5H_2O^{2-} + N_3^- \rightarrow Co(CN)_5N_3^{3-} + H_2O \tag{4.2}$$

or (b) by simple interchange in which there is a replacement of one ligand by another without the direct intervention of solvent, e. g.

$$Pt(dien)Br^+ + Cl^- \rightleftharpoons Pt(dien)Cl^+ + Br^- \tag{4.3}$$

Indirect substitution of the type indicated in (4.1) and (4.2) appears to be the method much preferred by octahedral complexes,[4] while direct substitution is more relevant with square-planar complexes. This situation could perhaps be predicted in view of the more crowded conditions with octahedral than with planar complexes. For other geometries both routes are used.

The substitution process permeates the whole realm of coordination chemistry. It is frequently the first step in a redox reaction[5] and in the dimerization or polymerization of a metal ion, the details of which in many cases are still rather scanty (e. g. for Cr(III)[6,7]). An understanding of the kinetics of substitution can be important for defining the best conditions for a preparative or analytical procedure.[8] Substitution pervades the behavior of metal or metal-activated enzymes. The production of apoprotein (demetalloprotein and the regeneration of the protein, as well as the interaction of substrates and inhibitors with metalloproteins are important examples[9].

4.1.1 Solvated Metal Ion

Before we consider substitution processes in detail, the nature of the metal ion in solution will be briefly reviewed.[10] A metal ion has a primary, highly structured, solvation sheath which comprises solvent molecules near to the metal ion. These have lost their translational degrees of freedom and move as one entity with the metal ion in solution. There is a secondary solvation shell around the metal ion, but the solvent molecules here have essentially bulk dielectric properties.[10,11] The (primary) solvation number n in $M(S)_n^{m+}$ of many of the labile and inert metal ions has been determined, directly by x-ray or neutron diffraction of *concentrated solutions,*[10,12] from spectral and other considerations and by examining the exchange process

$$M(S)_n^{m+} + {}^*S \rightleftharpoons M(S)_{n-1}({}^*S)^{m+} + S \tag{4.4}$$

From the ratio of the areas of nmr peaks due to coordinated and free solvent, or from simple isotopic analyses, the value of n can be determined.[13] It may be necessary to slow the exchange process (4.4) by lowering the temperature of the solution. A variety of solvation numbers n is observed, with four and six being the most prevalent. As we have noted already, there is a wide range of labilities associated with the solvent exchanges of metal ions (Fig. 4.1).

4.1.2 Representation of Substitution Mechanisms

The simplest type of replacement reaction is the exchange of a coordinated ligand by an identical free ligand, an important example of which arises when the ligand is a solvent molecule. [1, 14] The mechanisms we can visualize are presented in schematic form in Fig. 4.2. [1, 14] The larger circle represents the total coordination sphere of the metal ion (of any geometry) and the small circle labelled E and L represents an (identical) entering and leaving ligand molecule. If we can deduce by kinetics or other tests that there is an intermediate of higher or lower coordination number than in the reactant, the mechanisms are of the extreme types, denoted associative A or dissociative D respectively. [15] When the interchange is concerted and there is partial, and equal, association and dissociation of the entering and leaving groups, the mechanism is termed I. This will rarely occur and more likely there will be a preference for I_a or I_d, in which the entering and leaving groups are either firmly (I_a) or weakly (I_d) enbedded in the coordination sphere of the metal (Fig. 4.2). Microscopic reversibility considerations (Sec. 2.3.6) require that the activated complex be identical in both directions for this exchange reaction. As we have already implied, substitution in octahedral complexes is dissociatively activated, whereas with square planar complexes, associative activation is favored. Other factors, e. g. ligand crowding in the reactant, may modify these generalizations. Distinguishing mechanisms, I_d from D, I_a from A and particularly I_d from I_a, can be very difficult and it will be noted in this chapter that it is a preoccupation of the workers in this area. It is necessary to emphasize that these classifications are necessarily approximate and that there is a continuous range of behavior. [16]

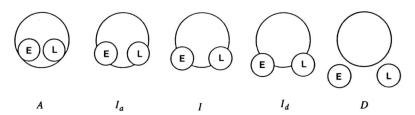

$$A \qquad\qquad I_a \qquad\qquad I \qquad\qquad I_d \qquad\qquad D$$

Fig. 4.2 Schematic representation of the mechanisms for substitution reactions. Based on Ref. 1.

4.2 Substitution in Octahedral Complexes [17]

The replacement of one unidentate ligand by another (particularly if the two ligands are identical) is the simplest substitution to envisage and has been used extensively to establish the rules of substitution mechansims. This becomes handy information in order to understand substitution involving chelates and macrocycles of increased complexity. From what has been already stated, at least one of the unidentate ligands will be a solvent molecule. In (4.1) the forward direction is variously referred to as solvation, solvolysis or dissociation and the reverse reaction is termed a formation (or anation, if the entering group is anionic) process. Mechanistic information can be obtained by studying the reaction in either direction (since these are intimately related (Sec. 2.3.6, Prob. 1) and particularly if the results are combined with those from solvent exchange studies. Substitution reactions of both labile and inert metal complexes have been investigated mainly, but not exclusively, [18] in aqueous solution. The earlier studies of the Werner-type complexes have been augmented by investigations of organometallic complexes, particularly of the metal carbonyls and related derivatives [19] although we shall not deal specifically with these.

4.2.1 Solvent Exchange with Metal Ions

Valuable information on mechanisms has been obtained from data on solvent exchange (4.4). [1, 14, 16, 20, 21] The rate law, one of the most used mechanistic tools, is not useful in this instance, unfortunately, since the concentration of one of the reactants, the solvent, is invariant. Sometimes the exchange can be examined in a "neutral" solvent, although this is difficult to find. [22] The reactants and products are however identical in (4.4), there is no free energy of reaction to overcome, and the activation parameters have been used exclusively, with great effect, to assign mechanism. This applies particularly to volumes of activation, since solvation differences are approximately zero and the observed volume of activation can be equated with the intrinsic one (Sec. 2.3.3).

(a) Divalent Metal Ions

Kinetic parameters are shown in Table 4.1 [23-25] for the exchange of the first-row (and one second-row [25]) divalent transition metal ions in water. Since $\Delta V^{\ddagger}$ for a D mechanism is $V_{MS_5} + V_S - V_{MS_6}$ and usually $V_{MS_5} < V_{MS_6}$, then $\Delta V^{\ddagger}$ will be less than V_S, the molar volume of the released solvent molecule.

Table 4.1 Kinetic Parameters for Water Exchange of Divalent Transition Metal Ions, $M(H_2O)_6^{2+}$ at $25\,^{\circ}C$ Refs. 23–25

	V^{2+}	Mn^{2+}	Fe^{2+}	Co^{2+}	Ni^{2+}	Ru^{2+}
k, s^{-1}	89	2.1×10^7	4.4×10^6	3.2×10^6	3.2×10^4	1.8×10^{-2}
$\Delta H^{\ddagger}$, kJ mol^{-1}	62	33	41	47	57	88
$\Delta S^{\ddagger}$, J K^{-1}mol^{-1}	-0.4	$+6$	$+21$	$+37$	$+32$	$+16$
$\Delta V^{\ddagger}$, cm^{3}mol^{-1} [a]	-4.1	-5.4	$+3.8$	$+6.1$	$+7.2$	-0.4
Electronic Config.	t_{2g}^3	$t_{2g}^3 e_g^2$	$t_{2g}^4 e_g^2$	$t_{2g}^5 e_g^2$	$t_{2g}^6 e_g^2$	t_{2g}^6
Ionic Radius, Å [b]	0.79	0.83	0.78	0.74	0.69	0.73

[a] On Basis of $\Delta \beta^{\ddagger} = 0$; [b] R.D. Shannon, Acta Crystallogr. Sect. A: Cryst. Phys. Diff. Theo. Gen. Crystallogr. **A32**, 751 (1976).

In aqueous solution therefore $\Delta V^{\ddagger}$ will be less than $+18$ cm^3 mol^{-1} and probably near $+9$ to $+11$ cm^3 mol^{-1} for a D mechanism, independent of ionic charge. [11, 16, 21] For an A mechanism $\Delta V^{\ddagger}$ will be negative to the extent of about -11 cm^{3}mol^{-1}. The increasingly positive values for $\Delta V^{\ddagger}$ from V^{2+} to Ni^{2+} (Table 4.1) signify therefore an increasingly dissociative mode for substitution. It appears that the designations I_a (V^{2+}, Mn^{2+}) $I(Fe^{2+})$ and $I_d(Co^{2+}$, $Ni^{2+})$ are most appropriate for water exchange with these metal ions. A strikingly similar pattern holds for exchange in MeOH[26] (we can add the I_d designation for Cu^{2+}) and CH$_3$CN,[27,28] Tables 4.2 and 4.3. The increasingly dissociative activation mode of exchange from Mn to Ni is accompanied in all three solvents by slower rates and increasingly positive $\Delta S^{\ddagger}$ and larger $\Delta H^{\ddagger}$ values. Since it is anticipated that the entering and leaving solvent molecules will be along the three-fold axes of the octahedral complex, it might be expected that the greater the t_{2g} electron density (see Table 4.1) the less likely the associative path will be electrostatically favored, which is precisely what occurs.[1,16] The relative inertness (low k and large $\Delta H^{\ddagger}$) for V^{2+} stems from its stable t_{2g}^3 electronic configuration. The high lability (large k and low $\Delta H^{\ddagger}$) for Cu^{2+} results from its tetragonal distortion.[26] When the size of the ligand (solvent) molecule increases, e. g. as in M(dmf)$_6^{2+}$, an associative path for exchange is obviously less favored on steric grounds, all $\Delta V^{\ddagger}$ values are positive (even Mn^{2+} [29,30]) and a dissociative activation mode is favored for all four ions. Perhaps even an extreme D mechanism is operative for Ni(dmf)$_6^{2+}$, Table 4.4.[27,29] Finally, the sequence of solvent exchange rate constants H$_2$O > dmf > CH$_3$CN > CH$_3$OH is metal-ion independent. Attempts to explain this sequence have so far been unsuccessful, but a relationship between $\Delta H^{\ddagger}$ for solvent exchange and ΔH_d, the heat of dissociation of solvent molecules S from MS$_6^{2+}$ has been suggested.[31]

Table 4.2 Kinetic Parameters for Methanol Exchange of First-Row Divalent Transition Metal Ions, M(CH$_3$OH)$_6^{2+}$ at 25 °C Ref. 26

	Mn^{2+}	Fe^{2+}	Co^{2+}	Ni^{2+}	Cu^{2+}
k, s^{-1}	3.7×10^5	5.0×10^4	1.8×10^4	1.0×10^3	3.1×10^7
$\Delta H^{\ddagger}$, kJ mol^{-1}	26	50	58	66	17
$\Delta S^{\ddagger}$, J K^{-1}mol^{-1}	-50	$+13$	$+30$	$+34$	-44
$\Delta V^{\ddagger}$, cm^{3}mol^{-1}	-5.0	$+0.4$	$+8.9$	$+11.4$	$+8.3$

Table 4.3 Kinetic Parameters for Acetonitrile Exchange of First-Row Divalent Transition Metal Ions, M(CH$_3$CN)$_6^{2+}$ at 25 °C Refs. 27, 28

	Mn^{2+}	Fe^{2+}	Co^{2+}	Ni^{2+}
k, s^{-1}	1.4×10^7	6.6×10^5	3.4×10^5	2.8×10^3
$\Delta H^{\ddagger}$, kJ mol^{-1}	30	41	50	64
$\Delta S^{\ddagger}$, J K^{-1}mol^{-1}	-9	$+5$	$+27$	$+37$
$\Delta V^{\ddagger}$, cm^{3}mol^{-1}	-7.0	$+3.0$	$+6.7$	$+7.3$

Table 4.4 Kinetic Parameters for Dimethylformamide Exchange of First-Row Divalent Transition Metal Ions, M(dmf)$_6^{2+}$ at 25 °C Refs. 27, 29

	Mn^{2+}	Fe^{2+}	Co^{2+}	Ni^{2+}
k, s^{-1}	2.2×10^6	9.7×10^5	3.9×10^5	3.8×10^3
$\Delta H^{\ddagger}$, kJ mol^{-1}	35	43	57	63
$\Delta S^{\ddagger}$, J K^{-1}mol^{-1}	-7	$+14$	$+53$	$+34$
$\Delta V^{\ddagger}$, cm^{3}mol^{-1}	$+2.4$	$+8.5$	$+6.7$	$+9.1$

(b) Trivalent Metal Ions

The exchange of the trivalent ions of the metals Cr, Fe, Ru and Ga in water is governed by the rate law

$$k_{exch} = a + b[H^+]^{-1} \tag{4.5}$$

from which exchange data for $M(H_2O)_6^{3+}$ and $M(H_2O)_5OH^{2+}$ may be extracted (Sec. 2.1.7(b)), Tables 4.5 and 4.6.[25,32,33] The negative signs of $\Delta V^{\ddagger}$ indicate an associative mechanistic mode for the transition metal ions $M(H_2O)_6^{3+}$, with $Ti(H_2O)_6^{3+}$ probably A, and the others I_a-controlled. In contrast, a dissociative mechanism I_d for $M(H_2O)_5OH^{2+}$ is supported by the pressure measurements. A strong labilizing effect of coordinated OH^- in $M(H_2O)_5OH^{2+}$, presumably on the trans H_2O, leads to a 10^2–10^3 fold enhanced rate for the hydroxy- over the hexaaqua ion and portends an important effect of coordinated ligands on the lability of other bound ligands (Sec. 4.3.3). Again, the greater the t_{2g} electron density, the less effective is associative activation. This is shown by increasingly less negative $\Delta V^{\ddagger}$s from Ti^{3+} to Fe^{3+}. The behavior of Ga^{3+} is included for comparative purposes.[32] The trend is however much less pronounced than with the bivalent metal ions. A comparison between iron and ruthenium is interesting. The stable low-spin configurations of Ru(II) and Ru(III) compared with their high-spin iron counterparts lead to dramatic reductions in exchange rates.[33] The two e_g electrons in Fe(II) and Fe(III) are missing in Ru(II) and Ru(III) and this favors more associative character for exchange with the latter ions. This is shown by more negative $\Delta V^{\ddagger}$'s for reaction of $Ru(H_2O)_6^{2+}$ and $Ru(H_2O)_6^{3+}$. General relationships of reactivity with

Table 4.5 Kinetic Parameters for Water Exchange of Trivalent Transition Metal Ions, $M(H_2O)_6^{3+}$ at 25°C Refs. 25, 32, 33

	Ti^{3+}	V^{3+}	Cr^{3+}	Fe^{3+}	Ru^{3+}	Ga^{3+c}
k, s^{-1}	1.8×10^5	5.0×10^2	2.4×10^{-6}	1.6×10^2	3.5×10^{-6}	4.0×10^2
$\Delta H^{\ddagger}, kJ\ mol^{-1}$	43	49	109	64	90	67
$\Delta S^{\ddagger}, J\ K^{-1}mol^{-1}$	+1	−28	+12	+12	−48	+30
$\Delta V^{\ddagger}, cm^3mol^{-1a}$	−12.1	−8.9	−9.6	−5.4	−8.3	+5.0
Electronic Config.	t_{2g}^1	t_{2g}^2	t_{2g}^3	$t_{2g}^3 e_g^2$	t_{2g}^5	$t_{2g}^6 e_g^4$
Ionic Radius, Å[b]	0.67	0.64	0.61	0.64	0.68	0.62

[a] On basis of $\Delta \beta^{\ddagger} = 0$; [b] R.D. Shannon, Acta Crystallogr. Sect. A: Cryst. Phys. Diffr. Theo. Gen. Crystallogr. **A32**, 751 (1976); [c] Included for comparative purposes.

Table 4.6 Kinetic Parameters for Water Exchange of Trivalent Transition Metal Ions, $M(H_2O)_5OH^{2+}$ at 25°C Refs. 25, 32, 33

	$CrOH^{2+}$	$FeOH^{2+}$	$RuOH^{2+}$	$GaOH^{2+}$
k, s^{-1}	1.8×10^{-4}	1.2×10^5	5.9×10^{-4}	$(0.6 - 2.0) \times 10^{5b}$
$\Delta H^{\ddagger}, kJmol^{-1}$	110	42	96	59
$\Delta S^{\ddagger}, JK^{-1}mol^{-1}$	+55	+5	+15	−
$\Delta V^{\ddagger}, cm^3mol^{-1a}$	+2.7	+7.0	+0.9	+6.2

[a] On basis of $\Delta \beta^{\ddagger} = 0$, [b] Range arises from the uncertainty of pK of $Ga(H_2O)_6^{3+}$

electronic configuration have been understood for some time, mainly on the basis of ligand reactions (next Section).[2,34] Crystal field considerations for example indicate kinetic inertness associated with the d^3, low spin d^4, d^5 and d^6 and d^8 electronic configurations.[2,B25] Undoubtedly, a much clearer picture of the intimate mechanism of reactions has emerged from the studies in the past decade of solvent exchange reactions.

4.2.2 The Interchange of Different Unidentate Ligands

Now we examine the replacement of a coordinated unidentate ligand L by a different unidentate ligand L_1. Either L or L_1 will be H_2O. The reaction can be examined in either direction. Figure 4.2

$$ML + L_1 \rightleftharpoons ML_1 + L \tag{4.6}$$

is still generally useful for depicting mechanism. We can consider the mechanism associatively activated if the reaction characteristics (activation parameters, steric effects, etc.) are more sensitive to a change of the entering group, then they are to the leaving group.[16] We can now obtain a meaningful rate law, but kinetic parameters are likely to be composite and less easy to evaluate than those from solvent exchange.

The *D* mechanism is represented by the scheme

$$ML \rightleftharpoons M + L \qquad k_1, \quad k_{-1} \tag{4.7}$$

$$M + L_1 \rightleftharpoons ML_1 \qquad k_2, \quad k_{-2} \tag{4.8}$$

in which a five-coordinated intermediate represented by M is generated with a sufficient lifetime to discriminate between L and L_1. The full rate law governing this mechanism has been referred to in Sec. 1.6.5. If we use L_1 in excess over ML and the reaction is irreversible $(k_{-2} \approx 0)$ then

$$-d(ML)/dt = d(ML_1)/dt = \frac{k_1 k_2 [ML][L_1]}{k_{-1}[L] + k_2[L_1]} \tag{4.9}$$

In the interchange mechanism, there is an interchange of L and L_1 perhaps within an outer-sphere complex $(ML \cdots L_1)$ which is very rapidly formed from the reactants

$$ML + L_1 \rightleftharpoons ML \cdots L_1 \qquad K_0, \text{ fast} \tag{4.10}$$

$$ML \cdots L_1 \rightarrow ML_1 \cdots L \qquad k_3 \tag{4.11}$$

$$ML_1 \cdots L \rightarrow ML_1 + L \qquad \text{fast} \tag{4.12}$$

The extent of influence of L_1 on the k_3 process, will dictate the applicable designation I_d, I or I_a. For this reaction scheme,

$$d(ML_1)/dt = \frac{k_3 K_0 [ML]_0 [L_1]_0}{1 + K_0 [L_1]_0} \tag{4.13}$$

where the subscripts 0 indicate the total (starting) concentrations of the species. For an A mechanism, in which there is a seven-coordinated intermediate or activated complex, a second-order rate law obtains,

$$ML + L_1 \overset{k}{\rightleftharpoons} [MLL_1]^{\ddagger} \rightleftharpoons ML_1 + L \tag{4.14}$$

$$d[ML_1]/dt = k[ML][L_1] \tag{4.15}$$

4.2.3 Outer Sphere Complexes

Before considering the kinetics associated with the various mechanisms, a discussion of the outer-sphere complex is necessary, since it features so prominently in the interchange mechanism (4.10)–(4.12). The secondary interaction of an inner-sphere complex with ligands in solution to give an outer-sphere complex as depicted in (4.10) is most effective between oppositely charged species (ion pairs). The presence of an outer-sphere complex, a term first coined by Alfred Werner in 1913, is easily demonstrated in a number of systems. Rapid spectral changes in the 200–300 nm region, which can be ascribed to outer-sphere complexing, occur on addition of a number of anions to $M(NH_3)_5H_2O^{3+}$, where M = Cr or Co,[35] *long* before final equilibration to the inner-sphere complex occurs. For example,

$$Co(NH_3)_5H_2O^{3+} + N_3^- \overset{K_0}{\rightleftharpoons} Co(NH_3)_5H_2O^{3+} \cdots N_3^-$$
$$\text{Outer-sphere complex}$$
$$\downarrow$$
$$Co(NH_3)_5N_3^{2+} + H_2O \tag{4.16}$$
$$\text{Inner-sphere complex}$$

The separation of the two stages is easier to discern when the rates of the two processes are so different, but it can also be seen in the ultrasonic spectra of metal-sulfate systems (Sec. 3.4.4). Ultrasonic absorption peaks can be attributed to formation of outer-sphere complexes (at higher frequency, shorter τ) and collapse of outer-sphere to inner-sphere complexes (at lower frequency). In addition to uv spectral and ultrasonic detection, polarimetry and nmr methods have also been used to monitor and measure the strength of the interaction. There are difficulties in assessing the value of K_0, the outer-sphere formation constant. The assemblage that registers as an ion pair by conductivity measurements may show a blank spectroscopically.[16] The value of K_0 at T K may be estimated using theoretically deduced expressions:[36]

$$K_0 = \frac{4\pi N a^3}{3000} \exp\left(-\frac{U(a)}{kT}\right) \tag{4.17}$$

where $U(a)$ is the Debye-Hückel interionic potential

$$U(a) = \frac{z_1 z_2 e^2}{aD} - \frac{z_1 z_2 e^2 \kappa}{D(1 + \kappa a)} \tag{4.18}$$

$$\kappa^2 = \frac{8\pi N e^2 \mu}{1000\, D k T}$$ (4.19)

and

N	=	Avogadro's number.
a	=	distance of closest approach of two ions (cm).
k	=	Boltzmann's constant (erg).
e	=	charge of an electron in esu units.
D	=	bulk dielectric constant.
μ	=	ionic strength.
z_1, z_2	=	charge of reactants.

To give some idea of the value of K_0 — it is approximately 14 M^{-1} for interaction of $2+$ and $2-$ charged reactants at $\mu = 0.1$ M and 0.15 M^{-1} for interaction between a cation and a zero charged species.[37] It is necessary to reaffirm the point that the occurrence of outer-sphere complexes, which can be observed in the studies of the $Fe(III)-Br^-$, $Ni(II)-CH_3PO_4^{2-}$ and the ultrasonics of a number of $M^{2+}-SO_4^{2-}$ systems,[38] does not necessitate their being in the direct pathway for the formation of products (Sec. 1.6.4). Their appearance does not help in the deciphering of the mechanism.

4.2.4 Characteristics of the Various Mechanisms

Although the rate laws derived from the three mechanisms assume distinctly different forms, (4.9), (4.13) and (4.15), the assignment of the mechnisms on the basis of these alone is difficult. If reaction (4.6) is studied using an excess of L_1, the rate laws shown in Table 4.7 will be observed for different concentrations of L_1 for the D and I mechnisms. For the A mechanism the second-order rate law holds for all concentrations of L_1. At low $[L_1]$, *all* mechanisms give second-order rate behavior with composite rate constants for D and I mechanisms. Only if the mechanism is D, is the rate slowed down by L (mass law retardation). This can be used for diagnosing a D mechanism for L, L_1 interchange reactions in nonaqueous solvents (see however Ref. 19) but for the ligand replacement of coordinated solvents, commonly studied, [L], the solvent, is constant and the ability to use this feature to differentiate amongst the mechnisms is lost. For D and I mechanisms. the linear dependence of k_{obs} on $[L_1]$ at low $[L_1]$ may be replaced by an independence at high $[L_1]$. The limiting rate constants will be k_1 for the breakage of the ML bond or k_3, for the interchange within the outer-sphere complex. The experimental limiting first-order rate constant is unlikely to be useful for distinguishing k_1 and k_3 (in an I_d mechanism) but in any case its value should be close to the value of k_{exch}.[16,39,40]

$$M—H_2O + H_2{}^*O \rightleftharpoons M—H_2{}^*O + H_2O \qquad k_{exch} \qquad (4.20)$$

Unfortunately even a slight deviation from linearity of k_{obs} with $[L_1]$ is rarely observed.

Table 4.7 Rate Laws for Substitution Mechanisms with Different Conditions

	Low $[L_1]$	Medium $[L_1]$	High $[L_1]$
D	$k_2[L_1] < k_{-1}[L]$ M scavenged by L preferentially	$k_2[L_1] \sim k_{-1}[L]$	$k_2[L_1] > k_{-1}[L]$ M scavenged by L_1 preferentially
Rate	$\dfrac{k_1 k_2 [ML][L_1]}{k_{-1}[L]}$	$\dfrac{k_1 k_2 [ML][L_1]}{k_{-1}[L] + k_2[L_1]}$	$k_1[ML]$
I	$K_0[L_1]_0 < 1$ Small build up of outer- sphere complex	$K_0[L_1]_0 \sim 1$	$K_0[L_1] > 1$ Formation of outer-sphere complex complete
Rate	$k_3 K_0 [ML]_0 [L_1]_0$	$\dfrac{k_3 K_0 [ML]_0 [L_1]_0}{1 + K_0[L_1]_0}$	$k_3[ML]_0$

4.2.5 The Limiting First-Order Rate Constant

Only in a relatively few systems are deviations from constancy for the function $k_{obs}/[L_1]$ observed and in even fewer, is a point reached in which k_{obs} is a constant, independent of $[L_1]$.

In the replacement of H_2O by SCN^- in the Co(III) porphyrin **1** abbreviated CoP^{5+}

$$Co(tmpyp)(H_2O)_2^{5+} + SCN^- \rightarrow Co(tmpyp)(H_2O)(SCN)^{4+} + H_2O \qquad (4.21)$$

the pseudo first-order rate constant k_{obs} vs. $[SCN^-]$ (used in excess) is shown in Figure 4.3.[41] The further replacement of the water in the product by SCN^- is very rapid. The rate law

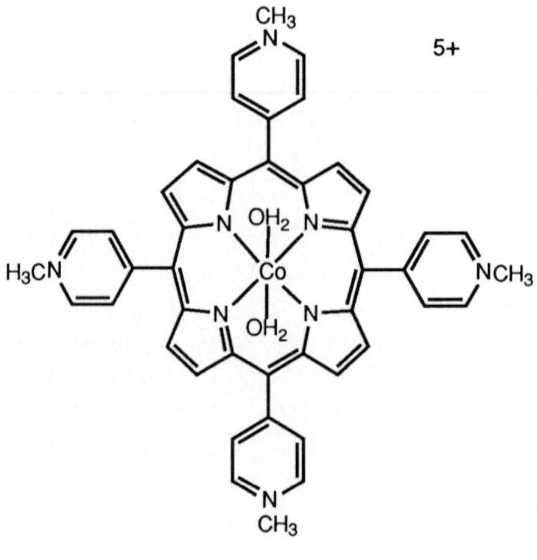

1

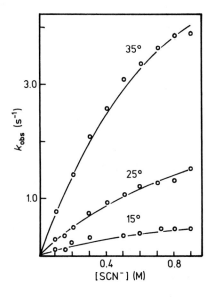

Fig. 4.3 The dependence of the pseudo first-order rate constants upon anion concentration for the anation of $Co(tmpyp)(H_2O)_2^{5+}$ by SCN^- at 15, 25, and 35 °C in 0.1 M H^+, $\mu = 1.0$ M.[41] The solid lines conform to Eqn. (4.22). Reproduced with permission from K. R. Ashley, M. Berggren and M. Cheng, J. Amer. Chem. Soc. **97**, 1422 (1975) © (1975) American Chemical Society.

(4.22) therefore applies and an A mechanism is eliminated. For a D mechanism, $a = k_1$ and $b = k_{-1}/k_2$. For an I mechanism,

$$-d[CoP^{5+}]_{total}/dt = \frac{a[SCN^-]}{b + [SCN^-]}[CoP^{5+}]_{total} \qquad (4.22)$$

$a = k_3$ and $b = K_0^{-1}$ (Table 4.7, Medium [L_1] Condition). The value of K_0 which results from this analysis (0.83 M^{-1} at 25 °C) appears small for a $+5$, -1 reaction pair in an I mechanism, although the effective charge is probably markedly reduced from $+5$ since it is smeared over the whole porphyrin.[41] Better evidence for the D mechanism than simply by default, comes from studies of the pressure effect on (4.21) using the high-pressure stopped-flow technique (Sec. 3.3.3).[42] On the basis of the rate law (4.22), which was confirmed, the variation of k_{obs}, the pseudo first-order rate constant, with pressure P will be given by (Sec. 2.3.3)

$$k_{obs} = \frac{a\exp(-P\Delta V_a^{\ddagger}/RT)[SCN^-]}{b\exp(-P\Delta V_b^{\ddagger}/RT) + [SCN^-]} \qquad (4.23)$$

For a D mechanism, (4.24) and (4.25) the value of $\Delta V_a^{\ddagger} (= \Delta V_1^{\ddagger}) = 14 \pm 4$ cm^3 mol^{-1} is reasonable for the loss of one H_2O molecule in the activated complex. The value of

$$Co(tmpyp)(H_2O)_2^{5+} \rightleftharpoons Co(tmpyp)(H_2O)^{5+} + H_2O \quad \Delta V_1^{\ddagger}, \Delta V_{-1}^{\ddagger} \qquad (4.24)$$

$$Co(tmpyp)(H_2O)^{5+} + SCN^- \rightarrow Co(tmpyp)(H_2O)SCN^{4+} \quad \Delta V_2^{\ddagger} \qquad (4.25)$$

$\Delta V_b^{\ddagger}(= \Delta V_{-1}^{\ddagger} - \Delta V_2^{\ddagger}) = 5 \pm 4$ cm^{3}mol^{-1} is also plausible. On the other hand for an I_d mechanism (4.26) and (4.27) the large value of $\Delta V_a^{\ddagger} (= \Delta V_3^{\ddagger})$ for the interchange step and the negative value for $\Delta V (= -\Delta V_b^{\ddagger})$ are quite unlikely.[42,43] See Ref. 44.

$$\text{Co(tmpyp)(H}_2\text{O)}_2^{5+} + \text{SCN}^- \rightleftharpoons \text{Co(tmpyp)(H}_2\text{O)}_2^{5+} \cdots \text{SCN}^- \qquad \Delta V \qquad (4.26)$$

$$\text{Co(tmpyp)(H}_2\text{O)}_2^{5+} \cdots \text{SCN}^- \rightarrow \text{Co(tmpyp)SCN}^{4+} + \text{H}_2\text{O} \qquad \Delta V_3^{\ddagger} \qquad (4.27)$$

All the kinetic features expected for a D mechanism and rate law (4.9) i.e. marked effects of L and L_1 on the rate constants, are shown in the comprehensive studies in nonaqueous solution of substitution in low-spin Fe(II) complexes of the type FeN_4XY where N_4 are planar porphyrins, phthalocyanins and macrocycles and X and Y are neutral ligands, CO, R_3P, pyridines etc. Small discrimination factors (k_{-1}/k_2) suggest that the five-coordinated intermediate in these systems is very reactive.[45,46] There have been problems in the confirmation of curvature in the plots of $k_{obs}/[L_1]$ for "classical" reactions of a number of aquapentammine complexes.[47]

4.2.6 Second-Order Rate Constants

We have to nearly always use the activation parameters from the second-order rate law to differentiate between mechanisms. For a single reaction, the kinetic parameters are of little use and it is usually necessary to compare the behavior of a number of reaction systems. The ploy then is to deduce with which mechanism the kinetic data are most consistent. Some values for k_3 for reactions of Ni^{2+} ion with a variety of unidentate ligands, calculated from the experimental rate constant $(k_3K_0$, see Table 4.7) using an estimation of K_0, are contained in Table 4.8[40]. The values of k_3 are reasonably constant, close to the water exchange rate constant (Table 4.1), and these results represent strong support for an I_d mechanism. It was this type of evidence that Eigen used to propose the ion-pair mechanism for the reactions of a number of bivalent metal ions[48]. It is difficult to distinguish an I_d from a D mechanism, although $\Delta V^{\ddagger}$ values for solvent exchange and ligation by neutral ligands of $\text{Ni(H}_2\text{O)}_6^{2+}$ also support an I_d mechanism.[49] The situation appears different with the bulkier dmf ligand. The computed value of $k_3 > 1.4 \times 10^4\text{s}^{-1}$ for reaction of Ni(dmf)_6^{2+} with SCN^- and Et_2dtc^- is substantially greater[50] than the solvent exchange value $k_{exch} = 3.8 \times 10^3\text{s}^{-1}$ (Table 4.4). An I_d mechanism is ruled out and a D mechanism favored by default.[51] The values of $\Delta V^{\ddagger}$ $(+8.8$ to $+12.4 \text{ cm}^3\text{mol}^{-1})$ for reaction of a number of ligands with Ni(dmf)_6^{2+} also support a dissociative mechanism.[51]

Table 4.8 Computed Values for k_3 from Second-Order Rate Constants (K_0k_3) for the Formation of Nickel(II) Complexes from Unidentate Ligands at 25°C Ref. 40

L^{n-}	$10^{-3} \times K_0k_3$ $M^{-1}s^{-1}$	K_0 M^{-1}	$10^{-4} \times k_3$ s^{-1}
$\text{CH}_3\text{PO}_4^{2-}$	280	40[a]	0.7[a]
CH_3CO_2^-	300	20[a]	1.5[a]
NH_3	4.5	0.15[b]	3.0
$\text{C}_5\text{H}_5\text{N}$	3.6	0.15[b]	2.0
$\text{NH}_2(\text{CH}_2)_2\text{NMe}_3^+$	0.4	~ 0.02[b]	~ 2

[a] These values are directly determined from relaxation data.
[b] Estimated values (see text).

Associative mechanisms are indicated by the second-order rate constant showing a decided dependence on the nucleophilicity (or basicity) of the entering ligand. We have already noted this in the reactions of $Fe(H_2O)_6^{3+}$ (Sec. 2.1.7) and find that a similar situation holds for $Cr(H_2O)_6^{3+}$,[52] $V(H_2O)_6^{3+}$ [53] and $Ti(H_2O)_6^{3+}$ [33] ions. The I_a assignment for substitution in these metal ions is satisfyingly consistent with their activation parameters for water exchange (Table 4.5). It has been suggested that in replacement of coordinated H_2O by SCN^- and Cl^- ions in any complex, if the ratio of their second-order rate constants (r) exceeds 10, $k(SCN^-)/k(Cl^-) > 10$, an I_a mechanism is indicated.[54] This is a useful rule. Values of r exceeding unity are usually associated with metal ions which have negative values for $\Delta V^{\ddagger}$ for water exchange again supporting associative activation. The selectivity indicated in the r value declines as the rates become faster.[16] Thus, r decreases in the order $Cr(H_2O)_6^{3+}$, $Fe(H_2O)_6^{3+}$ and probably $Mn(H_2O)_6^{2+}$, even though solvent exchange data indicate an I_a mechanism is operative in all cases. This mild selectivity for entering ligands has caused interpretation problems in the assignment of mechanism for the most labile metal ions.[16]

Examining the relationship between the hydrolysis rate constants (k_1) and the equilibrium constant (K_1) for a series of reactions of the type (4.28) and (4.29) involving charged ligands X^{n-} has been very helpful in delineating the type of I mechanism.

$$M(NH_3)_5X^{(3-n)+} + H_2O \rightleftharpoons M(NH_3)_5H_2O^{3+} + X^{n-} \quad k_1, k_{-1}, K_1 \tag{4.28}$$

$$M(H_2O)_5X^{(3-n)+} + H_2O \rightleftharpoons M(H_2O)_6^{3+} + X^{n-} \quad\quad k_1, k_{-1}, K_1 \tag{4.29}$$

For these reactions the general LFER holds

$$\log k_1 = a \log K_1 + b \tag{4.30}$$

With an I_d mechanism, the rate constants for anation (k_{-1}) are approximately constant and any differences in the formation constants K_1 reside in differing k_1 values since $K_1 = k_1/k_{-1}$. This requires that $a = 1.0$ as we have already seen for $Co(NH_3)_5X^{2+}$ (Sec. 2.4).[55] Simple reasoning indicates that for an I_a mechanism $a = 0.5$.[20,56] The decreasing values of a for the hydrolyses of $Co(NH_3)_5X^{2+}$ (1.0), $Cr(NH_3)_5X^{2+}$ (0.69) and $Cr(H_2O)_5X^{2+}$ (0.58) suggest increasing associative character (I_a) for these reactions. This is supported by the values of $\Delta V^{\ddagger}$ ($+1.2$, -5.8 and -9.3 cm^{3}mol^{-1} respectively) for the water exchange of the corresponding aqua ion.[20] See also Ref. 57.

4.2.7 Summary

Since we shall not obtain the comparable amount of detailed information on the mechanisms of substitution in octahedral complexes from the studies of more complicated substitutions involving chelation and macrocycle complex formation (Secs. 4.4 and 4.5) it is worthwhile summarizing the salient features of substitution in Werner-type complexes.

Associative (A) mechanisms are extremely rare and it is uncertain whether an authentic example exists.[58] Dissociative (D) mechanisms are more common although difficult to establish. Some examples were cited in Secs. 4.2.5 and 4.2.6. Thus interchange (I) mechanisms dominate the scene. This leads to the following generalizations:

(a) Relatively small influence of an entering group on the rate or rate law.

(b) Parallel rate constants for substitution and water exchange for a large number of complexes. Equality of k_3 and k_{exch}, or even better of the associated parameters $\Delta H_3^{\ddagger}$ and $\Delta H_{exch}^{\ddagger}$, is strong evidence for an I_d mechanism.

(c) Correlation of hydrolysis rate with the binding tendencies of the leaving group, leading to a variety of LFER involving activation and reaction parameters.

(d) Decrease of rate with an increase in the charge of the complex (Fig. 4.1) since bond rupture is an important component, even of an I_a mechanism.

(e) Steric acceleration and deceleration effects (Sec. 2.4).

Attempts to improve the simple mechanistic classification have been made using More-O'Ferall diagrams, [16] or transition state bond order variations. [59]

4.3 Accelerated Substitution of Unidentate Ligands

Reagents such as H^+, OH^-, metal ions and ligands may alter the rate of replacement of one ligand by another. These reagents act either by modifying the structure of one of the reactants, or by direct participation in the transition state (and the difference may be a subtle one and difficult to diagnose; see Sec. 2.3.6). It is important to establish that these reagents are promoting another reaction pathway and not just producing a medium effect (see Sec. 2.9.2). If the reagent is not used up in the reaction, the accelerating effect is termed *catalytic*. If on the other hand it is consumed, perhaps ending up in the product, the terms *reagent-accelerated* or *-assisted* are more appropriate. We shall deal in the next section with the catalytic and accelerated replacement of unidentate ligands and the ideas developed will be then incorporated in the discussion of chelation and macrocycle complex formation (Sec. 4.4 and 4.5).

4.3.1 H^+-Assisted Removal

Studies on the removal of unidentates in acid media have been made mainly with inert complexes. It might be expected that the removal of ligands that retain some basicity, even when coordinated, would be acid-promoted, e.g.

$$CrX^{2+} + H^+ \rightleftharpoons CrXH^{3+} \qquad\qquad K \qquad\qquad (4.31)$$

$$CrX^{2+} + H_2O \rightarrow Cr^{3+} + X^- \qquad\qquad k_0 \qquad\qquad (4.32)$$

$$CrXH^{3+} + H_2O \rightarrow Cr^{3+} + HX \qquad\qquad k_1 \qquad\qquad (4.33)$$

for which,

$$-d[CrX^{2+} + CrXH^{3+}]/dt = k_{obs}[CrX^{2+} + CrXH^{3+}] \qquad\qquad (4.34)$$

where

$$k_{obs} = \frac{k_0 + k_1 K[H^+]}{1 + K[H^+]} \qquad\qquad (4.35)$$

Sometimes appreciable amounts of $CrXH^{3+}$ build up and the full rate law (4.35) is applicable, as with $X = CH_3CO_2^-$, Ref. 60. Normally however, $K \ll [H^+]$, and

$$k_{obs} = k_0 + k_1 K[H^+] \tag{4.36}$$

Placing a proton on the X group presumably weakens the Cr-X bond. The enhanced lability is due to a reduced enthalpy of activation, compared with that associated with the k_0 step.[60] Normally, the removal of unidentate ammonia or amine ligands from metal complexes is not accelerated by acid, since the nitrogen is coordinately saturated. This situation changes when we consider multidentates (Sec. 4.4.2).

4.3.2 Metal Ion-Assisted Removal

Metal ions can function much like protons, and coordinated ligands whose removal are accelerated by H^+ (Sec. 4.3.1), are often ones (N_3^-, CN^-) whose loss are also metal ion-assisted. Metal ions can also speed up the removal of an additional type of ligand (NCS^-, Cl^-) that shows a strong tendency to form bridged binuclear complexes. The catalytic efficiency of a metal ion depends on a number of factors. There is a close correlation of the rate of accelerated aquation by M^{n+} with the complexing ability of M^{n+}. Hard metal ions (Be^{2+}, Al^{3+}), like H^+, readily remove the hard ligands, such as F^-. Soft metal ions (Hg^{2+}, Ag^+) are most effective when the leaving ligand is soft (Cl^-, Br^-).[61] Substitution-inert metal ions or complexes are usually ineffective.

The phenomenon has been mainly explored using inert Co(III) and Cr(III) complexes with Hg(II) and Tl(III) as the accelerating ions, and the leaving groups are usually halides, pseudohalides, alkyls and carboxylates.[61] The majority of these induced aquations follow simple second-order kinetics. At high inducing metal ion concentration, deviations from second-order behavior might be expected with the (rapid) appearance of an adduct (exactly as might be observed with H^+ catalysis) e. g.

$$cis\text{–}Co(en)_2Cl_2^+ + Hg^{2+} \xrightarrow{K_1} Co(en)_2Cl_2Hg^{3+} \tag{4.37}$$

with k_3 and k_2 pathways leading to:

$$cis\text{–}Co(en)_2(H_2O)Cl^{2+} + HgCl^+$$

$$V = \frac{a[Hg^{2+}][Co(III)]}{1 + K_1[Hg^{2+}]} \tag{4.38}$$

Fruitful interaction might occur via the adduct ($a = k_2K_1$) or be extraneous to adduct formation ($a = k_3$).[62] The two mechanisms are not easily distinguished, since they lead to the same kinetics (Sec. 1.6.4). See also Ref. 63.

The Hg(II) assisted aquation of Co(III)-chloro complexes has been throughly studied to gain insight into the effects of solvent, ionic strength and polyelectrolytes on reaction rates and equilibria.[61] For the two reactions in 1.0 M $HClO_4$ (4.39) and (4.40), $(N)_5$ representing five nitrogen donors in unidentates or multidentates or mixtures thereof,

$$Co(N)_5Cl^{2+} + Hg^{2+} \rightarrow Co(N)_5H_2O^{3+} + HgCl^+ \qquad k_{Hg^{2+}} \qquad (4.39)$$

$$Co(N)_5Cl^{2+} + H_2O \rightarrow Co(N)_5H_2O^{3+} + Cl^- \qquad k_{H_2O} \qquad (4.40)$$

a LFER for $\log k_{Hg^{2+}}$ vs $\log k_{H_2O}$ (slope 0.96) is constructable for 34 complexes, Fig. 2.6. This relationship suggests a similar 5-coordinated species as an intermediate in both these dissociatively-activated reactions.[64] When the removal of coordinated halides is speeded up with metal ions, products not normally obtained in aquation may result.[65] Mercury(II) is useful for producing chelated esters for hydrolytic examination (Sec. 6.3.1) and to probe for intermediates (Sec. 2.2.1 (b)).

4.3.3 Ligand-Assisted Removal

Anions can promote hydrolysis of complex cations by producing ion pairs of enhanced reactivity (see 2.178). Usually however, ligands accelerate the removal of a coordinated ligand by entering the metal coordination sphere with it and thereby labilizing it towards hydrolysis. We have already seen the effect of coordinated OH^- on the enhanced labilities of Fe(III) and Cr(III). Dissociative mechanisms and considerable acceleration are promoted by CH_3, CN^-, SO_3^{2-} and other groups on inert Cr(III), Co(III) and Pt(IV) complexes.[66] Nitrate ions, for example, reduce the half-life for replacement of water in $Cr(H_2O)_6^{3+}$ by dmso from ~ 380 h to 10 s![67]

In some cases, the unidentate ligand is liberated at the end of the reaction. Usually, however, the ligand is found in both the reactant and the product. The effect has been most systematically examined for Ni(II).[21,40] Coordinated NH_3 and polyamines have the largest accelerating influence. The rate acceleration induced by macrocycles resides primarily in reduced $\Delta H^{\ddagger}$ values (by 15–26 kJ mol^{-1}). The 6- and 5-coordination of solvated tetramethylcyclam complexes is controlled by the conformation at the 4 N-centers, **2** and **3**. These complexes exchange by I_d and I_a mechanisms, respectively, as indicated by positive and negative $\Delta S^{\ddagger}$ values (Table 4.9). Also Sec. 4.9.

Table 4.9 Water Exchange Rate Constants[a] for a Number of Nickel-(II) Complexes at 25°C. Refs. 21 and 40.

Complex	$10^{-5}k$, s^{-1}	Complex	$10^{-5}k$, s^{-1}
$Ni(H_2O)_6^{2+}$	0,32	$Ni(12[ane]N_4)(H_2O)_2^{2+}$	200
$Ni(NH_3)(H_2O)_5^{2+}$	2.5	$Ni(Me_4cyclam)(D_2O)_2^{2+\,b}$	1600
$Ni(NH_3)_2(H_2O)_4^{2+}$	6.1	$Ni(Me_4cyclam)(D_2O)^{2+\,c}$	160
$Ni(NH_3)_5(H_2O)^{2+}$	43	$Ni(bpy)(H_2O)_4^{2+}$	0.49
$Ni(2,3,2\text{-tet})(H_2O)_2^{2+}$	40	$Ni(tpy)(H_2O)_3^{2+}$	0.52

[a] For exchange of single solvent molecule [b] For *RRSS* form **2** of macrocycle, $\Delta H^{\ddagger} = 37.4$ kJ mol^{-1} and $\Delta S^{\ddagger} = +38$ J K^{-1}mol^{-1} [c] For *RSRS* **3** which forms 5-coordinated solvated species $\Delta H^{\ddagger} = 27.7$ kJ mol^{-1} and $\Delta S^{\ddagger} = -24$ J K^{-1}mol^{-1}.

There are linear correlations between $\log k$ (formation) and certain properties of the ligand (number of nitrogen atoms[17] or electron-donor constant[68]). The enhanced rate resides largely

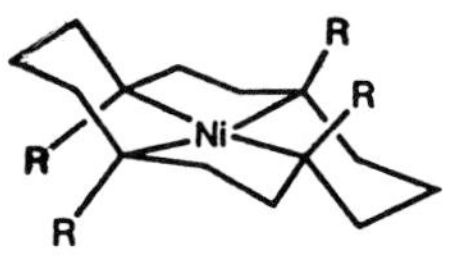

Trans III	Trans I
RRSS	RSRS
2	3

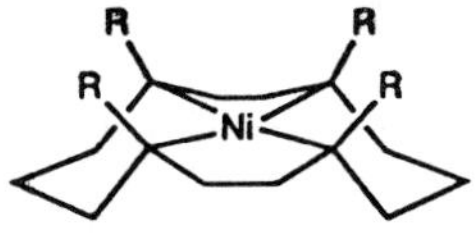

in the k_3 term in (4.11). Other coordinating groups such as aminocarboxylates and heterocycles, bpy, etc. have much less labilizing influence (Table 4.9). This behavior contrasts sharply with the pronounced effect that coordinated edta and related ligands have on the rates of substitution of the waters attached to Ti(III), Cr(III), Fe(III), Ru(II) and Ru(III), Co(III) and Os(III). Factors of $10^6–10^8$ enhanced substitution rates, compared with the hexaaqua ions have been reported.[69] Porphyrins also accelerate dramatically the substitution of the axial unidentate group in Cr(III), Fe(III) and Co(III) complexes.[70]

We have been concerned in this section with the formation of adducts and reactions of the type (S_x and S_y representing different ligand entities):

$$S_xML + L_1 \rightleftharpoons S_yMLL_1 \qquad k_1 \qquad\qquad (4.41)$$

We are interested in the effect of L, compared with S_x, on the rate constant (k_1) for the process. Ternary complex formation depicted in (4.41) has been actively studied, to a large extent because of the biological implications of the results.[71,72]

4.3.4 Base-Assisted Removal

The hydroxide ion can modify the reactivity of a system in acid medium. This has been known for a long time and an example is used in Section 1.1. The ability of hydroxide to modify a reactant is probably most important in the base-assisted hydrolysis of metal ammine and amine complexes.[73] The overwhelming bulk of these studies have been with Co(III), for example,

$$Co(NH_3)_5X^{2+} + OH^- \rightarrow Co(NH_3)_5OH^{2+} + X^- \qquad\qquad (4.42)$$

and these will be considered first. The kinetics are usually second-order,

$$V = k_{OH}[Co^{III}][OH^-] \qquad\qquad (4.43)$$

a rate law which is maintained in up to 1 M OH$^-$, at which point flow methods must be used to follow the rapid rates. Although a number of mechanisms have been suggested to explain these simple kinetics,[73] there is overwhelming support for a conjugate base mechanism for the majority of systems studied.

(a) The Conjugate Base Mechanism. As originally proposed by Garrick,[74] base removes a proton from the ammonia or amine ligand in a rapid preequilibrium to form a substitutionally labile amide complex

$$\text{Co(NH}_3)_5\text{X}^{2+} + \text{OH}^- \rightleftharpoons \text{Co(NH}_3)_4(\text{NH}_2)\text{X}^+ + \text{H}_2\text{O} \quad k_1, k_{-1}, K_1 \tag{4.44}$$

Unimolecular solvolysis of this conjugate base in steps (4.45) and (4.46) produces an aqua amide complex that rapidly converts to the final product (4.47):

$$\text{Co(NH}_3)_4(\text{NH}_2)\text{X}^+ \rightarrow \text{Co(NH}_3)_4\text{NH}_2^{2+} + \text{X}^- \quad k_2 \tag{4.45}$$

$$\text{Co(NH}_3)_4\text{NH}_2^{2+} + \text{H}_2\text{O} \rightarrow \text{Co(NH}_3)_4(\text{NH}_2)\text{H}_2\text{O}^{2+} \quad \text{fast} \tag{4.46}$$

$$\text{Co(NH}_3)_4(\text{NH}_2)\text{H}_2\text{O}^{2+} \rightarrow \text{Co(NH}_3)_5\text{OH}^{2+} \quad \text{fast} \tag{4.47}$$

This mechanism termed D_{cb} (formerly S_N1CB) was developed by Basolo and Pearson and their groups in the 1950's in the face of a good deal of healthy opposition from Ingold, Nyholm, Tobe and their co-workers who favored a straightforward A (S_N2) attack by OH^- ion on the complex.[75] For a D_{cb} mechanism, in general,

$$V = d\,[\text{Co(III)}]/dt = k_{\text{OH}}[\text{Co(III)}]\,[\text{OH}^-] = \frac{nk_1 k_2}{k_{-1} + k_2}[\text{Co(III)}]\,[\text{OH}^-] \tag{4.48}$$

where there are n equivalent amine protons in the cobalt reactant and assuming a steady-state concentration of the conjugate-base. Some of the evidence for the various steps proposed in the conjugate base mechanism will now be considered.

1. The base-catalyzed exchange of hydrogen between the cobalt amines and water demanded by equilibrium (4.44) has been amply demonstrated. Normally all the exchange will proceed by (4.44) but when k_2 and k_{-1} are similar in magnitude, the amount of H exchange between solvent and reactant will be less than the amount of exchange between solvent and product. This rarely has been observed.[76] If there is a build-up of conjugate base, i.e. $K_1[\text{OH}^-] \approx 1$, or of an ion-pair of the Co(III) substrate with OH^-, it is easy to show that there will be a deviation from linearity of the $V/[\text{OH}^-]$ plot. Again, this is rarely observed.[73,77] Normally, $k_{-1} \gg k_2$ and $k_1 > k_{\text{OH}}$. With these conditions, proton-transfer in (4.44) is a preequilibrium and

$$V = nK_1 k_2[\text{Co(III)}]\,[\text{OH}^-] \tag{4.49}$$

In the unlikely event that $k_2 \gg k_{-1}$, $k_{\text{OH}} = nk_1$ and the act of deprotonation becomes rate limiting, affording powerful evidence for the necessity of (4.44) in the base reaction.[78] Changes of rds with conditions show up in changing values of $\Delta H^{\ddagger}$ (but surprisingly not $\Delta V^{\ddagger}$) with temperature (Sec. 2.6).

2. An intermediate of the type $\text{Co(NH}_3)_4\text{NH}_2^{2+}$ is postulated in (4.45). The subsequent reactions of this intermediate should be independent of the nature of the X group in the starting material. The results of early experiments to verify this point have represented some of the most powerful support for the D_{cb} mechanism. Base hydrolysis of $\text{Co(NH}_3)_5\text{X}^{2+}$ in an $\text{H}_2^{16}\text{O}/\text{H}_2^{18}\text{O}$ mixture was found, as required, to give a constant proportion of $\text{Co(NH}_3)_5{}^{16}\text{OH}^{2+}$ and $\text{Co(NH}_3)_5{}^{18}\text{OH}^{2+}$, independent of X^- being Cl^-, Br^- or NO_3^-.[79] The competition experiments described in Sec. 2.2.1 (b) support a five-coordinate intermediate which is so short lived that it retains the original ion-atmosphere of $\text{Co(NH}_3)_5\text{X}^{(3-n)+}$ but has lost "memory" of the X group.[80]

3. There is strong evidence for a dissociative type of mechanism for base hydrolysis. There is an $\approx 10^5$-fold rate enhancement (steric acceleration) for base hydrolysis of Co(iso-BuNH$_2$)$_5$Cl^{2+} compared to Co(NH$_3$)$_5$Cl^{2+} (mainly residing in k_2[73]) while the corresponding factor for aquation is only $\approx 10^2$, emphasizing the different degrees of dissociation (D vs I_d).[81] There is, incidentally, a LFER for log k_{OH^-} vs log k_{H_2O}, slope 1.0, for reactions of a series of Co(III) complexes.[82] Finally, on the basis of a D_{cb} mechanism, the estimated properties of the conjugate base Co(NH$_3$)$_4$NH$_2^{2+}$ such as heat content[83] and partial molar volume[84] (see Prob. 15, Chap. 2) are constants independent of its source. Some other characteristics of the 5-coordinated intermediate will be discussed in Sec. 4.3.5.

There is no reason to believe that the conjugate base mechanism does not apply with the other metal ions studied. Complexes of Cr(III) undergo base hydrolysis, but generally rate constants are lower, often $10^3 - 10^4$ less than for the Co(III) analog,[73, 85] Table 4.10.[86] The lower reactivity appears due to both lower acidity (K_1) and lower lability of the amido species (k_2) in (4.49) (provided k_{-1} can be assumed to be relatively constant). The very unreactive Rh(III) complexes are as a result of the very low reactivity of the amido species. The complexes of Ru(III) most resemble those of Co(III) but, as with Rh(III), base hydrolyses invariably takes place with complete retention of configuration.[73]

4.3.5 The Quest for Five Coordinate Intermediates

Many experiments have been performed to throw light upon, and much been written about, the existence of an intermediate in substitution reactions. Most of the work has concerned Co(III) and often the complex ion Co(NH$_3$)$_5$X^{n+} has been the examining substrate of choice. Evidence rests largely on competition experiments. The existence of an intermediate Co(NH$_3$)$_5^{3+}$ in the replacement of X on Co(NH$_3$)$_5$X (charges omitted) by Y is strengthened if Co(NH$_3$)$_5$Y is formed *directly* (k_1 route):

$$\text{Co(NH}_3)_5\text{X} \xrightarrow{k_1(Y)} \text{Co(NH}_3)_5\text{Y} \tag{4.50}$$

$$\downarrow k_2 \qquad\qquad \uparrow k_3(Y)$$

$$\xrightarrow{} \text{Co(NH}_3)_5\text{H}_2\text{O}$$

$$(+\text{H}_2\text{O}) \qquad\qquad (-\text{H}_2\text{O})$$

Table 4.10 Rate Parameters for Base Hydrolysis and Exchange of Trans-M(*RRSS*-cyclam)Cl$_2^+$ at 0°C Ref. 86

M	k_{OH} $M^{-1}s^{-1}$	k_1 $M^{-1}s^{-1}$	k_2/k_{-1}
Co	4.1×10^3	2.5×10^{3a}	7.9×10^{-1}
Cr	1.2×10^{-2}	9.8^a	3.8×10^{-4}
Rh	3.8×10^{-8b}	6.5^{ab}	1.5×10^{-9b}
Ru	2.3	1.3×10^{7a}	4.3×10^{-8}
Co^c	4.5×10^5	5.8×10^{6d}	4.1×10^{-2}

[a] Proton exchanging is *cis* to Cl. [b] 20°C [c] Data for *cis*-Co(*RRRR*-cyclam)Cl$_2^+$ [d] Proton exchanging is *trans* to Cl.

Unfortunately, spontaneous aquation and anion-interaction with the aqua product represents another route for X-Y interchange, and it is normally difficult to separate the primary (k_1) and secondary routes (k_2 and k_3). This problem is eased if rapidly-leaving groups X, such as $CF_3SO_3^-$, ClO_4^- are employed. A comprehensive examination of 14 different complexes with $t_{1/2}$ (hydrolysis) ranging from < 1s to ≈ 1 hour, using SCN^- as a competitor, shows that there is $3-19\%$ *direct* anion capture (k_1) and that the values are leaving group dependent. Significantly, the ratio S-/N-bound thiocyanate in the products differs appreciably. These facts suggest that an I_d rather than a D mechanism operates.[4] With aquation induced by Hg^{2+} and NO^+, Ref. 87, generally dissociative activation is supported but the fine details, particularly the nature of the intermediate, has been a subject of some controversy.[4,87] Most work related to and the best evidence for, a 5-coordinate intermediate is in that generated in base-accelerated reactions.[73] The charge and nature of the leaving group X^{n-} in $Co(NH_3)_5X^{(3-n)+}$ only slightly affects the competition ratio (it may for example vary from 8.5 to 10.6%) and when SCN^- is used as a competitor the S/N bound isomer ratios are quite constant for a large number of leaving groups (Sec. 2.2.1(b)), in contrast with the acid hydrolysis results (above.)

The consensus is for a short-lived 5-coordinated intermediate of the type $Co(NH_3)_4NH_2^{2+}$ **4** which may react quicker than it can equilibrate with its solvent cage.[88,89] Experiments using competition with anions at concentrations as high as 1 M are complicated by ion-pairing.[87,90] Both non-aggregates and aggregates are reactive, but curiously the aggregate MY scavenges Y from solution and not from the second coordination sphere. It is not easy to arrive at any firm conclusions about the geometry of the five-coordinated intermediate of the conjugate base mechanism.[73] The base hydrolysis of a number of octahedral cobalt(III) and chromium(III) complexes, particularly of the type $M(en)_2XY^{n+}$ is accompanied by stereochemical change and it is reasonable to suppose that there is a rearranged trigonal-bipyramidal intermediate, although at which point along the reaction profile it appears, is uncertain.[73] The amido conjugate base is very labile and the manner in which the amido group labilizes is still highly speculative. It is also uncertain whether the amido group generated has to be in a specific position (*cis* or *trans* with respect to the leaving group).[73] It is however generally true that the presence of a meridional or "flat" sec-NH proton, *cis* to the leaving group, leads to high base hydrolysis rates. Finally, observations on the relative rates of base hydrolysis and loss of optical activity of cleverly conceived substrates have allowed reasonable conclusions about the symmetry of a relatively stable five-coordinated intermediate in base hydrolysis (Sec. 7.9).[91,92]

4

4.4 Replacement Reactions Involving Multidentate Ligands

There is no reason to believe that replacement of water by the donor groups of a chelating agent is fundamentally different from replacement when only unidentate ligands are involved. However, the multiplicity of steps may increase the difficulty in understanding the detailed mechanism, and mainly for this reason the simpler bidentate ligands have been most studied.

4.4.1 The Formation of Chelates

The successive steps in the replacement of two coordinated waters by a bidentate ligand L-L is represented as

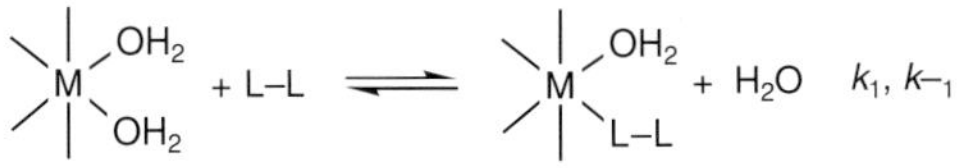

$$[M(OH_2)_2]$$

$$\tag{4.51}$$

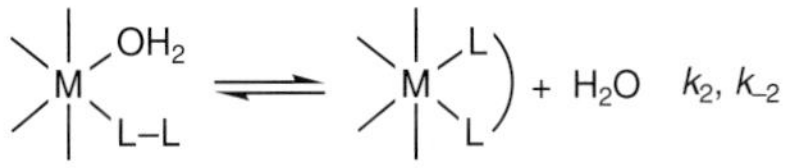

$$M(L_2)$$

$$\tag{4.52}$$

Assuming stationary-state conditions for the intermediate, in which $L-L$ is acting as a unidentate ligand, we find

$$d\,[M(L_2)]/dt = k_f[M(OH_2)_2]\,[L-L] - k_d[M(L_2)] \tag{4.53}$$

with

$$k_f = \frac{k_1 k_2}{(k_{-1} + k_2)} \qquad k_d = \frac{k_{-1}k_{-2}}{(k_{-1} + k_2)} \tag{4.54}$$

The function k_2/k_{-1} will dominate the kinetics of bidentate chelation.

(a) $k_2 \gg k_{-1}$. For this condition, $k_f = k_1$ and the overall rate of chelate formation will be determined by the rate of formation of the $M-L-L$ entity, a process we can assume is controlled by the same factors that apply with the entry of unidentates.[93] The relation $k_2 \gg k_{-1}$ is anticipated when the first bond formed is relatively strong and the tendency for the intermediate to bond-break (measured by k_{-1}) is much less than its ability to ring close (k_2). This behavior is heralded by a single discernible rate process with rate constants for complexing by bidentate (or multidentate) ligands resembling that of the appropriate unidentate ligand. The rate constant at 25 °C for complexing of Ni^{2+} with py, bpy and tpy are within a factor of three, namely 4×10^3, 1.5×10^3 and 1.4×10^3 $M^{-1}s^{-1}$ respectively.[40] Complexes containing no, one and two chelate rings are formed. The reactions of $Cr(H_2O)_6^{3+}$ with acetate, dicarboxylates, hydroxy- and amino acids have the common feature of a rate independent of the concentration of ligand. All reactions obey a single first-order process with similar energies of activation (75–97 kJ mol^{-1}) and in the suggested scheme

$$Cr(H_2O)_6^{3+} + H_3\overset{+}{N}CH_2CO_2^- \rightleftharpoons [Cr(H_2O)_6^{3+} \cdot H_3\overset{+}{N}CH_2CO_2^-] \rightarrow [Cr(H_2O)_4H_2NCH_2CO_2]^{2+}$$

$$(4.55)$$

(with glycine for example) the outer-sphere complexing is considered complete, the rate-determining step is the expulsion of one water from the Cr(III) coordination sphere and ring closure then is rapid.[94,95] The volumes of activation for complexing of Mn^{2+}, Fe^{2+}, Co^{2+} and Ni^{2+} with tpy are -3.4, $+3.5$, $+4.5$ and $+6.7$ cm^3 mol^{-1} respectively. Comparison with the water exchange data (Table 4.1) provides convincing evidence that the loss of the first H_2O in (4.56) is rate-determining and that the mechanisms of replacement resemble those deduced for water exchange.[96]

$$M(H_2O)_6^{2+} + tpy \rightarrow M(H_2O)_3 tpy^{2+} + 3 H_2O \tag{4.56}$$

The establishment of the first bond appears to signal rapid successive ring closures with most of the multidentate ligands examined. (However, consider the Ni(II)-fad system, Sec. 1.8.2) In certain cases the later steps in chelation can be shown to be more rapid than the earlier ones, by clever experiments involving laser photolysis (Table 3.4)[97] or pH-adjustments of solutions containing partially formed chelates.[98]

The dissociation rate constant is now composite, $k_d = k_{-1}k_{-2}/k_2$. Following the first bond rupture (k_{-2}) the competition between further bond rupture (k_{-1}) and reformation (k_2) which may lead to a small k_{-1}/k_2 is the basic reason for the high kinetic stability of the chelate. The problem of complete dissociation is intensified when complexes of ligands of higher dentate character are examined. The situation is altered when the successively released donor atom(s) can be prevented from reattachment (see subject of accelerated substitution).

(b) $k_2 \ll k_{-1}$. Now, $k_f = k_1k_2/k_{-1}$ and the rate-determining step is ring closure. The condition (b) is likely to arise in the following circumstances:

(i) If the first step is unusually rapid. The reaction of $Co(tren)(OH)_2^+$ and $Co(tren)(OH)H_2O^{2+}$ with CO_2 at pH 7–9 is biphasic (4.57). The first step is rapid because no $Co-O$ bond breakage is involved. This is followed by slower intramolecular chelation.[99] For other examples see Ref. 100.

$$Co(tren)(OH)H_2O^{2+} + CO_2 \rightarrow Co(tren)(H_2O)OCO_2H^{2+} \rightarrow Co(tren)CO_3^+ + H_3O^+$$

$$(4.57)$$

The steps in the complexing of Ni^{2+} by the terdentate dye sulfonated 2-pyridylazo-1-naphthol (5) have been carefully studied. At pH = 8.0 and low $[Ni^{2+}]$ (<0.4 mM) the formation of the first bond is the rds. At much higher $[Ni^{2+}] \approx 100$ mM, the initial substitution is now faster than the first ring closure. The final ring closure ($k \sim 25$ s^{-1} at 25 °C) is abnormally slow.[101]

(ii) If the closing of the ring is inhibited. This may arise with the formation of certain six-membered chelate rings involving diketones[102] or β-aminoacids. Steric hindrance or strain may then lead to "sterically-controlled" or "chelation-controlled" substitution, which is more

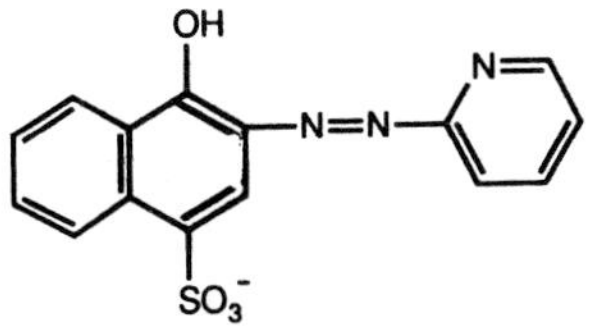

5

important with the more labile ions. [17,40] If the closing arm is protonated and a proton has to be lost prior to coordination then this may inhibit ring closure. This is likely only with the more reactive metal ions, for it is only with these that substitution rate constants can be comparable to those of protonation and deprotonation. [103] The difficulty of ring closure from this cause is however intensified when internal hydrogen bonds need to be broken. The rate constant for reaction of Ni^{2+} with the unprotonated 3,5-dinitrosalicylate, $dnsa^{2-}$ is 3.1×10^4 $M^{-1}s^{-1}$ (25°C, $I = 0.3$ M), that is, a normal value. However the corresponding value is 3.8×10^2 M^{-1} s^{-1} for $dnsaH^-$ (**6**) in which strong hydrogen bonding exists. For the scheme (4.58) the second-order formation rate constant (k_f) is given by

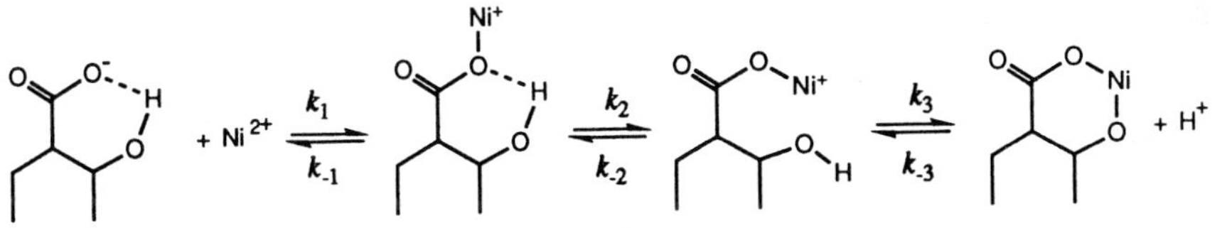

$$(4.58)$$

$$k_f = \frac{k_1 k_2 k_3}{k_{-1} k_{-2} + k_{-1} k_3 + k_{-2} k_3} \qquad (4.59)$$

on the assumption of steady state concentrations for **7** and **8**. However it is uncertain whether the rds is opening of the H-bond in **7** or closure of the chelate ring in **8**. [104,105] See also Ref. 106 and 107, and Prob. 4.

The dissociation rate constant k_d measures directly the value of k_{-2} in (4.52). The strain resident in multi-ring complexes is clearly demonstrated by some hydrolysis rate studies of nickel(II) complexes. The $\Delta H^{\ddagger}$ values for the *first* bond rupture for Ni(II)-polyamine complexes fall neatly into groups. They are highest for en, containing the most strain-free ring (84 kJ mol^{-1}), about 75 kJ mol^{-1} for complexes with terdentate ligands and only ~63 kJ mol^{-1} for complexes of quadridentate and quinquedentate amines and with NH_3 itself. [108] See also Ref. 109.

4.4.2 Effect of [H$^+$] on the Rates of Substitution in Chelate Complexes

Both the formation and hydrolysis rates of chelates will generally be pH-dependent. Let us compare the behavior of a chelating ligand L with its monoprotonated form LH$^+$. Each will exist in predominant amounts at particular pH's and usually have separate and distinct reactivities.

$$HL^+ \rightleftharpoons H^+ + L \qquad K_1 \tag{4.60}$$

$$M + L \rightleftharpoons ML \qquad k_2, k_{-2}, K_2 \tag{4.61}$$

$$M + LH^+ \rightleftharpoons ML + H^+ \qquad k_3, k_{-3}, K_3 \tag{4.62}$$

It is easy to show that

$$K_3 = K_1 K_2 \text{ and therefore } K_2 \gg K_3 \text{ since } K_1 \ll 1 \tag{4.63}$$

The smaller value of $K_3 (= k_3/k_{-3})$ than $K_2 (k_2/k_{-2})$ will reside in k_3 being less than k_2 and $k_{-3}[H^+]$ being smaller than k_{-2}. Most reported studies examine the hydrolysis of metal complexes of polyamines, particularly those of Ni(II)[108, 109] and Cr(III).[110]

In the reverse direction, a proton may be effective by aiding ring-opening directly or via a reactive protonated species. It may intervene with the ring-opened species. A splendid example of these effects is shown in the acid hydrolysis of ferrioxamine B (**9**). Four stages can be separated and the kinetics and equilibria have been characterized by stopped-flow and rapid-scan spectral methods.[111]

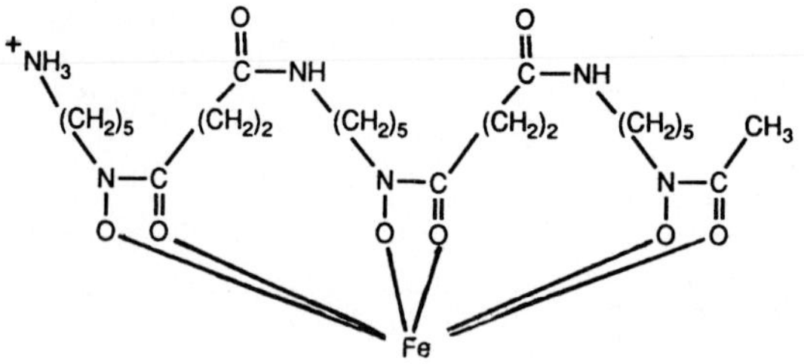

9

4.4.3 Metal Ion-Assisted Dechelation

Metal ions can assist in the dissociation (hydrolysis) of complexes containing multidentate ligands. The metal ion may not necessarily complex with the detached ligand, for example, in the metal-assisted acid-catalyzed aquation of $Cr(C_2O_4)_3^{3-}$. Ref. 112. Usually however the metal ion removes and complexes the ligand as in

$$ML + M_1 \rightarrow M_1L + M \tag{4.64}$$

The reactions have been followed spectrally, including luminescent changes,[113] polarographically (when M_1 is monitored) or by isotope exchange (M_1 being an isotopic form of M). They are commonly second-order reactions, with the second-order rate constant often dependent on $[H^+]$, $[M_1]$ and even $[M]$[17, 113]

It is supposed that binuclear intermediates involving M and M_1 occur in the various paths. The most studied systems involve edta complexes.[114]

4.4.4 Ligand-Assisted Dechelation

Interchange involving multidentate ligands is obviously even less likely to be direct than interchange involving unidentate ligands. When solvent is available on the complex (or part of the coordinated ligand is easily replaceable by solvent), the incoming ligand can gain a "coordination foothold". This can thereby lead to eventual complete ejection of the original multidentate ligand. [17, 115] The simplest example of this is shown in the exchanges of metal chelates of β-diketones with free ligand which have been studied in a variety of solvents using nmr line broadening, isotopic labelling with ^{14}C and spectral methods. [116] In the general scheme, where $\widehat{OO}$ represents the diketone and the asterisk distinguishes the interchanging diketones

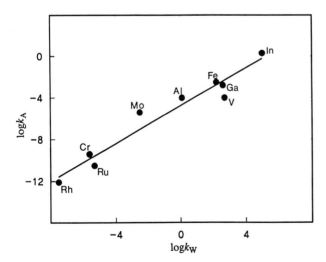

$$(4.65)$$

the rds can be (1) breaking of the $M-O$ bond, (2) formation of an intermediate containing dangling groups or (3) intramolecular proton transfer between the dangling groups. [116] For exchange of $M^{III}(acac)_3$ with acac in CH_3CN, rate laws suggest that the first step is the common rds. Because $M-O$ cleavage is involved in this step it is found that there is a LFER between ligand substitution rates of $M(acac)_3$ and water exchange of $M(H_2O)_6^{3+}$ and even $M(NH_3)_5H_2O^{3+}$, Fig.4.4. [117]

Fig. 4.4. LFER between exchange rate constants for $M(acac)_3$ k_A and $M(H_2O)_6^{3+}$ k_w at 25 °C. The values are both in s^{-1} units and the former exchanges are in acac. The data for Mo are second-order rate constants for anation. [117]

Of course, when highly dentated ligands have to be replaced, more complicated behavior is observed. In this category, are the well-studied ligand exchange reactions of edta and polyamine complexes, a typical one of which (charges omitted) is

$$Cu(edta) + trien \rightleftharpoons Cu(trien) + edta \tag{4.66}$$

This may be studied in either direction by adjusting the pH and reactant concentrations. Substantial evidence exists for a mechanism in which three nitrogen atoms of the polyamine bind to the metal (**10**) before the rds of $M-N$ cleavage, which leads to the final products.[118] See also Refs. 17 and 119.

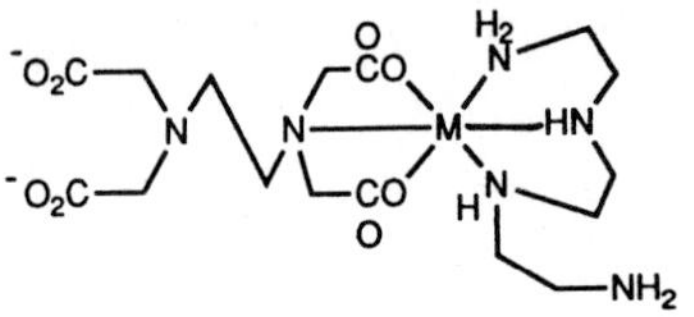

10

The incoming ligand (or metal ion in the previous section) simulates H^+ by preventing reclosing of the Ni-donor bonds as these are successively broken (see also Fig. 8.7)

4.5 Replacement Reactions Involving Macrocycles

The study of the complexing of macrocycle ligands should be considered for its intrinsic importance rather than for its value in illuminating the mechanism of substitution. Kinetic (but much more thermodynamic [120]) data are available for the reactions of the different macrocycle ligand types, shown in Fig. 4.5, including azamacrocycles,[121, 122] crown ethers and cryptands,[123, 124] and porphyrins.[70, 125]
Conventional, flow, temperature-jump, ultrasonic absorption, electric-field jump and nmr line broadening have all been used to measure the rates. UV-vis spectrophotometry and conductivity are the monitoring methods of choice. A variety of solvents have been used. The focus has been often on the dissociation since the dissociation rate constant appears in general to be the main controller of the overall stability.

The constraints imposed by a macrocycle on complex formation are well illustrated by a comparison of copper ion sequestering by flexible chain polyamines, 2,3,2-tet and Me_4trien, macrocycles, tetraazatetraamines, cyclam and (N-Me_4)cyclam, a bicyclic tetraamine cryptand $2_N 1_O 1_O$ and a rigid porphyrin.[121, 126–128] The reactions are carried out in a strongly basic medium, where ligand protonation is unimportant, and second-order rate constants for reaction of $Cu(OH)_3^-$ and $Cu(OH)_4^{2-}$ are shown in Table 4.11. Generally it is seen that an acyclic tetraamine allows easy stepwise replacement of coordinated H_2O and OH groups and reacts $\approx 10^8$ times faster than the rigid porphyrin, where simultaneous multiple desolvation of the metal ion is mandatory. With cyclam, twisting or folding of the ligand is possible but

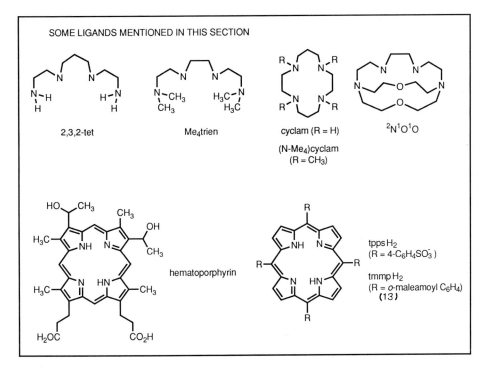

Fig. 4.5 Some ligands mentioned in this section (4.5).

(N-Me₄)cyclam and $2_N1_O1_O$ (compare Me₄trien) are more rigid and the rate constants reflect this condition. All rate constants are less than the diffusion-controlled limit and even of the anticipated axial water lability of Cu(II), Sec. 4.2.1(a). It is apparent that the mechnisms are complex and must differ appreciably (position of rds etc.) for the various types. Nevertheless there are also general similarities when the details are examined.

Table 4.11 Formation Rate Constants for Cu(II) Reactions [121, 126, 127]

Ligand	$k, \mathrm{Cu(OH)_3^-}$ $\mathrm{M^{-1}s^{-1}}$	$k, \mathrm{Cu(OH)_4^{2-}}$ $\mathrm{M^{-1}s^{-1}}$
2,3,2-tet	1.0×10^7	4.3×10^6
Me₄trien	4.1×10^6	4.2×10^5
cyclam	2.7×10^6	3.8×10^4
(N-Me₄)cyclam	3.1×10^3	< 10
$2_N1_O1_O$	6.6×10^4	3.8×10^3
hematoporphyrin	—	$\approx 2 \times 10^{-2}$

4.5.1 Azamacrocycles

The stepwise nature of complexing is illustrated in the suggested scheme (Fig. 4.6) for the reaction of $Cu(OH)_3^-$ and $Cu(OH)_4^{2-}$ with the cyclic tetraamines. [121, 127] When $Cu(OH)_3^-$ and $Cu(OH)_4^{2-}$ react with similar rate constants it may be supposed that they have a common rds, which will then depend on the ligand. Indirect arguments indicate that the rds is after the first substitution of axial water. It is almost certainly before formation of the third Cu-N bond, otherwise $Cu(OH)_4^{2-}$ reacting would lead to an unlikely 7-coordinate Cu(II) complex at some point. The Jahn-Teller inversion step is an additional complication with reactions of Cu(II) and may indeed be the rds following bond formation. Differences greater than a factor of 10 for reaction of $Cu(OH)_3^-$ over $Cu(OH)_4^{2-}$ suggest a shift of rds. This is believed to be a change from the first- to the second-bond formation, respectively. [121, 127] Formation rate constants vary little with the ring size of the tetraaza macrocycle, except that reaction with the 12-membered ring is particularly slow, perhaps because of the difficulty in folding such a ring, which would be expected to be helpful in the difficult insertion process. Conformational changes in the macrocycle can arise after metal ion incorporation. The problem of binding a metal ion to a macrocycle may be eased if the macrocycle has a pendent arm which contains a ligand center. The metal ion can bind then to the arm and from thence be pulled into the macrocycle. [129]

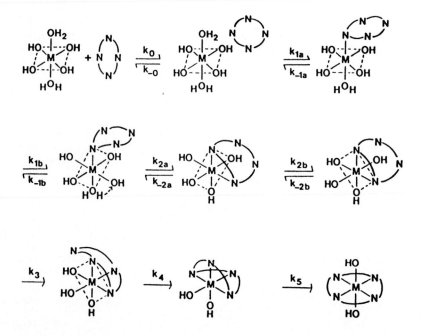

Fig. 4.6 Stepwise complexing of $Cu(OH)_4^-$ by a tetradentate macrocyclic ligand. The first Cu(II)-N bond is formed by replacement of an axial solvent molecule (k_{1a}) followed by a Jahn-Teller inversion (k_{1b}) which brings the coordinated nitrogen into an axial position. Second-bond formation follows a similar pattern (k_{2a} and k_{2b}). [121, 127] Reproduced with permisson from J. A. Drumhiller, F. Montavon, J. M. Lehn and R. W. Taylor, Inorg. Chem. **25**, 3751 (1986). © (1986) American Chemical Society.

(a) Effect of H^+

The macrocycles are generally characterized by their extreme resistance to dissociation, (see however Prob. 6). The kinetics of acid-promoted dissociation of an extensive series of Ni(II) and Cu(II) complexes, (ML) have been reported,[122] The general rate law is (see Prob. 2, Chap. 2):

$$V = \{k_1 + k_2[H^+]^n\} [ML] \tag{4.67}$$

where n can vary from 1 to 3, the value of n depending on the number of preequilibria involving H^+. "Acid-limiting" kinetics are interpreted either in terms of a mechanism:

$$ML + H^+ \rightleftharpoons MLH^+ \qquad K \quad \text{rapid} \tag{4.68}$$

$$MLH^+ \rightarrow \text{products} \qquad k_H \tag{4.69}$$

for which the first-order loss of ML is given by

$$k_{obs} = \frac{K k_H [H^+]}{1 + K[H^+]} \tag{4.70}$$

Alternatively, they can arise as a result of the rate-determining reaction of H^+ with a reactive form of ML (ML*) which might be an isomer originating, for example, from the chiral NH centers (Sec. 7.9.).

$$ML \underset{k_{-1}}{\overset{k_1}{\rightleftharpoons}} ML^* \xrightarrow[k_2]{H^+} \text{products} \tag{4.71}$$

$$k_{obs} = \left[\frac{k_1 k_2 [H^+]}{k_{-1}} \right] \left[1 + \frac{k_2 [H^+]}{k_{-1}} \right]^{-1} \tag{4.72}$$

The formation of $Cu(cyclam)^{2+}$ by reaction of cyclam with a large variety of copper complexes (CuL^{n+}) is interesting:

$$CuL^{n+} + cyclamH^+ \rightarrow Cu(cyclam)^{2+} + LH^{(n-1)+} \quad k \tag{4.73}$$

Provided that the stability of the CuL^{n+} is not very high ($\log K < 10$), the value of k is similar to that for Cu_{aq}^{2+}. There is an inverse relationship of $\log k$ with $\log K$ (CuL^{n+}) with more stable complexes (Fig. 8.7). It is also worthy of note that although the main cyclam species at the pH of the study (4–9) is $cyclamH_2^{2+}$, the reactive species is $cyclamH^+$. Ref. 130.

4.5.2 Crown-Ethers and Cryptands

Studies with these ligand systems mainly involve the Group One and Two metal ions. The suggested mechanism for the formation of complexes has a number of features in common with those proposed for porphyrins (Sec. 4.5.3). A possible rapid pre-equilibrium involving a ligand conformational change,

$$C \rightleftharpoons C^*$$
(4.74)

is frequently followed by a rds:

$$M^+_{solv} + C^* \rightarrow MC^{**+} + solv$$
(4.75)

and (rapidly)

$$MC^{**+} \rightarrow MC^+$$
(4.76)

where C, C* and C** represent different conformations of the crown ether or cryptand. With for example $2_o2_o2_o$, conformational changes involving endo-endo, exo-endo and exo-exo forms (**11**) have been established. Since in general there is little effect of solvents on the rates (see however Ref. 131) it is surmised that there is little desolvation in the transition state for (4.75), which thus resembles the reactants. As a corollary of this, there is a striking linear plot, slope −1 over 12 orders of magnitude, for log k(diss) vs log K(stab) for a variety of systems.[124] Outer sphere complexing may also precede the main step as has been postulated for the first of three steps observed in the complexing, in propylene carbonate, of UO_2^{2+} by [18]crown-6 to give a 1:1 complex.[132] With the cryptand complexes, dissociation is only proton-promoted when easily approached donor sites are available and probably occurs through the exo-endo isomer. Dissociation of the crown-ether[131] and cryptand[133] complexes is rarely metal-ion assisted. An intriguing example of ligand-aided removal of a transition metal ion from a macrocycle, involves complexes of catenends. These are interlocked as represented in **12**. The rate law for removal of Cu(I) by CN^- from the Cu(I) complex of **12**, CuC^+ (see Fig. 6.1) is

exo-exo   exo-endo   endo-endo
11

Conformers of the Cryptand (2.2.2)

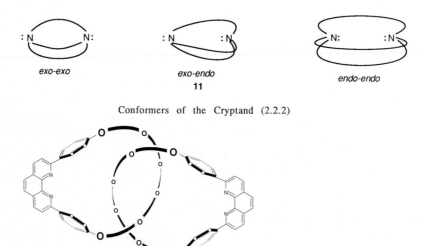

12

$$-d\,[CuC^+]/dt = k_1 + k_2[CN^-] \tag{4.77}$$

The k_1 term corresponds to spontaneous dissociation and the k_2 term arises from cyanide-assisted dissociation:

$$CuC^+ + CN^- \longrightarrow Cu(CN) + C \tag{4.78}$$

$$Cu(CN) + 3\,CN^- \xrightarrow{\text{fast}} Cu(CN)_4^{3-} \tag{4.79}$$

The difficulty in disengaging the two interlocked rings is seen when the values of k_1 ($2 \times 10^{-4}\,s^{-1}$) and k_2 ($0.16\,M^{-1}s^{-1}$) are compared with those for the corresponding breakdown of $Cu(dpp)_2^+$ by CN^-, $k_1 = 1.3\,s^{-1}$ and $k_2 = 14.6\,M^{-1}s^{-1}$, dpp = 2.9-diphenyl-1, 10-phenanthroline. [134(a)]

The formation of Li^+, Cd^{2+}, Zn^{2+} and Co^{2+} (but not Cu^+) complexes of **12** is biphasic. The first step (overall second-order) likely represents binding of the metal ion to one of the chelating subunits of **12**. The second step (first-order) is very slow ($k \approx 10^{-2} - 1\,s^{-1}$) and possibly corresponds to the gliding motion of one ring within the other while the second phenanthroline fragment attempts to bind to the metal center. [134(b)]

4.5.3 Porphyrins

Most of the studies have been carried out in nonaqueous solution. The important processes are (A) direct metal ion (M) interaction with porphyrin and (B) metal-ion M* assisted entry (transmetallation) shown schematically in (4.80). The reactions are usually slow, easily followed spectrally because of the high characteristic absorption coefficients of the complexes and free porphyrins, and attended by beautiful isosbestic points [135(a)] (Fig. 3.10). The free base in (4.80) is represented as H_2P and is the reactive species. [135(b)] The mono- and di-protonated forms are unreactive.

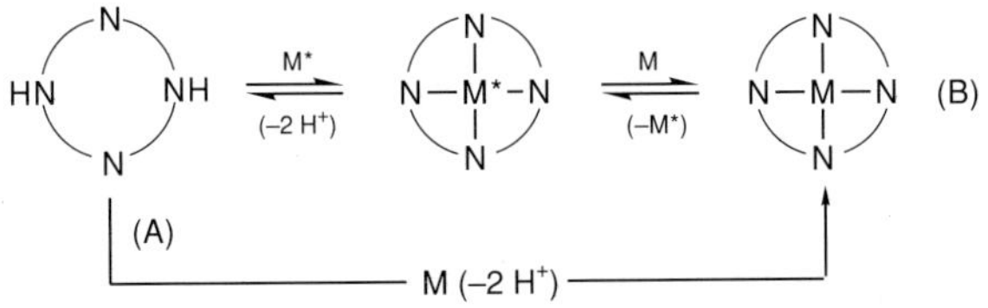

$$\tag{4.80}$$

(a) Direct Interaction

For route (A) the rate law is often second-order

$$V = k\,[M]\,[H_2P] \tag{4.81}$$

The insertion steps are probably preceded by deformation of H_2P and/or outer sphere complexing. [70, 135-137] If these are incomplete they will introduce rapid preequilibria constants K_1 and K_2 into the rate expression, and by reducing the

$$H_2P \rightleftharpoons H_2P^* \qquad K_1 \tag{4.82}$$

$$H_2P^* + MS_6 \rightleftharpoons MS_6 \cdots H_2P^* \quad K_2 \tag{4.83}$$

concentrations of the reactive species, lower markedly the overall rates of reaction with the solvento complex, MS_6. Ligand dissociation and the formation of the first bond then occurs:

$$MS_6 \cdots H_2P^* \rightleftharpoons MS_5 \cdot H_2P + S \tag{4.84}$$

Is this the rds? There is striking correlation of the exchange rate constants for MS_6 and the values of k in (4.81).[135, 136] In addition the volumes of activation are positive for solvent exchange and interaction of $M(dmf)_6^{2+}$ (M = Mn, Co, Ni, Zn and Cd) with N-Metpp in dmf (compare Table 4.4). In spite of these two facts however, it is considered that one of two further steps, probably the first, controls the overall rate. A sitting-atop (SAT) complex[138] is formed in which metal is attached by two bonds to the porphyrin and two N−H bonds remain intact.

$$MS_5 \cdot H_2P \rightleftharpoons MS_n = H_2P + (5-n)S \tag{4.85}$$

In the final step, the SAT complex collapses to the ultimate product with a concerted release of 2 protons

$$MS_n = H_2P \rightarrow MP + nS + 2H^+ \tag{4.86}$$

This segment of the mechanism and the concept of an SAT complex[138] have played an important role in the understanding of the complex mechanism by which metal ions react with porphyrins.

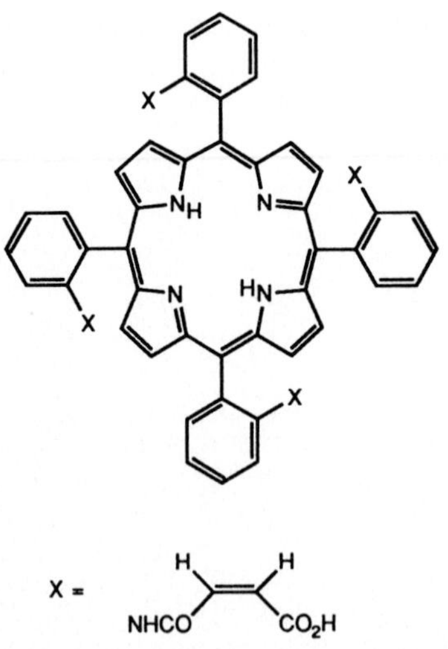

13

Appropriately placed groups on the porphyrin periphery can aid considerably the incorporation of metal ions into the ring. The porphyrin shown, H_2tmpp, **13** reacts rapidly with Co(II), Ni(II), Cu(II) in dmf/H_2O. The rate data indicate that an initial adduct is formed which allows the metal to more easily transfer into the porphine ring.[139] See also Ref. 129 for a similar behavior with an azamacrocycle.

(b) Metal-Ion Assisted Entry

Metal ions can assist, in a novel way, the formation of a metalloporphyrin, by route (B).[125] Since the free porphyrin does not appear in any amount during the M, M* interchange it is a truly associative process. The reaction of Mn^{2+} with $tppsH_2^{4-}$ in water is very slow.

$$Mn^{2+} + tppsH_2^{4-} \rightarrow Mn(tpps)^{4-} + 2H^+ \qquad \text{very slow} \qquad (4.87)$$

The incorporation is accelerated markedly by the presence of small amounts of Cd^{2+} ions acting as a catalyst. The following mechanism is suggested

$$Cd^{2+} + tppsH_2^{4-} \rightleftharpoons Cd(tpps)^{4-} + 2H^+ \qquad k_1, k_{-1} \qquad (4.88)$$

$$Cd(tpps)^{4-} + Mn^{2+} \rightarrow Mn(tpps)^{4-} + Cd^{2+} \qquad k_2 \qquad (4.89)$$

on the basis of the observed rate law, $[Mn^{2+}] \gg [tppsH_2^{4-}], [Cd^{2+}]$:

$$-d[tppsH_2^{4-}]/dt = k_{obs}[tppsH_2^{4-}] \qquad (4.90)$$

where

$$k_{obs} = \frac{k_1 k_2 [Mn^{2+}][Cd^{2+}]}{k_{-1}[H^+] + k_2[Mn^{2+}]} \qquad (4.91)$$

The reaction of Cd^{2+} with $tppsH_2^{4-}$ is rapid, because the larger metal ion forms an out-of-plane complex. This is easily attacked, probably from the other side of the ring, by the

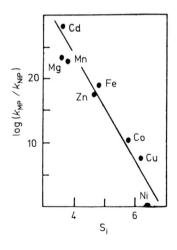

Fig. 4.7 Plot of the log of the relative rates of hydrolysis of MP and the corresponding Ni(II)P vs. the stability index S_i. The parameter S_i is defined as $100 zE_N/r_i$ where z = charge, r_i = radius and E_N = Pauling's electronegativity of the metal in the metalloporphyrin. The parameter is empirical but agrees generally with the order of stability, determined experimentally (J. W. Buchler, Ch. 5 in Porphyrins and Metalloporphyrins, ed. K. M. Smith, Elsevier, 1975).[70] Reproduced with permission from D. K. Lavallee, Coord. Chem. Revs. **61**, 55 (1985).

stronger binding Mn^{2+}.[140] At high pH, $k_{obs} = k_1[Cd^{2+}]$. At 25°C, $k_1 = 4.9 \times 10^2 \ M^{-1}s^{-1}$ and $k_{-1}/k_2 = 3.0 \times 10^{10} M^{-1}$ ($\mu = 0.1$ M). By studying (4.89) directly, $k_2 = 2 \times 10^2 M^{-1}s^{-1}$ and thus $k_{-1} = 6 \times 10^{12} \ M^{-2}s^{-1}$.[125] Reaction (4.89) is about 10^4 faster than (4.87).

Dissociation of most metal porphyrins, M(II)P, is very slow. As well as metal ions, H^+ accelerates the metal removal. The order of dependence on $[H^+]$ varies from 1–3 depending on the system, with a rate law

$$V = k[M(II)P][H^+]^2 \tag{4.92}$$

common.[141] This behavior suggests that a single metal-N bond is readily broken. The rate constant k parallels the instability of the prophyrin complex in a dramatic manner. Fig. 4.7.[70]

4.6 Substitution in Square-Planar Complexes

There are a number of metals, particularly with a low spin d^8 configuration, that form four-coordinate square-planar complexes. Of these, the Pt(II) complexes have been the most intensively investigated. They are therefore representatives of this geometry, much as Co(III) complexes epitomize octahedral behavior — and for precisely the same reasons, namely that they have been previously well characterized and studied, particularly by Russian workers, and that they react slowly. There have never been any problems therefore in the initiation and the rate measurement of these reactions. With the easier access to rapid reaction techniques, particularly flow, square-planar complexes of a number of other metals, Rh(I), Ir(I), Ni(II), Pd(II), Cu(II) and Au(III) are being increasingly examined. Many of these metal complexes, unlike most octahedral complexes, are soluble in aprotic solvents and these have been, as well as water, examining media. The ultraviolet and nmr spectral methods, and less commonly, conductivity and radioactive isotopic exchange have been the methods most commonly employed for monitoring the rates. In many respects, substitution in square-planar complexes is one of the best understood dynamic processes in chemistry.[142, 143]

4.6.1 The Kinetics of Replacement Involving Unidentate Ligands

Most studies have again been concerned with replacment of one unidentate ligand by another, and the rules and patterns of behavior that have evolved are based mainly on this simple type of substitution reaction. Consider the substitution scheme

$$-\overset{|}{\underset{|}{M}}-X + Y \rightleftharpoons -\overset{|}{\underset{|}{M}}-Y + X \tag{4.93}$$

The rate law governing substitution in planar complexes usually consists of two terms, one first-order in the metal complex (M) alone and the other first-order in both M and the entering ligand Y:

$$V = -d\,[MX]/dt = k_1\,[M] + k_2\,[M]\,[Y] \qquad (4.94)$$

The experiments are invariably carried out using excess Y and therefore with pseudo first-order conditions. The experimental first-order rate constant k is given by

$$V = k\,[M] \qquad (4.95)$$

Therefore

$$k = k_1 + k_2\,[Y] \qquad (4.96)$$

A plot of k vs [Y] will have an intercept k_1 and a slope k_2. An example is shown in Fig. 4.8.[144] For different nucleophiles reacting with the same complex, the value of k_1 is the same, whereas the value of k_2 usually will be different. Often, $k_1 \ll k_2\,[Y]$ and it is then difficult to measure k_1 accurately. Care has to be taken that a positive intercept in a plot of the type shown in Fig. 4.8 is not mistakenly assigned to k_1 in (4.96), when it may in fact represent the reverse rate constant of (4.93) (Sec. 1.5).[145, 146]

(a) Significance of k_1. The term containing k_1 resembles octahedral complexes in their substitution behavior where it represents ligand-ligand replacement via the solvated complex:

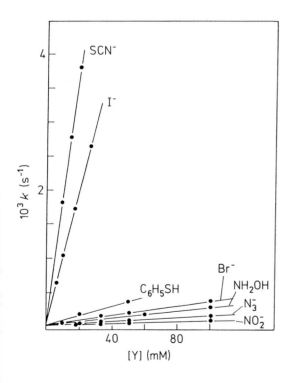

Fig. 4.8 Plots of pseudo first-order rate constants at 30°C vs [nucleophile, Y] for reaction of *trans*-Pt(py)$_2$Cl$_2$ in methanol.[144]

$$-\underset{|}{\overset{|}{M}}-X \underset{X\ \text{slow}}{\overset{S}{\rightleftharpoons}} -\underset{|}{\overset{|}{M}}-S \underset{S\ \text{fast}}{\overset{Y}{\rightleftharpoons}} -\underset{|}{\overset{|}{M}}-Y \qquad (4.97)$$

As with solvolysis reactions of octahedral complexes, the rate-determining step may be solvolytic or dissociative; in any case, it is independent of the concentration of Y:

$$
\begin{array}{ccc}
 & & \overset{|}{-}\text{M} + \text{X} \xrightarrow[+\text{Y}]{k_{\text{D}}^{\text{Y}}} \overset{|}{-}\text{M}\overset{|}{-}\text{Y} + \text{X} \\
\overset{|}{-}\text{M}\overset{|}{-}\text{X} & \overset{k_{\text{D}}}{\underset{k_{-\text{D}}}{\rightleftarrows}} & \\
 & \overset{k_{\text{S}}}{\underset{k_{-\text{S}}+\text{S}}{\rightleftarrows}} & \overset{|}{-}\text{M}\overset{|}{-}\text{S} + \text{X} \xrightarrow[+\text{Y}]{k_{\text{S}}^{\text{Y}}} \overset{|}{-}\text{M}\overset{|}{-}\text{Y} + \text{S} + \text{X}
\end{array}
\tag{4.98}
$$

Setting up stationary-state conditions for $\overset{|}{-}\text{M}$ and $\overset{|}{-}\text{M}\overset{|}{-}\text{S}$ yields

$$
k_1 = \frac{k_{\text{D}}^{\text{Y}} k_{\text{D}} [\text{Y}]}{k_{\text{D}}^{\text{Y}}[\text{Y}] + k_{-\text{D}}[\text{X}]} + \frac{k_{\text{S}}^{\text{Y}} k_{\text{S}} [\text{S}] [\text{Y}]}{k_{\text{S}}^{\text{Y}}[\text{Y}] + k_{-\text{S}}[\text{X}]}
\tag{4.99}
$$

The full equation (4.99) must be invoked when reversibility of any of the steps in (4.98) occurs.[142] When the step involving Y is fast, (4.99) reduces to

$$
k_1 = k_{\text{D}} + k_{\text{S}}[\text{S}]
\tag{4.100}
$$

On the basis of either interpretation, k_1 should equal the solvolysis rate constant and if MS is an isolatable intermediate, it should be shown to react rapidly with Y. Both consequences have been realized in certain systems. Although there has been some skepticism shown towards a dissociative mode of substitution (k_{D}, $k_{-\text{D}}$ in (4.98)), there is growing evidence for its existence in cases where the importance of the associatively activated pathways can be reduced. For the first step of the reaction

$$
cis\text{-Pt}(\text{C}_6\text{H}_5)_2(\text{Me}_2\text{SO})_2 + \text{L}\!-\!\text{L} \rightarrow cis\text{-Pt}(\text{C}_6\text{H}_5)_2(\text{Me}_2\text{SO})(\text{L}\!-\!\text{L}) + \text{Me}_2\text{SO}
$$

$$
\xrightarrow{\text{fast}} \text{Pt}(\text{C}_6\text{H}_5)_2(\text{L}\!-\!\text{L}) + \text{Me}_2\text{SO}
\tag{4.101}
$$

where L—L (a bidentate ligand) and Me_2SO are used in excess, the pseudo first-order rate constant k for the loss of the Pt(II) reactant is:

$$
k = \frac{a\,[\text{L}\!-\!\text{L}]}{b\,[\text{L}\!-\!\text{L}] + [\text{Me}_2\text{SO}]} + c\,[\text{L}\!-\!\text{L}]
\tag{4.102}
$$

This conforms to the usual two-term rate law (4.96), but in which the k_1 term is associated with a three-coordinated intermediate in (4.99). The values of $a = k_{\text{D}} k_{\text{D}}^{\text{Y}}/k_{-\text{D}}$, $b = k_{\text{D}}^{\text{Y}}/k_{-\text{D}}$ and $c = k_2$ (in (4.96)). Significantly the exchange of Me_2SO with $cis\text{-Pt}(\text{C}_6\text{H}_5)_2(\text{Me}_2\text{SO})_2$, in CDCl_3 followed by stopped flow-nmr, yields a value for k_{D}, 0.08 s^{-1} (27 °C) similar to that obtained by examining (4.101) using $\text{L}-\text{L} = \text{bpy}$, and other bidentate ligands in C_6H_6.[147(a)] Markedly different values of b for different entering ligands (L$-$L) support a D mechanism. The D mechanism might be favored in very poor nucleophilic solvents such as C_6H_6 or

$CHCl_3$, but its differentiation from the solvolytic path is still very difficult. [142, 148] The $\Delta V^{\ddagger}$ value of $+5.5 \pm 0.8$ cm^{3}mol^{-1} for the exchange of Me_2SO with $Pt(C_6H_5)_2(Me_2SO)_2$ also strongly supports a D mechanism which appears favored in complexes containing a Pt-C bond. [147(b)]

(b) Significance of k_2. It is generally accepted that the key rate term in substitution in square-planar complexes involves nucleophilic associative attack of the entering nucleophile Y on the metal complex. It follows from this that the bond-making as well as the bond-breaking process will be important, and it can be expected that there will be varying degrees of participation by both. All attempts, however, to observe an intermediate in the bimolecular reactions of Pt(II) have failed, even with the most favorable situation, that is, using a strong entering ligand and a weak leaving one. [145]

4.6.2 Activation Parameters

Substitution in Pt(II) and Pd(II) complexes is invariably attended by large negative values of $\Delta S^{\ddagger}$ and negative values of $\Delta V^{\ddagger}$ (Table 4.12). [149-152] This is consistent with an associative mechanism and a net increase in bonding in an ordered and charged transition state. These considerations apply to both the k_1 and the k_2 terms. [149] Although there is strong evidence for an absolute A mechanism, the small value of $\Delta V^{\ddagger}$ for the simple $M(H_2O)_4^{2+} - H_2O$ exchanges (Table 4.12) [151], M = Pd(II) or Pt(II), might suggest otherwise and support an I_a process (see also $Pt(Me_2SO)_4^{2+}$, Me_2SO exchange (Sec. 7.4.1)). However the formation of a fifth (equatorial) metal-water bond in the trigonal-bipyramidal intermediate or transition state (Sec. 4.7.5) *may* be accompanied by lengthening of the two (axial) metal-water bonds offsetting somewhat the anticipated larger negative $\Delta V^{\ddagger}$. [151] More data are obviously required.

Table 4.12 Activation Parameters for Substitution in Some Square-Planar Complexes in Water at 25°C Refs. 149–152.

Complex	Reagent	k M^{-1}s^{-1}	$\Delta H^{\ddagger}$ kJ mol^{-1}	$\Delta S^{\ddagger}$ J K^{-1}mol^{-1}	$\Delta V^{\ddagger}$ cm^{3}mol^{-1}
$Pd(H_2O)_4^{2+}$	H_2O	5.6×10^{2a}	50	-26	-2.2
$Pt(H_2O)_4^{2+}$	H_2O	3.9×10^{-4a}	90	-9	-4.6
$Ni(CN)_4^{2-}$	CN^-	$>5 \times 10^5$	–	–	–
$Pd(CN)_4^{2-}$	CN^-	1.2×10^{2b}	17	-178	–
$Pt(CN)_4^{2-}$	CN^-	26^b	26	-143	–
$Au(CN)_4^-$	CN^-	3.9×10^{3b}	28	-100	–
$Pt(dien)Br^+$	H_2O	1.4×10^{-4a}	84	-63	-10
	Cl^-	6×10^{-4}	75	-46	–
	Br^-	6×10^{-3}	67	-63	–
	N_3^-	6.4×10^{-3}	65	-71	-8.5
	py	2.8×10^{-3}	46	-136	-7.7
	NO_2^-	1.4×10^{-3}	72	-56	-6.4
	I^-	0.32	46	-104	–
	NCS^-	0.68	39	-112	–
	$SC(NH_2)_2$	1.3	35	-121	–

^{a}s^{-1} ^{b}No [CN^-]-independent term

4.7 Ligand Effects on the Rate

One of the consequences of an associative mechanism is the decided importance of all the ligands — entering, leaving and remaining — on the rate of the process. This arises because all the ligands involved are present in the five-coordinate activated complex and can therefore affect its stability and the activation energy for its production. This feature distinguishes planar from octahedral substitution. There have therefore been, mainly in the 60's and early 70's a large number of studies in which systematic variations are made in the character of all the ligands. [149] The interpretation of the results has been helped by the fact that complete retention of configuration during substitution has been consistently observed with the type of Pt(II) complexes studied. [142]

4.7.1 Effect of Entering Ligand

There have been extensive studies of the influence of an entering ligand on its rate of entry into a Pt(II) complex. [144, 153] The rate constants for reaction of a large number and variety of ligands with $trans$-Pt(py)$_2$Cl$_2$ have been measured (Table 4.13). The large range of reactivities is a feature of the associative mechanism and differentiates it from the behavior of octahedral complexes. The rate constants may be used to set up quantitative relationships. For a variety of reactions of Pt complexes in different solvents (Sec. 2.5.4):

$$\log k_y = s n_{Pt} + \log k_s \tag{4.103}$$

Table 4.13 Rate Constants (k, $M^{-1}s^{-1}$) for Reaction of $trans$-Ptpy$_2$Cl$_2$ with a Number of Nucleophiles in CH$_3$OH [144, 153]

Nucleophile	$10^3 \times k$	n_{Pt}	Nucleophile	$10^3 \times k$	n_{Pt}
CH$_3$OH	0.00027	0.0	I$^-$	107[a]	5.46
CH$_3$O$^-$	Very Slow	<2.4	(CH$_3$)$_2$Se	148	5.70
Cl$^-$	0.45[a]	3.04	SCN$^-$	180[a]	5.75
NH$_3$	0.47[a]	3.07	SO$_3^{2-}$	250[a]	5.79
C$_5$H$_5$N	0.55[a]	3.19	Ph$_3$Sb	1,810	6.79
NO$_2^-$	0.68[a]	3.22	Ph$_3$As	2,320	7.68
C$_3$H$_3$N$_2$	0.74	3.44	CN$^-$	4,000	7.14
N$_3^-$	1.6[a]	3.58	SeCN$^-$	5,150[a]	7.11
N$_2$H$_4$	2.9[a]	3.86	SC(NH$_2$)$_2$	6,000[a]	7.17
Br$^-$	3.7[a]	4.18	Ph$_3$P	249,000	8.93
(CH$_3$)$_2$S	21.9	4.87			

[a] Kinetic data at 30°C.

The term s is the nucleophile discrimination factor. The values of n_{Pt}, the nucleophilic reactivity constant, is useful for correlating kinetic data for other Pt(II) complexes. [144, 149, 153]

4.7.2 Effect of Leaving Group

Reactions of the type

$$Pt(dien)X^+ + py \rightarrow Pt(dien)py^{2+} + X^- \tag{4.104}$$

in aqueous solution have been used to study the effect of X on the rates of the reactions.[149] The members of the series differ in rate constants by as much as 10^6 from the slowest

$$CN^- < NO_2^- < SCN^- < N_3^- \ll I^- < Br < Cl^- < H_2O < NO_3^- \tag{4.105}$$

(CN^-) to the fastest. Generally, the second-order rate constant increases with decreasing basicity of the leaving group and this gives rise to LFER. It is therefore fairly clear from these observations that metal-ligand bond breakage must be significant, even in a predominantly associative reaction.

4.7.3 Effect of Ligands Already Present

The group *trans* to the leaving ligand appears to have a more pronounced influence than the two *cis* to it, on the rate of its departure.[149] It has been known for many years that a ligand can be assigned an order of *trans effect* which denotes its tendency to direct an incoming group in the position *trans* to itself. In Pt(II) complexes, this power decreases approximately in the order[149]

$$CN^-, C_2H_4, CO, NO > R_3P \approx H^- \approx SC(NH_2)_2 > CH_3^- > C_6H_5^- >$$

$$SCN^- > NO_2^- > I^- > Br^- > Cl^- > NH_3 > OH^- > H_2O \tag{4.106}$$

The effect has played an important role in preparing Pt(II) complexes of specific geometry. The greater *trans* effect appears to be associated with a larger rate constant for the elimination of the *trans* ligand, from a limited number of studies.[154] Comparison of (4.106) with (4.105) and Table 4.12 shows that the groups with a pronounced ability to trans-labilize are replaced the least easily and are the more powerful nucleophiles. This might to anticipated since in the five-coordinate intermediate (see 4.109) the entering nucleophile E, *trans* ligand T, and the departing ligand D, all occupy positions in the trigonal plane and all may influence the energetics of the transition state in similar ways.

In an associative mechanism, increasing the steric hindrance in the complex should lead to decreasing rates (steric retardation). This has been realised with a number of systems, that of $Pd(R_5dien)X^{n+}$ being the most explored.[155, 156] Activation parameters for the reaction

$$Pd(R_5dien)Cl^+ + I^- \rightarrow Pd(R_5dien)I^+ + Cl^- \tag{4.107}$$

are shown in Table 4.14.[156(a)] For R = H, i.e. dien itself, the usual two term rate law (4.94) is observed and the value of k_1 is shown in Table 4.14. The rate constant k_1, decreases with increasing steric hindrance as a result of increases in the value of $\Delta H^{\ddagger}$. With the heavily hindered complexes the rate constant k_1 is almost 10^5 less in value than for the dien complex and now the k_2 term is difficult to detect [155] unless large concentrations of a strong entering nucleophile are used. [143, 156(a)] The values of $\Delta V^{\ddagger}$ however throughout the series remain negative at -12 ± 2, although rising to -3 ± 1 cm^3 mol when the replaced group is neutral in Pd(Et$_4$dien)NH$_3^{2+}$. One is forced to conclude that the solvolyses remain associative in character even in the hindered complexes although its extent may change. Similar conclusions are reached from an examination of exchange reactions of Pd(R$_5$dien)H$_2$O^{2+} Ref. 156(b). In some respects then these hindered complexes, which have also been examined with Pt(II) and Au(III), [149] resemble their octahedral analogs and they have been termed pseudo-octahedral complexes. [157] They result when bulky ligands in the plane of a four-coordinate metal complex spill over and hinder the apical positions and they resemble the octahedral complexes in reactivity characteristics, while at the same time retaining some features of a square planar complex.

Table 4.14 Activation Parameters for the Solvolysis of Pd(R$_5$dien)Cl$^+$ in Aqueous Solution using Reaction (4.107) at 25 °C. From Ref. 156(a))

R$_5$dien	k_1 s^{-1}	$\Delta H^{\ddagger}$ kJ mol^{-1}	$\Delta S^{\ddagger}$ J K^{-1}mol^{-1}	$\Delta V^{\ddagger}$ cm^{3}mol^{-1}
dien	44	43	-69	-10.0
1,4,7-Et$_3$dien	10	41	-86	-10.8
1,1,7,7-Et$_4$dien	2.2×10^{-3}	66	-74	-14.9
1,1,4,7,7-Me$_5$dien	0.28	50	-88	-10.9
1,1,4,7,7-Et$_5$dien	7.2×10^{-4}	59	-106	-12.8

4.7.4 Effect of Solvent

The solvent is the reaction medium and as such, by solvating the ground and activated states, will influence the energetics of the activation process. In addition it acts as a nucleophile in the reaction path represented by k_1. A large value of k_1 relative to k_2 is observed in solvents capable of coordinating strongly to the metal so that *generally* the order

$$(CH_3)_2SO > MeNO_2, H_2O > ROH \tag{4.108}$$

is observed. [149] In solvents that are poor coordinators such as C$_6$H$_6$ and CCl$_4$, the k_2 value dominates. The order of nucleophilicities does not however change in different solvents. A linear plot of $\Delta V^{\ddagger}$ vs the solvent electrostriction parameter for reaction of *trans*-Pt(py)$_2$(Cl)NO$_2$ with py in a variety of solvents, attests to the importance of solvation in these reactions. [158]

4.7.5 Reaction Pathways

There is naturally an overriding interest in the geometry of the five-coordinate intermediate, or activated complex. General considerations of the shape in which there will be least mutual interaction of five ligands and of the available orbitals [149] support a trigonal bipyramid. Much replacement behavior can be rationalized on this basis. It has been suggested, however, that both square-pyramid and trigonal-bipyramid geometries are developed in the course of the replacement: [142, 149]

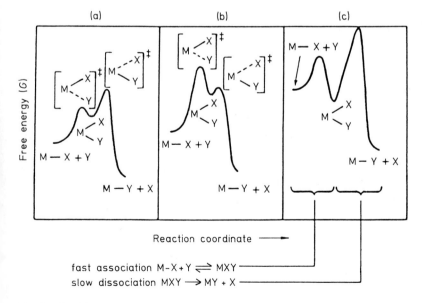

T,D and E in trigonal plane   (4.109)

The departing ligand D is replaced by the entering E. The ligands C_1 and C_2 are *cis*, and T is *trans* to D. The full scheme (4.109) should therefore be represented by a reaction profile (Sec. 2.3.6) which contains 3 minima, corresponding to the 3 intermediates and 4 peaks, corresponding to the activated complexes for the four steps. [142] For the sake of simplicity, the reaction profile can be represented as shown in Fig. 4.9. A number of cases are depicted, exaggerating those likely to be encounted in practice. In Fig. 4.9(a) bond breaking is rate-determining and in Fig. 4.9(b) bond making is rate-determining. In both cases, the energy of

Fig. 4.9 Simplified reaction profiles for various situations in the associative mechanism for substitution in square planar complexes, focusing attention on the replacement $M-X + Y \rightarrow M-Y + X$ (4.93).

the intermediate is above the energies of reactant and products. The highest energy level determines the *major* transition state. If a five-coordinate intermediate is detected as in certain Rh(I) and Ni(II) reactions, the energy corresponding to this intermediate will be between the energies of reactant and products and Fig. 4.9(c) results. [149, 154] With Au(III) all the data suggest that bond formation and bond breaking are synchronous, the pentacoordinated species is therefore the activated complex and the peaks in the reaction profile are replaced by one major peak (Chap. 8. Au(III)). [159] The order of reactivity is

$$Ni(II) \gg Au(III) > Pd(II) > Pt(II) \qquad (4.110)$$

(See Table 4.12) and is possibly related to the stability of 5-coordinate complexes and therefore the ease with which this is reached in the transition state.

4.7.6 Chelation in Square-Planar Complexes

As with substitution in octahedral complexes, in chelation in square planar complexes the formation of the first bond is usually rate-determining (see (4.101)). [159]

Compare the ring closure shown in (4.111) with the corresponding process involving only a unidentate ligand in (4.112), rate constants at 25 °C.

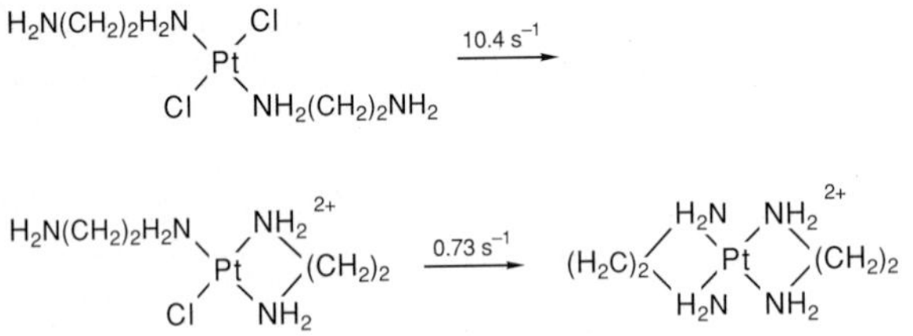

$$(4.111)$$

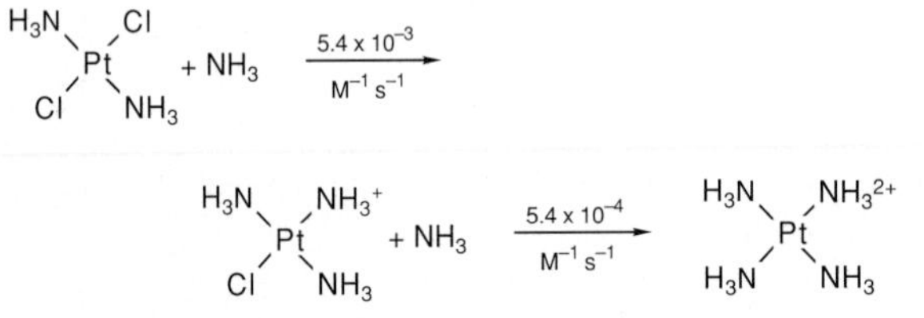

$$(4.112)$$

The comparison has to be made between a first-order rate constant for the unimolecular process and a second-order rate constant for the corresponding intermolecular reaction (Sec. 6.1.1). One may arbitrarily decide on a 1 M concentration of reagent NH_3, in which case the pseudo first-order rate constant for the intermolecular process is $5.7 \times 10^{-3} s^{-1}$ (but the procedure is far from satisfactory). On this basis ring closures are over 10^3 times faster than unidentate replacement in 1 M NH_3. When a slightly greater nucleophilicity of ethylenediamine over ammonia is allowed for, a nearly 10^3 fold enhancement in rate attends chelation. A large part of this effect resides in a higher effective concentration of $-NH_2$ nearer the replaced chloride in (4.111) than in (4.112). [160] Ring opening of chelates appears to occur at only a slightly slower rate than cleavage of unidentates from the metal (about a factor of 10) so that the enhanced stability of Pt(en)$_2^{2+}$ over Pt(NH$_3$)$_4^{2+}$ resides largely in enhanced formation rates. [160] A proton may aid in the removal of chelating ligands in a manner similar to that of octahedral complexes, although second-order reaction paths are possible:

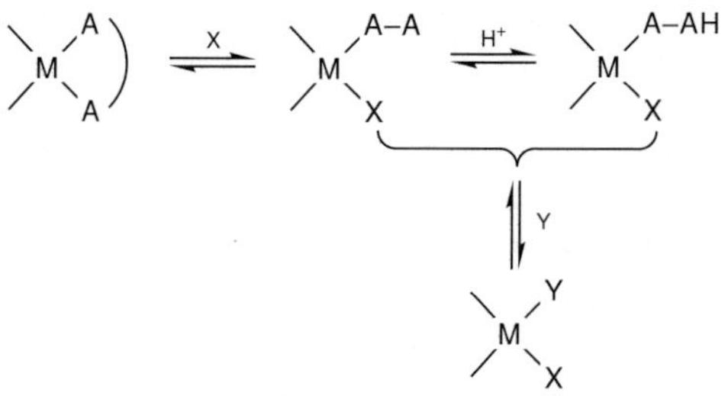

$$(4.113)$$

X and Y may be H$_2$O or halide groups. A number of reactions of Pd(II) appear to fit this scheme. [161] Ring opening and the subsequent slower displacement of single bonded amine are kinetically separable with M = Au(III), and each step proceeds via a normal nucleophilic attack. [159]

(a) Coordination Chain Reactions

The interaction of two metal chelates, MA$_2$ with NB$_2$, poses interesting mechanistic problems. Depending on the system, nature of chelating molecules A and B etc., complete exchange may occur,

$$\text{MA}_2 + \text{NB}_2 \ \rightarrow\ \text{MB}_2 + \text{NA}_2 \qquad (4.114)$$

or mixed species may be partially or completely formed [162-164]

$$\text{MA}_2 + \text{NB}_2 \ \rightleftharpoons\ \text{MAB} + \text{NAB} \qquad (4.115)$$

As an example, consider the interchange between Cu(mnt)$_2^{2-}$ and Ni(Et$_2$dsc)$_2$

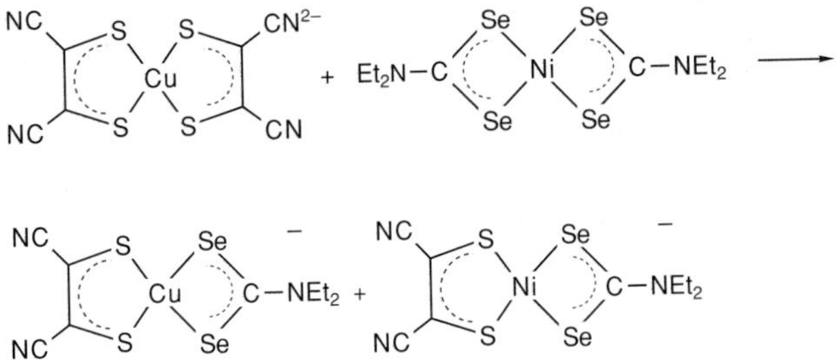

$$(4.116)$$

The rate has been measured in 1:1 acetone/CHCl$_3$ by a stopped-flow/epr method. [163] The epr signals of Cu(mnt)$_2^{2-}$ and Cu(mnt)(Et$_2$dsc)$^-$ are sufficiently different to allow analysis. (Fig. 4.10). A second-order rate law is usually observed, but this belies a simple mechanism,

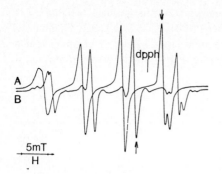

dpph

A
B

5mT
H

Fig. 4.10 X-band epr spectra of $Cu(mnt)_2^{2-}$ (A) and $Cu(mnt)(Et_2dsc)^-$ (B) at 25 °C. The reaction is monitored at the field strengths indicated by the arrows. Full details for the mixing and accumulation of epr signals are given in Ref. 163. Reproduced with permission from J. Stach, R. Kironse, W. Dietzsch, G. Lassmann, V. K. Belyaeva and I. N. Marov Inorg. Chim. Acta **96**, 55 (1985).

since with some conditions an S-shaped formation of product with time is observed. In addition the rate is retarded by added ligand (Na_2mnt) and accelerated by Cu(II). Although a full kinetic study was not reported, the observations were consistent with a chain mechanism, in which there are solvated monochelate intermediates:[162, 163]

$$Cu(mnt)_2^{2-} \rightleftharpoons Cu(mnt) + mnt^{2-} \quad \left.\right\} \text{chain initiation} \qquad (4.117)$$
$$Ni(Et_2dsc)_2 \rightleftharpoons Ni(Et_2dsc)^+ + Et_2dsc^- \qquad (4.118)$$

$$Cu(mnt)_2^{2-} + Ni(Et_2dsc)^+ \rightleftharpoons Ni(mnt)(Et_2dsc)^- + Cu(mnt) \quad \left.\right\} \text{chain propagation} \qquad (4.119)$$
$$Ni(Et_2dsc)_2 + Cu(mnt) \rightleftharpoons Cu(mnt)(Et_2dsc)^- + Ni(Et_2dsc)^+ \qquad (4.120)$$

$$Cu(mnt) + Et_2dsc^- \rightleftharpoons Cu(mnt)(Et_2dsc)^- \quad \left.\right\} \text{chain termination} \qquad (4.121)$$
$$Ni(Et_2dsc)^+ + mnt^{2-} \rightleftharpoons Ni(mnt)(Et_2dsc)^- \qquad (4.122)$$

Solid evidence for this type of chain mechanism (differing only in the chain propagation steps) had been earlier obtained by Margerum and his co-workers[165, 166] in the study of "coordination chain reactions" between two metal complexes each containing *one* multidentate ligand, edta, trien and so on, e.g.,

$$Ni(trien)^{2+} + Cu(edta)^{2-} \rightarrow Ni(edta)^{2-} + Cu(trien)^{2+} \qquad (4.123)$$

A number of cross reactions of the type (4.115) have been studied and, when sufficiently slow, an elegant HPLC batch method (Sec. 3.11(a)) has been used.[164]

4.8 Substitution in Tetrahedral Complexes

The exchange reactions (M = Fe, Co, and Ni; Ar = Ph or *p*-tolyl)

$$M(PAr_3)_2Br_2 + PAr_3 \rightleftharpoons M(PAr_3)_2Br_2 + PAr_3 \qquad (4.124)$$

studied in $CDCl_3$ by nmr linewidth techniques are all second-order (Table 4.15).[167] The lability trend Fe > Ni > Co resides mainly in a $\Delta H^{\ddagger}$ effect.

Table 4.15 Rate Constants and Activation Parameters for Ligand Exchange of $M(PPh_3)_2Br_2$ in $CDCl_3$ at 25 °C [167]

M	k $M^{-1}s^{-1}$	$\Delta H^{\ddagger}$ $kJ\,mol^{-1}$	$\Delta S^{\ddagger}$ $J\,K^{-1}mol^{-1}$
Fe	2.0×10^5	16	-92
Co	8.7×10^2	32	-79
Ni	6.9×10^3	20	-104

Ligand field arguments indicate that the tetrahedral d^6 ground state and five-coordinate d^6 transition state are both stabilized to a lesser extent than the d^7 and d^8 counterparts. These effects would make Fe(II) more reactive and less reactive, respectively, than Co(II) and Ni(II), so that presumably ground-state destabilization is the more important. For the $Co(PPh_3)_2Br_2$ exchange with Ph_3P in $CDCl_3$, a value for $\Delta V^{\ddagger} = -12.1$ cm^{3}mol^{-1} at 30 °C as well as the highly negative $\Delta S^{\ddagger}$, are strong supportive evidence for an associative mechanism.[168] See also Ref. 169 for substitution in the tetrahedral $FeCl_4^{2-}$ ion.

The situation is quite different with tetrahedral complexes of Ni(0), Pd(0) and Pt(0). We might anticipate that an associative mechanism would be deterred, because of strong mutual repulsion of the entering nucleophile and the filled d orbitals of the d^{10} system. Thus a first-order rate law for substitution in Ni(0) carbonyls,[170] and $M^0(P(OC_2H_5)_3)_4 M = Ni$, (Sec. 1.4.1) Pd and Pt,[171] as well as a positive volume of activation ($+8$ cm^{3}mol^{-1}) for the reaction of $Ni(CO)_4$ with $P(OEt)_3$ in heptane[172] support an associative mechanism.

4.9 Substitution in Five-Coordinate Complexes

Five-coordinate complexes may have either a square-pyramidal or a trigonal-bipyramidal geometry. The replacement reactions of the complexes of mainly Co(II), Ni(II) and Cu(II) have been studied in both aqueous and nonaqueous media. As might be expected, the mechanisms of substitution may be associative or dissociative in character.[173] Axial methyl groups in macrocycles such as cyclam and tet b can impose five-coordination on metal complexes, usually square pyramidal geometry, and associative mechanisms. The complex $Ni(RSRSMe_4cyclam)(D_2O)^{2+}$ (3) exchanges with D_2O with a negative entropy of activation $(-24 \text{ J K}^{-1}mol^{-1})$ whereas for $trans$-$Ni(RRSSMe_4cyclam)(D_2O)_2^{2+}$, **2**, $\Delta S^{\ddagger} = +38$ J K^{-1}mol^{-1} for the corresponding exchange[174] (Table 4.9). For the $Co(Me_4cyclam)CH_3CN^{2+}$, CH_3CN exchange, $k = 6.4 \times 10^5$ s^{-1}, $\Delta H^{\ddagger} = 18.7$ kJ mol^{-1}, $\Delta S^{\ddagger} = -71$ J K^{-1}mol^{-1} and $\Delta V^{\ddagger} = -9.6$ cm^{3}mol^{-1} all evidence for an I_a mechanism. However, for the corresponding Ni complex, a slightly positive value for $\Delta V^{\ddagger}$ $(+2.3$ cm^3 mol$^{-1})$ is interpreted in terms of a D mechanism. Significantly, an equilibrium between a 5- and 4-coordinate Ni(II) − $Me_4cyclam$ complex in CH_3CN has been established.[175] LFER for the reactions (for tet a and tet b see Sec. 7.9):

red or *blue* $Cu(tet\,b)H_2O^{2+} + Y^- \rightleftharpoons$ *red* or *blue* $Cu(tet\,b)Y^+ + H_2O \quad k_1, k_{-1}$ (4.125)

support the associative character for the anation (Sec. 2.5). [176] The value of k_1 depends on the nucleophilicity of Y^- but since the effect is not as pronounced as with Pt(II), an I_a mechanism may hold.

Table 4.16 Activation Parameters for the Exchange of $M(Me_6tren)dmf^{2+}$ with dmf at 25°C Ref. 177

M	k_{exch} s^{-1}	$\Delta H^{\ddagger}$ kJ mol^{-1}	$\Delta S^{\ddagger}$ J K^{-1} mol^{-1}	$\Delta V^{\ddagger}$ cm^3 mol^{-1}
Mn	2.7×10^6	18	-61	-6
Co	51	52	-36	-2.7
Cu	5.6×10^2	43	-47	$+6.5$

Substitution behavior in a trigonal-bipyramidal structure has been confined to an examination of the exchange of $M(Me_6tren)dmf^{2+}$, M = Mn(II), Co(II) and Cu(II), with dmf (Table 4.16). [177] The values of $\Delta V^{\ddagger}$ suggest an increasing dissociative character for the exchange with progressive filling of the d_{xz} and d_{yz} orbitals, from Mn to Cu. It is suggested that there is thereby increasing hindering of an approaching dmf to any of the three faces of the trigonal bipyramid which are adjacent to the axially-coordinated dmf. [177] The dissociative character for substitution in $Cu(Me_6tren)dmf^{2+}$ is supported by saturation kinetics for anation by SCN^- and an associated $K_0 = 157$ M^{-1} and (as expected) $k_3 = 5.5 \times 10^2$ s^{-1}. Eqns. (4.10) and (4.11). Ref. 178.

4.10 Substitution in Organized Surfactant Systems

The rate of metal complex formation is often modified (usually enhanced) by the presence of a charged interface in the aqueous phase. This may be provided by ionic micelles, e.g., SDS, [179, 180] microemulsions [179] or polyelectrolytes. [181, 182] The reactions of Ni^{2+} and Co^{2+} with hydrophobic ligands pan, pap and pad **14–16** are popular ones for examining effects, since they are well characterized in the bulk water. The simple model (4.126)

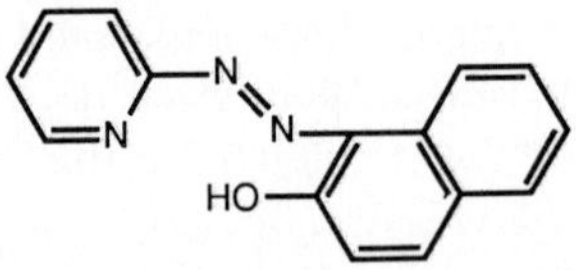

14
(pan)

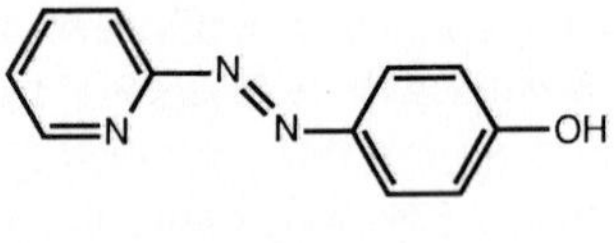

15
(pap)

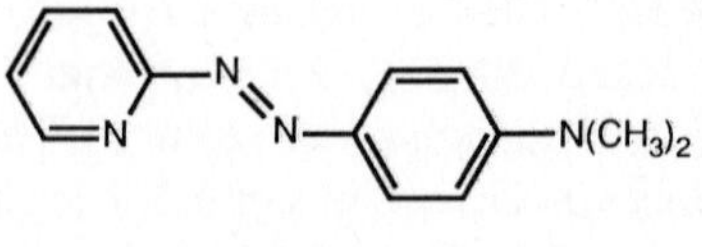

16
(pad)

$$M^{2+}_{sol} + L_{sol} \underset{k^{sol}_{-1}}{\overset{k^{sol}_{1}}{\rightleftharpoons}} ML^{2+}_{sol}$$

$$\updownarrow \qquad \updownarrow \qquad\qquad \updownarrow$$

$$M^{2+}_{sur} + L_{sur} \underset{k^{sur}_{-1}}{\overset{k^{sur}_{1}}{\rightleftharpoons}} ML^{2+}_{sur} \qquad\qquad (4.126)$$

presupposes rapid interchange of species between the bulk solvent (sol) and the pseudo phase associated with surfactant (sur), compared with the complexation rate. The usual equation for complex formation is modified to allow for micelle surface concentration of M^{2+}

$$k_{obs} = k^{sur}_1 \frac{[M^{2+}]_T}{([SDS]_T - cmc)AN} + k^{sur}_{-1} \qquad\qquad (4.127)$$

cmc = critical micelle concentration
A = surface area per SDS head group in m^2
N = Avogadro's Number

The rate constants k^{sur}_1 and k^{sur}_{-1} represent rate constants for a surface reaction and have units $m^2 mol^{-1} s^{-1}$ and s^{-1} respectively. The accelerative effects are about 10^2–10^3 fold. They indicate that both reactants are bound at the surface layer of the micelle (surfactant-water interface) and the enhanced rates are caused by enhanced reactant concentration here and *there are no other significant effects*.[179] Similar behavior is observed in an inverse micelle, where the water phase is now dispersed as micro-droplets in the organic phase. With this arrangement, it is possible to study anion interchange in the tetrahedral complexes $CoCl_4^{2-}$ or $CoCl_2(SCN)_2^{2-}$ by temperature-jump. A dissociative mechanism is favored, but the interpretation is complicated by uncertainty in the nature of the species present in the water-surfactant boundary, a general problem in this medium.[183]

If the polyelectrolyte can coordinate strongly to a metal ion, marked deceleration effects can be noted, as, for example, in the reactions of Ni^{2+} and Co^{2+} with pad in the presence of polyphosphates.[181, 182] Modifications of equilibria constants in these micelles must also be recognized as contributing to rate change, e.g., ligand pK or keto-enol equilibria may be altered.

4.11 Substitution in Metalloproteins

Undoubtedly the most complicated mileau for a substitution process is that of a protein. However, the principles developed in this chapter for substitution in metal complexes also apply to metalloproteins.[184] Allowance for a role for the protein, particularly near the site, must always be made. The formation and dissociation of a metalloprotein (PM) may be represented in an undoubtedly simplified form as:

$$P + ML \rightleftharpoons PML \rightleftharpoons PM + L \qquad\qquad (4.128)$$

where P is a demetallated (apo) protein and ML is a metal-ligand complex. The incorporation of metal ions into proteins appears to occur post-translationally and so the forward direction probably represents the last step in the biosynthesis of many metalloproteins. The occurrence of a ternary intermediate or intermediates is deduced by the observation of "saturation kinetics" or, in rare cases, directly observed. Conformational changes may occur after the metal ion is placed in the metal site. The reverse direction has also been studied, ligand L removing metal ion M from the protein to produce the apo form. From the latter a variety of metalloforms of the protein may be prepared for useful structural analysis. For references and the study of two systems see Refs. 185 and 186. In general, formation rate constants are smaller than those for simpler ligands.

Multisites in proteins are not uncommon. The removal of metal ions from such centers is likely to be involved. This is in fact illustrated by the iron removal from serotransferrin, see also Sec. 2.6. This protein is bilobal and each lobe contains an iron-binding site. These are 35 Å apart and it is believed that direct interaction between the sites is absent. The two Fe's, designated a and b, are different and their removal is biphasic, although not markedly so.

$$
\begin{array}{ccc}
 & \mathrm{TfFe_b} & \\
 & \nearrow^{k_1} \quad \searrow^{k_3} & \\
\mathrm{Fe_aTfFe_b} & & \mathrm{Tf} \\
 & \searrow_{k_2} \quad \nearrow_{k_4} & \\
 & \mathrm{Fe_aTf} &
\end{array}
\qquad (4.129)
$$

Simultaneous rate equations are complex, but solvable.[187] Simplifications are possible e. g. $k_1 = k_4$ and $k_2 = k_3$, if the sites are non-cooperative. Strong binding ligands such as edta or synthetic sidereophores effect iron removal and the two rate constants associated with the biphasic Fe removal are both curved towards saturation when plotted against [ligand].

Such behavior has been encountered in Chapter 1 and conforms to either a mechanism involving formation of a ternary species (considering just one iron):

$$
\mathrm{FeTf} + \mathrm{L} \rightleftharpoons \mathrm{FeTfL} \rightarrow \mathrm{FeL} + \mathrm{Tf} \qquad (4.130)
$$

or rate-determining transformation of a "closed" form (FeTf) to a reactive "open" one, FeTf*:

$$
\mathrm{FeTf} \rightleftharpoons \mathrm{FeTf^*} \underset{-L}{\overset{+L}{\rightleftharpoons}} \mathrm{FeTfL} \rightarrow \mathrm{FeL} + \mathrm{Tf} \qquad (4.131)
$$

(see Sec. 1.6.4).[188-190]

References

1. Y. Ducommun and A. E. Merbach in B 17, Chap. 2.
2. H. Taube, Chem. Revs. **50**, 69 (1952).
3. M. H. M. Abou El-Wafa, M. G. Burnett and J. F. McCullagh, J. Chem. Soc. Dalton Trans. 1059 (1987). Careful analysis shows no direct substitution of Cl^- by N_3^- in $Co(CN)_5Cl^{3-}$.
4. W. G. Jackson, B. C. McGregor and S. S. Jurisson, Inorg. Chem. **26**, 1286 (1987). The reactions of 14 different complexes of the type $Co(NH_3)_5X^{n+}$ with SCN^- show 3-19 % *direct* anion capture.

5. For example reaction of reducing ligands with tervalent metal complexes are often accompanied by complex formation.

6. J. Springborg, Adv. Inorg. Chem. **32**, 55 (1988).

7. F. P. Rotzinger, H. Stunzi and W. Marty, Inorg. Chem. **25**, 489 (1986).

8. H. A. Mottola, Kinetic Aspects of Analytical Chemistry, Wiley-Interscience, NY, 1988; D. P.-Bendito and M. Silva, Kinetic Methods in Analytical Chemistry, Ellis-Horwood, Chichester 1988.

9. A. S. Mildvan, Metals in Enzyme Catalysis, in the Enzymes, Chap. 9.

10. J. P. Hunt and H. L. Friedman, Prog. Inorg. Chem. **30**, 359 (1983). This review is concerned with the structure of hydration complexes of ions and includes a discussion of the x-ray or neutron diffraction method for determining structure of ions in solution.

11. T. W. Swaddle, Inorg. Chem. **22**, 2663 (1983); T. W. Swaddle and M. K. S. Mak, Can. J. Chem. **61**, 473 (1983).

12. J. E. Enderby, S. Cummings, G. J. Herdman, G. W. Neilson, P. S. Salmon and N. Skipper, J. Phys. Chem. **91**, 5851 (1987); G. W. Neilson and J. E. Enderby, Adv. Inorg. Chem. **34**, 195 (1989).

13. S. F. Lincoln, Coordn. Chem. Revs. **6**, 309 (1971); B28, Chap. 2.

14. A. E. Merbach, Pure Appl. Chem. **54**, 1479 (1982); **59**, 161 (1987).

15. Terms used by C. H. Langford and H. B. Gray in B31. They replace older ones, S_{N^2} or S_{N^1}, used in B25.

16. T. W. Swaddle, Substitution Reactions of Divalent and Trivalent Metal Ions, in Adv. Inorg. Bioinorg. Mechs. **2**, 95 (1983).

17. D. W. Margerum, G. R. Cayley, D. C. Weatherburn and G. K. Pagenkopf in B32, Chap. 1. A complete coverage of substitution and rate data up to 1977.

18. J. F. Coetzee, Pure Appl. Chem. **49**, 27 (1977); E. F. Caldin, Pure Appl. Chem. **51**, 2067 (1979).

19. D. Darensbourg, Adv. Organomet. Chem. **21**, 113 (1982); F. Basolo, Polyhedron, **9**, 1503 (1990).

20. L. Mønsted and O. Mønsted, Coordn. Chem. Revs. **94**, 109 (1989).

21. P. Moore, Pure Appl. Chem. **57**, 347 (1985).

22. L. S. Frankel, Inorg. Chem. **10**, 2360 (1971). The exchange rate constant of $Ni(dmf)_6^{2+}$ with dmf in CH_3NO_2 solution is independent of the composition of the solvent mixture. Since an I_d mechanism would require an abnormally large K_0, a D mechanism is favored.

23. Y. Ducommun, K. E. Newman and A. E. Merbach, Inorg. Chem. **19**, 3696 (1980).

24. Y. Ducommun, D. Zbinden and A. E. Merbach, Helv. Chim. Acta **65**, 1385 (1982).

25. P. Bernhard, L. Helm, I. Rapaport, A. Ludi and A. E. Merbach, Inorg. Chem. **27**, 873 (1988). All exchange k's are for a particular solvent molecule (Sec. 1.9).

26. L. Helm, S. F. Lincoln, A. E. Merbach and D. Zbinden, Inorg. Chem. **25**, 2550 (1986) and references therein.

27. F. K. Meyer, K. E. Newman and A. E. Merbach, Inorg. Chem. **18**, 2142 (1979).

28. M. J. Sisley, Y. Yano and T. W. Swaddle, Inorg. Chem. **21**, 1141 (1982).

29. C. Cossy, L. Helm and A. E. Merbach, Helv. Chim. Acta **70**, 1516 (1987).

30. L. Fielding and P. Moore, J. Chem. Soc. Chem. Communs. 49 (1988).

31. A. Hioki, S. Funahashi, M. Ishii and M. Tanaka, Inorg. Chem. **25**, 1360 (1986).

32. D. H-Cleary, L. Helm and A. E. Merbach, J. Amer. Chem. Soc. **109**, 4444 (1987) and references therein.

33. A. D. Hugi, L. Helm and A. E. Merbach, Inorg. Chem. **26**, 1763 (1987).

34. J. Bjerrum and K. G. Poulsen, Nature London, **169**, 463 (1952).

35. T. W. Swaddle and G. Guastalla, Inorg. Chem. **8**, 1604 (1969).

36. R. M. Fuoss, J. Amer. Chem. Soc. **80**, 5059 (1958); M. Eigen, Z. Phys. Chem. (Frankfurt) **1**, 176 (1954).

37. By equating the exponential term to unity and setting $a = 4$ Å arbitrarily, $K_0 = 2.5 \times 10^{21} a^3 \simeq 0.15$ M^{-1}. See also J. E. Prue, J. Chem. Soc. 7534 (1965); D. B. Rorabacher, Inorg. Chem. **5**, 1891 (1966).

38. D. W. Carlyle and J. H. Espenson, Inorg. Chem. **8**, 575 (1969); H. Brintzinger and G. G. Hammes, Inorg. Chem. **5**, 1286 (1966); P. Hemmes and S. Petrucci, J. Phys. Chem. **72**, 3986 (1968).

39. The evidence available indicates that outer-sphere complexing of the $M-H_2O$ entity does not markedly alter the value of k_{exch}. For literature see Refs. 16 and 40.

40. R. G. Wilkins, Acc. Chem. Res. **3**, 408 (1970); Comments Inorg. Chem. **2**, 187 (1983).

41. K. R. Ashley, M. Berggren and M. Cheng, J. Amer. Chem. Soc. **97**, 1422 (1975); K. R. Ashley and J. G. Leipoldt, Inorg. Chem. **20**, 2326 (1981).

42. S. Funahashi, M. Inamo, K. Ishihara and M. Tanaka, Inorg. Chem. **21**, 447 (1982).

43. ΔV is estimated as ≈ 0 cm^{3}mol^{-1} for outer-sphere complex formation involving an uncharged species and ≈ 3 cm^{3}mol^{-1} for reactants whose charge product is -2.^{42}

44. J. G. Leipoldt, R. van Eldik and H. Kelm, Inorg. Chem. **22**, 4147 (1983).

45. X. Chen and D. V. Stynes, Inorg. Chem. **25**, 1173 (1986) and references therein.

46. D. V. Stynes, Pure Appl. Chem. **60**, 561 (1988).

47. R. van Eldik, D. A. Palmer and H. Kelm, Inorg. Chem. **18**, 1520 (1979).

48. M. Eigen and K. Tamm, Ber. Bunsenges. Phys. Chem. **66**, 107 (1962), see also M. Eigen and R. G. Wilkins, Adv. Chem. Ser. **49**, 55 (1965).

49. $\Delta V^{\ddagger}$ values for reactions of Ni^{2+} with neutral ligands range from $+5$ to $+11$ cm^{3}mol^{-1}, Ref. 14.

50. A value of k_3/k_{exch} as low as 0.2 might arise in the I_d mechanism if the complexed anion is held randomly around the octahedral complex and only loss of water near to the anion leads to anion entry.

51. P. J. Nichols, Y. Fresard, Y. Ducommun and A. E. Merbach, Inorg. Chem. **23**, 4341 (1984).

52. J. H. Espenson, Inorg. Chem. **8**, 1554 (1969).

53. A. D. Hugi, L. Helm and A. E. Merbach, Helv. Chim. Acta **68**, 508 (1985).

54. Y. Sasaki and A. G. Sykes, J. Chem. Soc. Dalton Trans. 1048 (1975).

55. A. Haim, Inorg. Chem. **9**, 426 (1970). A similar pattern holds for Co(NH$_3$)$_5$X$^+$.

56. N. Agmon, Int. J. Chem. Kinetics **13**, 333 (1981).

57. N. J. Curtis, G. A. Lawrance and R. van Eldik, Inorg. Chem. **28**, 329 (1989).

58. Associative substitution (A) in organometallic octahedral complexes involving π-type ligands is well established but not common. F. Basolo, Inorg. Chim. Acta **100**, 33 (1985). See Also Ref. 19.

59. S. J. Formosinho, J. Chem. Soc. Faraday Trans. I, **83**, 431 (1987).

60. E. Deutsch and H. Taube, Inorg. Chem. **7**, 1532 (1968).

61. R. Banerjee, Coordn. Chem. Revs. **68**, 145 (1985); G. A. Lawrance, Adv. Inorg. Chem. **34**, 145 (1989).

62. C. Bifano and R. G. Linck, Inorg. Chem. **7**, 908 (1968).

63. W. Weber, D. A. Palmer and H. Kelm, Inorg. Chem. **21**, 1689 (1982).

64. D. A. House, Coordn. Chem. Revs. **23**, 223 (1977); Inorg. Chim. Acta **51**, 273 (1981).

65. B. F. Anderson, J. D. Bell, D. A. Buckingham, P. J. Cresswell, G. J. Gainsford, L. G. Marzilli, G. B. Robertson and A. M. Sargeson, Inorg. Chem. **16**, 3233 (1977).

66. A. L. Crumbliss and W. K. Wilmarth, J. Amer. Chem. Soc. **92**, 2593 (1970); D. E. Clegg, J. R. Hall, and N. S. Ham, Aust. J. Chem. **23**, 1981 (1970) and references.

67. S.-J. Wang and E. L. King, Inorg. Chem. **19**, 1506 (1980).

68. S. Yamada, T. Kido and M. Tanaka, Inorg. Chem. **23**, 2990 (1984).

69. H. Ogino and M. Shimura, Adv. Inorg. Bioinorg. Mechs. **4**, 107 (1986).

70. D. K. Lavallee, Coordn. Chem. Revs. **61**, 55 (1985).

71. K. J. Butenhof, D. Cochenour, J. L. Banyasz and J. E. Stuehr, Inorg. Chem. **25**, 691 (1986) studied the rates of reaction of bpy with various Ni(II) species, Ni(fad)$^{2-}$, Ni(fadH)$^-$ and Ni(fad)OH^{3-}, see Eqn. (8.121).

72. H. Sigel, R. M.-Balakrishnan and U. K. Häring, J. Amer Chem. Soc. **107**, 5137 (1985).

73. M. L. Tobe, Adv. Inorg. Bioinorg. Mechs. **2**, 1 (1983) gives an authoritative and detailed account covering all aspects of base hydrolysis.

74. F. J. Garrick, Nature London **139**, 507 (1937).

75. R. G. Pearson, J. Chem. Educ. **55**, 720 (1978).

76. G. Marangoni, M. Panayotou and M. L. Tobe, J. Chem. Soc. Dalton Trans. 1989 (1973); J. Lichtig, M. E. Sosa and M. L. Tobe, J. Chem. Soc. Dalton Trans. 581 (1984).

77. D. A. Buckingham, C. R. Clark and T. W. Lewis, Inorg. Chem. **18**, 2041 (1979); D. A. Buckingham, C. R. Clark and W. S. Webley, Aust. J. Chem. **33**, 263 (1980).

78. C. Blakeley and M. L. Tobe, J. Chem. Soc. Dalton Trans. 1775 (1987).

79. M. Green and H. Taube, Inorg. Chem. **2**, 948 (1963).

80. N. E. Brasch, D. A. Buckingham, C. R. Clark and K. S. Finnie, Inorg. Chem. **28**, 4567 (1989).

81. D. A. Buckingham, B. M. Foxman and A. M. Sargeson, Inorg. Chem. **9**, 1790 (1970).

82. D. A. House, Coordn. Chem. Revs. **23**, 223 (1977).

83. D. A. House and H. K. J. Powell, Inorg. Chem. **10**, 1583 (1971).

84. Y. Kitamura, R. van Eldik and H. Kelm, Inorg. Chem. **23**, 2038 (1984).

85. D. A. House, Inorg. Chem. **27**, 2587 (1988).

86. M. E. Sosa and M. L. Tobe, J. Chem. Soc. Dalton Trans. 427 (1986).

87. W. G. Jackson and B. H. Dutton, Inorg. Chem. **28**, 525 (1989). The case has been made that all "induced" reactions are spontaneous hydrolysis of modified reactants e. g. $(NH_3)_5CoN_3^{2+}$ is modified to $(NH_3)_5CoN_4O^{3+}$ in the NO^+ reaction.

88. N. E. Dixon, W. G. Jackson, W. Marty and A. M. Sargeson, Inorg. Chem. **21**, 688 (1982).

89. M. J. Gaudin, C. R. Clark and D. A. Buckingham, Inorg. Chem. **25**, 2569 (1986).

90. F. P. Rotzinger, Inorg. Chem. **27**, 772 (1988).

91. D. A. Buckingham, P. A. Marzilli and A. M. Sargeson, Inorg. Chem. **8**, 1595 (1969).

92. P. Comba and W. Marty, Helv. Chim. Acta **63**, 693 (1980). Sec. also A. A. Watson, M. R. Prinsep and D. A. House, Inorg. Chim. Acta **115**, 95 (1986).

93. Higher rate constants for reactions of aliphatic diamines than expected has long been recognized and explained in terms of a conjugate base mechanism. [17]

94. R. E. Hamm, R. L. Johnson, R. H. Pekins, and R. E. Davis, J. Amer. Chem. Soc. **80**, 4469 (1958).

95. M. A. Abdullah, J. Barrett and P. O'Brien, J. Chem. Soc. Dalton Trans. 1647 (1984); Inorg. Chim. Acta **96**, L35 (1985).

96. B. Mohr and R. van Eldik, Inorg. Chem. **24**, 3396 (1985).

97. B. H. Robinson and N. C. White, J. Chem. Soc. Faraday Trans. I **74**, 2625 (1978).

98. R. L. Wilder, D. A. Kamp, and C. S. Garner, Inorg. Chem. **10**, 1393 (1971).

99. T. P. Dasgupta and G. M. Harris, J. Amer. Chem. Soc. **97**, 1733 (1975).

100. R. van Eldik, Adv. Inorg. Bioinorg. Mechs. **3**, 275 (1984).

101. R. L. Reeves, G. S. Calabrese and S. A. Harkaway, Inorg. Chem. **22**, 3076 (1983); R. L. Reeves, Inorg. Chem. **25**, 1473 (1986).

102. M. J. Hynes and B. D. O'Regan, J. Chem. Soc. Dalton Trans. 162 (1979).

103. V. S. Sharma and D. L. Leussing, Inorg. Chem. **11**, 138 (1972).

104. H. Diebler, F. Secco and M. Venturini, J. Phys. Chem. **91**, 5106 (1987).

105. S. Chopra and R. B. Jordan, Inorg. Chem. **22**, 1708 (1983).

106. E. Mentasti, F. Secco and M. Venturini, Inorg. Chem. **19**, 3528 (1980).

107. B. Perlmutter-Hayman and R. Shinar, Inorg. Chem. **15**, 2932 (1976).

108. G. Melson and R. G. Wilkins, J. Chem. Soc. 2662 (1963).

109. G. Schwarzenbach, H.-B. Burgi, W. P. Jensen, G. A. Lawrance, L. Mønsted and A. M. Sargeson, Inorg. Chem. **22**, 4029 (1983). For a number of Ni(II) amine complexes, the rate constant for $Ni-N$ bond rupture decreases as the $Ni-N$ bond length shortens.

110. D. K. Lin and C. S. Garner, J. Amer. Chem. Soc. **91**, 6637 (1969); L. Mønsted, Acta Chem. Scand. **A30**, 599 (1976).

111. B. Monzyk and A. L. Crumbliss, J. Amer. Chem. Soc. **104**, 4921 (1982); M. Biruš, Z. Bradić, G. Krznarić, N. Kujundžic, M. Pribanić, P. C. Wilkins and R. G. Wilkins, Inorg. Chem. **26**, 1000 (1987).

112. H. Kelm and G. M. Harris, Inorg. Chem. **6**, 1743 (1967).

113. P. J. Breen, W. DeW. Horrocks, Jr. and K. A. Johnson, Inorg. Chem. **25**, 1968 (1986) and extensive refs. mostly before 1980.

114. D. W. Margerum, D. L. Jones and H. M. Rosen, J. Amer. Chem. Soc. **87**, 4463 (1965).

115. D. B. Rorabacher and D. W. Margerum, Inorg. Chem. **3**, 382 (1964).

116. W.-S. Jung, H. Tomiyasu and H. Fukutomi, Inorg. Chem. **25**, 2582 (1986) and references therein.

117. H. Kido and K. Saito, J. Amer. Chem. Soc. **110**, 3187 (1988).

118. J. D. Carr, R. A. Libby and D. W. Margerum, Inorg. Chem. **6**, 1083 (1967).

119. K. Kumar and P. C. Nigam, Inorg. Chem. **20**, 1623 (1981).

120. R. M. Izatt, J. S. Bradshaw, S. A. Nielsen, J. D. Lamb, J. J. Christensen and D. Sen, Chem. Revs. **85**, 271 (1985).

121. J. A. Drumhiller, F. Montavon, J. M. Lehn and R. W. Taylor, Inorg. Chem. **25**, 3751 (1986) and references therein.

122. N. F. Curtis and S. R. Osvath, Inorg. Chem. **27**, 305 (1988) and extensive references.

123. H. Diebler, M. Eigen, G. Ilgenfritz, G. Maass and R. Winkler, Pure Appl. Chem. **20**, 93 (1969).

124. J. C. Lockhart, Adv. Inorg. Bioinorg. Mechs. **1**, 217 (1982), extensive Tables.

125. M. Tanaka, Pure Appl. Chem. **55**, 151 (1983).

126. D. K. Cabbiness and D. W. Margerum, J. Amer. Chem. Soc. **92**, 2151 (1970).

127. C. T. Lin, D. B. Rorabacher, G. R. Cayley and D. W. Margerum, Inorg. Chem. **14**, 919 (1975).

128. F.-T. Chen, C.-S. Lee and C.-S. Chung, Polyhedron **2**, 1301 (1983).

129. F. McLaren, P. Moore and A. M. Wynn, J. Chem. Soc. Chem. Communs. 798 (1989). E. Kimura, Y. Kotake, T. Koike, M. Shionoya and M. Shiro, Inorg. Chem. **29**, 4991 (1990).

130. Y. H. Wu and T. A. Kaden, Helv. Chim. Acta **68**, 1611 (1985).

131. Dissociation of Na^+ complex of dibenzo-[24]-crown-6 is first-order alone in acetonitrile but Na^+-assisted in nitromethane (from nmr exchange data). A full discussion of unwrapping mechanisms is given, A. Delville, H. D. H. Stöver and C. Detellier, J. Amer. Chem. Soc. **109**, 7293 (1987).

132. P. Fux, J. Lagrange and P. Lagrange, J. Amer. Chem. Soc. **107**, 5927 (1985).

133. B. G. Cox, J. Garcia-Rosas and H. Schneider, J. Amer. Chem. Soc. **104**, 2434 (1982).

134. (a) A.-M. Albrecht-Gary, Z. Saad, C. O. Dietrich-Buchecker and J.-P. Sauvage, J. Amer. Chem. Soc. **107**, 3205 (1985) (b) A.-M. Albrecht-Gary, C Dietrich-Buchecker, Z. Saad and J.-P. Sauvage, J. Amer. Chem. Soc. **110**, 1467 (1988). J.-P. Sauvage, Acc. Chem. Res. **23**, 319 (1990).

135. (a) C. Grant Jr. and P. Hambright, J. Amer. Chem. Soc. **91**, 4195 (1969) (b) J. Turay and P. Hambright, Inorg. Chem. **19**, 562 (1980); A. Shamin and P. Hambright, Inorg. Chem. **22**, 694 (1983).

136. S. Funahashi, Y. Yamaguchi and M. Tanaka, Inorg. Chem. **23**, 2249 (1984).

137. R. F. Pasternack, G. C. Vogel, C. A. Skowronek, R. K. Harris and J. G. Miller, Inorg. Chem. **20**, 3763 (1981). Cu(II) incorporation into tetraphenylporphyrin in dmso shows saturation kinetics. An outer-sphere complex without appreciable distortion of the porphyrin ring is proposed to explain the kinetics.

138. E. B. Fleischer and J. H. Wang, J. Amer. Chem. Soc. **82**, 3498 (1960).

139. D. A. Buckingham, C. R. Clark and W. S. Webley, J. Chem. Soc. Chem. Communs. 192 (1981).

140. The importance of this deformation in easing metal entry is illustrated by the fact that the porphyrin N-Metetraphenylporphyrin is deformed and reacts rapidly and simply with many metal ions.[70, 125]

141. J. Nwaeme and P. Hambright, Inorg. Chem. **23**, 1990 (1984).

142. R. J. Cross, Chem. Soc. Revs. **14**, 197 (1985); Adv. Inorg. Chem. **34**, 219 (1989).

143. M. Kotowski and R. van Eldik, in B 17, Chap. 4.

144. U. Belluco, L. Cattalini, F. Basolo, R. G. Pearson and A. Turco, J. Amer. Chem. Soc. **87**, 241 (1965).

145. P. Haake, S. C. Chan, and V. Jones, Inorg. Chem. **9**, 1925 (1970).

146. J. K. Beattie, Inorg. Chim. Acta **76**, L69 (1983).

147. (a) S. Lanza, D. Minniti, P. Moore, J. Sachinidis, R. Romeo and M. L. Tobe, Inorg. Chem. **23**, 4428 (1984); (b) U. Frey, L. Helm, A. E. Merbach and R. Romeo, J. Amer. Chem. Soc. **111**, 8161 (1989).

148. D. Minniti, G. Alibrandi, M. L. Tobe and R. Romeo, Inorg. Chem. **26**, 3956 (1987).

149. L. Cattalini, Mechanism of Square-Planar Substitution, in B30, p. 266.

150. U. Belluco, R. Ettorre, F. Basolo, R. G. Pearson, and A. Turco, Inorg. Chem. **5**, 591 (1966).

151. L. Helm, L. I. Elding and A. E. Merbach, Helv. Chim. Acta **67**, 1453 (1984); Inorg. Chem. **24**, 1719 (1985).

152. J. J. Pesek and W. R. Mason, Inorg. Chem. **22**, 2958 (1983).

153. R. G. Pearson, H. Sobel, and J. Songstad, J. Amer. Chem. Soc. **90**, 319 (1968).

154. B31

155. J. B. Goddard and F. Basolo, Inorg. Chem. **7**, 963; 2456 (1968).

156. (a) M. Kotowski and R. van Eldik, Inorg. Chem. **25**, 3896 (1986); J. J. Pienaar, M. Kotowski and R. van Eldik, Inorg. Chem. **28**, 373 (1989); (b) J. Berger, M. Kotowski, R. van Eldik, U. Frey, L. Helm and A. E. Merbach, Inorg. Chem. **28**, 3759 (1989).

157. W.H. Baddley and F. Basolo, J. Amer. Chem. Soc. **86**, 2075 (1964).

158. M. Kotowski, D. A. Palmer and H. Kelm, Inorg. Chem. **18**, 2555 (1979).

159. L. H. Skibsted, Adv. Inorg. Bioinorg. Mechs. **4**, 137 (1986).

160. M. J. Carter and J. K. Beattie, Inorg. Chem. **9**, 1233 (1970).

161. J. S. Coe and J. R. Lyons, J. Chem. Soc. A 829 (1971) and refs.

162. I. N. Marov, M. N. Vargaftig, V. K. Belyaeva, G. A. Evtikova, E.Hoyer, R. Kirmse and W. Dietzsch, Russ. J. Inorg. Chem. **23**, 101 (1980).

163. J. Stach, R. Kironse, W. Dietzsch, G. Lassmann, V. K. Belyaeva and I. N. Marov, Inorg. Chim. Acta **96**, 55 (1985).

164. M. Moriyasu and Y. Hashimoto, Bull. Chem. Soc. Japan **53**, 3590 (1980); **54**, 3374 (1981).

165. D. C. Olson and D. W. Margerum, J. Amer. Chem. Soc. **85**, 297 (1963).

166. D. W. Margerum and J. D. Carr, J. Amer. Chem. Soc. **88**, 1639, 1645 (1966).

167. L. H. Pignolet, D. Forster, and W. DeW. Horrocks, Jr., Inorg. Chem. **7**, 828 (1968).

168. F. K. Meyer, W. L. Earl and A. E. Merbach, Inorg. Chem. **18**, 888 (1979).

169. G. P. Algra and S. Balt, Inorg. Chem. **20**, 1102 (1981); Acta Cryst. **B40**, 582 (1984).

170. L. S. Meriwether and M. L. Fiene, J. Amer. Chem. Soc. **81**, 4200 (1959).

171. M. Meier, F. Basolo, and R. G. Pearson, Inorg. Chem. **8**, 795 (1969).

172. K. R. Brower and T. S. Chen, Inorg. Chem. **12**, 2198 (1973).

173. D. A. Sweigart, Inorg. Chim. Acta **18**, 179 (1976).

174. P. Moore, J. Schinidis and G. R. Willey, J. Chem. Soc. Dalton Trans. 1323 (1984).

175. L. Helm, P. Meier, A. E. Merbach and P. A. Tregloan, Inorg. Chim. Acta **73**, 1 (1983).

176. D.-T. Wu and C.-S. Chung, Inorg. Chem. **25**, 4841 (1986).

177. S. F. Lincoln, A. M. Hounslow, D. L. Pisaniello, B. G. Doddridge, J. H. Coates, A. E. Merbach and D. Zbinden, Inorg. Chem. **23**, 1090 (1984).

178. S. F. Lincoln, J. H. Coates, B. G. Doddridge and A. M. Hounslow, Inorg. Chem. **22**, 2869 (1983).

179. P. D. I. Fletcher and B. H. Robinson, J. Chem. Soc. Faraday Trans. I **80**, 2417 (1984) and refs.

180. S. Diekmann and J. Frahm, J. Chem. Soc. Faraday Trans. I **75**, 2199 (1979).

181. C. Tondre, J. Chem. Soc. Faraday Trans. I **78**, 1795 (1982).

182. N. Sbiti and C. Tondre, J. Chem. Soc. Faraday Trans. I **78**, 1809 (1982).

183. A. Yamagishi, T. Masui and F. Watanabe, Inorg. Chem. **20**, 513 (1981).

184. Methods in Enzymology, Vol. 158, Metallobiochemistry, Part A, eds. J. F. Riordan and B. L. Vallee Academic Press, 1988. Contains authoritative accounts of all aspects of metalloproteins.

185. J.A. Blaszak, D.R. McMillin, A. T. Thornton and D. L. Tennent, J. Biol. Chem. **258**, 9886 (1983) (Cu(II) + apoazurin).

186. J. Hirose and R. G. Wilkins, Biochemistry **23**, 3149 (1984) (Co(II) + apoarsanilazotyrosine-248 carboxypeptidase A).

187. For a full analysis see D. Baldwin, Biochim. Biophys. Acta **623**, 183 (1980).

188. S. A. Kretchmar and K. N. Raymond, J. Amer. Chem. Soc. **108**, 6212 (1986).

189. I. Bertini, J. Hirose, H. Kozlowski, C. Luchinat, L. Messori and A. Scozzafava, Inorg. Chem. **27**, 1081 (1988).
190. P. K. Bali and W. R. Harris, J. Amer. Chem. Soc. **111**, 4457 (1989).

Selected Bibliography (Generally relevant to Chaps. 4–8)

B24. C. F. Baes, Jr. and R. E. Mesmer, The Hydrolysis of Cations, Wiley-Interscience, NY, 1976.
B25. F. Basolo and R. G. Pearson, Mechanisms of Inorganic Reactions, Wiley-Interscience, NY, 1967.
B26. D. Benson, Mechanisms of Inorganic Reactions in Solution, McGraw-Hill, London, 1968.
B27. J. Burgess, Metal Ions in Solution, Ellis Horwood, Chichester, 1978.
B28. J. Burgess, Ions in Solution, Wiley-Interscience, NY, 1988.
B29. J. O. Edwards, Inorganic Reaction Mechanisms, Benjamin, NY, 1964.
B30. J. O. Edwards (ed) Inorganic Reaction Mechanisms, Wiley-Interscience, NY, 1970 (Part I); 1972 (Part 2).
B31. C. H. Langford and H. B. Gray, Ligand Substitution Processes, Benjamin, NY, 1965.
B32. A. E. Martell (ed) Coordination Chemistry Vol. 2, American Chemical Society, Washington, 1978.
B33. A. G. Sykes, Kinetics of Inorganic Reactions, Pergamon, Oxford, 1966.
B34. M. L. Tobe, Inorganic Reaction Mechanisms, Nelson, London, 1972.
B35. G. Wilkinson, R. D. Gillard and J. A. McCleverty (eds) Comprehensive Coordination Chemistry, Vol. 1. Theory and Background, Pergamon, Oxford, 1987. Chapter 7.1, M. L. Tobe (Substitution); Chapter 7.2, T. J. Meyer and H. Taube (Electron Transfer); Chap. 7.4 D. St. C. Black (Reactions of Coordinated Ligands).

Problems

1. For the reaction

$$Cr(H_2O)_6^{3+} + SCN^- \rightleftharpoons Cr(H_2O)_5NCS^{2+} + H_2O$$

the forward direction is governed by the rate law

$$V = (k_1 + k_2[H^+]^{-1} + k_3[H^+]^{-2})[Cr^{3+}][SCN^-]$$

Using microscopic reversibility considerations, write down the rate law for the reverse direction and deduce the relationship between the various rate constants and thermodynamic parameters for the system.
C. Postmus and E. L. King, J. Phys. Chem. **59**, 1216 (1955).

2. For a large number of hydrolyses, X being anions and neutral ligands

$$Co(NH_3)_5X^{n+} + H_2O \rightarrow Co(NH_3)_5H_2O^{3+} + X^{(n-3)+}$$

$$\Delta V^{\ddagger} = 0.48 \, \Delta V + 1.5$$

where $\Delta V^{\ddagger}$ = activation volume and ΔV = reaction volume. Discuss the significance of this equation with respect to the mechanism of the hydrolyses.
Y. Kitamura, K. Yoshitani and T. Itoh, Inorg. Chem. **27**, 996 (1988).

3. The tautomeric equilibrium

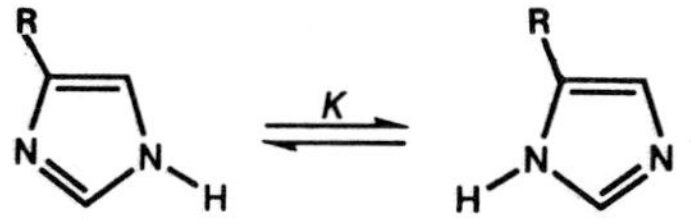

is established rapidly ($\tau < 10^{-4}$s). For histamine, $R = -CH_2CH_2NH_3^+$, the tautomeric equilibrium constant K is estimated to be 0.26 from ^{15}N nmr chemical shift experiments. How might the tautomerism influence the kinetics of chelation of Ni^{2+} with histamine? P. Dasgupta and R. B. Jordan, Inorg. Chem. **24**, 2721 (1985).

4. (a) Show that the open form of **6** present in $\approx 0.1\%$ in dnsaH$^-$ cannot account for the reduced rate of reaction of dnsaH$^-$ (Sec. 4.4.1).
   (b) Account for the higher rate constant for reaction of tsa$^-$ with Ni^{2+} ($9.4 \times 10^3 M^{-1}s^{-1}$) compared with that for dhba$^-$ ($65\ M^{-1}s^{-1}$)

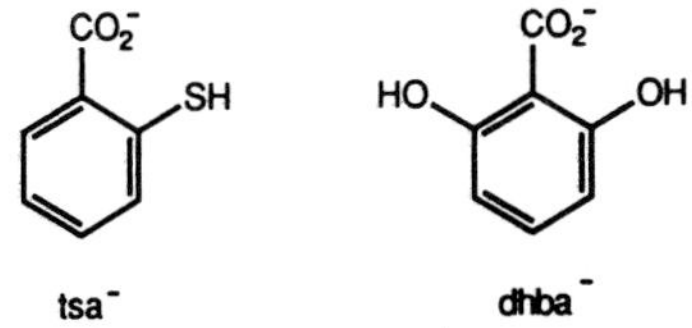

tsa$^-$ dhba$^-$

H. Diebler, F. Secco and M. Venturini, J. Phys. Chem. **91**, 5106 (1987).

5. There are two likely pathways for the replacement of chelates in complexes of the type $M(CO)_4(L_2)$ by an incoming unidentate nucleophile, L_1

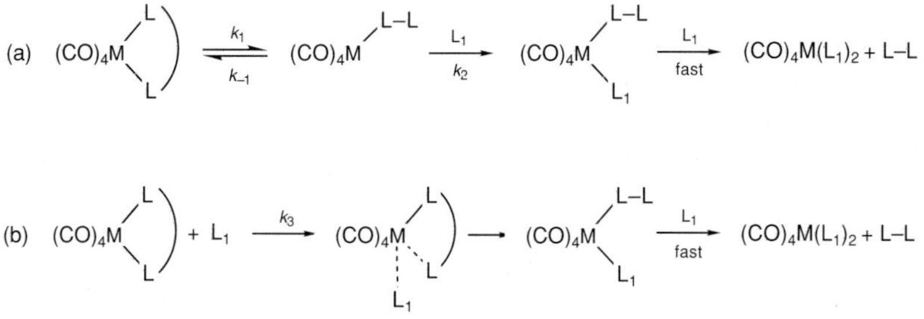

Derive the rate laws for (a) and (b) with L_1 in excess, and with assumption of a steady-state concentration of the 5-coordinate species in (a) and a rds associated with k_3 in (b). Under what conditions will the rate laws for (a) and (b) be identical? With these conditions in 1,2-dichloroethane $\Delta V^{\ddagger} = 14.7\ cm^3 mol^{-1}$ when $M = Cr$, $L-L = 3,6$-dithiaoctane and $L_1 = P(OC_2H_5)_3$. Which mechanism is supported?
H.-T. Macholdt, R. van Eldik and G. R. Dobson, Inorg. Chem. **25**, 1914 (1986).

6. (a) The acid hydrolyses of both $Cu(12[ane]N_4)^{2+}$ and $Cu(14[ane]N_4)^{2+}$ are acid-dependent

$$V = k[Cu(II)][H^+]$$

Account for the value of k being $\approx 10^4$ larger for $Cu(12[ane]N_4)^{2+}$ than for $Cu(14[ane]N_4)^{2+}$

R. W. Hay and M. P. Pujari, Inorg. Chim. Acta **100**, L1 (1985).

(b) The rates of hydrolyses of $Cu(2_N2_O2_O)^{2+}$ and $Cu(trans\text{-}Me_6[18]dieneN_4)^{2+}$ are both strongly base dependent with $n = 3$ and 2 respectively in

$$V = k[Cu(II)][OH^-]^n$$

Account for these rate laws.

J. A. Drumhiller, F. Montavon, J. M. Lehn and R. W. Taylor, Inorg. Chem. **25**, 3751 (1986); R. W. Hay and R. Bembi, Inorg. Chim. Acta **62**, 89 (1982).

7. Rationalize the following rate behavior for Ni(II) complexing with an excess of bidentate ligand, $XY = $ en, gly or phen:

  Ni^{2+}, overall second order at all concentrations of XY
  $Ni(trien)(H_2O)_2^{2+}$, second-order at low concentrations of XY and first-order at high concentrations of XY (independent of [XY])
  $Ni(14[ane]N_4)^{2+}$ first-order at all concentrations of XY

  E. J. Billo, Inorg. Chem. **23**, 2223 (1984) and references therein.

8. There have been a number of studies of the kinetics of interaction of Ni(II) complexes with CN^- ion.

(a) The square-planar complex NiA_2^{2+} reacts with CN^- to give $Ni(CN)_4^{2-}$ via a discernable intermediate $NiA(CN)_2$, A = 2,3-diamino-2,3-dimethylbutane

$$NiA_2^{2+} + 2\,CN^- \rightarrow NiA(CN)_2 + A \qquad \text{Stage I}$$
$$NiA(CN)_2 + 2\,CN^- \rightarrow Ni(CN)_4^{2-} + A \qquad \text{Stage II}$$

At high pH, all the cyanide is present as CN^-. The rate laws are

  Stage I $d[NiA(CN)_2]/dt = k_1[NiA_2^{2+}][CN^-]$
  Stage II $d[Ni(CN)_4^{2-}]/dt = k_2[NiA(CN)_2][CN^-]^2$

The pH-dependence of Stage II was determined from pH 10.7 to 6.3. The order in total cyanide ion remains 2 and the rate law is

$$d[Ni(CN)_4^{2-}]dt = (k_2[CN^-]^2 + k_3[CN^-][HCN] + k_4[HCN]^2 + k_5)[NiA(CN)_2]$$

Suggest mechanisms for these rate laws.

J. C. Pleskowicz and E. J. Billo, Inorg. Chim. Acta **99**, 149 (1985).

(b) The rate law for the reaction between Ni(edda) and CN^- at pH 10.8 is

$$V = k[\text{Ni(edda)}][CN^-]^n$$

where n varies from 3 at low $[CN^-]$ to 1 at high $[CN^-]$. Suggest a mechanism, and a reason why the similar reaction between Ni(cydta)^{2-} and CN^- is much slower.

L. C. Coombs, D. W. Margerum and P. C. Nigam, Inorg. Chem. **9**, 2081 (1970).

(c) For the reaction of the binuclear chelate Ni_2L with CN^- to form Ni(CN)_4^{2-} in one observable step, at high pH,

$$V = \{k_1 + k_2[CN^-]\}[\text{Ni}_2\text{L}]$$

where L = diethylenetriaminepentaacetate. Suggest a mechanism.

K. Kumar, H. C. Bajaj and P. C. Nigam, J. Phys. Chem. **84**, 2351 (1980).

9. (a) Account for the following two rate laws which have been observed for the incorporation of M(II) into PH_2 (porphyrin)

  (a) $V = k[M]^2[\text{PH}_2]$

  (b) $V = k_1[\text{PH}_2]$

J. Weaver and P. Hambright, Inorg. Chem. **8**, 167 (1969); C. Grant, Jr. and P. Hambright, J. Amer. Chem. Soc. **91**, 4195 (1969).

(b) The structure of the porphyrin ring of both the ligand and metal complexes of N-substituted porphyrins is very similar. How would you expect the dissociation of N-substituted and unsubstituted porphyrin complexes to differ with respect to rate law, effect of H^+, rates, etc?

D. K. Lavallee, The Chemistry and Biochemistry of N-Substituted Porphyrins, VCH 1987 p. 112.

10. Using scheme (4.109) show how *cis*-MA_2B_2 could be transformed into *trans*-MA_2B_2 in the presence of free ligand A. See Sec. 7.7.2.

R. J. Cross, Chem. Soc. Revs. **14**, 197 (1985).

11. (a) What form will the rate law for substitution in square-planar complexes take if the solvolysis of the complex is rapid compared with ligand substitution? (This occurs in reactions of PtCl_4^{2-} with $^*Cl^-$, bpy and phen).

F. A. Palocsay and J. V. Rund, Inorg. Chem. **8**, 524 (1969).

(b) Predict the effects of the concentrations of I^- and OH^- on the rate constants for replacement of Cl^- in Pd(dien)Cl^+, $\text{Pd(1,4,7-Me}_3\text{dien)Cl}^+$, $\text{Pd(1,1,7,7-Me}_4\text{dien)Cl}^+$ and $\text{Pd(1,1,4,7,7-Me}_5\text{dien)Cl}^+$.

J. B. Goddard and F. Basolo, Inorg. Chem. **7**, 936 (1968); E. L. J. Breet and R. van Eldik, Inorg. Chem. **23**, 1865 (1984).

12. Give a reasonable explanation for the fact that the lability of dmf in $M(Me_6tren)dmf^{2+}$ is some 10^5 less than in $M(dmf)_6^{2+}$ for both $M = Co$ and Cu.

S. F. Lincoln, J. H. Coates, B. G. Doddridge and A. M. Hounslow, Inorg. Chem. **22**, 2869 (1983).

Chapter 5

Oxidation-Reduction Reactions

5.1 General Characteristics

Oxidation-reduction (redox) reactions of the transition metal complexes are probably the best understood of the types of processes we are concerned with. In redox reactions, the oxidation state of at least two reactants changes. A variety of such reactions are shown in Table 5.1.[1-7]

Table 5.1 Some Types of Redox Reactions

Redox Reaction	Characteristics	Ref.
1. $*Co(NH_3)_6^{2+} + Co(NH_3)_6^{3+} \rightleftharpoons$ $*Co(NH_3)_6^{3+} + Co(NH_3)_6^{2+}$	Extremely slow, $k = (8 \pm 1) \times 10^{-6} M^{-1}s^{-1}$ at 40°C, $\mu = 2.5$ M	1
2. $*Fe(bpy)_3^{2+} + Fe(bpy)_3^{3+} \rightleftharpoons$ $*Fe(bpy)_3^{3+} + Fe(bpy)_3^{2+}$	$k = 3 \times 10^8 M^{-1}s^{-1}$	2
3. $*Cr^{2+} + CrCl^{2+} \rightleftharpoons$ $*CrCl^{2+} + Cr^{2+}$	$k = 9 M^{-1}s^{-1}$ at 0°C	3
4. $Cr^{2+} + Co(NH_3)_5Cl^{2+} + 5 H^+$ $\rightarrow CrCl^{2+} + Co^{2+} + 5 NH_4^+$	$k = 6 \times 10^5 M^{-1}s^{-1}$, one of earliest examples of an inner-sphere redox reaction	4, 5
5. $2 Fe^{3+} + H_2A \rightarrow$ $2 Fe^{2+} + 2 H^+ + A$	H_2A = ascorbic acid. Fe(III) complexes formed. Mechanism complicated.	6
6. Horse cytochrome-c(II) + $Co(phen)_3^{3+} \rightleftharpoons$ horse cytochrome-c(III) + $Co(phen)_3^{2+}$	$k = 1.8 \times 10^3 M^{-1}s^{-1}$. Site on protein implicated for binding, different than that used by $Fe(CN)_6^{3-}$ $(k = 9 \times 10^5 M^{-1}s^{-1})$	7

A net chemical change does not necessarily occur as a result of the redox reaction. Reactions 1 and 2 (Table 5.1) involve an interchange of electrons between two similar metal complex ions. Such isotopic exchange reactions were the subject of a sizable number of studies in the late forties and fifties,[8] and the novelty at that time of working with radioactive isotopes attracted many physical chemists to inorganic reaction mechanisms. Reactions 1 and 2 emphasize the wide variation in rates encountered here, as in substitution. One of the challenges we must face is rationalizing these large differences in rate constants (14 orders of magnitude), as well as interpreting smaller more subtle disparities. Reaction 3 indicates that isotopic exchange may not involve merely electron transfer but also movement of atoms (chlorine in this case).

Redox reactions usually lead, however, to a marked change in the species, as reactions 4-6 indicate. Important reactions involve the oxidation of organic and metalloprotein substrates (reactions 5 and 6) by oxidizing complex ions. Here the substrate often has ligand properties, and the first step in the overall process appears to be complex formation between the metal and substrate species. Redox reactions will often then be phenomenologically associated with substitution. After complex formation, the redox reaction can occur in a variety of ways, of which a direct intramolecular electron transfer within the adduct is the most obvious.

Spectrophotometry has been a popular means of monitoring redox reactions, with increasing use being made of flow, pulse radiolytic and laser photolytic techniques. The majority of redox reactions, even those with involved stoichiometry, have second-order characteristics. There is also an important group of reactions in which first-order intramolecular electron transfer is involved. Less straightforward kinetics may arise with redox reactions that involve metal complex or radical intermediates, or multi-electron transfer, as in the reduction of Cr(VI) to Cr(III).[9] Reactants with different equivalences as in the noncomplementary reaction

$$2\,Fe^{II} + Tl^{III} \rightarrow 2\,Fe^{III} + Tl^{I} \tag{5.1}$$

often give rise to complicated kinetic rate laws.[10]

Proton-accelerated rates are often observed when the net reaction involves protons since some of these will have been lost or gained at the transition state. This is the situation with a large number of reactions of oxyions.

5.2 Classification of Redox Reactions

The most important single development in the understanding of the mechanisms of redox reactions has probably been the recognition and establishment of *outer-sphere* and *inner-sphere* processes.[4] Outer-sphere electron transfer involves intact (although not completely undisturbed) coordination shells of the reactants. In inner-sphere redox reactions, there are marked changes in the coordination spheres of the reactants in the formation of the activated complex.

Reaction 2 in Table 5.1 must qualify for an outer-sphere redox category since the bipyridine could not become detached, even by just one end of the bidentate ligand, from the inert iron(II) or iron(III) centers during the course of the rapid redox reaction. There is thus no bond breaking or making during the electron transfer, a situation making them ideal for treatment by the theoretical chemist (Sec. 5.4).

Reaction 4 in Table 5.1, on the other hand, was one of the first-established examples of an inner-sphere redox reaction.[4] The rapid reaction gives $CrCl^{2+}$ as a product, characterized spectrally after separation by ion exchange from the remainder of the species in solution. It is clear that since $CrCl^{2+}$ could not possibly be produced from Cr^{3+} and Cl^- ions during the brief time for reaction and ion exchanger manipulation, it must arise from the redox process *per se.* Thus an activated complex or intermediate of the composition shown in (5.2) must arise from the penetration of the chromium(II) ion by the coordinated chloride of the cobalt(III):

$$(NH_3)_5CoCl^{2+} + Cr(H_2O)_6^{2+} \rightarrow (NH_3)_5CoClCr(H_2O)_5^{4+} + H_2O \tag{5.2}$$

Within this species, an intramolecular electron transfer from Cr(II) to Co(III) must occur, producing Cr(III) and Co(II). The adduct then breaks up and the Cr(III) takes along the chloride as the species $CrCl^{2+}$:

$$(NH_3)_5CoClCr(H_2O)_5^{4+} \xrightarrow{H^+} Co^{2+} + 5\,NH_4^+ + Cr(H_2O)_5Cl^{2+} \tag{5.3}$$

This scheme implies that at no time does chloride ion break free of the influence of at least one of the metals, and in support of this there is no incorporation of ^{36}Cl in $CrCl^{2+}$ when the reaction takes place in the presence of $^{36}Cl^-$ ion.[4]

5.3 Characterization of Mechanism

The characterization of a redox reaction as inner-sphere or outer-sphere is a primary preoccupation of the redox kineticist. The assignment is sometimes obvious, but often difficult and in certain cases impossible!

(a) From the Nature of the Products. The *eventual* products from reaction 4, Table 5.1, are Cr^{3+}, $CrCl^{2+}$, Co^{2+}, Cl^-, and NH_4^+ ions. These could arise from an outer- or an inner-sphere process:[11]

Outer-sphere:

$$Cr^{2+} + Co(NH_3)_5Cl^{2+} \xrightarrow[\text{transfer}]{e^-} Cr^{3+} + Co(NH_3)_5Cl^+ \tag{5.4}$$

$$Co(NH_3)_5Cl^+ \xrightarrow[\text{rapid}]{H^+} Co^{2+} + 5\,NH_4^+ + Cl^- \tag{5.5}$$

$$Cr^{3+} + Cl^- \underset{\text{slow}}{\rightleftharpoons} CrCl^{2+} \tag{5.6}$$

Inner-sphere:

$$Cr^{2+} + Co(NH_3)_5Cl^{2+} \xrightarrow[\text{mechanism}]{\text{bridged}} CrCl^{2+} + Co(NH_3)_5H_2O^{2+} \tag{5.7}$$

$$Co(NH_3)_5H_2O^{2+} \xrightarrow[\text{rapid}]{H^+} Co^{2+} + 5\,NH_4^+ + H_2O \tag{5.8}$$

$$CrCl^{2+} \underset{\text{slow}}{\rightleftharpoons} Cr^{3+} + Cl^- \tag{5.9}$$

The inertness of $CrCl^{2+}$ and the labilities of Cr^{2+} and Co^{2+} (in part responsible for the rapidity of the intermediate formation and the breakup steps) were thus cleverly exploited to provide unambiguous proof for the operation of the inner-sphere process.[4] Since most redox reactions involving Cr^{2+} are rapid, and the hydrolyses of most Cr(III) complexes slow, it is not difficult to detect the intermediate CrX^{n+}, for example,

$$Cr^{2+} + M^{III}X^{n+} \rightarrow Cr^{III}X^{n+} + M^{2+} \tag{5.10}$$

and, in so doing, characterize the reaction as inner-sphere. This has been demonstrated in the Cr(II) reduction of a large number of Co(III), Cr(III), Fe(III) and recently Rh(III)[12] oxidants.[13]

The only other common reducing agents that can lead to products leisurely characterizable, because they hydrolyze extremely slowly, are $Co(CN)_5^{3-}$, $Fe(CN)_5H_2O^{3-}$ and $Ru(NH_3)_5H_2O^{2+}$. With the other common reducing agents, Fe^{2+}, V^{2+}, Eu^{2+}, and Cu^+, any product will hydrolyze rapidly; for example,

$$M^{2+} + Co(NH_3)_5X^{2+} \xrightarrow{H^+} MX^{2+} + Co^{2+} + 5NH_4^+ \tag{5.11}$$

$$MX^{2+} \rightleftharpoons M^{3+} + X^- \quad \text{rapid} \tag{5.12}$$

Table 5.2 Rate Parameters and Characteristics of Some Reactions of V(II) at 25°C, Refs. 15 and 16.

Reactant	k $M^{-1}s^{-1}$	$\Delta H^{\ddagger}$ kJ mol^{-1}	$\Delta S^{\ddagger}$ J K^{-1} mol^{-1}
Inner-sphere redox, intermediate detected			
$CrSCN^{2+}$	8.0	54	−46
VO^{2+}	1.6	51	−71
$Co(NH_3)_5SCN^{2+}$	30	69	+25
$Co(NH_3)_5C_2O_4^+$	45	51	−42
cis-$Co(en)_2(N_3)_2^+$	33	...	...
$Co(CN)_5N_3^{3-}$	112	...	...
$Co(CN)_5SCN^{3-}$	140	...	...
Probably inner-sphere redox			
Cu^{2+}	27	48	−59
$Co(NH_3)_5N_3^{2+}$	13	49	−59
$Co(NH_3)_5SO_4^+$	26	49	−54
$Co(NH_3)_5OCOR^{2+}$	1−21[a]	46−51[a]	−54 to −71[a]
$Co(CN)_5X^{3-}$	120−280[b]	...	...
Probably outer-sphere redox			
$Co(NH_3)_5Cl^{2+}$	10	31	−121
$Co(NH_3)_5H_2O^{3+}$	0.53	34	−134
$Co(NH_3)_6^{3+}$	0.004	38	−167
$RuCl^{2+}$	1.9×10^3	...	...
Fe^{3+}	1.8×10^4	...	...
FeX^{2+}	$(4.6−6.6) \times 10^5$ [c]		
Replacement reaction			
NCS^-	24	67	+5
H_2O	89[d]	62	−0.4

[a] Variety of R groups.   [b] X = Cl$^-$, Br$^-$, I$^-$ and H$_2$O.   [c] X = Cl$^-$, N$_3^-$ and NCS$^-$.
[d] First-order rate constant (s^{-1}) for water exchange.

It will be very difficult to detect Cu(II)X and Eu(III)X as intermediates because of their marked lability, and therefore hard to characterize Cu$^+$ and Eu^{2+} as inner-sphere reductants by product identification. It is easier to detect Fe(III) and V(III) species, by flow methods, and a number of reactions of Fe^{2+} with Co(III) complexes[14] and V^{2+} with V(IV), Co(III), and Cr(III) complexes, Table 5.2,[15,16] have been shown to progress via the intermediate required of an inner-sphere reaction.

Closer examination of the reaction between Fe^{2+} and Co(C$_2$O$_4$)$_3^{3-}$ (Sec. 3.7), for example, shows the formation and decay of an intermediate FeC$_2$O$_4^+$ ion,[14]

$$\text{Fe}^{2+} + \text{Co(C}_2\text{O}_4)_3^{3-} \xrightarrow{\text{H}^+} \text{FeC}_2\text{O}_4^+ + \text{Co}^{2+} + 2\,\text{H}_2\text{C}_2\text{O}_4 \qquad (5.13)$$

$$\text{FeC}_2\text{O}_4^+ \xrightarrow{\text{H}^+} \text{Fe}^{3+} + \text{H}_2\text{C}_2\text{O}_4 \qquad (5.14)$$

Obviously, for success in this approach, the rate of the redox step producing the intermediate must be at least as fast as the decomposition of the intermediate. This can be sometimes accomplished by increasing the reactant concentrations, since the first step is second order and the second step is first order.

(b) By the Detection of a Bridged Species. The detection of a bridged complex comparable to that in (5.2) does not prove (although it may suggest) that it is an intermediate in an inner-sphere redox process. The bridged species could be in equilibrium with the reactants, but the products form directly from reactants by an outer-sphere process. This apparently occurs in the reaction of Co(edta)$^{2-}$ with Fe(CN)$_6^{3-}$ (Sec. 1.6.4). The possible oxidation states of the metals in the bridged species in (5.2) are either Cr(II) and Co(III) or Cr(III) and Co(II). In both cases, one of the components is quite labile, and the binuclear species will respectively either return to reactants or dissociate to products rather than exist independently for any length of time. When both partners in the bridged intermediate are inert, however, there is every chance that it will be detected, or at least its presence inferred from the form of the rate law, or the magnitude of the activation parameter (Sec. 5.5). A number of such systems are shown in Table 5.3.[16] The oxidation states of the detected species are deduced from spectral or chemical considerations. In only the last two entries are the oxidation states of the metals in the bridged complex the same as the oxidation state of the reactants. Such bridged in-

Table 5.3 Some Bridged Species Arising from Redox Reactions.[16]

Reactants	Species
Cr(II) + Ru(III) chloro complexes	Cr(III) – Cl – Ru(II)
Co(CN)$_5^{3-}$ + IrCl$_6^{2-}$	Co(III) – Cl – Ir(III)
Co(edta)$^{2-}$ + Fe(CN)$_6^{3-}$	Co(III) – NC – Fe(II)
Cr(II) + V(IV)	Cr(III) – (OH)$_2$ – V(III)[a]
Fe(II) + Co(NH$_3$)$_5$nta	Fe(II) – nta – Co(III)[b]

Fe(CN)$_5$H$_2$O^{3-} + Co(NH$_3$)$_5$ N⟨☐⟩N^{3+} (NC)$_5$Fe^{II}N⟨☐⟩NCo^{III}(NH$_3$)$_5$[b]

[a] This is one of a number of examples in which a binuclear complex with an $-\text{O}-$ (or (OH)$_2$) or OH bridge results from the interaction of oxyions.[18]
[b] These species undergo intramolecular electron transfer at measurable rates (Sec. 5.8.1).

termediates are termed precursor complexes to distinguish them from the more commonly encountered successor complexes in which electron transfer has already taken place (Sec. 5.5). In order to obtain sizeable amounts of a precursor complex it is clear that there must be a very strong affinity by the bridging ligand for the reactant partners.

(c) From Rate Data. Both inner- and outer-sphere redox reactions are usually second-order

$$V = k \text{[oxidant] [reductant]} \tag{5.15}$$

Only in a limited number of instances will the value of k and its associated parameters be useful in diagnosing mechanism. When the redox rate is faster than substitution within either reactant, we can be fairly certain that an outer-sphere mechanism holds.[17] This is the case with Fe^{3+} and $RuCl^{2+}$ oxidation of V(II)[15] and with rapid electron transfer between inert partners. On the other hand, when the activation parameters for substitution and redox reactions of one of the reactants are similar, an inner-sphere redox reaction, controlled by replacement, is highly likely.[18] This appears to be the case with the oxidation by a number of Co(III) complexes of V(II),[15] confirmed in some instances by the appearance of the requisite V(III) complex, e. g.

$$Co(CN)_5N_3^{3-} + V^{2+} \rightarrow Co(CN)_5^{3-} + VN_3^{2+} \tag{5.16}$$

An $[H^+]^{-1}$ term in the rate law for reactions involving an aqua redox partner strongly suggests the participation of an hydroxo species and the operation of an inner-sphere redox reaction (Sec. 5.5(a)). Methods (a) and (b) are direct ones for characterizing inner-sphere processes, analyzing for products or intermediates which are kinetically-controlled. Method (c) is indirect. Other methods of distinguishing between the two basic mechanisms are also necessarily indirect. They are based on patterns of reactivity, often constructed from data for authentic inner-sphere and outer-sphere processes. They will be discussed in a later section.

5.4 Outer Sphere Reactions

Consider one of the most common types of outer-sphere reactions involving bivalent and trivalent metal complexes, ML^{2+} and $M_1L_1^{3+}$ where L and L_1 represent the total ligand structure and M and M_1 or/and L and L_1 may be different, as in the reactant pairs, $V(H_2O)_6^{2+}$ and $Fe(H_2O)_6^{3+}$, $Fe(H_2O)_6^{2+}$ and $Fe(bpy)_3^{3+}$, and $V(H_2O)_6^{2+}$ and $Ru(NH_3)_6^{3+}$. The outer-sphere process can be envisaged as

$$ML^{2+} + M_1L_1^{3+} \rightleftharpoons ML^{3+} + M_1L_1^{2+} \tag{5.17}$$

proceeding in three steps

$$ML^{2+} + M_1L_1^{3+} \rightleftharpoons ML^{2+} \cdots M_1L_1^{3+} \tag{5.18}$$

$$ML^{2+} \cdots M_1L_1^{3+} \rightleftharpoons ML^{3+} \cdots M_1L_1^{2+} \tag{5.19}$$

$$ML^{3+} \cdots M_1L_1^{2+} \rightleftharpoons ML^{3+} + M_1L_1^{2+} \tag{5.20}$$

A precursor complex is very rapidly formed in (5.18). It undergoes intramolecular electron transfer (5.19) to give a successor complex, which rapidly breaks down to products (5.20). In outer sphere reactions, it is noted that the two reactants do not share at any time a common atom or group. Such reactions then are particularly suitable as a basis for the calculation of rate constants since no bond breaking or making occurs during the electron transfer. The coordination shell and immediate environment for the reactants and for the products will differ as a result of a redox reaction. However internuclear distances and nuclear velocities cannot change during the electron transition of a redox reaction (Franck-Condon principle). Therefore some "common state" must be reached for each reactant prior to electron transfer. It is the free energy ΔG^* that is required to change the atomic coordinates from their equilibrium values to those in the activated complex, which must be calculated in any theory.

The work required to bring the ions 1 and 2 (charges z_1 and z_2) to the separation distance $r (= a_1 + a_2)$ is w_{12} where

$$w_{12} = \frac{z_1 z_2 e^2}{D_s r (1 + \beta r \mu^{1/2})} = \frac{4.25 \times 10^{-8} z_1 z_2}{r (1 + 3.29 \times 10^7 r \mu^{1/2})} \tag{5.21}$$

and β (Debye-Hückel constant) is given by

$$\beta = \left(\frac{8 \pi N^2 e^2}{1000 D_s RT} \right)^{1/2} = 0.329 \,\text{Å}^{-1} \text{ in } H_2O \text{ at } 25 °C \tag{5.22}$$

The reorganization terms, λ_o and λ_i are given by

$$\lambda_o = (\Delta e)^2 \left(\frac{1}{2 a_1} + \frac{1}{2 a_2} - \frac{1}{r} \right) \left(\frac{1}{n^2} - \frac{1}{D_s} \right) \tag{5.23}$$

$$\lambda_i = \sum_j \frac{f_j^r f_j^p (\Delta q j)^2}{f_j^r + f_j^p} = \frac{3 \bar{f}(\Delta d)^2}{2} \tag{5.24}$$

where N = Avogadro's number
D_s = Dielectric constant of the medium
n = Refractive index of the medium
(Δe) = Charge transferred from one reactant to another
f_j^r and f_j^p = jth normal mode force constants in the reactants and products respectively. Breathing vibrations are often employed and $\bar{f}$ = mean of the breathing force constants.
$\Delta q j$ = change in equilibrium value of the jth normal coordinate, and when breathing vibrations are employed
Δd = the difference of the metal-ligand distance between oxidized and reduced complex.

Several workers, particularly Marcus and Hush, tackled the calculation of ΔG^*; for an account and comparison of the various early attempts, the reader is referred to Refs. 19 and 20. This important area has been thoroughly reviewed and representative examples in Refs. 21–25 as well as in the Selected Bibliography give accounts of the theory in varying depths as well as an entry into the vast literature.

The free energy barrier ΔG^* is considered to consist of various components:

1. The work required to bring the reactants (assumed to be rigid spheres of radius a_1 and a_2) to their mean separation distance in the activated complex ($r = a_1 + a_2$) and then remove the products to infinity. These work terms are w^r and $-w^p$, respectively, and incorporate electrostatic and nonpolar contributions.

2. The free energy required to reorganize the solvent molecules around the reactants (the outer coordination shell) and to reorganize the inner coordination shell of the reactants. These are termed λ_o and λ_i, respectively.

3. The standard free energy of the reaction in the conditions of the experimental medium and when the reactants are far apart. The quantity $(\Delta G^\circ + w^p - w^r)$ is important since it is the standard free energy of the reaction at the separation distance (it is the work-corrected free energy of reaction). Both the w and λ_o terms can be fairly easily calculated. The term λ_i is quite difficult to estimate, requiring a knowledge of bond lengths and force constants of the reactants (see Inset).

Marcus has derived the expression (alternative forms are often seen)

$$\Delta G^* = \frac{w^r + w^p}{2} + \frac{\lambda_o + \lambda_i}{4} + \frac{\Delta G^\circ}{2} + \frac{(\Delta G^\circ + w^p - w^r)^2}{4(\lambda_o + \lambda_i)} \tag{5.25}$$

The free energy term ΔG^* is related to the free energy of activation $\Delta G^{\ddagger}$ by

$$\Delta G^* = \Delta G^{\ddagger} - RT \ln\left(\frac{hZ}{kT}\right) = (\Delta G^{\ddagger} - 2.8) \text{ kcal mol}^{-1} \tag{5.26}$$

and to the rate constant k by the expression

$$k = \kappa A r^2 \exp(-\Delta G^*/RT) \tag{5.27}$$

The transmission coefficient κ is approximately 1 for reactions in which there is substantial ($>4\,\text{kJ}$) electronic coupling between the reactants (adiabatic reactions). Ar^2 is calculable if necessary[25,26] but is usually approximated by Z, the effective collision frequency in solution, and assumed to be $10^{11}\,\text{M}^{-1}\text{s}^{-1}$. Thus it is possible in principle to calculate the rate constant of an outer-sphere redox reaction from a set of *nonkinetic* parameters, including molecular size, bond length, vibration frequency and solvent parameters (see inset). This represents a remarkable step. Not surprisingly, exchange reactions of the type

$$ML_6^{2+} + ML_6^{3+} \rightleftharpoons ML_6^{3+} + ML_6^{2+} \tag{5.28}$$

(5.28) rather than (5.17) have been examined, since ΔG° (≈ 0) in (5.25) can be ignored and the properties of only one redox couple ($ML^{2+/3+}$) and not two ($ML^{2+/3+}$ and $M_1L_1^{2+/3+}$)

need be considered. Some results are contained in Table 5.4. See also Table II in Ref. 25. The details of the calculations and references to the experimental data obtained in a variety of ways are given in Refs. 23, 25, 27–31.

Table 5.4 Comparison of Observed Exchange Rate Constants with Values Calculated on the Basis of (5.25) at 25 °C

Couple	k_{obs} $M^{-1}s^{-1}$	k_{calc} $M^{-1}s^{-1}$	Refs.
$Ru(H_2O)_6^{2+} + Ru(H_2O)_6^{3+}$	20	60	27–29
$Ru(NH_3)_6^{2+} + Ru(NH_3)_6^{3+}$	7×10^3 (4°C)	1×10^5	27, 30
$Ru(bpy)_3^{2+} + Ru(bpy)_3^{3+}$	4×10^8	1×10^9	25, 27
$Tc(dmpe)_3^+ + Tc(dmpe)_3^{2+}$ [a]	$\sim 6 \times 10^5$	3×10^6	31

[a] dmpe = 1,2-bis(dimethylphosphino)ethane

Everything being considered, the agreement between the calculated and the observed rate constants is excellent. The rates tend to increase with size of the ligands (see Ru entries in Table 5.4). This arises from a decrease in the value of λ_o as the reactant size increases. For a given ligand type, ($\lambda_o \approx$ constant) rate constants increase with decreasing differences in the metal-ligand distances (smaller λ_i term) in the two oxidation states. This is strikingly illustrated by the $M(bpy)_3^{2+/3+}$ and $M(bpy)_3^{+/2+}$ couples shown in Table 5.5. [32] The transfer of σ^*d electrons between the two oxidation states leads to larger M-N bond distance changes (Δd), and slower rates than when only non-bonding πd electrons are involved. The varying self-exchange rate constants for a series of $M(sar)^{2+/3+}$ (Prob. 4) and Ru(II)-(III) complexes have been rationalized in a similar manner. [26,33,34]

Table 5.5 Electron Transfer Rate Constants and Differences in Metal-Ligand Distances between the Oxidation States (L = bpy) [32]

Couple	Electron Configuration	$\Delta d(\text{Å})$	k_{11}, s^{-1}
$NiL_3^{2+/3+}$	$(\pi d)^6(\sigma^*d)^2/(\pi d)^6(\sigma^*d)^1$	≈ 0.12	1.5×10^3
$CoL_3^{2+/3+}$	$(\pi d)^5(\sigma^*d)^2/(\pi d)^6$	0.19	18
$CoL_3^{+/2+}$	$(\pi d)^6(\sigma^*d)^2/(\pi d)^5(\sigma^*d)^2$	-0.02	1×10^9
$FeL_3^{2+/3+}$	$(\pi d)^6/(\pi d)^5$	0.00	3×10^8
$RuL_3^{2+/3+}$	$(\pi d)^6/(\pi d)^5$	0.00	4×10^8
$CrL_3^{2+/3+}$	$(\pi d)^4/(\pi d)^3$	0.00	2×10^9

A good deal of data is required for these calculations and the theoretical ideas developed have been more usefully applied to the estimation of rate constants for net chemical changes (5.17) in terms of the free energy change, ΔG^0 and the rate constants for related reactions (LFER, Sec. 2.5). Equation (5.25) can be written

$$\Delta G^* \approx w^r + \frac{\lambda_o + \lambda_i}{4}\left(1 + \frac{\Delta G^0 + w^p - w^r}{\lambda_o + \lambda_i}\right)^2 \tag{5.29}$$

If

$$(\Delta G^0 + w^p - w^r)(\lambda_o + \lambda_i)^{-1} < 1 \tag{5.30}$$

then

$$\Delta G^* \sim \frac{w^r + w^p}{2} + \frac{\lambda_o + \lambda_i}{4} + \frac{\Delta G^0}{2} \tag{5.31}$$

In the redox reactions of a series of *related* reagents with one *constant* reactant (so that ΔG^0 is the only important variable), a plot of ΔG^* vs ΔG^0 would be expected to be linear with slope 0.5.

We can take the analysis still further: Consider the "cross reaction" (5.17) with the various parameters subscripted 12 (forward direction)

$$ML^{2+} + M_1L_1^{3+} \rightleftharpoons ML^{3+} + M_1L_1^{2+} \qquad k_{12}, K_{12}, (\lambda_o + \lambda_i)_{12}, w_{12}, \kappa_{12} \tag{5.17}$$

and the related isotopic exchange ("self-exchange") reactions, (5.28) and (5.32) with the subscripts 11 and 22

$$ML_6^{2+} + ML_6^{3+} \rightleftharpoons ML_6^{3+} + ML_6^{2+} \qquad k_{11}, (\lambda_o + \lambda_i)_{11}, w_{11}, \kappa_{11} \tag{5.28}$$

$$M_1L_1^{2+} + M_1L_1^{3+} \rightleftharpoons M_1L_1^{3+} + M_1L_1^{2+} \qquad k_{22}, (\lambda_o + \lambda_i)_{22}, w_{22}, \kappa_{22} \tag{5.32}$$

In the first instance, we can ignore work terms. If, in addition, we can assume that the sum of λ's for the cross reaction $(\lambda_o + \lambda_i)_{12}$ is given by

$$(\lambda_o + \lambda_i)_{12} = 1/2\,[(\lambda_o + \lambda_i)_{11} + (\lambda_o + \lambda_i)_{22}] \tag{5.33}$$

then combining this with (5.29) yields

$$\Delta G_{12}^{\ddagger} = 0.50\Delta G_{11}^{\ddagger} + 0.50\Delta G_{22}^{\ddagger} + 0.50\Delta G_{12}^{\circ} - 1.15\,RT \log f_{12} \tag{5.34}$$

or

$$k_{12} = (k_{11}k_{22}K_{12}f_{12})^{1/2} \tag{5.35}$$

where

$$\log f_{12} = \frac{(\log K_{12})^2}{4\log (k_{11}k_{22}/Z^2)} \tag{5.36}$$

The value of K_{12} is the equilibrium constant for (5.17) in the prevailing medium, and may be determined directly (e. g. spectrally) or calculated from a knowledge of the oxidation potentials for the two self-exchanges. A relationship of the form (5.35) has been derived from a simplified nonrigorous statistical mechanical derivation[35] and by simple thermodynamic cycle and detailed balance considerations.[36] First we examine a series of reactions between a common reactant ML^{2+} (constant k_{11}) and a number of closely related complexes $M_1L_1^{3+}$ (there might be slight changes in the ligand structure L_1) so that k_{22} doesn't much vary either. In addition, if the equilibrium constant for the cross-reaction K_{12} does not deviate much from unity, i.e. $\log f_{12} \approx 0$, then it is easily seen from (5.35) that there should be a linear relationship between $\log k_{12}$ and $\log K_{12}$ with a slope 0.5. In a number of systems including metalloproteins[37] and electronically excited reactants[38,39] this simple relationship has been observed,[25] and an example has been already shown in Sec. 2.5. However with a large number of reactions (particularly between oppositely charged reactants and with large driving forces), work terms cannot be ignored and if the transmission coefficient κ is less than 1, i.e., there is an element of nonadiabaticity, then the modified Marcus expression is[32,40]

$$k_{12} = \left[\frac{k_{11} k_{22} K_{12} f_{12}}{\kappa_{11} \kappa_{22}} \right]^{1/2} \kappa_{12} W_{12} \tag{5.37}$$

$$\ln f_{12} = \frac{[\ln K_{12} + (w_{12} - w_{21})/RT]^2}{4 \left[\ln \dfrac{k_{11} k_{22}}{Z^2 \kappa_{11} \kappa_{22}} + \dfrac{w_{11} + w_{22}}{RT} \right]} \tag{5.38}$$

$$W_{12} = \exp \left[-(w_{12} + w_{21} - w_{11} - w_{22})/2RT \right] \tag{5.38a}$$

There are a number of ways of plotting the modified expression (5.37), using natural or decadic logarithms. Usually the self-exchange reactions are assumed adiabatic ($\kappa_{11} = \kappa_{22} = 1$), and rearrangement of (5.37) leads to

$$2 \ln k_{12} - \ln f_{12} - 2 \ln W_{12} - \ln k_{22} = \ln (k_{11} \kappa_{12}^2) + \ln K_{12} \tag{5.38b}$$

A plot of the left-hand side of (5.38b) versus $\ln K_{12}$ should be linear with a slope of unity and an intercept $= \ln (k_{11} \kappa_{12}^2)$.[41] Such a plot for the reactions of $Co(phen)_3^{3+}$ with $Cr(bpy)_3^{2+}$, $Cr(phen)_3^{2+}$ and their substituted derivatives yields a slope of 0.98 and an intercept of approximately -0.55. If k_{11}, the self-exchange rate constant for $Co(phen)_3^{3+}$ is 30 $M^{-1}s^{-1}$ this corresponds to $\kappa_{12} = 0.13$, indicating mild nonadiabaticity for reactions involving $Co(phen)_3^{3+}$. Ref. 41. See also Fig. 8.2.

We are now in a position to understand the full implications of the plot of Fig. 2.8. The value of $\log k_{12}$ at $\log K_{12} = 0$ is 2 and on the basis of the simple Marcus expression this equals $1/2(8.5 + x)$ where $8.5 = \log k_{11}$ (for PTZ/PTZ$^+$) and x is $\log k_{22}$ (for $Fe^{2+/3+}$). This leads to a value of $x \approx -4$, i.e. $k_{22} = 10^{-4} M^{-1}s^{-1}$. This is much smaller that the experimentally determined value of 4 $M^{-1}s^{-1}$. This difference may be taken care of by using the fuller expression (5.38b) and a value of $\kappa_{12} \approx 10^{-2}$. See also Ref. 42.

Plots for a reaction series have been however much less used than the application of these equations, both simple (Prob. 1) and complicated, to isolated reaction systems.[21,25] Table 5.6

Table 5.6 Calculated Values for the Self-Exchange Rate Constant for $Ru(H_2O)_6^{2+/3+}$ using (5.35) and Data for a Number of Cross-Reactions (from Ref. 43)

Reactions	ΔE° V	k_{12} $M^{-1}s^{-1}$	k_{22} $M^{-1}s^{-1}$	k_{11} $M^{-1}s^{-1}$
$Ru(H_2O)_6^{2+} + Ru(NH_3)_5py^{3+}$	0.082	1.1×10^4	4.7×10^5	15
$Ru(NH_3)_6^{2+} + Ru(H_2O)_6^{3+}$	0.150	1.4×10^4	2×10^4	32
$Ru(H_2O)_6^{2+} + Ru(NH_3)_5isn^{3+}$	0.167	5.5×10^4	4.7×10^5	14
$V(H_2O)_6^{2+} + Ru(H_2O)_6^{3+}$	0.47	2.8×10^2	1×10^{-2}	0.37
$Ru(H_2O)_6^{2+} + Co(phen)_3^{3+}$	0.15	53	40	0.24

shows the second-order rate constants k_{12} determined for a number of cross reactions involving the $Ru(H_2O)_6^{2+/3+}$ couple at 25 °C and $\mu = 1.0$ M. The difference in oxidation potentials for the two reactants, $\Delta E°$, allows us to calculate $\log K_{12}$ ($= 16.9 \times \Delta E°$) which together with k_{22}, permits an estimation of k_{11} for the $Ru(H_2O)_6^{2+/3+}$ couple using (5.35).[43] Inclusion of the work terms has only a small effect on the final results.[40] The values calculated for k_{11} using data for the highly exothermic reactions are lower than those derived from reactions with small driving forces. The higher values ($14-32$ M^{-1}s^{-1}) are satisfyingly close to those measured and calculated in Table 5.4.

If the reactants are oppositely charged, the collision complex in (5.18) takes the form of an outer-sphere complex with discernable stability. For the outer sphere redox reaction between $Co(NH_3)_5L^{n+}$ and $Fe(CN)_6^{4-}$, L being a series of pyridine or carboxylate derivatives, saturation kinetics are observed, with the pseudo first-order rate constant (k_{obs}), Fe(II) in excess, being given by

$$k_{obs} = \frac{k_{et} K_{os} [Fe(CN)_6^{4-}]}{1 + K_{os} [Fe(CN)_6^{4-}]} \tag{5.39}$$

This behavior is consistent with the mechanism (Note however Sec. 1.6.4).

$$Co(NH_3)_5L^{n+} + Fe(CN)_6^{4-} \xrightleftharpoons{fast} Co(NH_3)_5L^{n+} \,|\, Fe(CN)_6^{4-} \quad K_{os} \tag{5.40}$$

$$Co(NH_3)_5L^{n+} \,|\, Fe(CN)_6^{4-} \longrightarrow Co(NH_3)_5L^{(n-1)+} \,|\, Fe(CN)_6^{3-} \quad k_{et} \tag{5.41}$$

$$Co(NH_3)_5L^{(n-1)+} \,|\, Fe(CN)_6^{3-} \xrightleftharpoons{fast} Co(NH_3)_5L^{(n-1)+} + Fe(CN)_6^{3-} \tag{5.42}$$

At 25 °C and $\mu = 0.1$ M, the values of K_{os} are 10^2-10^4 M^{-1} and those of k_{et} are $10^{-4}-10^{-1}$s^{-1} depending on the identity of L.[44] The internal electron transfer rate in an outer-sphere complex can thus be analyzed[45,46] without considering work terms or, what is equivalent, the equilibrium controlling the formation of precursor complex.[47] This favorable situation is even improved when the metal centers are directly bridged. The relative orientation of the two metal centers in a well-established geometry can be better treated than in the outer-sphere complex (Sec. 5.8).

It is hardly possible to overestimate the impact that the Marcus-Hush ideas and their experimental exploitation, mainly by Sutin and his coworkers, have made on the study of redox reactions. The treatment here is necessarily brief. Other aspects have been discussed in the references cited. Some important points we have not considered are

(a) the application of the derived equations to other activation parameters.[23]

(b) the necessity to sometimes (not often) allow for the preexponential factor κ being non-unity, has been briefly alluded to. The non-adiabaticity becomes more pronounced when the standard free energy of the reaction increases (Table 5.6). Its assessment can be difficult (and controversial).[48]

(c) the occurrence of an inverted region where ΔG^* increases (rate constant decreases) as $\Delta G°$ becomes more negative. In this region, very large driving forces are involved.[23,49]

(d) the effect of solvents.[50]

5.4.1 The Applications of the Marcus Expression

Occasionally, the successful application of the Marcus expressions (5.35) and (5.37) to a reaction can support its designation as outer-sphere. The reduction of a series of substituted benzenediazonium salts by $Fe(CN)_6^{4-}$ and $(Me_5cp)_2Fe$ conforms to the simple Marcus expression and represents supporting evidence for the formulation of these reactions as outer sphere (or non-bonded electron transfer in organic systems)

$$ArN_2^+ + Fe^{II} \rightarrow ArN_2^{\cdot} + Fe^{III} \tag{5.43}$$

rather than inner sphere (or bonded electron transfer)[51]

$$ArN_2^+ + Fe^{II} \rightarrow ArN=N-Fe^{II} \rightarrow ArN_2^{\cdot} + Fe^{III} \tag{5.44}$$

The pattern for outer-sphere oxidation by $Co(NH_3)_6^{3+}$ compared with $Co(en)_3^{3+}$ (usually it is ≈ 10 times slower) towards inorganic reductants can be used[52] to support an estimate of the proportion of electron transfer (Marcus-dependent) and charge transfer which *$Ru(bpy)_3^{2+}$ displays towards these oxidants (45 and 11%, respectively), Sec. 2.2.1(b). Finally, Eqn. 5.35 can be used to determine K_{12} for a reaction in which the other kinetic parameters are known. The value of K_{12} can be used, in turn, to estimate the oxidation potential of one couple, which is normally inaccessible.[53] Thus the potentials of the o-, m- and p-benzene diol radicals $H_2A^{\cdot}$ were determined from kinetic data for the oxidation of the diols (H_2A) by $Fe(phen)_3^{3+}$ (5.45):[53]

As might be anticipated, there are exceptions to the Marcus equations (Prob. 3).[25]

$$Fe(phen)_3^{3+} + H_2A \rightleftharpoons Fe(phen)_3^{2+} + H_2A^{\cdot} \tag{5.45}$$

5.5 Inner Sphere Redox Reactions

Just as we did with outer-sphere reactions, we can dissect an inner sphere redox process into individual steps. Specifically, let us examine the reaction of $(H_2O)_6Cr^{II}$ with $Co^{III}(NH_3)_5L$. The first step is the formation of the precursor complex[19,20,54-56].

$$(H_2O)_6Cr^{II} + LCo^{III}(NH_3)_5 \rightleftharpoons (H_2O)_5Cr^{II}LCo^{III}(NH_3)_5 + H_2O \quad k_1, k_{-1}, K_1 \tag{5.46}$$

The product of intramolecular electron transfer within the precursor complex is the successor complex

$$(H_2O)_5Cr^{II}LCo^{III}(NH_3)_5 \rightleftharpoons (H_2O)_5Cr^{III}LCo^{II}(NH_3)_5 \qquad k_2, k_{-2} \tag{5.47}$$

which then undergoes dissociation into products

$$(H_2O)_5Cr^{III}LCo^{II}(NH_3)_5 \rightleftharpoons (H_2O)_5Cr^{III}L + Co^{II}(NH_3)_5 \qquad k_3, k_{-3} \tag{5.48}$$

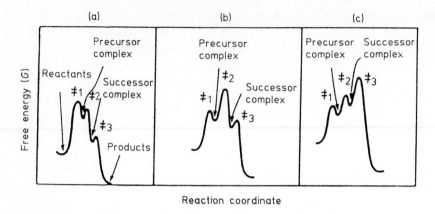

Fig. 5.1 Reaction profiles for inner-sphere redox reactions illustrating three types of behavior (a) prercursor complex formation is rate-limiting (b) precursor-to-successor complex is rate-limiting and (c) breakdown of successor complex is rate-limiting. The situation (b) appears to be most commonly encountered.

Of course the $Co^{II}(NH_3)_5$ breaks down rapidly in acid into Co^{2+} and $5\,NH_4^+$. Precursor complex formation, intramolecular electron transfer, or successor complex dissociation may severally be rate limiting. The associated reaction profiles are shown in Fig. 5.1. A variety of rate laws can arise from different rate-determining steps.[22] A second-order rate law is common, but the second-order rate constant k_{obs} is probably composite. For example, (Fig. 5.1(b)) if the observed redox rate constant is less than the substitution rate constant, as it is for many reactions of Cr^{2+}, Eu^{2+}, Cu^+, Fe^{2+} and other ions, and if little precursor complex is formed, then $k_{obs} = (k_1 k_2 k_{-1}^{-1})$. In addition, the breakdown of the successor complex would have to be rapid ($k_3 \gg k_{-2}$). This situation may even give rise to *negative* $\Delta H_{obs}^{\ddagger}$ ($= \Delta H_1^{\circ} + \Delta H_2^{\ddagger}$) since enthalpies of formation of precursor complexes (ΔH_1°) may be negative. Such negative values of $\Delta H_{obs}^{\ddagger}$ in turn, constitutes good evidence for the existence of precursor complexes.[57] A good deal of effort has gone into attempting to isolate the data for the electron transfer step (5.47).[22,58] One way of accomplishing this might be by bulding up large concentrations of precursor complex. If this is not possible, then the experimental rate constants for reaction of a series of related complexes may still parallel the corresponding values for the intramolecular rate constants, k_2, if the values of k_1/k_{-1} remain sensibly constant. However, the possibility that this is not the case should be always recognized.

5.6 The Bridging Ligand in Inner-Sphere Redox Reactions

The early work of Taube and his co-workers opened several interesting avenues of approach, most of which have been fully exploited.[21,22,48,B40] One of the most obvious is to examine the requirements for a good bridging group and determine the effects of this bridge on the rate of the inner-sphere redox reaction. Hundreds of different bridges have been examined since

Taube's original discovery.[4] Attendant changes of redox rates by many orders of magnitude have been observed. Much data have been obtained on reactions of the type

$$Cr^{2+} + Co(NH_3)_5L^{n+} \xrightarrow{5\,H^+} CrL^{n+} + Co^{2+} + 5\,NH_4^+ \qquad (5.49)$$

and these, together with reduction of the Co(III) complexes by other metal ions and complexes, are useful for discussion purposes (Table 5.7).[16] Oxidation by Cr(III) and Ru(III) also provides useful information, and isotopic exchanges of the type

$$*Cr^{2+} + CrL^{2+} \rightleftharpoons *CrL^{2+} + Cr^{2+} \qquad (5.50)$$

which cannot be outer-sphere, were early explored[3] (Table 5.8).

Table 5.7 Rate Constants (k, $M^{-1}s^{-1}$) for the Reduction of $Co(NH_3)_5L^{n+}$ by a Variety of Reductants at 25 °C

L	Cr^{2+}	V^{2+}	Fe^{2+} [d]	Eu^{2+}
NH_3	8.0×10^{-5}	3.7×10^{-3}		2×10^{-2}
py	4.1×10^{-3}	0.24		
H_2O	$\leqslant 0.1$	0.53		0.15
$OCOCH_3$	0.35	1.2	$< 5 \times 10^{-5}$	0.18
OCOCOOH	1.0×10^2	12.5	3.8×10^{-3}	
F^-	2.5×10^5	2.6	6.6×10^{-3}	2.6×10^4
Cl^-	6×10^5	10	1×10^{-3}	3.9×10^2
Br^-	1.4×10^6	25	7.3×10^{-4}	2.5×10^2
I^-	3×10^6	1.2×10^2		1.2×10^2
OH^-	1.5×10^6	< 4		$< 2 \times 10^3$
N_3^-	$\sim 3 \times 10^5$	13	8.8×10^{-3}	1.9×10^2
NCS^-	19^b	0.3	$< 3 \times 10^{-6}$	0.05
$\underline{S}CN^-$	$1.9 \times 10^{5\,b}$ $0.8 \times 10^{5\,c}$	30	0.12	3.1×10^3

L	Cu^+	$Co(CN)_5^{3-}$	$Cr(bpy)_3^{2+}$	$Ru(NH_3)_6^{2+}$	Ti^{3+}
NH_3		$8 \times 10^{4\,a}$	6.9×10^2	1.1×10^{-2}	
py				1.2	
H_2O	1.0×10^{-3}		$5 \times 10^{4\,e}$	3.0	
$OCOCH_3$		$1.1 \times 10^{4\,a}$	1.2×10^3	1.8×10^{-2}	
OCOCOOH				0.50	
F^-	1.1	1.8×10^3	1.8×10^3		2×10^2
Cl^-	4.9×10^4	$\sim 5 \times 10^7$	8×10^5	2.6×10^2	13
Br^-	4.5×10^5		5×10^6	1.6×10^3	2
I^-				6.7×10^3	4
OH^-	3.8×10^2	9.3×10^4	$1 \times 10^{3\,e}$	0.04	
N_3^-	1.5×10^3	1.6×10^6	4.1×10^4	1.8	
NCS^-	~ 1	1.1×10^6	1.1×10^4	0.74	
$\underline{S}CN^-$		$< 10^7$	2.0×10^6	3.8×10^2	

[a] $M^{-2}s^{-1}$, outer-sphere reduction by $Co(CN)_6^{3-}$. [b] Remote attack. [c] Adjacent attack. [d] J. H. Espenson, Inorg. Chem. **4**, 121 (1665). [e] 4 °C.

Examination of the data for (5.49) and (5.50) in Tables 5.7 and 5.8 shows that there is some general order of reactivity for the various ligands L. Containing an unshared electron pair *after coordination* appears a minimum requirement for a ligand to be potential bridging group, for it has to function as a Lewis base towards two metal cations. Thus $Co(NH_3)_6^{3+}$ and $Co(NH_3)_5py^{3+}$ oxidize Cr^{2+} by an outer-sphere mechanism, giving Cr^{3+} as the product, at a much slower rate than for the inner-sphere reactions.

The bridging group is often supplied by the oxidizing agent because this is invariably the inert reactant. In these cases, the bridging ligand normally transfers from oxidant to reductant during the reaction. This however is not an essential feature of an inner-sphere redox reaction. The cyanide bridge is supplied by $Fe(CN)_6^{4-}$ in some reductions and remains with the iron after electron transfer and breakup.[59] Such reactions, which proceed without ligand transfer, can only be shown to be inner-sphere directly, i.e. by the demonstration of a bridged intermediate.

Table 5.8 Rate Parameters for Cr(II) − Cr(III) Exchange Reactions (5.20) at 25 °C. Ref. 16.

Exchange Partners	k $M^{-1}s^{-1}$	$\Delta H^{\ddagger}$ kJ mol^{-1}	$\Delta S^{\ddagger}$ J K^{-1}mol^{-1}
$Cr^{2+} + Cr^{3+}$	$\leqslant 2 \times 10^{-5}$	...	...
$Cr^{2+} + CrOH^{2+}$	0.7	54	−67
$Cr^{2+} + CrNCS^{2+}$	1.4×10^{-4}	...	...
$Cr^{2+} + CrSCN^{2+}$	40	...	...
$Cr^{2+} + CrN_3^{2+}$	6.1	40	−96
$Cr^{2+} + CrF^{2+}$	2.4×10^{-3} (0 °C)	57	−84
$Cr^{2+} + CrCl^{2+}$	9 (0 °C)	...	...
$Cr^{2+} + CrBr^{2+}$	> 60	...	...
$Cr^{2+} + CrCN^{2+}$	7.7×10^{-2}	39	−134
$Cr^{2+} + cis\text{-}Cr(N_3)_2^{+}$	60	...	...

The different types of bridging ligands will now be discussed. Many of the varying patterns and theoretical bases have been built up by using the simpler bridging ligands on which we shall first concentrate. This will lead into the larger organic and protein bridges.

(a) Hydroxide and Water. With oxidants containing a coordinated water group, for example $Co(NH_3)_5H_2O^{3+}$, a term in the rate law containing an $[H^+]^{-1}$ dependency for their reaction is often found. This may make a significant contribution to the rate, and mask any $[H^+]$-independent term.[12] The inverse term is usually ascribed to reduction of the hydroxy species, for example $Co(NH_3)_5OH^{2+}$, offering a very effective OH bridge in an inner-sphere process.[60] Reactions in which the aqua and hydroxy forms have similar reactivities and in which no other bridging group is present are probably outer-sphere,[18] and assignment of mechanism on this basis is illustrated in Table 5.7.[61]

There has been a continuing discussion without resolution[22] of whether the aqua group acts as a (weak) bridge in the situations where the corresponding hydroxo complex reacts inner-sphere.

(b) Halides. The reduction of halide complexes has featured prominently in the development of redox chemistry. Rates vary monotonically from F to I but not in a consistent manner. In examining Table 5.7 it is seen that in inner-sphere reductions of $Co(NH_3)_5X^{2+}$ by Cr^{2+}, the rate increases with increasing size of the halogen.[62] The order is inverted when reduction by Eu^{2+} is considered, even although it is probable that this is inner-sphere also. The difference has been rationalized by calculating formal equilibrium constants for halide interchange in the transition state for various redox reactions.[22,64,65,B36]

The equilibrium constant for

$$[(NH_3)_5CoFCr^{4+}]^{\ddagger} + I^- \rightleftharpoons [(NH_3)_5CoICr^{4+}]^{\ddagger} + F^- \qquad K_1 \qquad (5.51)$$

can be calculated knowing the rate and equilibrium constants for:

$$Co(NH_3)_5F^{2+} + Cr^{2+} \rightleftharpoons [(NH_3)_5CoFCr^{4+}]^{\ddagger} \qquad k_2 \qquad (5.52)$$

$$Co(NH_3)_5I^{2+} + Cr^{2+} \rightleftharpoons [(NH_3)_5CoICr^{4+}]^{\ddagger} \qquad k_3 \qquad (5.53)$$

$$Co(NH_3)_5H_2O^{3+} + F^- \rightleftharpoons Co(NH_3)_5F^{2+} + H_2O \qquad K_4 \qquad (5.54)$$

$$Co(NH_3)_5H_2O^{3+} + I^- \rightleftharpoons Co(NH_3)_5I^{2+} + H_2O \qquad K_5 \qquad (5.55)$$

since $K_1 = k_3 K_5 k_2 K_4 = 0.064.$[65]

The value of $K_1 < 1$ for this Cr(II) reduction, as with the reactions of Eu(II), indicates that the substitution of bridging F^- by I^- is unfavorable in the bridged transition complex in both cases. The two sets of reactivity patterns noted above thus disappear. It has been noted that $K_1 < 1.0$ when both metal centers are hard acids, whereas $K_1 > 1$ when one reactant is soft e.g., Cu^+.[B36] These relationships have been rationalized.[22] The much better bridging properties of chloride than water are shown by the data in Table 5.7 and Table 5.9.

Table 5.9 Ratio (R) of Reactivities of $Co(edta)Cl^{2-}$ and $Co(edta)H_2O^-$ towards Reductants[5]

Reductants	R	Mechanism
Cr^{2+}	$>3 \times 10^2$	inner sphere
Fe^{2+}	2×10^3	inner sphere
$Fe(CN)_6^{4-}$	33	outer sphere
Ti(III)	31	outer sphere

(c) Ambidentate Ligands. The use in the oxidant of a polyatomic bridging ligand that presents more than one potential donor site towards the reducing metal ion introduces the concept of *remote* and *adjacent* attack. An authentic example of adjacent attack is rare but illustrated by the scheme[66]

$$
\begin{array}{l}
\text{remote attack by} \\
\xrightarrow{\hspace{1cm}} [(NH_3)_5CoSCNCr^{4+}]^{\ddagger} \rightarrow CrNCS^{2+} \text{ stable (purple)} \\
\quad Cr^{2+} \qquad\qquad\qquad\qquad\qquad\qquad\qquad\qquad\uparrow \quad \text{form} \\
Co(NH_3)_5SCN^{2+} \qquad\qquad\qquad\qquad\qquad\qquad\qquad\qquad\qquad (5.56) \\
\quad\uparrow \\
\text{adjacent attack by} \quad \left[(NH_3)_5CoSCr^{4+}\right]^{\ddagger} \rightarrow CrSCN^{2+} \rightarrow Cr^{3+} + SCN^- \\
\qquad Cr^{2+} \qquad\qquad\qquad\qquad\; C \qquad\qquad\quad \text{unstable linkage} \\
\qquad\qquad\qquad\qquad\qquad\qquad\qquad\; N \qquad\qquad\qquad \text{(green) isomer}
\end{array}
$$

Analysis for $CrSCN^{2+}$ and $CrNCS^{2+}$ in the products can be made by ion-exchange separation and spectral identification. This procedure indicates that about 30% of the reaction goes by the adjacent attack path in 1 M^+H^+ at 25 °C. Reduction of $Co(NH_3)_5NCS^{2-}$ by Cr^{2+}, in contrast, proceeds much more slowly and quantitatively by remote attack, leading to the unstable isomer $CrSCN^{2+}$. In the N-bound thiocyanate complex, the only lone pairs of electrons available for attack by the Cr^{2+} are on the sulfur. In the S-bound $Co(NH_3)_5SCN^{2+}$ both S and N have lone pairs.[22] It is generally found that $MSCN^{2+}$ is about 10^4 times more reactive than $MNCS^{2+}$, $M = Co(NH_3)_5$ and $Cr(H_2O)_5$, in its reaction with Fe(II) and Cr(II), (Table 5.7) as well as with a number of other reducing ions which go by inner-sphere. With outer-sphere reductants the ratio is less, about 10^2, and these ratios have been rationalized.[22] With V(II), the ratio is also much less, and this supports the idea that the V(II)-$Co(NH_3)_5SCN^{2+}$ reaction is substitution-controlled (Table 5.7).[67] Redox reactions of the type outlined above have been used to prepare linkage isomers (Sec. 7.4).

The azide bridging ligand cannot offer the interesting dual possibilities of the thiocyanate group. Because it is symmetrical and presents a nitrogen donor atom, which is favored over sulfur for most incipient tervalent metal centers, $Co(NH_3)_5N_3^{2+}$ is likely to be a more effective oxidant than $Co(NH_3)_5NCS^{2+}$ if the reaction goes by an inner-sphere mechanism; it is not likely to be much different in an outer-sphere reaction. This has been a useful diagnostic tool[22] (see Table 5.7).

Double bridges have been established for example, in the inner-sphere reaction of Cr^{2+} with *cis*-$Cr(N_3)_2^+$.[68] Surprisingly, the double bridge does not offer a markedly faster route than the single bridge (a factor of only 31 in enhanced rate for the example cited[68]). If a chelate site presents itself to an attacking metal ion, a chelate product can result.[21] The oxidation of $Co(en)_2(H_2O)_2^{2+}$ by $Co(C_2O_4)_3^{3-}$ yields $Co(en)_2(C_2O_4)^+$, probably via a double bridge, and with a very small amount of chiral discrimination (Sec. 5.7.4). Using nmr and ^{13}C-enriched free oxalate ion it can be shown that there is no enriched oxalate in the $Co(en)_2(C_2O_4)^+$ product, which must therefore arise from an inner-sphere process.[69]

(d) Non-Bridging Ligands. We might wonder what happens to the ligands that are not involved in the bridging act during the redox process, and what influence they might have as a result on the rates of such reactions.[70] This is an area where theoretical predictions preceeded experimental results. Orgel first drew attention to a model in which electronic states in the activated complex are matched by changing bond distances and therefore the ligand fields of the reactant ions.[71] For the reduction of Co(III) and Cr(III) complexes, for example, an electron from the reducing agent would appear in an unoccupied e_g say d_{z^2} orbital directed towards X (the bridging group) and Y in *trans*-$Co(NH_3)_4XY$. The energy of this orbital will obviously be more sensitive to changes in the *trans* Y group than to changes in the *cis* NH_3 ligands, and the orbital will be stabilized by a weak field ligand Y.[71] In general, the consequences of these happenings have been confirmed in both rate experiments and isotopic fractionation experiments. It has been shown, for example, that towards Cr(II), *trans*-$Cr(NH_3)_4(H_2^{16}O)Cl^{2+}$ reacts 1.6% faster than *trans*-$Cr(NH_3)_4(H_2^{18}O)Cl^{2+}$ (Sec. 3.12.2). This strongly supports the idea of stretching of the bond in the *trans*-position during the redox reaction. Stretching of the bonds in the *cis* postition is less important, although not negligible, judged by the value for $k_{16_O}/k_{18_O} = 1.007$ for the *cis* isomer.[72]

(e) The Estimation of Rate Constants for Inner-Sphere Reactions. There is evidence that a Marcus-type relationship may be applied to inner-sphere as well as outer-sphere redox reactions.[25,73] Varying A in *cis*-$Co(en)_2(A)Cl^{?+}$ has the same effect on the rate in the outer-sphere reduction by $Ru(NH_3)_6^{2+}$ as in the inner-sphere reduction by Fe^{2+} (constant Cl bridges). Since there is a nice correlation of $\log k$ vs ΔG for the outer-sphere reductions, it follows that a similar LFER must also apply to the inner-sphere process. [74]

5.7 Some Other Features of Redox Reactions

We now consider some other aspects of redox reactions which might be outer- and/or inner-sphere in nature.

5.7.1 Mixed Outer- and Inner-Sphere Reactions

As might be foreseen, there are a (limited) number of systems where the energetics of the outer- and inner-sphere reactions are comparable and where therefore both are paths for the reaction. An interesting example of this behavior is the reaction of $Cr(H_2O)_6^{2+}$ with $IrCl_6^{2-}$ which has been studied by a number of groups and is now well understood. [22] At 0°C, most of the reaction proceeds via an outer-sphere mechanism. The residual inner-sphere process utilizes a binuclear complex, which can undergo both $Cr-Cl$ and $Ir-Cl$ cleavage:

$$Cr(H_2O)_6^{2+} + IrCl_6^{2-} \begin{array}{c} \overset{71\%}{\nearrow} Cr(H_2O)_6^{3+} + IrCl_6^{3-} \\ \underset{29\%}{\searrow} (H_2O)_5CrClIrCl_5 \end{array} \qquad (5.57)$$

$$(H_2O)_5CrClIrCl_5 \begin{array}{c} \overset{39\%}{\nearrow} Cr(H_2O)_6^{3+} + IrCl_6^{3-} \\ \underset{61\%}{\searrow} Cr(H_2O)_5Cl^{2+} + IrCl_5H_2O^{2-} \end{array} \qquad (5.58)$$

The outer-sphere rate constant for the $Cr(H_2O)_6^{2+}/IrCl_6^{2-}$ reaction can be estimated, using Marcus' equation, as $\approx 10^9 M^{-1}s^{-1}$. A value of this magnitude can obviously be competitive with that for the inner-sphere path, which is more usual with the highly labile $Cr(H_2O)_6^{2+}$ ion. [22]

A rather involved, but interesting, example of a reaction which proceeds by both inner- and outer-sphere pathways is summarized in the scheme

$$Ru(NH_3)_5pz^{3+} + Co(edta)^{2-} \underset{k_{-1}}{\overset{k_1}{\rightleftharpoons}} [(NH_3)_5Ru^{II}pzCo^{III}(edta)]^+ \qquad (5.59)$$

$$Ru(NH_3)_5pz^{3+} + Co(edta)^{2-} \underset{k_{-2}}{\overset{k_2}{\rightleftharpoons}} Ru(NH_3)_5pz^{2+} + Co(edta)^- \qquad (5.60)$$

The outer-sphere pathway (k_2) produces the final products directly, as shown by a rapid increase in absorbance at 474 nm, which is a maximum for $Ru(NH_3)_5pz^{2+}$. At the same time, a rapid inner-sphere (k_1) production of the binuclear complex takes place. A slower absorbance increase at 474 nm arises from the back electron transfer in the binuclear intermediate (k_{-1}). This produces the original reactants which then undergo outer-sphere reaction (k_2). The values of the rate constants at 25°C and $\mu = 0.1$ M are 2.5×10^3 M^{-1}s^{-1} (k_1), 16.9 s^{-1} (k_{-1}) 1.0×10^3 M^{-1}s^{-1} (k_2) and 3.2 M^{-1}s^{-1} (k_{-2}).[75]

5.7.2 Two-Electron Transfer

In most of the discussions so far, we have been concerned with reactants undergoing one-electron transfer processes. When one or both of the participants of a redox reaction has to undergo a change of two in the oxidation state, the point arises as to whether the two-electron transfer is simultaneous or nearly simultaneous, a question that has been much discussed.[21] The Tl(I)–Tl(III) second-order exchange (k_{exch}) proceeds by a two-electron transfer. One would need to postulate the equilibrium (5.61) if Tl(II) was involved in the

$$Tl(III) + Tl(I) \rightleftharpoons 2\,Tl(II) \qquad k_1, k_{-1}, K_1 \tag{5.61}$$

exchange. The value of k_{-1} can be determined by pulse radiolytic (one-electron) reduction of Tl(III) and observation of the subsequent disproportionation of the resultant Tl(II). The value of k_{-1} is 1.9×10^8 M^{-1}s^{-1}. By using data for the Fe^{2+}, Tl^{3+} reaction, involved but sound reasoning, which is well worth examining, allows the estimation of $K_1 = 4 \times 10^{-33}$. It follows that $k_1 = 7.6 \times 10^{-25}$ M^{-1}s^{-1}, and since this is very much less than k_{exch} (1.2×10^{-4} M^{-1}s^{-1}) the exchange cannot proceed via a Tl(II) species which is free in solution.[76]

The immediate product of the reaction of Tl(III) with Cr(II) is the dimer $Cr_2(OH)_2^{4+}$. This is likely to result only from an interaction of Cr(II) with Cr(IV), produced in the redox step with Tl(III). If Cr(III) (and Tl(II)) resulted directly from Cr(II) and Tl(III), it would undoubtedly be in the form of a mononuclear Cr(III) species, since this is the product of most of the oxidations of Cr(II).[77] Other examples are in B36 and Ref. 78.

5.7.3 Redox Catalyzed Substitution

Certain substitutions can be catalyzed by the operation of a redox process. It is most easily detected with inert Cr(III), Co(III) and Pt(IV). Hydrolysis, anation and anion interchange all have been accelerated in complexes of these metals by the presence of the lower oxidation state (which is more labile).

Chromium(II) catalyzes the ligation of Cr(III) by X by a mechanism:

$$Cr^{II} + X \rightleftharpoons Cr^{II}X \tag{5.62}$$

$$Cr^{II}X + Cr^{III} \rightarrow Cr^{III}X + Cr^{II} \tag{5.63}$$

based on a third-order rate law [79]

$$V = k \,[\mathrm{Cr}^{II}]\,[\mathrm{Cr}^{III}]\,[X] \tag{5.64}$$

Chromium(II) must catalyze the aquation of $\mathrm{Cr}^{III}X$ by the reverse of the two steps (5.62) and (5.63). [80] Traces of lower oxidation state may be the cause of apparent lability in the higher one e.g. Fe(II) labilising Fe(III). [81,82] Exchange of $\mathrm{Co(NH_3)_6^{3+}}$ with $\mathrm{NH_3(aq)}$, catalyzed by Co(II) has been examined using $^{15}\mathrm{N}$-labelled $\mathrm{NH_3}$ and nmr monitoring. [1] The treatment is complicated by side reactions.

$$\mathrm{Co(II)} + 6\,\mathrm{NH_3} \rightleftharpoons \mathrm{Co(NH_3)_6^{2+}} \tag{5.65}$$

$$\mathrm{Co(NH_3)_6^{2+}} + \mathrm{Co(NH_3)_6^{3+}} \rightleftharpoons \mathrm{Co(NH_3)_6^{3+}} + \mathrm{Co(NH_3)_6^{2+}} \tag{5.66}$$

5.7.4 Stereoselectivity

One might anticipate that there would be a rate difference for the reaction of enantiomers with a chiral compound. The first demonstration of stereoselectivity in an outer-sphere electron transfer was as recent as 1980. [83] Since then such asymmetric induction has been established with a number of examples, nearly all involving outer-sphere redox reactions. Thus, consider the two reactions [83–85]

$$\Delta\text{-Co(edta)}^- + \Delta\text{-Co(en)}_3^{2+} \xrightarrow{k_1} (\pm)\text{-Co(edta)}^{2-} + \Delta\text{-Co(en)}_3^{3+} \tag{5.67}$$

$$\Delta\text{-Co(edta)}^- + \Lambda\text{-Co(en)}_3^{2+} \xrightarrow{k_2} (\pm)\text{-Co(edta)}^{2-} + \Lambda\text{-Co(en)}_3^{3+} \tag{5.68}$$

We are looking for a difference in k_1 and k_2. Since the Co(II) complex is a labile racemic mixture ($\pm$), such a difference can only be demonstrated by a competition approach, searching for a preponderance in the product of one of the optical forms of $\mathrm{Co(en)_3^{3+}}$. In this instance a small (11%) excess of $\Lambda\text{-Co(en)}_3^{3+}$ is observed in the product, meaning that $k_2/k_1 \approx 1.2$. With this typical result it is clear that the kinetic approach, when applicable, is less likely to be as sensitive as that in which the product estimation is assessed by cd and optical rotation. If (5.67) and (5.68) are considered as outer-sphere redox reactions, which is highly likely, the observed rate constant k_1 or k_2 is composite, a product of a precursor complex formation constant and an intramolecular electron transfer rate constant (Sec. 5.4). Since there is an ion pairing selectivity between $\mathrm{Co(edta)}^-$ and $\mathrm{Co(en)}_3^{3+}$ which is also $\Delta\Lambda$, this suggests that precursor ion pair formation between $\mathrm{Ce(edta)}^-$ and $\mathrm{Co(en)}_3^{2+}$ is an important component of the observed electron transfer stereoselectivity. The stereoselectivity increases to 32% $\Delta\Lambda$ preference in dmso. [85] From a detailed examination of the variety of stereochemical products of reduction of $\mathrm{Co(edta)}^-$ by a number of Co(II) complexes of the type $\mathrm{Co(N)_6^{2+}}$, it has been possible to make deductions about the detailed interactions, for example, that there is strong hydrogen bonding of the pseudo-C_3 carboxylate face of $\mathrm{Co(edta)}^-$ with amine hydrogens on the reductant. [84–86]

The reductant may be optically stabilized by using optically active forms of the coordinated ligand. Such ligands may impose stereochemical restraints even with labile oxidation states

(Chap. 3, Structure **10**). The *SS* form of the ligand alamp (chirality relating to the two asymmetric C atoms) *imposes* a Λ-configuration at the Fe(II) center in the complex Λ-[Fe(*SS*)-alamp], **1**. Treatment of this complex with the racemic mixture ($\pm$)-Co(bamap)H$_2$O$^+$ (see **2**) shows a change of cd (λ = 367 nm) with time indicated in Fig. 5.2. The sign of the cd signal

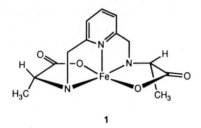

1

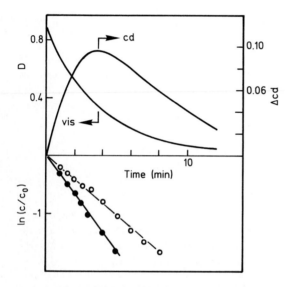

Time (min)

Fig. 5.2 (a) Change in absorbance at 502 nm and in cd intensity at 367 nm on mixing ($\pm$)−Co(bamap) (H$_2$O)$^+$ (2.5 mM) and Fe(*SS*-alamp) (from 2.5 mM Fe^{2+} and 10 mM (*SS*)-alamp). The solution contains 0.1 M ascorbic acid.[87] (b) ln(c/c_0) vs t for reaction of (−) Co(bamap)(H$_2$O)$^+$ (upper curve) and (+) Co(bamap)(H$_2$O)$^+$. The rate constants are 3.2 × 10^{-3}s^{-1} and 5.5 × 10^{-3}s^{-1} respectively.[87] pH = 4.0 and 25 °C. Reproduced with permission from K. Bernauer, P. Pousaz, J. Porret and A. Jean-guenat, Helv. Chim. Acta **68**, 1611 (1985).

corresponds to Λ-(−)$_{436}$-[Co(*SS*)-bamap(H$_2$O)]$^+$. This must therefore be the enantiomer building up in the residue and therefore be the one with the lower rate. This is confirmed when the two Co(III) complexes, Λ-(−)$_{436}$-[Co(*SS*)-bamap(H$_2$O)]$^+$ and Δ-(+)$_{436}$-[Co(*RR*)-bamap(H$_2$O)]$^+$ are separately examined in their rates with Λ-[Fe(*SS*)-alamp]. The ratio of rate constants is 0.6 from the cd measurements and 0.53 measured directly at 25 °C and pH = 4.0. The chiral faces of the Fe(II) and Co(III) complexes fit better in a precursor complex when they show opposite chirality (**2**) and thus the Λ-Δ couple might be expected to react faster. An inner-sphere mechanism with a H$_2$O bridging is proposed, although it is admitted that such a bridge is unusual (Sec. 5.5 (a)).[87]

   Both outer- and inner-sphere pathways are observed when Co(C$_2$O$_4$)$_3^{3-}$ reacts with Co^{2+} in the presence of en ligand.[69] The outer-sphere pathway is favored in higher en concentrations and produces Co(en)$_3^{3+}$. The ΔΛ combination is preferred (9% stereospecificity). The inner sphere pathway leads to $Co(en)_2(C_2O_4)^+$ in only 1.5% ($\Delta\Delta$) preference. It would perhaps be anticipated that inner-sphere redox processes, which lead to more intimate interactions, might be more stereoselective but in this case the bridge is extended and maintains the two cobalts some distance (5 Å) apart. There is a detailed discussion of the various interactions in the two pathways and it is clear that asymmetric induction studies have the potential for probing structural details of the mechanisms of redox reactions.[69]

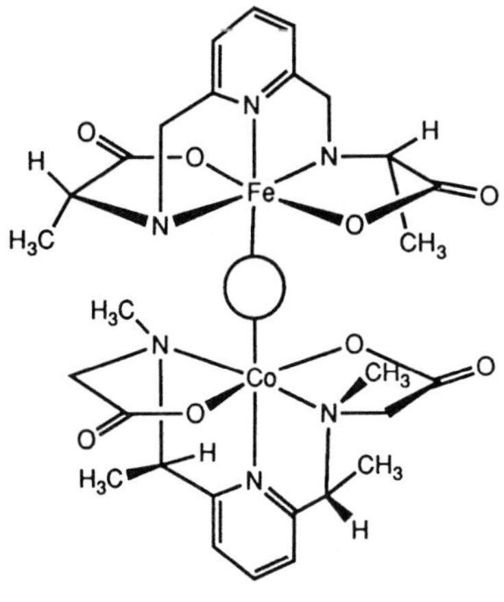

2

5.8 Intramolecular Electron Transfer

A major effort has been made to determine the rate of electron transfer between two well defined sites in a molecule and thus to assess the effect, on the rate, of the distance and the nature of the medium separating the points, the potential drive and so on.[88] Again, the pioneering work on simple complexes has laid the basis for understanding the behavior of larger molecules, especially protein systems which are attracting so much attention.

5.8.1 Between Two Metal Centers

The pioneering studies of Taube and his co-workers established remote attack by Cr^{2+} on an extended organic ligand attached to M = Co(III), Cr(III) or Ru(III)[89,90]

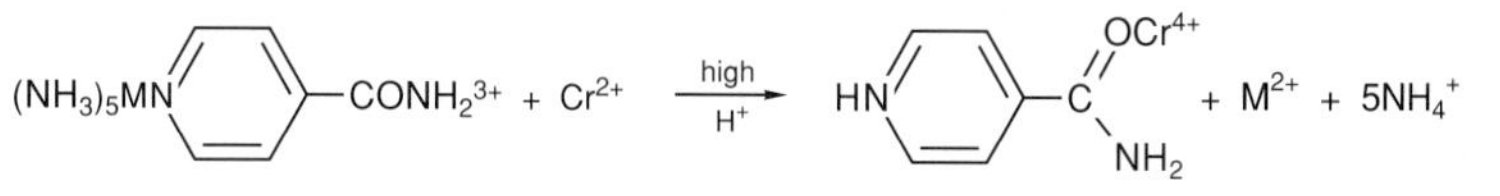

$$(NH_3)_5MN \underset{H^+}{\overset{high}{\longrightarrow}} \text{—CONH}_2^{3+} + Cr^{2+} \quad \longrightarrow \quad HN \text{—C} \begin{matrix} OCr^{4+} \\ \\ NH_2 \end{matrix} + M^{2+} + 5NH_4^+ \quad (5.69)$$

The result made it possible to begin to answer the important question as to how electron transfer from one redox center to another occurred. Was it (i) by the passage of electrons to the bridging group to give a radical ion, which passes an electron further to the oxidant center?

$$M^{III}L + Cr^{II} \rightarrow M^{III} \cdot L \cdot Cr^{II} \rightarrow M^{III} \cdot L^- \cdot Cr^{III} \tag{5.70}$$

$$M^{III} \cdot L^- \cdot Cr^{III} \rightarrow M^{II} \cdot L \cdot Cr^{III} \rightarrow M^{II} + Cr^{III}L \tag{5.71}$$

This is called a chemical, radical or stepwise mechanism. Or was it (ii) by the action of the bridging group to increase the probability of electron transfer by tunneling, termed resonance transfer? [18,56,91]

The chemical mechanism was supported for M = Co and Cr, whereas a resonance transfer was favored by Ru(III) and these differences were rationalized. [89,92-94] As important as these results are, they do not allow us to observe the actual electron transfer between the metal centers. Subsequent strategy has been to produce a very stable precursor complex, preferably with a versatile bridging system, and to observe the subsequent internal electron transfer. It is essential that the system is analyzed carefully kinetically, so as to obtain the internal electron tranfer rate constant separated from other possible reactions also occurring (break up of precursor complex to constituents or outer-sphere electron transfer (Sec. 1.6.4(d)). For such examples see Refs. 22, 75. Examples of the types of bridging systems which have been examined are shown in Chart 5.1. [95-99]

Chart 5.1 Bridged Ligand Systems Within Which Electron Transfer Occurs

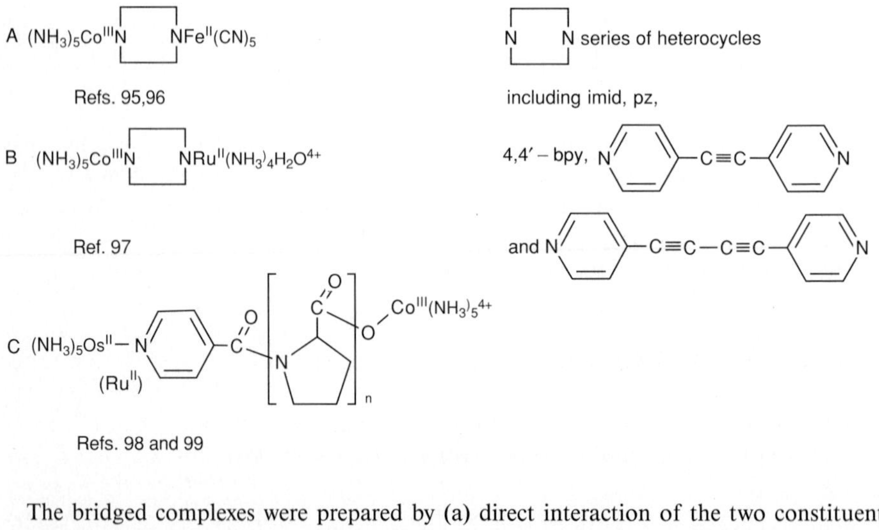

A $(NH_3)_5Co^{III}N \qquad NFe^{II}(CN)_5$

Refs. 95,96

B $(NH_3)_5Co^{III}N \qquad NRu^{II}(NH_3)_4H_2O^{4+}$

Ref. 97

N   N series of heterocycles

including imid, pz,

$4,4' - bpy$, N⎓⎓⎓$-C\equiv C-$⎓⎓⎓N

and N⎓⎓⎓$-C\equiv C-C\equiv C-$⎓⎓⎓N

C $(NH_3)_5Os^{II}-N$
   (Ru^{II})

$Co^{III}(NH_3)_5^{4+}$

Refs. 98 and 99

The bridged complexes were prepared by (a) direct interaction of the two constituents or (b) one electron reduction of the fully oxidized bridged complex (it would be the $Co^{III}-Ru^{III}$ complex in one example shown above[97]). The speed with which the reduction must be carried out depends on the subsequent electron transfer rate. Both chemical reductants or rapidly generated reducing radicals have been used. The latter approach (b) has been an effective one for investigating electron transfer within proteins (Sec. 5.9). A special approach (c) involves optical (picosecond) excitation to produce an unstable isomer, and examination of the reverse reaction [100, 101]

$$(NH_3)_5Ru^{II}-NN-Ru^{III}(edta)^+ \underset{\substack{(dark)\\ (8 \times 10^9 s^{-1})}}{\overset{h\nu}{\rightleftharpoons}} (NH_3)_5Ru^{III}-NN-Ru^{II}(edta)^+ \tag{5.72}$$

An important goal using these types of bridged complexes is to determine the factors which will control the rate constants. From the previous discussions, we can guess that the important ones will include driving force, distances between redox centers and type of bridging ligand. Other more subtle influences are expected. The rate constant (k_{et}) for intramolecular electron transfer within a ligand bridged binuclear complex (as in an inner-sphere mechanism) or in an outer-sphere complex, and in which the metal centers are separated by a distance r, is given by the expression (compare (5.27))

$$k_{et} = \nu_n \kappa_{el} \kappa_n \tag{5.73}$$

ν_n is the effective nuclear vibration frequency that destroys the activated complex configuration and is $\approx 10^{13} \text{ s}^{-1}$. κ_{el} is the electronic transmission coefficient and is approximately one for an adiabatic electron transfer, which occurs when the electronic coupling of the two redox sites is relatively strong. The value of κ_{el} will be, otherwise, a function of the separation and relative orientation of the two redox centers. κ_n is a nuclear factor [23] and given by (5.74); see (5.27) and (5.25), with w^r and w^p irrelevant ($= 0$).

$$\kappa_n = \exp(-\Delta G^*/RT) \tag{5.74}$$

$$\Delta G^* = \frac{(\lambda_o + \lambda_i + \Delta G^o)^2}{4(\lambda_o + \lambda_i)} \tag{5.75}$$

ΔG^o is the standard free energy change for intramolecular electron transfer. Both κ_{el} and κ_n are contributary factors to a dependence of k on r, and (5.73) can be written

$$k_{et} = \nu_n \exp[-\beta(r - r_0)] \kappa_n^0 \exp(-\gamma r) \tag{5.76}$$

The term r_0 is defined as that in which the value of κ_{el} ($= \exp(-\beta(r - r_0))$) is unity when $r = r_0$, i.e. it is then an adiabatic reaction.

We can now consider some of the limited results which have emerged so far. In Haim's bridging systems (Chart 5.1), all reactions appear to undergo adiabatic electron transfer by a resonance mechanism with $\kappa_{el} = 1$. A linear plot of ΔG^* (or $\Delta G^{\ddagger}$) against $1/r$ is obtained over a range of r from 6 to 16 Å for the five bridges shown. This indicates that only the solvent reorganizational term λ_o in (5.75) is a variable in the κ_n factor. [22, 95, 96] Other effects of the ligand structure are discussed (see Prob. 12). The Os(II)–L–Co(III) polyproline system has about an 0.6 V increased potential drive over the Ru(II)–L–Co(III) system (C in Chart 5.1) which is an analogous one in every other respect. A substantial increase in the rate of electron transfer results from this factor (Table 5.10). For the slower reactions ($n = 3$ or 4) equilibration of *trans*- to *cis*-proline interferes with the analysis. With the Os(II)–L–Ru(III) system, both the nuclear ($\gamma = 0.9$ Å^{-1}) and the electronic ($\beta = 0.68$ Å^{-1}) factors are shown to contribute to the (exponential) dependence on r of k_{et}.[98,99,102]

Table 5.10 Intramolecular Electron Transfer Across Oligoprolines[98]

n in System C (Chart 5.1)	k_{Os} s^{-1}	k_{Ru} s^{-1}
0	1.9×10^5	1.2×10^{-2}
1	2.7×10^2	1.0×10^{-4}
2	0.74	6.0×10^{-6}
3	0.09	5.6×10^{-5}
4	0.09	1.4×10^{-4}

5.8.2 Metal Complexes with Reducible Ligands

Strong evidence for the feasibility of the chemical mechanism (Sec. 5.8.1) has been afforded by the production of a transient nitrophenyl radical attached to a Co(III) complex

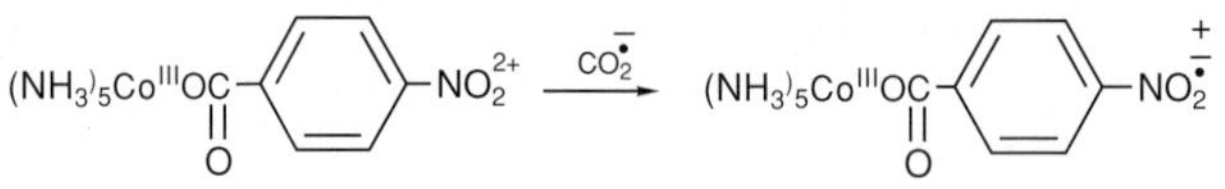

$$(5.77)$$

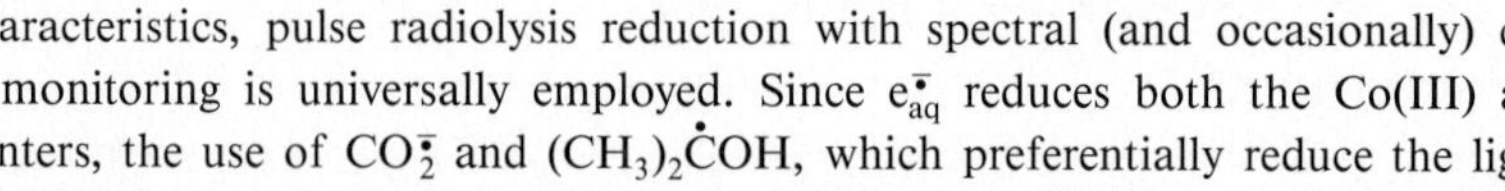

$$(5.78)$$

The radical undergoes first-order decay (k_{et}) via electron transfer to Co(III).[103] This approach has been much employed to investigate the internal electron transfer process. Invariably Co(III) complexes have been used and since the transient radicals have strong absorbance characteristics, pulse radiolysis reduction with spectral (and occasionally) conductometric monitoring is universally employed. Since e_{aq}^- reduces both the Co(III) and the ligand centers, the use of CO_2^- and $(CH_3)_2\dot{C}OH$, which preferentially reduce the ligand, is preferred. In the types of systems so far examined (Table 5.11)[104–106] decay of the coordinated radical ligand usually occurs as in (5.78) by an intramolecular electron transfer (k_1) leading to the cobalt(II) species.

$$\text{Co}^{III}-L \xrightarrow{CO_2^-} \text{Co}^{III}-L^- \xrightarrow{k_1} \text{Co}^{II}-L \tag{5.79}$$

This is indicated by the observation of first-order loss of radical signal, with the rate constant k_1 being independent of the concentration, both of radical complex (i.e., independent of the dose of reductant radical used) and of Co^{III}–L complex used (in excess). Only rarely is the first-order loss of Co^{III}–L$^-$ directly dependent on the concentration of Co^{III}–L present, and this is interpreted in terms of reaction (5.80). Sometimes the loss of $Co^{III}-L^{\bar{\cdot}}$ is second-order,[104]

$$Co^{III}-L^{\bar{\cdot}} + Co^{III}-L \rightarrow Co^{III}-L + Co^{II}-L \qquad k_2 \qquad (5.80)$$

and this is attributed to a disproportionation of the protonated adduct in acid (5.81).

$$2\,Co^{III}-L^{\cdot}H \rightarrow Co^{III}-L + Co^{II}-LH_2 \qquad 2k_3 \qquad (5.81)$$

In all cases (5.79)–(5.81), the cobalt(II) product is expected to rapidly dissociate to constituents. The presence of (5.80) makes it difficult to determine the value of k_1, which is a small intercept on a k_{obs} vs $[Co^{III}-L]$ plot.[106] The occurrence of (5.81) is usually signalled by mixed first- and second-order kinetics since as the concentration of $Co^{III}-L^{\bar{\cdot}}$ (or $Co^{III}-L^{\cdot}H$) decreases, the importance of (5.81) is superseded by (5.79) and k_1 results from the (first-order) end of the decay (see Fig. 1.6). Most unusually, the complex $(NH_3)_5Co(N\text{-Mebpy}^{\cdot})^{3+}$ undergoes all three modes of decay (Table 5.11).[106]

Table 5.11 Rate Constants for Decay of $Co(NH_3)_5(L^{\bar{\cdot}})^{n+}$ at 23–25 °C

$-L$	k_1, s^{-1}	$k_2, M^{-1}s^{-1}$	$2k_3, M^{-1}s^{-1}$	Ref.
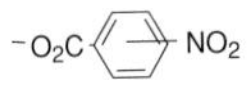 $-O_2C$—⟨ring⟩—NO_2	4.0×10^5 (o) 1.5×10^2 (m) 2.6×10^3 (p)			104
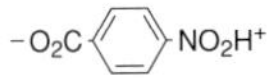 $-O_2C$—⟨ring⟩—NO_2H^+	5		1.5×10^8	104
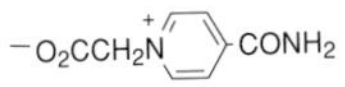 $-O_2CCH_2\overset{+}{N}$—⟨ring⟩—$CONH_2$	$<2.0\times10^4$	1.5×10^9		105
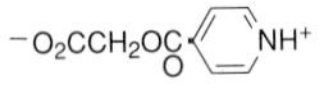 $-O_2CCH_2O\underset{O}{C}$—⟨ring⟩—$NH^+$	2.0×10^4	$<6\times10^6$		105
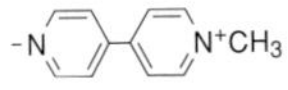 $-N$⟨ring⟩—⟨ring⟩—N^+CH_3	8.7×10^2	5.4×10^7	2.4×10^8	106

There has been nothing like the enthusiasm for the application to these systems of the theoretical equations, which we have noted in the previous sections and will encounter in the next. Nevertheless, a number of features are present which are qualitatively consistent with the discussions in Sec. 5.8.1 and which are in part illustrated in Table 5.11. There is a correlation of rate constant with the driving force of the internal electron transfer.[104-108] The p-nitrophenyl derivative is a poorer reducing agent when protonated and k_1 is much less than for the unprotonated derivative. Consequently disproportionation $(2k_3)$ becomes important. Although there are not marked effects of structural variation on the values of k_1, the associated activation parameters may differ enormously and this is ascribed to the operation of different mechanisms.[104] The "resonance-assisted through-chain" operates with the p-nitrophenyl derivative and has large Franck-Condon requirements, hence $\Delta H^{\ddagger} = 71$ kJ mol^{-1} and $\Delta S^{\ddagger} = +59$ J K^{-1} mol^{-1}. "Direct ligand by-pass" is possible for the o-nitrophenyl derivative since orbital overlap of the donor and acceptor sites is now feasible. The much smaller $\Delta H^{\ddagger}$ values (~ 16 kJ mol^{-1}) are consistent with this, but the advantage is offset by a large negative $\Delta S^{\ddagger}$ ($= -100$ J K^{-1} mol^{-1}) which is attributed to a greatly reduced configurational flexibility which accompanies activation. Two other ways of electron transfer are suggested, namely flexible "indirect overlap" and nonadiabatic transfer with very little coupling between the centers. [104]

Finally, attention should be drawn to the elegant studies of Meyer and his co-workers. Metal complexes have been designed which contain both an excitable (e. g. bpy) and a quencher (e. g. N-Mebpy$^+$) group. Following excitation, intramolecular electron and energy transfer occurs and the dependence of rate on distances, metal and so on, can be assessed. [108]

5.8.3 Induced Electron Transfer

Internal ligand-to-metal electron transfer may be initiated by the action of an external oxidant on the ligand. This phenomenon of induced electron transfer has received rather scant attention. In the complex **3** shown in (5.82) the one-electron oxidizing center of the Co(III) and the two-electron reducing ligand 4-pyridylcarbinol can coexist because of their redox "incompatability". The complex is therefore relatively stable. This situation is upset when a strong one-electron oxidant such as Ce(IV) or Co(III) is added to a solution of the Co(III) complex. The oxidant attacks the carbinol function to generate an intermediate or intermediates; the intermediate in this case is oxidized *internally* by the Co(III) center; for example,

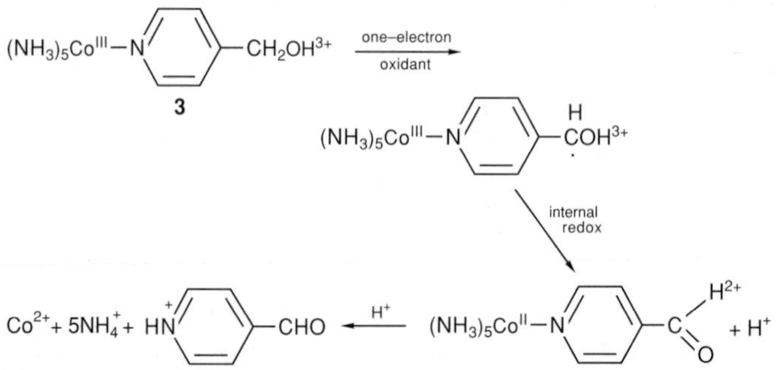

$$(5.82)$$

Thus, one equivalent of an external oxidant and one of the Co(III) complex are consumed in oxidizing one equivalent of the alcohol to the aldehyde. Two-equivalent oxidants, Cl$_2$, Cr(VI), give no such radical intermediate, and therefore no Co(II), and only the Co(III) complex. [109]

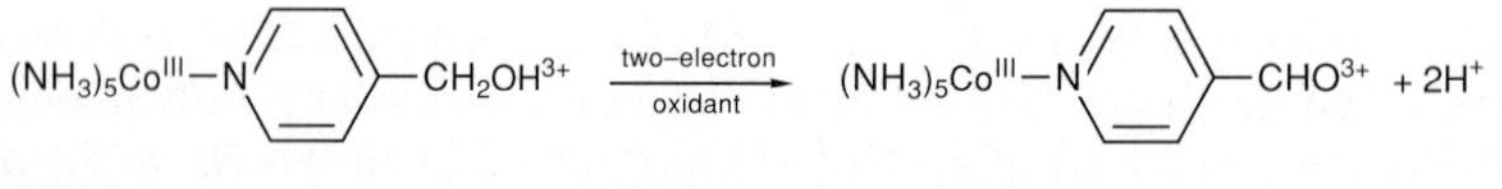

$$(5.83)$$

For more recent examples involving induced electron transfer linked to cobalt(III) complexes of α-hydroxy acids and hydroquinone esters see Refs. 110, 111.

5.9 Electron Transfer in Proteins

Early attempts at observing electron transfer in metalloproteins utilized redox-active metal complexes as external partners. The reactions were usually second-order and approaches based on the Marcus expression allowed, for example, conjectures as to the character and accessibility of the metal site. [112,113] The agreement of the observed and calculated rate constants for cytochrome c reactions for example is particularly good, even ignoring work terms. [25] The observations of deviation from second-order kinetics ("saturation" kinetics) allowed the dissection of the observed rate constant into the components, namely adduct stability and first-order electron transfer rate constant (see however Sec. 1.6.4). [113] Now it was a little easier to comment on the possible site of attack on the proteins, particularly when a number of modifications of the proteins became available.

Protein surface "patches" — Certain sites on the surface of proteins appear to act as conduits for electron passage. Thus in the protein azurin *(Pseudomonas aeruginosa)* a hydrophobic patch around the partly exposed His 117 (which is attached to Cu) is believed to be involved in electron transfer with nitrite reductase and in the self exchange of the Cu(I) and Cu(II) forms of azurin. The effect of specific labelling at the patch on the kinetic characteristics helps confirm this patch as playing a role in the electron transfer. Earlier, chemical modifications were employed (O. Farver, Y. Blatt and I. Pecht, Biochemistry **21**, 3556 (1982)). More recently the powerful technique of site-directed mutagenesis has been used. Met 44 located next to His 117 is replaced by Lys with minimum overall structural changes in the protein. The electron self-exchange measured by nmr line broadening is slower and pH-dependent for the modified protein compared with the native (wild type) protein ($\approx 10^6 M^{-1} s^{-1}$, which is also pH-independent). The patch is therefore implicated in the electron transfer process. (M. van de Kamp, R. Floris, F. C. Hali and G. W. Canters, J. Amer. Chem. Soc. **112**, 907 (1990)).

This general approach has, however, serious limitations. The position of the site for attack (and therefore the electron transfer distance involved) is very conjectural. In addition, the vexing possibility, which we have encountered several times, of a "dead-end" mechanism (Sec. 1.6.4) is always present. [113] One way to circumvent this difficulty, is to bind a metal complex to the protein *at a specific site,* with a known (usually crystallographic) relationship to the metal site. The strategy then is to create a metastable state, which can only be alleviated by a discernable electron transfer between the labelled and natural site. It is important to establish that the modification does not radically alter the structure of the protein. A favorite technique is to attach $(NH_3)_5 Ru^{3+}$ to a histidine imidazole near the surface of a protein. Exposure of this modified protein to a deficiency of a powerful reducing agent, will give a concurrent (partial) reduction of the ruthenium(III) and the site metal ion e.g. iron(III) heme in cytochrome c

$$(NH_3)_5Ru(His\text{-}33)^{3+} \quad cyt\ c(Fe^{3+}) \longrightarrow \begin{array}{l} \text{rapid one-} \\ \text{electron} \\ \text{reduction} \end{array} \longrightarrow \begin{array}{c} (NH_3)_5Ru(His\text{-}33)^{2+} \quad cyt\ c(Fe^{3+}) \\ \Big\downarrow k_{et} \\ (NH_3)_5Ru(His\text{-}33)^{3+} \quad cyt\ c(Fe^{2+}) \end{array} \tag{5.84}$$

The intramolecular electron transfer k_{et}, subsequent to the rapid reduction, must occur because the Ru(III)–Fe(II) pairing is the stable one. It is easily monitored using absorbance changes which occur with reduction at the Fe(III) heme center. Both laser-produced $*Ru(bpy)_3^{2+}$ and radicals such as CO_2^- (from pulse radiolysis (Prob. 15)) are very effective one-electron reductants for this task (Sec. 3.5). [114,115] In another approach, [116,117] the Fe in a heme protein is replaced by Zn. The resultant Zn porphyrin (ZnP) can be electronically excited to a triplet state, $^3ZnP*$, which is relatively long-lived ($\tau = 15$ ms) and is a good reducing agent ($E^\circ = -0.62$ V). Its decay via the usual pathways (compare (1.32)) is accelerated by electron transfer to another metal (natural or artificial) site in the protein e.g.,

$$(NH_3)_5Ru^{3+} \quad ZnP \xrightarrow{h\nu} (NH_3)Ru^{3+} \quad ^3ZnP* $$
$$\qquad\qquad {}_{k'_{et}}\searrow \qquad\qquad\qquad \swarrow{}_{k_{et}} \tag{5.85}$$
$$(NH_3)_5Ru^{2+} \quad ZnP^{\pm}$$

The $ZnP^{\pm}$ accepts an electron from the Ru^{2+} to return the system to its initial state. Although $k'_{et} > k_{et}$, both can be measured. [118] Examples of these approaches with iron and copper proteins are shown in Table 5.12. There are a number of excellent short reviews of this subject. [121-124]

Table 5.12 Intramolecular Electron Transfer Rate Constants in Metalloproteins at $\approx 25°C$

Protein	Redox Centers Involved	Initiating Mode	k_{et} s^{-1}	Distance[a] Å	ΔE Volts	Ref.
Cytochrome c (horse heart)	$(NH_3)_5Ru^{II}(His\text{-}33)$; Fe(III)heme	$*Ru(bpy)_3^{2+}$	30^b	11.8	0.19	114
		CO_2^-	53^c	11.8	0.19	115
Plastocyanin (A. variabilis)	$(NH_3)_5Ru^{II}(His\text{-}59)$; Cu(II)	CO_2^-	$\leqslant 0.08$	11.9	0.26	119
Azurin (P. aeruginosa)	$(NH_3)_5Ru^{II}(His\text{-}83)$; Cu(II) (Sec. 3.5.1)	$*Ru(bpy)_3^{2+}$ CO_2^-	1.9 2.5	11.8 11.8	0.28 0.28	120 119
Cytochrome c (horse heart)	$(NH_3)_5Ru^{III}(His\text{-}33)$; Zn(II)	$^3ZnP*$ $(h\nu)$	7.7×10^5	11.8	0.72	118

[a] The shortest distance separating the delocalized metal centers, e.g. heme edge.
[b] $\Delta H^{\ddagger} = 8$–16 kJ mol^{-1}.   [c] $\Delta H^{\ddagger} = 15$ kJ mol^{-1}.

Electron transfer within polypeptide material is surprisingly fast even over relatively long distances. The rate constants however are much less than the typical values of 10^6 to $10^9 s^{-1}$ for electron transfer between biphenyl radical anions and organic acceptors held about $10 Å$ apart by a steroid frame.[125] In the latter case, there is generally a larger potential drive and electron transfer may be "through-bond" rather than the "through space" in proteins, although this does not now appear likely.[126]

We have seen that two important factors control the value of k_{et}. These are (a) the distance of separation of the two sites (r) and (b) the driving force ΔG° for the electron transfer. Recalling (5.73) to (5.76) and setting $r_0 = 3 Å$, $\lambda = \lambda_o + \lambda_i$:

$$k_{et} = \nu_n \kappa_{el} \kappa_n = \nu_n \exp\left[-\beta(r-3)\right] \exp\left[\frac{-(\Delta G^\circ + \lambda)^2}{4\lambda RT}\right] \qquad (5.86)$$

For a series of Ru modified (all at His-33 so that r is constant), Fe or Zn substituted cytochrome c derivatives, κ_{el} and λ are relatively constant and ΔG° is variable. The plot of $\ln k_{et}$ (which varies from 30 to $3.3 \times 10^6 s^{-1}$) vs ΔG° (-0.18 to -1.05 V) shows the semblance of the expected parabolic shape for (5.86) and leads to $\lambda = 1.10$ eV.[118,127] Experiments directed towards the distance dependence of $\ln k_{et}$ lead to β values of 0.9-$1.0 Å^{-1}$.[127] Similar results are obtained with Co(III) cage modified cytochrome-c.[128]

More subtle factors that might affect k_{et} will be the sites structures, their relative orientation and the nature of the intervening medium.[122] That these are important is obvious if one examines the data for the two copper proteins plastocyanin and azurin. Despite very similar separation of the redox sites and the driving force (Table 5.12), the electron transfer rate constant within plastocyanin is very much the lesser (it may be zero).[119] See Prob. 16. In striking contrast, small oxidants are able to attach to surface patches on plastocyanin which are more favorably disposed with respect to electron transfer to and from the Cu, which is about $14 Å$ distant. It can be assessed that internal electron transfer rate constants are $\approx 30 s^{-1}$ for Co(phen)$_3^{3+}$, $>5 \times 10^3 s^{-1}$ for Ru(NH$_3$)$_5$imid^{2+} and $3.0 \times 10^6 s^{-1}$ for *Ru(bpy)$_3^{2+}$, Refs. 119 and 129. In the last case the excited state *Ru(bpy)$_3^{2+}$ is believed to bind about 10-$12 Å$ from the Cu center. Electron transfer occurs both from this remote site as well as by attack of *Ru(bpy)$_3^{2+}$ adjacent to the Cu site.[129] At high protein concentration, electron transfer occurs solely through the remote pathway.

Experimental evidence for long-range electron transfer in polypeptides and proteins had been early accrued.[130] The value of using a metal center as a marker is apparent from the above. The approach can be extended to electron transfer between two proteins which are physiological partners.[25] Metal substitution (e. g. Zn for Fe) can be used to alter the value of ΔG° and permit photoinduced initiation. The parabolic behavior predicted by (5.86) has been verified for the electron transfer rate constant vs ΔG° within the adduct between cyt c and cyt b$_5$.[117]

5.10 The Future

Of all the areas covered in this book, that of oxidation-reduction reactions has attracted the most attention by a variety of chemists with substantial results. Many of the future problems delineated by Taube in his book *Electron Transfer Reactions of Complex Ions in Solution* published 20 years ago, have been tackled with success. The detailed arrangement of reactants in the activated complex for both outer-sphere and inner-sphere reactions is better understood as well as the controlling factors. The calculations of outer-sphere self-exchange rate constants is likely now to be a successful exercise and the application of cross-reaction equations almost routine. Deviations are rare and lead to interesting concepts. The assessment of solvent and medium effects as well as volumes of activation still requires a good deal of examination. The observation of very short-lived transients is increasingly easier and with it comes an appreciation of an increasingly detailed mechanism, small molecule interactions with heme proteins being the best example of this. Finally, the transfer of electrons in proteins will undoubtedly attract much time and talent. The approaches that are likely to be successful have been well established in the past decade.

References

1. A. Hammershøi, D. Geselowitz and H. Taube, Inorg. Chem. **23**, 979 (1984).
2. I. Ruff and M. Zimonyi, Electrochim. Acta **18**, 515 (1973).
3. H. Taube and E. L. King, J. Amer. Chem. Soc. **76**, 4053, 6423 (1954); D. L. Ball and E. L. King, J. Amer. Chem. Soc. **80**, 1091 (1958).
4. H. Taube, H. Myers and R. L. Rich, J. Amer. Chem. Soc. **75**, 4118 (1953); H. Taube and H. Myers, J. Amer. Chem. Soc. **76**, 2103 (1954).
5. H. Ogino, E. Kikkawa, M. Shimura and N. Tanaka, J. Chem. Soc. Dalton Trans. 894 (1981).
6. J. Xu and R. B. Jordan, Inorg. Chem. **29**, 4180 (1990); M. J. Hynes and D. F. Kelly, J. Chem. Soc. Chem. Communs. 849 (1988).
7. P. L. Drake, R. T. Hartshorn, J. McGinnis and A. G. Sykes, Inorg. Chem. **28**, 1361 (1989).
8. D. R. Stranks and R. G. Wilkins, Chem. Rev. **57**, 743 (1957).
9. J. H. Espenson, Acc. Chem. Res. **3**, 347 (1970).
10. K. G. Ashurst and W. C. E. Higginson, J. Chem. Soc. 3044 (1953).
11. The eventual proportions of Cr^{3+} and $CrCl^{2+}$ formed will depend on the concentrations and formation constant of $CrCl^{2+}$.
12. E. F. Hills, M. Moszner and A. G. Sykes, Inorg. Chem. **25**, 339 (1986).
13. For tables and references see T. J. Williams and C. S. Garner, Inorg. Chem. **9**, 2058 (1970); D. W. Carlyle and J. H. Espenson, J. Amer. Chem. Soc. **91**, 599 (1969).
14. A. Haim and N. Sutin, J. Amer. Chem. Soc. **88**, 5343 (1966).
15. A. G. Sykes and M. Green, J. Chem. Soc. **A**, 3221 (1970); M. Green, R. S. Taylor and A. G. Sykes, J. Chem. Soc. **A**, 509 (1971); K. M. Davies and J. H. Espenson, J. Amer. Chem. Soc. **91**, 3093 (1969).
16. The data for this Table are taken from the compilations in B36 and in the Selected Bibliography in Chap. 4.
17. The reaction of Fe^{2+} with $Fe(phen)_3^{3+}$ is considered to be an outer-sphere reaction because of the inertness of the Fe(III) complex. Marked effects of anions on the rate have been interpreted however

in terms of nucleophilic attack of the anion on the phenanthroline ligand. Even the unaccelerated reaction has been considered as an inner-sphere process involving a water bridge, N. Sutin and A. Forman, J. Amer. Chem. Soc. **93**, 5274 (1971); R. Schmid and L. Han, Inorg. Chim. Acta **69**, 127 (1983).

18. N. Sutin, Acc. Chem. Res. **1**, 225 (1968).

19. R. A. Marcus, J. Chem. Phys. **24**, 966 (1956); Ann. Rev. Phys. Chem. **15**, 155 (1964); R. A. Marcus and P. Siders in B38, Chap. 10.

20. N. S. Hush, Trans. Faraday Soc. **57**, 557 (1961); B38, Chap. 13.

21. D. E. Pennington in B32, Chap. 3.

22. A. Haim, Prog. Inorg. Chem. **30**, 273 (1983).

23. N. Sutin, Acc. Chem. Res. **15**, 275 (1982); Prog. Inorg. Chem. **30**, 441 (1983). N. Sutin, B. S. Brunschwig, C. Creutz and J. R. Winkler, Pure App. Chem. **60**, 1817 (1988).

24. M. D. Newton and N. Sutin, Ann. Rev. Phys. Chem. **35**, 437 (1984).

25. R. A. Marcus and N. Sutin, Biochim. Biophys. Acta **811**, 265 (1985).

26. G. M. Brown and N. Sutin, J. Amer. Chem. Soc. **101**, 883 (1979).

27. B. S. Brunschwig, C. Creutz, D. H. Macartney, T.-K. Sham and N. Sutin, Faraday Disc. Chem. Soc. **74**, 113 (1982).

28. M. Kozik and L. C. W. Baker, J. Amer. Chem. Soc. **112**, 7604 (1990).

29. P. Bernhard, L. Helm, I. Rapaport, A. Ludi and A. E. Merbach, J. Chem. Soc. Chem. Communs. 302 (1984); P. Bernhard, L. Helm, A. Ludi and A. E. Merbach, J. Amer. Chem. Soc. **107**, 312 (1985).

30. P. J. Smolenaers and J. K. Beattie, Inorg. Chem. **25**, 2259 (1986).

31. M. N. Doyle, K. Libson, M. Woods, J. C. Sullivan and E. Deutsch, Inorg. Chem. **25**, 3367 (1986).

32. D. J. Szalda, D. H. Macartney and N. Sutin, Inorg. Chem. **23**, 3473 (1984). D. H. Macartney and N. Sutin, Inorg. Chem. **22**, 3530 (1983).

33. P. Bernhard and A. M. Sargeson, Inorg. Chem. **26**, 4122 (1987).

34. P. Bernhard and A. M. Sargeson, Inorg. Chem. **27**, 2582 (1988).

35. T. W. Newton, J. Chem. Educ. **45**, 571 (1968).

36. M. A. Ratner and R. D. Levine, J. Amer. Chem. Soc. **102**, 4898 (1980).

37. T. E. Meyer, C. T. Przysiecki, J. A. Watkins, A. Bhattacharyya, R. P. Simondson, M. A. Cusanovich and G. Tollin, Proc. Natl. Acad. Sci. USA **80**, 6740 (1983).

38. M. Z. Hoffman, F. Bolleta, L. Moggiand and G. L. Hug, J. Phys. Chem. Ref. Data **18**, 219 (1989).

39. T. J. Meyer, Prog. Inorg. Chem. **30**, 389 (1983).

40. J. T. Hupp and M. J. Weaver, Inorg. Chem. **22**, 2557 (1983).

41. U. Fürholz and A. Haim, J. Phys. Chem. **90**, 3686 (1986); K. Zahir, J. H. Espenson and A. Bakač, Inorg. Chem. **27**, 3144 (1988).

42. W. H. Jolley, D. R. Stranks and T. W. Swaddle, Inorg. Chem. **29**, 1948 (1990).

43. W. Böttcher, G. M. Brown and N. Sutin, Inorg. Chem. **18**, 1447 (1979).

44. E. Kremer, G. Cha, M. Morkevicius, M. Seaman and A. Haim, Inorg. Chem. **23**, 3028 (1984).

45. H.-M. Huck and K. Wieghardt, Inorg. Chem. **19**, 3688 (1980).

46. A large positive $\Delta V^{\ddagger}$ is associated with the internal electron transfer within the outer-sphere complex between $Co(NH_3)_5X^{3+}$ and $Fe(CN)_6^{4-}$. I. Krack and R. van Eldik, Inorg. Chem. **25**, 1743 (1986); Y. Sasaki, K. Endo, A. Nagasawa and K. Saito, Inorg. Chem. **25**, 4845 (1986). This value is not easy to interpret.

47. A. Haim, Comments Inorg. Chem. **4**, 113 (1985).

48. U. Fürholz and A. Haim, Inorg. Chem. **24**, 3091 (1985).

49. J. R. Miller, J. V. Beitz and R. K. Huddleston, J. Amer. Chem. Soc. **106**, 5057 (1984). H. B. Gray, L. S. Fox, M. Kozik and J. R. Winkler, Science, **247**, 1069 (1990).

50. M. J. Weaver and G. E. McManis, Acc. Chem. Res. **23**, 294 (1990).

51. M. P. Doyle, J. K. Guy, K. C. Brown, S. N. Mahapatro, C. M. VanZyl and J. R. Pladziewicz, J. Amer. Chem. Soc. **109**, 1536 (1987). The application of Marcus ideas to organic and organometallic reactions is well covered by L. Eberson, Electron Transfer Reactions in Organic Chemistry, Springer-Verlag,

Berlin 1987 and J. K. Kochi, Angew. Chem. Int. Ed. Engl. **27**, 1227 (1988), where detailed accounts and comprehensive lists of references are contained.

52. K. Zahir, W. Böttcher and A. Haim, Inorg. Chem. **24**, 1966 (1985).
53. M. Kimura, S. Yamabe and T. Minato, Bull. Chem. Soc. Japan **54**, 1699 (1981).
54. H. Taube, Adv. Inorg. Chem. Radiochem. **1**, 1 (1959).
55. N. Sutin, Ann. Rev. Nucl. Sci. **12**, 285 (1962).
56. J. Halpern and L. E. Orgel, Disc. Faraday Soc. **29**, 32 (1960).
57. R. C. Patel, R. E. Ball, J. F. Endicott and R. G. Hughes, Inorg. Chem. **9**, 23 (1970). $\Delta H_{obs}^{\ddagger}$ is negative for the reduction of a number of complexes of the type *cis*-Co(en)$_2$XY^{n+} by Cr^{2+}.
58. A. Haim, Pure Appl. Chem. **55**, 89 (1983).
59. J. P. Birk, Inorg. Chem. **9**, 125 (1970).
60. R. K. Murmann, H. Taube and F. A. Posey, J. Amer. Chem. Soc. **79**, 262 (1957). Oxygen-18 tracer experiments have indicated that transfer of OH$^-$ to Cr^{2+} from Co(NH$_3$)$_5$OH^{2+} is quantitative.
61. J. F. Ojo, O. Olubuyide and O. Oyetunji, Inorg. Chim. Acta **119**, L5 (1986).
62. J. P. Candlin and J. Halpern, Inorg. Chem. **4**, 766 (1965); M. C. Moore and R. N. Keller, Inorg. Chem. **10**, 747 (1971). The effect is very small probably because of the extreme reactivity of Cr(II) towards these oxidants.
63. J. P. Candlin, J. Halpern and D. L. Trimm, J. Amer. Chem. Soc. **86**, 1019 (1964).
64. A. Haim, Inorg. Chem. **7**, 1475 (1968).
65. A. related approach can be made to this calculation, B39, p. 51.
66. C. Shea and A. Haim, J. Amer. Chem. Soc. **93**, 3055 (1971).
67. D. P. Fay and N. Sutin, Inorg. Chem. **9**, 1291 (1970).
68. R. Snellgrove and E. L. King, J. Amer. Chem. Soc. **84**, 4609 (1962); A. Haim, J. Amer. Chem. Soc. **88**, 2324 (1966).
69. R. A. Marusak, P. Osvath, M. Kemper and A. G. Lappin, Inorg. Chem. **28**, 1542 (1989).
70. J. E. Earley, Prog. Inorg. Chem. **13**, 243 (1970).
71. L. E. Orgel, Report of the Tenth Solvay Conference, Brussels, 1956, p. 286.
72. Sr. M. J. DeChant and J. B. Hunt, J. Amer. Chem. Soc. **90**, 3695 (1968).
73. A. Haim and N. Sutin, J. Amer. Chem. Soc. **88**, 434 (1966).
74. R. C. Patel and J. F. Endicott, J. Amer. Chem. Soc. **90**, 6364 (1968); D. P. Rillema, J. F. Endicott, and R. C. Patel, J. Amer. Chem. Soc. **94**, 394 (1972).
75. G. C. Seaman and A. Haim, J. Amer. Chem. Soc. **106**, 1319 (1984).
76. R. Dodson in B38, p. 132.
77. M. Ardon and R. A. Plane, J. Amer. Chem. Soc. **81**, 3197 (1959).
78. U. Fürholz and A. Haim, Inorg. Chem. **26**, 3243 (1987).
79. R. D. Cannon and J. E. Earley, J. Chem. Soc. A, 1102 (1968); D. E. Pennington and A. Haim, Inorg. Chem. **6**, 2138 (1967), and references therein.
80. J. Doyle, A. G. Sykes and A. Adin, J. Chem. Soc. A, 1314 (1968).
81. E. M. Sabo, R. E. Shepherd, M. S. Rau and M. G. Elliott, Inorg. Chem. **26**, 2897 (1987).
82. A. D. James, R. S. Murray and W. C. E. Higginson, J. Chem. Soc. Dalton Trans. 1273 (1974).
83. D. A. Geselowitz and H. Taube, J. Amer. Chem. Soc. **102**, 4525 (1980).
84. P. Osvath and A. G. Lappin, Inorg. Chem. **26**, 195 (1987).
85. D. A. Geselowitz, A. Hammershøi and H. Taube, Inorg. Chem. **26**, 1842 (1987).
86. Even the effects of chelate ring conformation in the reductants on the stereoselectivity can be assessed. Four conformational isomers are possible with M(en)$_3^{n+}$ $\Delta(\lambda\lambda\lambda)$, $\Delta(\delta\lambda\lambda)$, $\Delta(\delta\delta\lambda)$ and $\Delta(\delta\delta\delta)$, in which Δ refers to the chirality of the metal center and δ and λ to the chelate conformation (Secs. 7.1 and 7.6). There is a preference for $\Delta\Delta$ interactions for $\Delta(\lambda\lambda\lambda)$ isomers and $\Delta\Lambda$ interactions for $\Delta(\delta\delta\delta)$ isomers.[84]
87. K. Bernauer, P. Pousaz, J. Porret and A. Jeanguenat, Helv. Chim. Acta **71**, 1339 (1988).

88. S. Isied, Prog. Inorg. Chem. **32**, 443 (1984).

89. F. Nordmeyer and H. Taube, J. Amer. Chem. Soc. **90**, 1162 (1968).

90. R. G. Gaunder and H. Taube, Inorg. Chem. **9**, 2627 (1970).

91. H. Taube and E. S. Gould, Acc. Chem. Res. **2**, 321 (1969).

92. M. K. Loar, Y.-T. Fanchiang and E. S. Gould, Inorg. Chem. **17**, 3689 (1978).

93. C. Norris and F. R. Nordmeyer, J. Amer. Chem. Soc. **93**, 4044 (1971).

94. C. A. Radlowski, P.-W. Chum, L. Hua, J. Heh and E. S. Gould, Inorg. Chem. **19**, 401 (1980).

95. A. P. Szecsy and A. Haim, J. Amer. Chem. Soc. **103**, 1679 (1981) and previous work.

96. G.-H. Lee, L. D. Ciana and A. Haim, J. Amer. Chem. Soc. **111**, 2535 (1989).

97. S. K. S. Zawacky and H. Taube, J. Amer. Chem. Soc. **103**, 3379 (1981).

98. S. S. Isied, A. Vassilian, R. H. Magnuson and H. A. Schwarz, J. Amer. Chem. Soc. **107**, 7432 (1985).

99. S. S. Isied, A. Vassilian, J. F. Wishart, C. Creutz, H. A. Schwarz and N. Sutin, J. Amer. Chem. Soc. **110**, 635 (1988).

100. C. Creutz, P. Kroger, T. Matsubara, T. L. Netzel and N. Sutin, J. Amer. Chem. Soc. **101**, 5442 (1979).

101. C. Creutz, Prog. Inorg. Chem. **30**, 1 (1983).

102. In a series of polypeptides of the type [TrpH-(proline)$_n$-TyrOH] it is possible to preferentially oxidize (with N$_3^\bullet$) the trpH residue and produce [$\bullet$Trp-(proline)$_n$-TyrOH]. The rate constant for the subsequent change to [TrpH-(proline)$_n$-TyrO$\bullet$] correlates with ΔG°, but is only slightly sensitive to the value of n, M. Faraggi, M. R. DeFelippis and M. H. Klapper, J. Amer. Chem. Soc. **111**, 5141 (1989).

103. M. Z. Hoffman and M. Simic, J. Amer. Chem. Soc. **94**, 1757 (1972).

104. K. D. Whitburn, M. Z. Hoffman, N. V. Brezniak and M. G. Simic, Inorg. Chem. **25**, 3037 (1986) and previous references.

105. J. V. Beitz, J. R. Miller, H. Cohen, K. Wieghardt and D. Meyerstein, Inorg. Chem. **19**, 966 (1980).

106. K. Tsukahara and R. G. Wilkins, Inorg. Chem. **28**, 1605 (1989).

107. H. Cohen, E. S. Gould, D. Meyerstein, M. Nutkovich and C. A. Radlowski, Inorg. Chem. **22**, 1374 (1983). H. Cohen, M. Nutkovich, D. Meyerstein and K. Wieghardt, J. Chem. Soc. Dalton Trans. 943 (1982).

108. T. J. Meyer, Pure Appl. Chem. **62**, 1003 (1990).

109. B39, Chap. 4. The situation presented above is oversimplified and some puzzling features remain to be resolved. J. E. French and H. Taube, J. Amer. Chem. Soc. **91**, 6951 (1969).

110. V. S. Srinivasan and E. S. Gould, Inorg. Chem. **20**, 208 (1981).

111. R. A. Holwerda and J. D. Clemmer, Inorg. Chem. **21**, 2103 (1982).

112. S. Wherland and H. B. Gray in Biological Aspects of Inorganic Chemistry, A. W. Addison, W. R. Cullen, D. Dolphin and B. R. James, eds. Wiley, NY, 1977, p. 289.

113. A. G. Sykes, Chem. Soc. Revs. **15**, 283 (1986).

114. J. R. Winkler, D. G. Nocera, K. M. Yocom, E. Bordignon and H. B. Gray, J. Amer. Chem. Soc. **104**, 5798 (1982); **106**, 5145 (1984).

115. S. S. Isied, G. Worosila and S. J. Atherton, J. Amer. Chem. Soc. **104**, 7659 (1982); S. S. Isied, C. Kuehn and G. Worosila, J. Amer. Chem. Soc. **106**, 1722 (1984).

116. S. E. Peterson-Kennedy, J. L. McGourty, J. A. Kalweit and B. M. Hoffman, J. Amer. Chem. Soc. **108**, 1739 (1986).

117. G. McLendon, Acc. Chem. Res. **21**, 160 (1988), who also gives a short understandable account of long-distance electron transfer.

118. H. Elias, M. H. Chou and J. R. Winkler, J. Amer. Chem. Soc. **110**, 429 (1988).

119. M. P. Jackman, J. McGinnis, R. Powls, G. A. Salmon and A. G. Sykes, J. Amer. Chem. Soc. **110**, 5880 (1988). See also O. Farver and I. Pecht, Inorg. Chem. **29**, 4855 (1990) who find $k_1 = 0.07 \pm 0.01$ s^{-1} and $k_2 = 0$ for similar experiments with stellacyanin.

120. N. M. Kostić, R. Margalit, C.-M. Che and H. B. Gray, J. Amer. Chem. Soc. **105**, 7765 (1983).

121. H. B. Gray, Chem. Soc. Revs. **15**, 17 (1986).

122. S. L. Mayo, W. R. Ellis, Jr., R. J. Crutchley and H. B. Gray, Science **233**, 948 (1986).
123. A. G. Sykes, Chem. in Britain **24**, 551 (1988).
124. R. A. Scott, A. G. Mauk and H. B. Gray, J. Chem. Educ. **62**, 932 (1985).
125. G. L. Closs and J. R. Miller, Science **240**, 440 (1988).
126. D. N. Beratan, J. N. Onuchic, J. N. Betts, B. E. Bowler and H. B. Gray, J. Amer. Chem. Soc. **112**, 7915 (1990).
127. A. W. Axup, M. Albin, S. L. Mayo, R. J. Crutchley and H. B. Gray, J. Amer. Chem. Soc. **110**, 435 (1988). T. J. Meade, H. B. Gray and J. R. Winkler, J. Amer. Chem. Soc. **111**, 4353 (1989).
128. D. W. Conrad and R. A. Scott, J. Amer. Chem. Soc. **111**, 3461 (1989).
129. B. S. Brunschwig, P. J. DeLaive, A. M. English, M. Goldberg, H. B. Gray, S. L. Mayo and N. Sutin, Inorg. Chem. **24**, 3743 (1985).
130. M. Faraggi, M. R. DeFelippis and M. H. Klapper, J. Amer. Chem. Soc. **111**, 5141 (1989) for references.

Selected Bibliography

B36. R. D. Cannon, Electron Transfer Reactions, Butterworth, London, 1980.
B37. W. L. Reynolds and R. W. Lumry, Mechanisms of Electron Transfer, Ronald Press, NY, 1966.
B38. D. B. Rorabacher and J. F. Endicott (eds.) Mechanistic Aspects of Inorganic Reactions, ACS Symposium Series, 198, American Chemical Society, Washington, D.C. 1982.
B39. H. Taube, Electron Transfer Reactions of Complex Ions in Solution, Academic, NY, 1970.
B40. J. J. Zuckerman, Ed. Inorganic Reactions and Methods. Vol. 15. Electron Transfer and Electrochemical Reactions; Photochemical and other Energized Reactions, VCH, Weinheim, 1986.

Problems

1. Calculate the rate constants for the reactions

$$Ce^{IV} + Fe(CN)_6^{4-} \rightarrow Ce^{III} + Fe(CN)_6^{3-}$$

and $\quad MnO_4^- + Fe(CN)_6^{4-} \rightarrow MnO_4^{2-} + Fe(CN)_6^{3-}$

on the basis of the Marcus equation, (5.35). Use the following information for the isotopic exchange reactions

	E_0, V	$k, M^{-1}s^{-1}$
$*Ce^{IV} + Ce^{III} \rightleftharpoons Ce^{IV} + *Ce^{III}$	+1.44	4.6
$*Fe(CN)_6^{3-} + Fe(CN)_6^{4-} \rightleftharpoons Fe(CN)_6^{3-} + *Fe(CN)_6^{4-}$	+0.36	3×10^2
$*MnO_4^- + MnO_4^{2-} \rightleftharpoons MnO_4^- + *MnO_4^{2-}$	+0.56	3.6×10^3

All values at 25 °C.

2. Using the Marcus expression, the self-exchange for the couples CO_2^-/CO_2 and SO_2^-/SO_2 have been estimated as approximately $10^{-5}M^{-1}s^{-1}$ and $10^4 M^{-1}s^{-1}$ respectively. Suggest a reason for such a tremendous difference. H. A. Schwarz, C. Creutz and N. Sutin, Inorg. Chem. **24**, 433 (1985); R. J. Balahura and M. D. Johnson, Inorg. Chem. **26**, 3860 (1987).

3. The electron self-exchange rate constants evaluated by the Marcus expressions (using cross-reaction data) and those determined experimentally differ in the following cases. Give possible reasons for these differences.

Couple	Evaluation $M^{-1}s^{-1}$	Experimental $M^{-1}s^{-1}$	Refs.
Cu(II) peptides, Cu(III) peptides	$\sim 10^{8}$ [a]	6×10^{4}	e
$Fe(H_2O)_6^{2+}$, $Fe(H_2O)_6^{3+}$	10^{-3} [b]	4	f, g
O_2^{-}, O_2	$10^{-8}-10^{5}$ [b]	4.5×10^{2}	h, i
$Cu(phen)_2^{+}$, $Cu(phen)_2^{2+}$	50^{c-} 5×10^{7} [d]	7×10^{4}	j

[a] Using $IrCl_6^{2-}$ [b] Large number of metal complex oxidants or reductants
[c] Using cyt-c as partner   [d] Using $Co(edta)^{-}$ as partner   [e-j] Refs. below

(e) G. D. Owens and D. W. Margerum, Inorg. Chem. **20**, 1446 (1981).
(f) J. T. Hupp and M. Weaver, Inorg. Chem. **22**, 2557 (1983).
(g) P. Bernhard and A. M. Sargeson, Inorg. Chem. **26**, 4122 (1987).
(h) M. S. McDowell, J. H. Espenson and A. Bakač, Inorg. Chem. **23**, 2232 (1984).
(i) J. Lind, X. Shen, G. Merényi and B. Ö. Jonsson, J. Amer. Chem. Soc. **111**, 7654 (1989).
(j) C.-W. Lee and F. C. Anson, Inorg. Chem. **23**, 837 (1984).

4. The electron self-exchange rate constants (k) have been assessed for the following couples at 25°C and $\mu = 0.1$ M:

Couple	$k, M^{-1}s^{-1}$	E_0
$Mn(sar)^{2+/3+}$	17	0.52
$Fe(sar)^{2+/3+}$	6.0×10^{3}	0.09
$Ru(sar)^{2+/3+}$	1.2×10^{5}	0.29
$Co(sar)^{2+/3+}$	2.1	-0.43
$Co(en)_3^{2+/3+}$	3.4×10^{-5}	-0.18
$Ni(sar)^{2+/3+}$	1.7×10^{3}	0.86

Structure sar — Chap. 6, Structure *19*

Correlate these data with the electronic configurations of the couples or alternatively deduce structural differences for the couples from the kinetic data.
I. I. Creaser, A. M. Sargeson and A. W. Zanella, Inorg. Chem. **22**, 4022 (1983).
P. Bernhard and A. M. Sargeson, Inorg. Chem. **26**, 4122 (1987).

5. Rationalize the relatively slow self-exchange rates for the couples involving the aqua ions, $Ti^{3+/4+} \geq 3 \times 10^{-4} M^{-1}s^{-1}$; $TiOH^{2+/3+} \approx 10^{-2} M^{-1}s^{-1}$; $V^{2+/3+} = 10^{-2} M^{-1}s^{-1}$.
D. H. Macartney, Inorg. Chem. **25**, 2222 (1986) and references therein.

6. a. Explain why the rate constants for a number of Cr^{2+} reductions, although inner-sphere, do not vary much.

J. P. Candlin and J. Halpern, Inorg. Chem. **4**, 766 (1965); R. C. Patel, R. E. Ball, J. F. Endicott and R. G. Hughes, Inorg. Chem. **9**, 23 (1970); N. Sutin, Acc. Chem. Res. **1**, 225 (1968).

b. $Co(NH_3)_5NH_2CHO^{3+}$ reacts rapidly with Cr^{2+} to give $Cr(H_2O)_5OCHNH_2^{3+}$, with a rate law:

$$V = k\,[Cr^{II}]\,[Co^{III}]\,[H^+]^{-1}$$

whereas the linkage isomer $Co(NH_3)_5OCHNH_2^{3+}$ only slowly reacts with Cr^{2+}, with no $[H^+]$ dependency in the rate law. Expain.
R. J. Balahura and R. B. Jordan, J. Amer. Chem. Soc. **92**, 1533 (1970).

c. Discuss the probable mechanisms for Cu^+ reductions from the data of Table 5.7.
O. J. Parker and J. H. Espenson, J. Amer. Chem. Soc. **91**, 1968 (1969); E. R. Dockal, E. T. Everhart and E. S. Gould, J. Amer. Chem. Soc. **93**, 5661 (1971).

d. The reduction of nicotinic or isonicotinic acid complexes of the pentamminecobalt(III) moiety are much faster by V^{2+} than by Cr^{2+}, and this is unusual. Suggest how this might arise, bearing in mind the discussion in Sec. 5.8.1.
C. Norris and F. R. Nordmeyer, Inorg. Chem. **10**, 1235 (1971).

7. Comment on the rate constants, $M^{-1}s^{-1}$ at 25 °C, for the following pairs of reactions considering particularly the relative values:

reductant	oxidant	k	oxidant	k
Cr^{2+}	CrN_3^{2+}	6.1	FeN_3^{2+}	$\sim 3 \times 10^7$
	$CrNCS^{2+}$	1.5×10^{-4}	$FeNCS^{2+}$	3×10^7
V^{2+}	FeN_3^{2+}	5.2×10^5	$Co(CN)_5N_3^{3-}$	1.1×10^2
	$FeNCS^{2+}$	6.6×10^5	$Co(CN)_5SCN^{3-}$	1.4×10^2
$Co(CN)_5^{3-}$	$Co(NH_3)_5N_3^{2+}$	1.6×10^6 [a]		
	$Co(NH_3)_5NCS^{2+}$	1.0×10^6 [a]		
$TiOH^{2+}$	$Co(CN)_5N_3^{3-}$	1.5		
	$Co(CN)_5NCS^{3-}$	$< 10^{-3}$		

[a] Inner-sphere.

O. Oyetunji, O. Olubuyide and J. F. Ojo, Bull. Chem. Soc. Japan **63**, 601 (1990) and references therein.

8. The reduction of a number of complexes $Co(NH_3)_5X^{(3-n)+}$ by $Co(CN)_5^{3-}$ in solutions containing CN^- ion has been examined. With $X^{n-} = Cl^-$, N_3^-, NCS^-, and OH^-, the redox reactions are second-order, with a wide range of values for the second-order rate constant, and a product $Co(CN)_5X^{3-}$. The rate law is different with $X^{n-} = NH_3$, PO_4^{3-}, CO_3^{2-} and SO_4^{2-},

$$V = k\,[Co^{III}]\,[Co^{II}]\,[CN^-]$$

with k similar for these reductions, and the product $Co(CN)_6^{3-}$. Give an explanation for this behavior.
J. P. Candlin, J. Halpern and S. Nakamura, J. Amer. Chem. Soc. **85**, 2517 (1963).

9. A strong autocatalysis is observed in the reaction of Co(bamap)(H$_2$O)$^-$ (see **2**) with Fe^{2+} in the presence of ascorbic acid, pH < 3.5 (Sec. 5.7.4). The autocatalysis is eliminated if a large excess of Zn^{2+} is present and the second-order rate constant for the Fe^{2+} reaction can then be determined (as 9.0 × 10^{-4}M^{-1}s^{-1}). What might be happening?
   K. Bernauer, P. Pousaz, J. Porret and A. Jeanguenat, Helv. Chim. Acta **71**, 1339 (1988).

10. Suggest why the oxidation

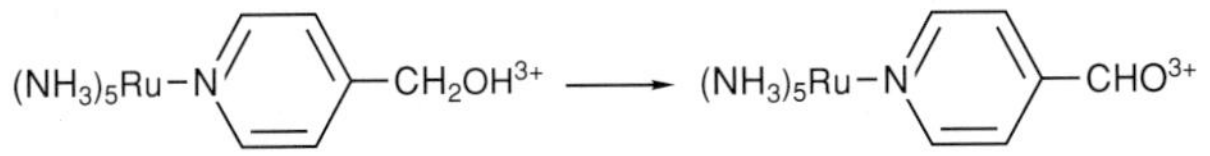

    gives solely one product, whether a one- or a two-electron oxidant is used, compared with the behavior of the Co(III) analog (Sec. 5.8.3).
    H. Taube, Electron Transfer Reactions of Complex Ions in Solution, Chap. 4.

11. When the yellow-orange ion Co(NH$_3$)$_5$py^{3+} is mixed with Fe(CN)$_6^{4-}$ (in the presence of edta to prevent the precipitation of products) three color changes are observed. The solution goes orange very rapidly (within milliseconds), then turns yellow in a period of seconds and finally over hours becomes purple. Interpret these changes.
    A. J. Miralles, A. P. Szecsy and A. Haim, Inorg. Chem. **21**, 697 (1982).

12. Give a reasonable explanation for the intramolecular electron transfer rate constants (k) at 25 °C for the following precursor complexes:

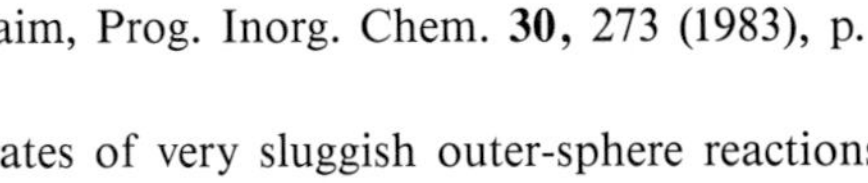

    X = CH$_2$, k < 6 × 10^{-4}s^{-1}
    X = (CH$_2$)$_2$, k = 2.1 × 10^{-3}s^{-1}
    X = −CH=CH−, k = 1.4 × 10^{-3}s^{-1}
    A. Haim, Prog. Inorg. Chem. **30**, 273 (1983), p. 340.

13. The rates of very sluggish outer-sphere reactions of Eu^{2+} (also V(II) and V(III)) with Co(NH$_3$)$_6^{3+}$ and Co(NH$_3$)$_5$py^{3+} are increased markedly by addition of small (mM) amounts of 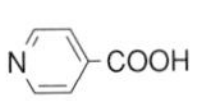 and inhibited by Eu^{3+}. The reaction products remain unchanged. The 3-pyridine carboxylic acid is without effect. Account for this behavior.
    E. S. Gould, Acc. Chem. Res. **18**, 22 (1985).

14. The second-order rate constants for reaction of $Co(NH_3)_5L^{n+}$ (k_H) and $Co(ND_3)_5L^{n+}$ (k_D) with V^{2+} and Cr^{2+} have been measured. Explain the results shown

Ligand, L	$k_H/k_D\,(V^{2+})$	$k_H/k_D\,(Cr^{2+})$
py	1.54	1.48
H_2O	1.54	–
N_3^-	1.04	–
NCS^-	1.41	1.34
⬡–$CONH_2$ (N)	1.16	1.07
⬡ $CONH_2$ (N)	1.44	1.45

M. M. Itzkowitz and F. R. Nordmeyer, Inorg. Chem. **14**, 2124 (1975).

15. The absorbance changes shown below occur for the reaction of the radicals with penta-ammine(histidine-33)ruthenium(III) ferricytochrome c, $PFe^{III} - Ru^{III}$ (see (5.84)). The *final* product is $PFe^{II}Ru^{III}$. Absorbance increases at 550 nm are largely as a result of the step $PFe^{III} \rightarrow PFe^{II}$. Interpret the changes (particularly the relative absorbances associated with the very fast and slower absorbances).

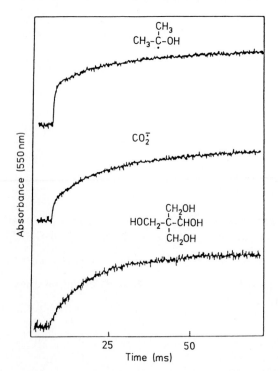

Problem 15. Reduction of Ru(III)-cytochrome-c-Fe(III) by pulse radiolysis-generated radicals. Reproduced with permission from S. S. Isied, C. Kuehn and G. Worosila, J. Amer. Chem. Soc. **106**, 1722 (1984).

S. S. Isied, C. Kuehn and G. Worosila, J. Amer. Chem. Soc. **106**, 1722 (1984).

16. The reduction of the ruthenated plastocyanin protein $PCu^{II}Ru^{III}$ by CO_2^- results in 72% $PCu^{I}Ru^{III}$ and 28% $PCu^{II}Ru^{II}$. This very fast stage is followed by a slower one in which $PCu^{II}Ru^{II}$ is converted by a first order process into $PCu^{I}Ru^{III}$. For this conversion

$$d[PCu^{I}Ru^{III}]/dt = \{k_1 + k_2[PCu^{II}Ru^{III}]\}[PCu^{II}Ru^{II}]$$

$k_1 = 0.024 \pm 0.058 \text{ s}^{-1}$ and $k_2 = 1.2 \times 10^5 M^{-1}s^{-1}$. Comment on the values and the significance of k_1 and k_2.

M. P. Jackman, J. McGinnis, R. Powls, G. A. Salmon and A. G. Sykes, J. Amer. Chem. Soc. **110**, 5880 (1988).

Chapter 6

The Modification of Ligand Reactivity by Complex Formation

Coordination modifies the properties of the metal ion; equally important, however, is the impact that the metal has on the behavior of the coordinated ligand towards chemical reagents.[1-4] The reactivity of the ligand can be enhanced or the ligand can be deactivated. Activation is important in the catalytic effects of metal ions, and deactivation is useful when "masking" of a reaction center is required. The product of the reaction may remain coordinated to the metal, and may be a weaker or a stronger ligand than the original reactant. Alternatively, the product may break away from the metal ion, which is then able to coordinate with more reactant and function in a catalytic manner. Chemical reaction may occur at a point within a chelate ring, in an adjacent ring, or at the side chain of a chelate ring, with variable results. In general, the further the reaction site from the metal center, the less the influence of the latter, unless conjugative effects are present.

We shall consider the ways in which a metal may influence a reaction. These are listed in Table 6.1. The effects of the metal and the reactivity of the coordinated ligand are interrelated since invariably at least one of the reactants becomes coordinated to the metal during a catalyzed reaction.

Table 6.1 Functions of Metal Ion Center in Altering Reaction Characteristics

Function	Examples of use
Serve as a collecting point for reactants, Sec. 6.1	Neighboring group and template effects.
Promote electron shifts in the metal-ligand system, Secs. 6.2–6.5	Promotion of nucleophilic substitution. Enhanced acidity of coordinated ligand.
Protect a coordinated function from reaction, Sec. 6.6.	Masking of normally reactive groups.
Force a reaction to completion, Sec. 6.7.	Promotion of macrocycle and Schiff base formation.
Alter the strain within or the conformational characteristics of the reacting ligand, Sec. 6.8.	Catalyzed rearrangement of highly strained polycyclics.

6.1 The Metal as a Collecting Point for Reactants

The metal, in acting as a collection point to bring reactants together, is likely at the very least to promote enhanced rates of reaction by operation of the *neighboring group effect*. If, in addition, the transmission of electronic effects of the metal (Secs. 6.2–6.5) also occurs, as is usually the case, then large overall rate enhancements may be encountered.

6.1.1 Neighboring Group Effects

The ability of a substituent in one part of an organic molecule to influence a reaction by partially or completely bonding to the reaction center in another part of that molecule, thereby leading to an intramolecular reaction, is well recognized in organic chemistry.[5] These neighboring group effects, or anchimeric effects, often give rise to a rate 10^5 to 10^6 times faster than the rate for the "unassisted" reaction. It is surprising, when one considers the possibilities for juxtaposing reactants within the coordination sphere of the metal, that the effect has not been more exploited in transition metal chemistry. We shall be concerned in this section with the interaction of a coordinated nucleophile with a reaction center in a metal complex, and are interested in how much acceleration might result compared with the situation when the metal is not present.

One obvious area in which anchimeric effects might materialize is in chelation reactions; this is the underlying reason that chelation is dominated by the first step (Sec. 4.4.1). We have seen it operate in the formation of Pt(II) chelates (Sec. 4.7.6). The accompanying values of $\Delta S^{\ddagger}$ are less negative than is usual for the associative substitution in Pt(II), consistent with the intramolecular mechanism.[6] The rate constant for the ring-closure reaction (6.1)

$$(NH_3)_4Co \overset{\underset{\displaystyle N}{H_2}}{\underset{\underset{\displaystyle OH_2}{|}}{\diagup}} \overset{Co(NH_3)_4^{4+}}{\underset{\underset{\displaystyle Cl}{|}}{}} \xrightarrow[k]{- Cl^-} (NH_3)_4Co \overset{\underset{\displaystyle N}{H_2}}{\diagup}\overset{\diagdown}{\underset{\underset{\displaystyle H_2}{O}}{}} Co(NH_3)_4^{5+} \qquad (6.1)$$

is some 10^2 times faster than the intermolecular aquation of the corresponding mononuclear complex,[7]

$$Co(NH_3)_5Cl^{2+} + H_2O \rightarrow Co(NH_3)_5H_2O^{3+} + Cl^- \qquad (6.2)$$

Since the bridged complex reacts without formation of the diaqua intermediate (this can be separately prepared and shown to be much less reactive than would be required), the anchimeric effect of the adjacent coordinated water is established. See also Ref. 8.

Using a Co(III) complex as a scaffold, it is possible to place a coordinated nucleophile *cis* to an incipient substrate. As an example (of many (see later sections)) the amidolysis of the phosphate ester, which is normally extremely slow, is quite markedly promoted in (6.3).[9] The value of $k(20\ s^{-1})$ is an enhancement of about 10^6 compared with the attack of (1 M) NH_3 on the free dinitrophenylphosphate.[9] In addition to holding the reactants together, the metal may also provide a driving force in producing a stable chelate ring. Finally, the shift of electrons away from the ligand, induced by the positive charge of the metal, will lead to accelerated nucleophilic attack at the ligand. These effects are even more dramatic in the intramolecular attack by OH^- on esters, amides etc. (Secs. 6.3.1 and 6.3.2).

$$(NH_3)_4Co \overset{\underset{\displaystyle NH_2}{\diagup}}{\underset{\underset{\underset{\displaystyle O}{|}}{O-P-O-C_6H_3(NO_2)_2}}{\overset{\displaystyle O}{\diagdown}}} \xrightarrow{k} (NH_3)_4Co \overset{\underset{\displaystyle N}{H_2}}{\diagup}\overset{\overset{\displaystyle O \quad O}{\diagdown P \diagup}}{\underset{\underset{\displaystyle O-C_6H_3(NO_2)_2}{O}}{}} \qquad (6.3)$$

Comparison of intra- vs inter-molecular reactions – The problems in assessing the accelerating effects of intramolecular processes (which we do constantly in this chapter) have been analyzed by W. P. Jencks (Catalysis in Chemistry and Enzymology, McGraw-Hill, NY, 1969, p 8). It may be difficult to estimate the magnitude of the neighboring group effect because the analogous intermolecular reaction may occur by a different mechanism or may not take place at a conveniently measurable rate. Even if the two rates can be measured the vexing question of units arises. The intramolecular reaction is first-order with the usual units $k_{intra} = s^{-1}$ whereas the intermolecular counterpart is second-order, k_{inter} typically $M^{-1}s^{-1}$. The ratio k_{intra}/k_{inter} is *not* therefore dimensionless, but will depend on units of concentration employed, here M. The ratio represents the concentration (in M) of one of the reactants which must be used in the intermolecular reaction to produce the same (pseudo) first-order rate constant, as observed for the intramolecular reaction. Alternatively, the ratio may be regarded as the "effective" concentration of one of the entities in the intramolecular reaction. This assessed large "local" concentration may not account for the effects observed, signifying that other factors are contributing to the neighboring group effect. An entropic advantage for intra-over inter-molecular reactions of about 140 $JK^{-1}mol^{-1}$ has been assessed. This amounts to $\Delta G^{\ddagger}$ ($= -T\Delta S^{\ddagger}$) of 42 $kJmol^{-1}$ at 25 °C, or a 10^8 enhanced rate factor (K. N. Houk, J. A. Tucker and A. E. Dorigo, Accs. Chem. Res. **23**, 107 (1990)).

6.1.2 Template Chemistry

The use of metals for prearranging reaction centers as neighboring groups has a special value in the production of macrocycles (template effect).[2, 10, 11] Although these ligands can be sometimes prepared directly, the addition of metal ion during the synthesis will often increase the yield, modify the stereochemical nature of the product, or even be essential in the buildup of the macrocycle.[10, 11] There have been few mechanistic studies of these processes. The alkali and alkaline-earth metal ions can promote the formation of benzo[18]crown-6 in methanol:

(6.4)

(6.5)

The value of k_2/k_1 varies from 1.2×10^3 (Sr^{2+}) to 17 (Cs$^+$). There are two opposing factors influencing this ratio. There is a rate-enhancing effect due to the increased proximity of the chain ends in (6.5). This is offset by a rate retardation which arises because of the reduced inherent nucleophilicity of ArO$^-$ which ion-pairs with M^{n+}.[12] The formation of the interlocked catenands under the influence of Cu$^+$ is a striking example of the template effect, Fig. 6.1.[13]

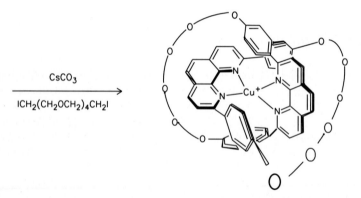

Fig. 6.1 Synthesis of a cuprocatenane. High dilution conditions for the second step allow the intremolecular condensation of the terminal OH groups on a single phenanthroline. Removal of the metal from the copper(I) complex by treatment with Me$_4$N$^+$CN$^-$ in acetonitrile/water affords the catenand.[13] Reproduced with permission from C. O. D. Buchecker, J.-P. Sauvage and J.-M. Kern, J. Amer. Chem. Soc. **106**, 3043 (1984). © (1984) American Chemical Society.

This particular function, and also others of the metal, are also beautifully utilized in the work on the chemical synthesis of corrins (Fig. 6.2). In the synthesis of **2**, a metal, Co(II), Ni(II) or Pd(II), is required to stabilize the precursor **1**, which would otherwise be extremely labile configurationally and constitutionally. As added bonuses, the metal ion helps to activate the methylene carbon for its attack on the iminoester carbon, and also forces the four nitrogens into a planar conformation, thereby bringing the condensation centers of rings A and B close together. This is strikingly shown by X-ray structural determination of **1**. With all this help from the metal, the final ring closure occurs smoothly. However, it is impossible to remove the metal from **2** without its complete destruction. To obtain the metal-free corrin, it is necessary to synthesize the zinc analog by a route somewhat more involved than that used with the nickel complex. The zinc can be replaced by cobalt, or the ring closure can be made using the cobalt(III) complex directly. These are key types of steps in the mammoth synthesis of vitamin B_{12} that has been completed. [14]

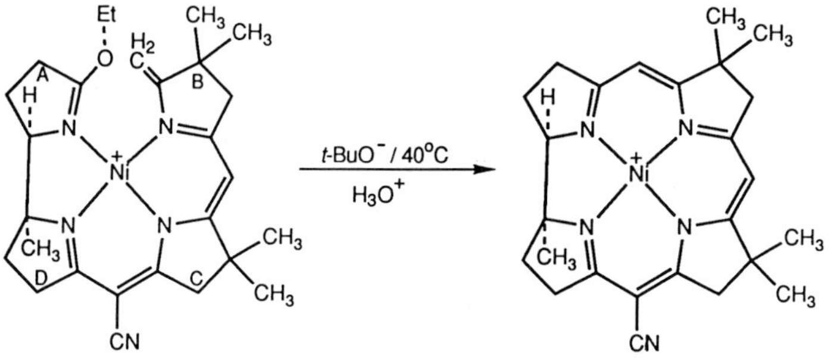

Fig. 6.2 Utilization of template and other effects in corrin syntheses. [14]

6.1.3 Collecting Reactant Molecules

In the examples above, one or both of the reaction centers are already attached to the metal center. In many cases, the reactants are free before reaction occurs. If a metal ion or complex is to promote reaction between A and B, it is obvious that at least one species must coordinate to the metal for an effect. It is far from obvious whether both A and B enter the coordination sphere of the metal in a particular instance. A number of metal-oxygen complexes can oxygenate a variety of substrates (SO_2, CO, NO, NO_2, phosphines) in mild conditions. Probably the substrate and O_2 are present in the coordination sphere of the metal during these so-called autoxidations. In the reaction of oxygen with transition metal phosphine complexes, oxidation of metal, of phosphine or of both, may result. [15] The initial rate of reaction of O_2 with $Co(Et_3P)_2Cl_2$ in tertiary butylbenzene,

$$-d\,[Co(II)]/dt \;=\; k\,[Co(II)]\,[O_2] \tag{6.6}$$

is consistent with the mechanism

$$Co(Et_3P)_2Cl_2 \,+\, O_2 \;\rightleftharpoons\; Co(Et_3P)_2(O_2)Cl_2 \tag{6.7}$$

$$Co(Et_3P)_2(O_2)Cl_2 \;\rightarrow\; Co(OPEt_3)_2Cl_2 \tag{6.8}$$

in which O−O bond rupture occurs within a oxygen adduct and two P−O bonds result. In the reaction of $Co(Et(OEt)_2P)Cl_2$ with O_2 the 1:1 dioxygen adduct can be isolated at lowered temperatures ($-46\,°C$) and its decomposition (6.8) studied. [15] Reaction cannot be occurring via dissociated free phosphine since this reacts with oxygen via radical reactions to give products such as $Et_n(OEt)_{3-n}PO$, and these are not observed. In contrast, the formation of phosphine oxide by the $Pt(R_3P)_3$-catalyzed oxidation of R_3P by O_2 occurs by more involved steps with major mechanistic differences in protic and aprotic media.[16,17]

Another example, where both reactants coordinate for an effective neighboring group participation, is in the Cu(II) catalysis of the reactions (Prob. 2, Chap. 1):

$$2 H_2O_2 \rightarrow 2 H_2O + O_2 \tag{6.9}$$

$$2 H_2O_2 + N_2H_4 \rightarrow N_2 + 4 H_2O \tag{6.10}$$

The metal appears to function by complexing with H_2O_2 or HO_2^- (sometimes this is detected); the resulting species then interacts with another molecule of H_2O_2 or other H donor, such as N_2H_4. Copper(II) complexes with no coordinated water appear inactive, and an interesting comparison of the catalytic activity of 1:1 Cu complexes with en, dien, and trien is shown in Fig. 6.3. The type of detailed mechanism envisaged for (6.10) is

$$\tag{6.11}$$

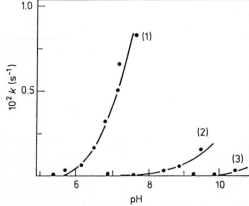

Fig. 6.3 Catalytic activity of Cu(II) chelates toward H_2O_2 decomposition. The plot illustrates the decrease in catalytic activity with decreasing number of free H_2O coordination sites on the metal ion: $Cu(en)(H_2O)_2^{2+}$ (1); $Cu(dien)(H_2O)^{2+}$ (2); $Cu(trien)^{2+}$ (3). Ref. 18.

N−N might represent bipyridine, which will retain the metal in solution, even in alkaline conditions.[18]

The iron enzymes catalase and peroxidase[19] promote these reactions very effectively, but it is unlikely that both reactants are coordinated to the metal during the reaction. It is generally true that the metal site is restricted in enzymes, allowing coordination of only one reactant. The other reactant is however often held close by the protein structure.[20, 21]

There has been an increasing interest in the mediation by metal ions of reaction between two coordinated organic reactants. Consider the reaction between pyridine-2-carbaldoxime anion and phosphoryl imidazole in the presence of Zn(II). A ternary complex **3** or **4** is formed and it is believed that reaction occurs within this framework. The first-order rate constant for reaction of the ternary complex is at least 10^4 times larger than the second-order rate constant for the metal-free reaction. The zinc appears to function in this example, as in a number of a similar type,[22] as (a) a collector of the two reactants in the correct orientation for an intramolecular reaction (the so-called *proximity effect*) and (b) a shield between the negative charges of the reactants.[23, 24] An important condensation that metal ions catalyze involves pyridoxal and amino acids (Fig. 6.4). In the formation of Schiff bases from salicylaldehyde, glycine, and metal ions, there is, as metal-dependent terms in the rate law indicate, a direct participation of the metal in the condensation. In this, a kinetic template mechanism is believed operative, involving a rapid preequilibrium between reactants and metal ion to form a ternary complex, within which a rate-determining reaction occurs.[25]

3   **4**

The Schiff base can undergo a variety of reactions in addition to transamination, shown in Fig. 6.4; for example, racemization of the amino acid via the α-deprotonated intermediate. Many of these reactions are catalyzed by metal ions and each has its equivalent nonmetallic enzyme reaction, each enzyme containing pyridoxal phosphate as a coenzyme. Many ideas of the mechanism of the action of these enzymes are based on the behavior of the model metal complexes.[26]

6.2 Promotion of Reaction within the Metal-Bound Ligand

The shift of electrons away from the ligand, which is usually induced by the positive charge of the metal, will lead to accelerated nucleophilic attack at the ligand. Since nucleophiles donate electrons they thus act to redress the balance. This effect may well reinforce the neighboring group effect. Some examples of its operation are shown in Table 6.2. All the reactions are second-order.

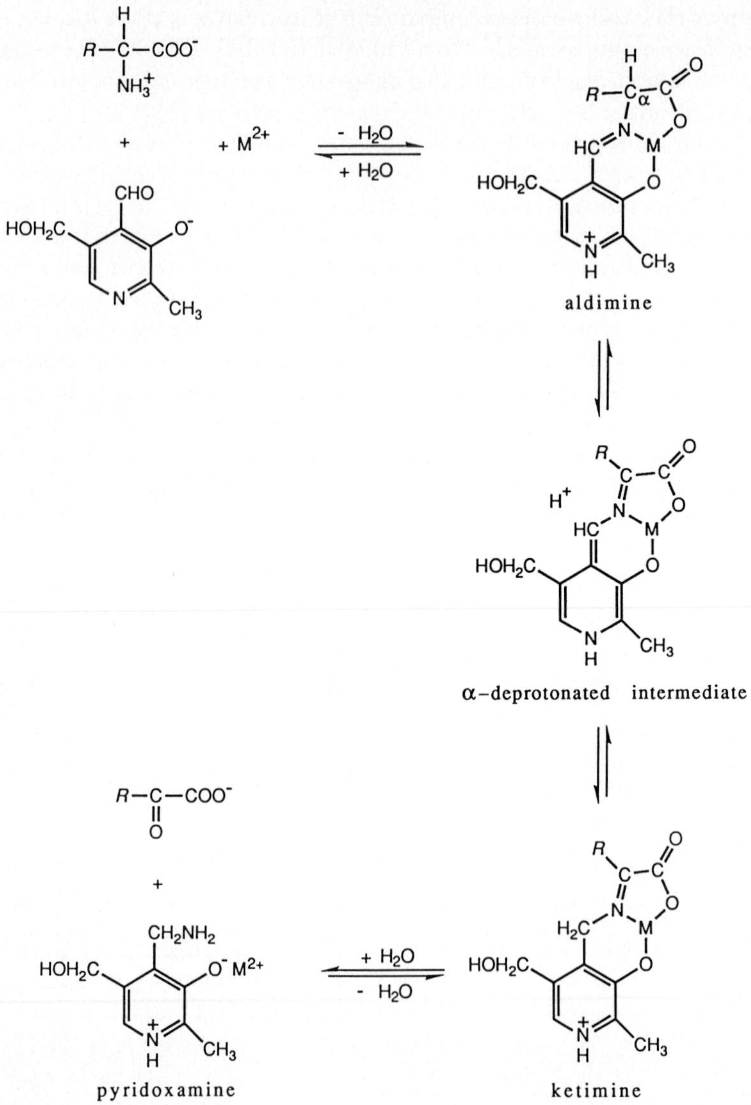

Fig. 6.4 Reversible interconversion of amino acid and keto acid. Conjugation of the imine bond in the aldimine with the electron sink of the pyridine ring plus protonation of the pyridine nitrogen as well as the metal ion — all this results in weakening of the C−H bond of the amino acid residue. Thus, also catalyzed is α-proton exchange, racemization of a chiral center at the α-carbon atom and decarboxylation of the appropriate amino acid. [26]

Table 6.2 Some Examples of Metal-Promoted Reactions

Substrate	Nucleophile	Products	Comments	Ref.
$Fe(CN)_5NO^{2-}$	OH^-	$Fe(CN)_5NO_2^{4-}$ via $Fe(CN)_5NO_2H^{3-}$	metal polarization causes ligand to be present as NO^+ in substrate	a
cis-$Ru(bpy)_2(NO)Cl^{2+}$	N_3^-	$Ru(bpy)_2(H_2O)Cl$ $+ N_2 + N_2O$		b
$Ru(NH_3)_5N_2O^{2+}$	Cr^{2+}	$Ru(NH_3)_5N_2^{2+}$ (no sign of intermediates)	Ru weakens N—O bond $k = 8 \times 10^2 M^{-1}s^{-1}$ compared with $k = 6.5 \times 10^{-6}M^{-1}s^{-1}$ for $N_2O + Cr^{2+}$ (25°C)	c
$PtCl(Ph_3P)_2CO^+$	ROH	$PtCl(Ph_3P)_2C(O)OR$	Rare attack of nucleophile on carbon rather than on metal	d
$Co(NH_3)_5NCC_6H_5^{3+}$	OH^-	$Co(NH_3)_5NHCOC_6H_5^{2+}$	$k = 18.8$ $M^{-1}s^{-1}$; $\Delta H^{\ddagger} = 69 kJ\ mol^{-1}$; $\Delta S^{\ddagger} = 11 J\ K^{-1}mol^{-1}$ compared with $k = 8.2 \times 10^{-6}\ M^{-1}s^{-1}$; $\Delta H^{\ddagger} = 83 kJ\ mol^{-1}$ $\Delta S^{\ddagger} = -64 J\ K^{-1}mol^{-1}$ for free C_6H_5CN	e, 47
$Co(NH_3)_5NCCH_3^{3+}$	N_3^-	$Co(NH_3)_5N_4CCH_3^{3+}$ (5-methyltetrazole) see (6.24)	complete in 2 hours at room temperature; 25 hours at 150°C for metal-free system	48

a J. H. Swinehart and P. A. Rock, Inorg. Chem. **5**, 573 (1966); G. Stochel, R. van Eldik, E. Hejmo and Z. Stasicka, Inorg. Chem. **27**, 2767 (1988).
b F. J. Miller and T. J. Meyer, J. Amer. Chem. Soc. **93**, 1294 (1971).
c J. N. Armor and H. Taube, J. Amer. Chem. Soc. **93**, 6476 (1971).
d J. E. Byrd and J. Halpern, J. Amer. Chem. Soc. **93**, 1634 (1971); H. C. Clark and W. J. Jacobs, Inorg. Chem. **9**, 1229 (1970).
e D. Pinnell, G. B. Wright and R. B. Jordan, J. Amer. Chem. Soc. **94**, 6104 (1972); D. A. Buckingham, F. R. Keene and A. M. Sargeson, J. Amer. Chem. Soc. **95**, 5649 (1973) note a similar enhancement for CH_3CN complexes.

The metal ion or complex, which is often termed a superacid, resembles the proton in being able to produce this electron shift, so that it is quite usual to compare the proton- and metal-ion-assisted reactivities. For example, the values for the second-order rate constants at 25 °C for the reactions shown

$$HNCO + H_3O^+ \rightarrow NH_4^+ + CO_2 \qquad k = 0.12\ M^{-1}s^{-1} \qquad (6.12)$$

$$M(NH_3)_5NCO^{2+} + H_3O^+ \rightarrow M(NH_3)_6^{3+} + CO_2 \qquad \begin{array}{l} k = 0.16\ M^{-1}s^{-1}\ (M = Co) \\ k = 0.62\ M^{-1}s^{-1}\ (M = Rh) \\ k = 0.06\ M^{-1}s^{-1}\ (M = Ru) \end{array} \qquad (6.13)$$

indicate that, although both H_3O^+ and metal complexes promote reaction, there is not a large difference in their effects here, see also Table 6.3. It should however be remembered that

(a) the larger coordination number of the metal ion might allow more flexibility in positioning reactants for intramolecular grouping and (b) a larger concentration of metal ion and therefore increased rates is tolerable at neutral and basic pH.[1,2]

The polarization effects of metals have a substantial impact particularly in two areas: the promotion of the hydrolysis and other nucleophilic reactions of chelated ligands and the enhanced ionization of coordinated acidic ligands. Their importance has encouraged extensive attacks on the effects.

6.3 Hydrolysis of Coordinated Ligands

The ability of metal ions to accelerate the hydrolysis of a variety of linkages has been a subject of sustained interest. If the hydrolyzed substrate remains attached to the metal, the reaction becomes stoichiometric and is termed metal-ion promoted. If the hydrolyzed product does not bind to the metal ion, the latter is free to continue its action and play a catalytic role. The *modus operandi* of these effects is undoubtedly as a result of metal-complex formation, and this has been demonstrated for both labile and inert metal systems. Reactions of nucleophiles other than H_2O and OH^- will also be considered.

6.3.1 Carboxylate Esters: $-CO_2R \rightarrow -CO_2H$

The discovery that metal ions such as Cu^{2+} catalyzed the hydrolysis of amino acid esters was first reported nearly 40 years ago.[27] Table 6.3 shows that in the absence of direct interaction of the ester grouping with the metal ion, the primary cause of any increased rates experienced

Table 6.3 Effect of Metal Coordination on Rate Constants for Base Hydrolysis of Amino Acid Esters at 25 °C

Substrate	k $M^{-1}s^{-1}$	Ref.
$NH_2CH_2CH(NH_2)CO_2Me$	0.73	28
$NH_3^+ CH_2CH(NH_2)CO_2Me$	57	28
$(H_2O)_2Cu \begin{smallmatrix} NH_2-CH_2^{2+} \\ NH_2-CHCO_2Me \end{smallmatrix}$	6.2×10^2	28
8	84 (30 °C)	30
$NH_2CH_2CO_2Et$	0.6	29
$NH_3^+ CH_2CO_2Et$	24	29
$(H_2O)_2Cu(NH_2CH_2CO_2Et)^{2+}$	7.4×10^4	29
$en_2Co(NH_2CH_2CO_2Pr^i)^{3+}$	1.5×10^6	34
$NH_2(CH_2)_2CO_2Pr^i$	0.02	35
$en_2Co(OH)(NH_2(CH_2)_2CO_2Pr^i)^{2+}$, **12**	$5.6 \times 10^{-6}(s^{-1})^a$	35
$en_2Co(NH_2(CH_2)_2CO_2Pr^i)^{3+}$, **11**	4×10^4	35

^{a} For intramolecular attack; 0.22 $M^{-1}s^{-1}$ for external OH^- attack

by the metal complex system resides mainly in a charge effect.[28] Association of the ester grouping to the metal on the other hand induces marked rate acceleration.[29]

There is a major problem in the interpretation of the results when labile metal ion complexes are used as promoters.[1] It is difficult to characterize the reactant species and therefore to distinguish between the following modes of attack by OH⁻ (and H_2O):

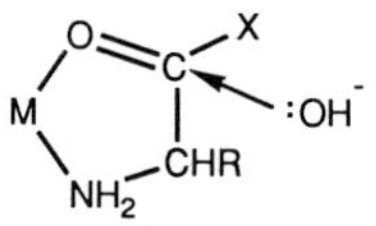

5 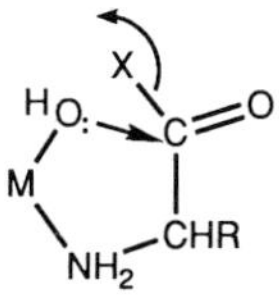   **6** 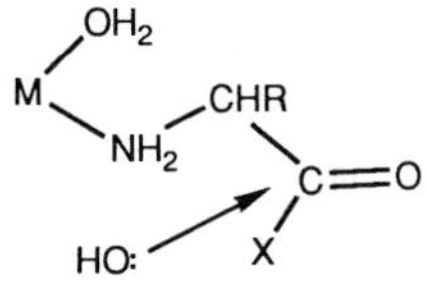   **7**

In **5**, coordination through O polarizes the C = O bond producing the incipient carbonium ion, which is more susceptible to nucleophilic attack. In **6**, there is intramolecular attack by coordinated OH⁻ on the N-bonded monodentate amino acid. In **7**, only a small rate enhancement would be anticipated from intermolecular attack on the N-bonded monodentate. In these representations, X is OR, but anticipating the next section, it might also be NHR. These interpretive problems are illustrated in the hydrolysis of complexes of the type **8**, where again any of three mechanisms corresponding to **5-7** could apply.

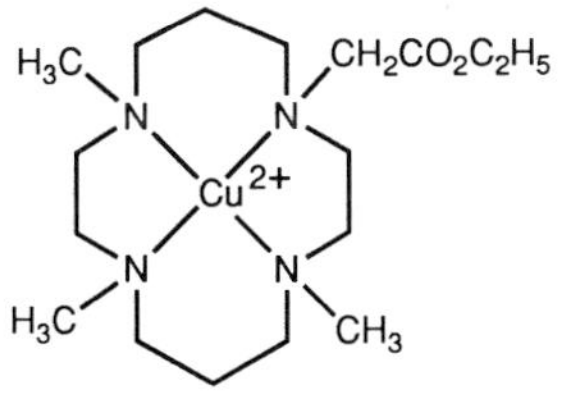

8

Absorption spectra suggest that the carbonyl O-atom is coordinated to Cu in **8** and therefore the attack corresponding to **5** is favored.[30] These problems are to a large extent overcome by using a *cis*-structure of the type **9** or **10** which we shall abbreviate CoN₄(X)Y. With this framework, there are only two available sites for reaction. The known inertness and well-characterized behavior of cobalt(III) allow the identification of reactants and intermedieates and in so doing more easily differentiate between alternative mechansisms, an approach not possible with the labile systems.[1, 31] There is some merit also in using a tetradentate ligand (N₄) which can only coordinate to leave *cis*-positions for X and Y, e. g. **10**. Although most studies use activated esters as substrates, recently strong catalysis of the hydrolysis of methyl acetate by Co(trpn)(H₂O)OH²⁺ has been reported. Complexation of the ester to the Co(III) complex is rate determining.[32]

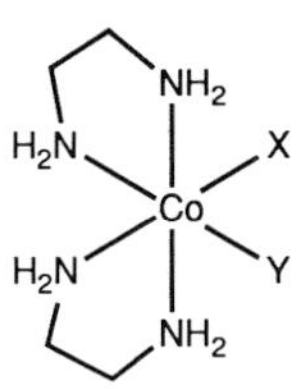

9 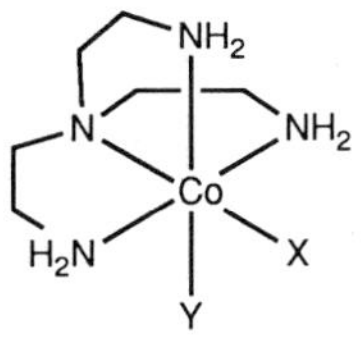   **10**

A complex of the type $CoN_4(NH_2(CH_2)_nCO_2R)X^{2+}$ in which the ester must be unidentate, linked only through the NH_2 grouping, is relatively stable towards hydrolysis. If we can rapidly remove the halide X from the coordination sphere, it may be possible to see what takes place prior to and during ester hydrolysis. Mercury(II) ion or OH^- can be used to effect removal. The pioneering work on these lines of Alexander and Busch,[33] Sargeson[34] and Buckingham[31, 34] forms a strong basis for a good understanding of the factors which influence the cobalt(III)-catalyzed hydrolysis of coordinated amino acid esters. The complexes **11** and **12** can be isolated and separately examined.[35] Both give the chelated $(Co(en)_2$ $(\beta\text{-alaO})^{2+}$ as product:

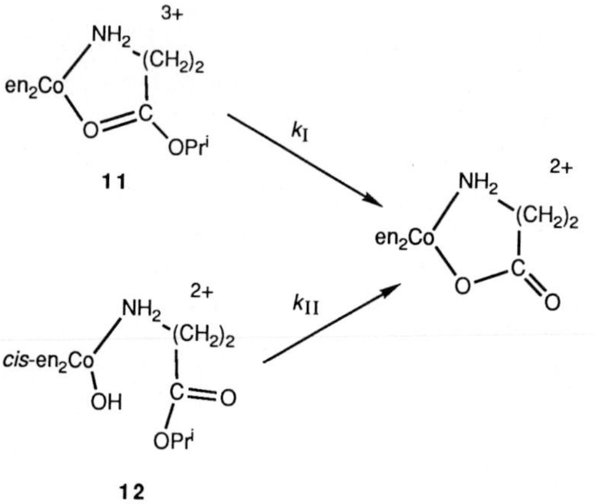

(6.14)

For **11** and **12** both OH^--unassisted and -assisted paths exist for hydrolysis with ring closure.

$$k_I = a[H_2O] + \frac{(b[OH^-] + c[OH^-]^2)}{1 + d[OH^-]}$$
(6.15)

$$k_{II} = e + f[OH^-]$$
(6.16)

The a term refers to rate-determining addition of H_2O to the chelate **11** ($a = 8.3 \times 10^{-7} M^{-1}s^{-1}$; $a_{H_2O}/a_{D_2O} = 2.5$). The value is little different from that for intramolecular attack by H_2O in **12** (in acid, **12** converts from Co-OH to Co-OH$_2$, $pK_a = 6.05$; k_{H_2O} $1.8 \times 10^{-6} s^{-1}$ and no ^2H isotope effect). The second term in (6.15) can be ascribed to the mechanism (6.17)–(6.18) with b, c and d composite rate constants.[36]

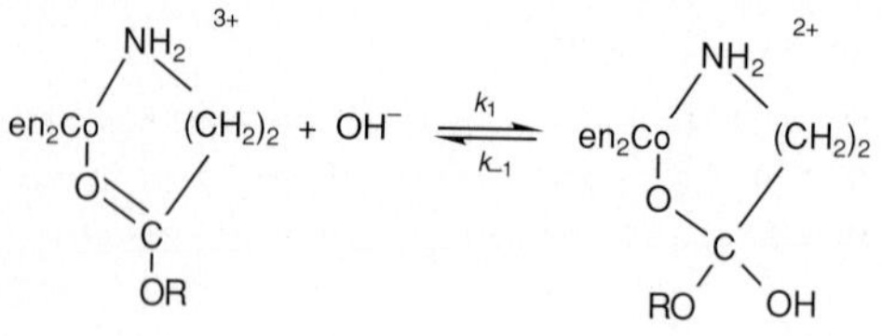

(6.17)

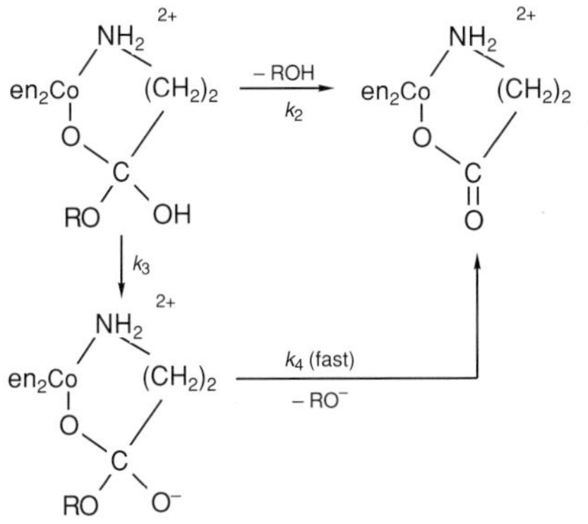

$$\text{(6.18)}$$

The value of k_1 extracted from the data for $R = Pr^i$ is shown in Table 6.3 and shows the very large rate enhancement (almost 10^8) due to attachment of the $-CO_2R$ moiety to Co(III) compared with that of **12** which is considerably slower. The effect resides mainly in a much more positive $\Delta S^{\ddagger}$. In the pH range 7–10, the rds is proton abstraction (k_3) from the addition intermediate. At higher pH, the rds is OH^- addition to the chelated ester (k_1). This is the first kinetic demonstration of a tetrahedral intermediate in amino acid ester hydrolysis. Tracer experiments show that **11** reacts without ring-opening and that in **12** the coordinated $-OH$ group is incorporated into the chelate. Trapping experiments indicate that a common tetrahedral intermediate features in the reactions of both **11** and **12** with OH^- ion.[36]

6.3.2 Amides and Peptides: $-CONHR \rightarrow -CO_2H$

The amide and peptide linkages are much more difficult to hydrolyze than the ester grouping. Both free and metal bound groups hydrolyze with second-order rate constants approximately 10^4–10^6 less than for the corresponding esters.[1,31] There are two potential sites for coordination in the -CONHR residue, namely at the carbonyl O in **13** and at the amide N in **14** where ionization of the amide proton is induced (Sec. 6.4.3).[37] Cu^{2+} promotes hydrolysis of glycinamide at low pH where it is present as **13**. However it inhibits hydrolysis at high pH, where it is **14**, to such a degree that hydrolysis cannot be observed.[29,37]

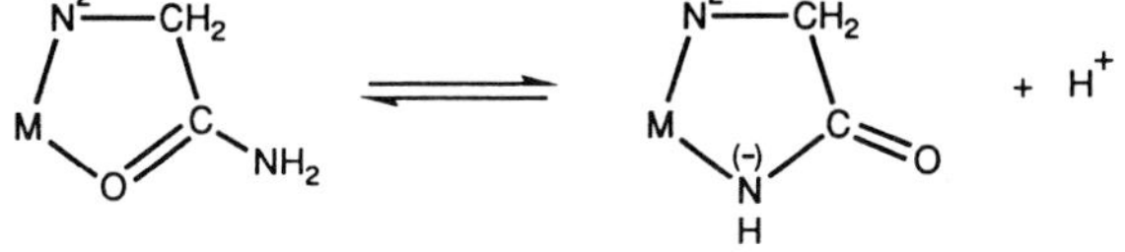

$$\text{(6.19)}$$

13   **14**

Once again, a better understanding of the basis of these effects has emerged from studies of the Co(III) complexes.[1,31] Hydrolysis of chelates of the type **13** occurs by rate-determining attack of OH^- by a mechanism similar to that shown in (6.17) and (6.18) with OR replaced by NHR.[38] The metal promotes hydrolysis by OH^- by a factor of about 10^4 (Table 6.4) and again this resides entirely in a different $\Delta S^{\ddagger}$. The acceleration is however about the same as that observed with singly O-bound amide in $Co(NH_3)_5(OCHNMe_2)^{3+}$ (Table 6.4) so that chelation is apparently unnecessary to achieve hydrolytic promotion.

Table 6.4 Effect of Cobalt(III) Coordination on Rate Constants for Base Hydrolysis of Amino Acid Amides

Substrate	k $M^{-1}s^{-1}$	$\Delta H^{\ddagger}$ kJ mol^{-1}	$\Delta S^{\ddagger}$ J K^{-1} mol^{-1}	Ref.
$NH_2CH_2CONH_2$	2.9×10^{-3}	58	-99	29
$en_2Co(OH)NH_2CH_2CONH_2^{2+}$	1.5×10^{-4} [a]	50	-150	38, 39
$en_2Co(H_2O)NH_2CH_2CONH_2^{3+}$	9.2×10^{-3} [a]	75	-33	38, 39
$en_2Co(NH_2CH_2CONH_2)^{3+}$	14	58	-30	38, 39
$(NH_3)_5Co(OCHNMe_2)^{3+}$	1.3	58	-50	31

[a] s^{-1}

Startling results are obtained with the unidentate species **15**. The intramolecular hydrolysis by coordinated H_2O in **15**, $X = H_2O$, is more marked than by coordinated OH^- **15**, $X = OH^-$ (Table 6.4)

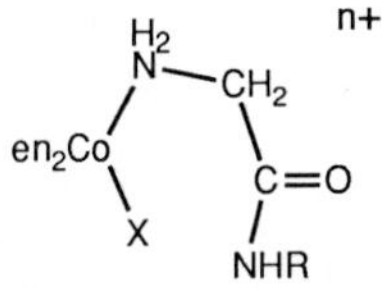

15

A bigger effect for H_2O than OH^- is very unusual and is a behavior certainly not shown by the uncoordinated amide. The effect is ascribed to a benefit from cyclization and concerted loss of protonated amide, without formation of the tetrahedral intermediate. [31] Although the coordinated OH^- is some 10^2 times less effective than coordinated H_2O (Table 6.4), it is still about 10^2 times faster with **15** than via external attack by OH^- at pH 7 on the chelated amide **13**. Early studies showed that complexes of the type $CoN_4(H_2O)OH^{2+}$ can promote the hydrolysis of esters, amides and dipeptides and that this probably arises via formation of ester, amide or peptide chelates. These then hydrolyze in the manner above. [31]

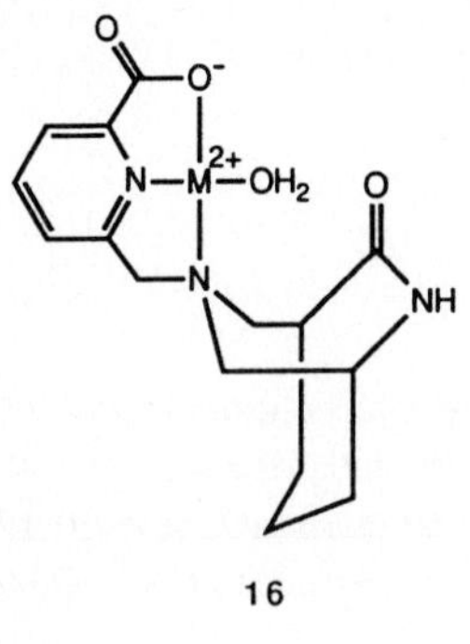

16

In ester hydrolysis, rate-limiting formation of the tetrahedral intermediate usually applies (Sec. 6.3.1) since the alkoxide group is easily expelled. In contrast, amide hydrolysis at neutral pH involves rate-limiting breakdown of the tetrahedral intermediate, because RNH^- is a poor leaving group. The catalytic effect of metal ions on amide hydrolysis has been ascribed to accelerated breakdown of the tetrahedral intermediate.[40]

The importance of correct juxtapositioning of metal and amide function is illustrated in recent studies by Groves and co-workers.[41] If the metal (Ni^{2+}, Cu^{2+} or Zn^{2+}) is forced to lie above the plane of the amide function in a specially designed complex **16**, 10^4–10^7 fold accelerations can be observed, in contrast to the meagre rate enhancements with previous models where this favorable geometry was not present.[41]

6.3.3 Nitriles: $-CN \rightarrow -CONH_2$

Considerable attention has been paid to this transformation (which is sometimes referred to as hydration[1]) in the past 15 years.[1,42,43] An early example of the effect was the marked acceleration of the base hydrolysis of 2-cyanophenanthroline by Ni^{2+}, Cu^{2+} and Zn^{2+} ions. The second-order rate constant is 10^7-fold higher for the Ni^{2+} complex than for the free ligand, residing mainly in a more positive $\Delta S^{\ddagger}$. An external OH^- attack on the chelate was favored but an internal attack by Ni(II) coordinated OH^- cannot be ruled out.[1,22,44] Nickel-ion catalysis of the hydrolysis of the phenanthroline-2-amide product is much less effective, being only about 4×10^2 times the rate for spontaneous hydrolysis.[44]

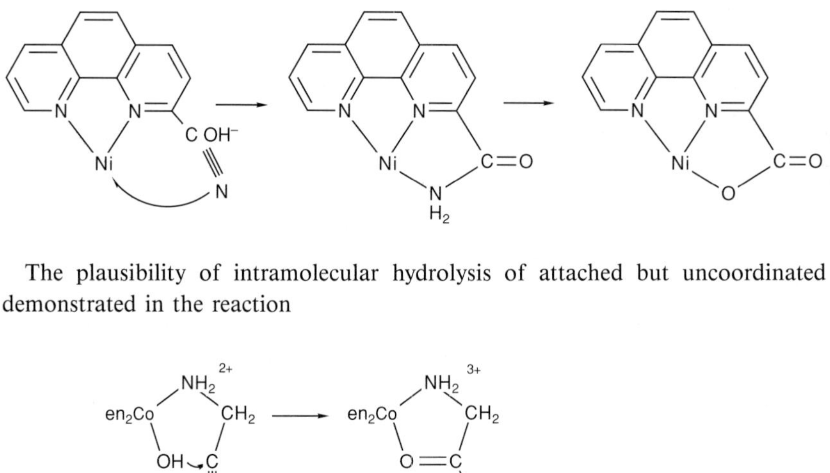

$$(6.20)$$

The plausibility of intramolecular hydrolysis of attached but uncoordinated nitrile is demonstrated in the reaction

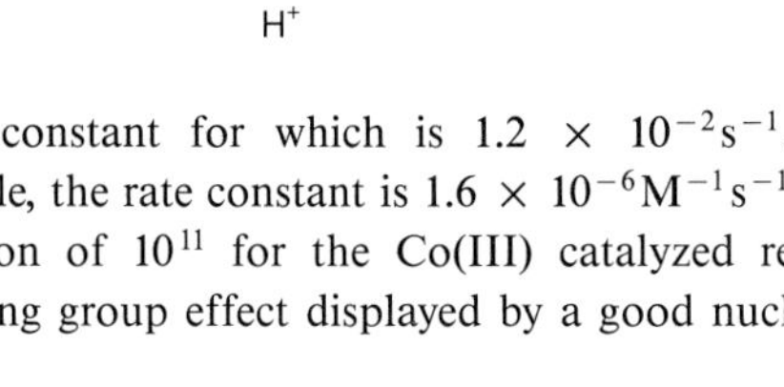

$$(6.21)$$

the rate constant for which is $1.2 \times 10^{-2} s^{-1}$. For intermolecular hydrolysis of free acetonitrile, the rate constant is $1.6 \times 10^{-6} M^{-1} s^{-1}$. Therefore at pH $= 7$, this represents an acceleration of 10^{11} for the Co(III) catalyzed reaction, undoubtedly as a result of the neighboring group effect displayed by a good nucleophile (coordinated OH^-).[45,46]

More attention has been paid to the behavior of nitriles which are coordinated to the metal e. g.,

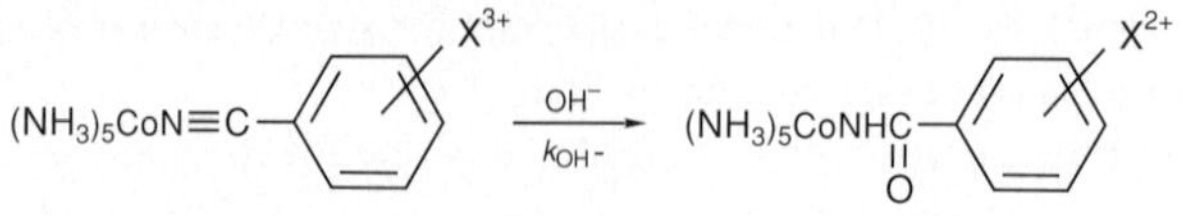

$$(NH_3)_5CoN{\equiv}C{-}\underset{}{\text{(ring)}}^{X^{3+}} \xrightarrow[k_{OH^-}]{OH^-} (NH_3)_5CoNHC{-}\underset{\overset{\|}{O}}{}\underset{}{\text{(ring)}}^{X^{2+}} \tag{6.22}$$

Rate accelerations of $\approx 10^6$, residing in more positive $\Delta S^{\ddagger}$, over the free ligand are observed. For 11 coordinated nitriles,

$$\log k_{OH^-} = 1.93\sigma + 1.30 \tag{6.23}$$

The slope ($\rho = 1.93$) is slightly smaller than that for uncoordinated nitriles ($\rho = 2.31$). Since ^{13}C nmr shows the effects of coordination on the nitrile carbon chemical shift are independent of substitutents, the observed difference in ρ must be an effect associated with the activated complex. [47]

Similar high rate enhancements on coordination are observed with the analogous $Rh(NH_3)_5^{3+}$- and $Ru(NH_3)_5^{3+}$- but not with $Ru(NH_3)_5^{2+}$-coordinated nitriles. This is an important finding. With ruthenium(II) complexes considerable metal-ligand π-bonding occurs. This results in back donation of electron density from the metal center to the $C{\equiv}N$ bond. The polarization by the metal of the nitrile, which is the basis for the enhanced effects with Co(III), Rh(III) and Ru(III), is therefore lost with Ru(II). [1]

A number of nucleophiles including CN^-, BH_4^-, amines and alcohols react with coordinated nitriles at $\geqslant 10^4$ enhanced rates compared with the free ligand. [1] One of the most novel examples involves N_3^- as a nucleophile. Acetonitrile and sodium azide form 5-methyltetrazole only after 25 hours at 150°C whereas the cobalt(III) coordinated acetonitrile reacts in 2 hours at room temperature:

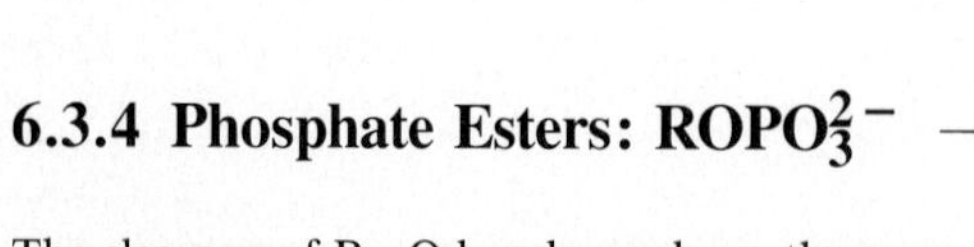

$$(NH_3)_5CoN{\equiv}CR^{3+} + N_3^- \longrightarrow (NH_3)_5Co{-}N{\underset{N{-}N}{\overset{\overset{R}{\underset{C}{|}}}{\langle\cdots\rangle}}}N^{2+} \tag{6.24}$$

The product undergoes linkage isomerism (Table 7.4). [48]

6.3.4 Phosphate Esters: $ROPO_3^{2-} \rightarrow PO_4^{3-}$

The cleavage of $P{-}O$ bonds produces the energy required for many biological processes, including synthesis of complex molecules, muscle functioning and ion transport. The catalytic modification of phophate chemistry by metalloenzymes is therefore very important. *E. coli* alkaline phosphatase is a Zn(II)-containing enzyme that catalyzes the hydrolysis of phosphate monoesters about 10^{11} times faster then in the absence of enzyme. [49] Using labile metal ions to probe these effects is rarely fruitful because of the difficulty in characterizing the species involved. Yet again Co(III) is the metal of choice for model investigations. A number of groups have examined the accelerated hydrolysis of phosphate monoesters by *cis*-$CoN_4(H_2O)_2^{3+}$ and the ionized forms.[50,51] Invariably $CoN_4(H_2O)OH^{2+}$ is the most reactive species, but the formation and dissociation of the phosphate complex and its hydrolytic breakdown may be mixed up kinetically and it is better to by-pass the formation altogether and study the isolated complex. ^{31}P nmr, ^{18}O isotopes and spectral monitoring have been used to interpret the attack by *cis* OH^- on the coordinated ester in (6.25). The nitrophenyl ester is a convenient one to employ because of the color of the 4-nitrophenolate product ($\varepsilon = 2.6 \times 10^3$ M^{-1}cm^{-1} at 400 nm and pH 6.3). There is believed to be a five-coordinated phosphorane intermediate which breaks down with a rate constant $k = 7.8 \times 10^{-4}$s^{-1} at 25 °C (pH 9–12). This is a 10^5 larger value than that for the uncoordinated ester.[52] The reaction (6.25) is accompanied by some *cis, trans* rearrangement and although this has little impact on the interpretation, the complication can be avoided by using as N_4, chelates which can only leave *cis*-positions (e. g. tren). There is a marked effect on the rate by the N_4 grouping used.[51,53]

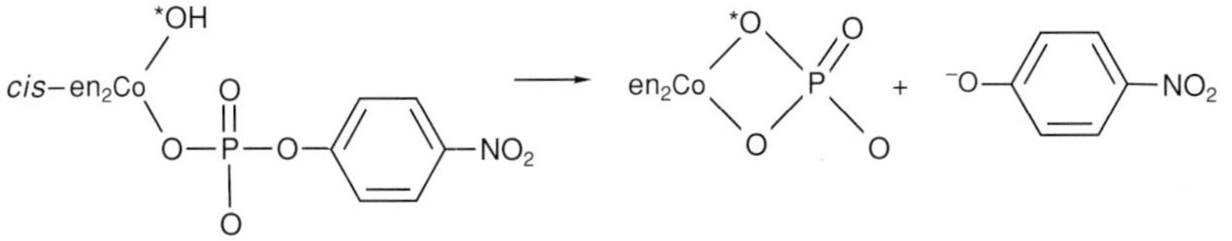

$$\text{(6.25)}$$

The phosphodiester in the complex $Co(en)_2(OH)(OP(O)(OC_6H_4NO_2)_2)^+$ is cleaved more slowly than the monoester (6.25) (this is usual), but the rate constant (2.7×10^{-3}s^{-1} at 50 °C) is still 10^7 times that for the unbound ester.[53]

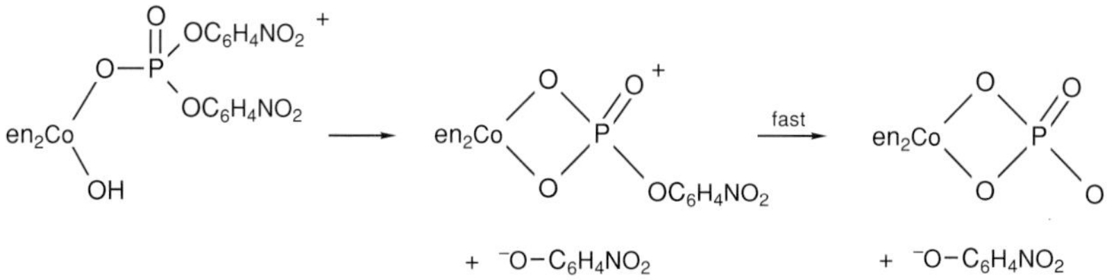

$$\text{(6.26)}$$

This accelerating factor also holds for the corresponding *cis*-Ir(III) diester complex, where stereochemical changes are also unimportant. Now however biphasic release of nitrophenolate is observed and the product of the first step is not the chelate but *cis*-$en_2Ir(OH)[OP(O)_2(OC_6H_4NO_2)]$. The rates are some 10^3 slower than for the Co(III) complexes. These important differences are ascribed to the larger size of Ir(III) which deters ring closure.[54] The strategy described in this section relies on the generation of a coordinated nucleophile *cis* to a reactive coordinated phosphate derivative. As well as OH^-, an NH_2^- group can be examined, generated from an NH_3 ligand as in the example (compare Sec. 1.6.2):

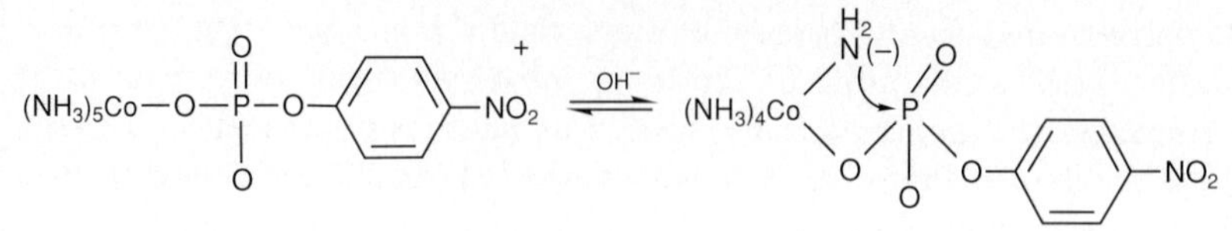

$(NH_3)_5Co-O-\overset{\overset{O}{\|}}{\underset{\underset{O}{|}}{P}}-O-\!\!\left\langle\!\!\bigcirc\!\!\right\rangle\!\!-NO_2 \rightleftharpoons (NH_3)_4Co$ ... $-NO_2$ (6.27)

$\longrightarrow (NH_3)_4Co$ ... $+ \ ^-O-\!\!\left\langle\!\!\bigcirc\!\!\right\rangle\!\!-NO_2$

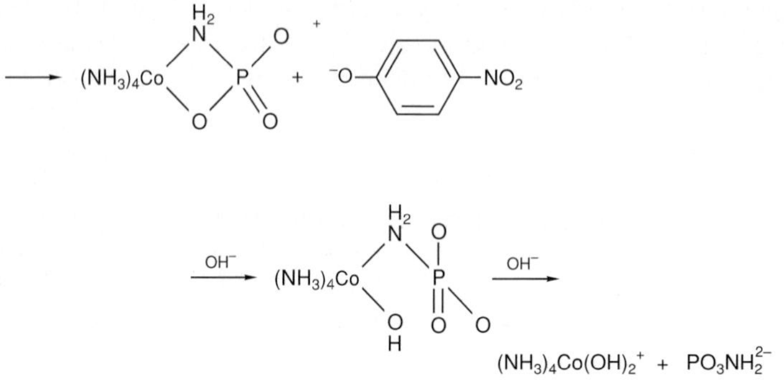

(6.28)

$\overset{OH^-}{\longrightarrow} (NH_3)_4Co$ ... $\overset{OH^-}{\longrightarrow}$

$(NH_3)_4Co(OH)_2^+ \ + \ PO_3NH_2^{2-}$

The first observed product is the hydroxo-N-bound phosphoramidate complex, although there are almost certainly other intermediates. Both ester hydrolysis (to nitrophenolate ion) and transfer of a phosphate residue from O to N occur. An acceleration of at least 10^8 fold can be assessed for both processes, compared with the reaction of the uncoordinated ester with NH_3 or OH^- ion.[1,55] The O to N transfer is a general biochemical occurrence, e.g. creatine kinase uses Mg^{2+} and creatine to transform ATP to ADP and form creatine phosphate.[1]

If a hydroxyester function is incorporated into a potential chelating system then the ability even of a labile metal ion to catalyze the hydrolysis can be assessed. The 8-hydroxyquinoline **17** and 2,2′-phenanthroline **18** frameworks have proved popular for this purpose.

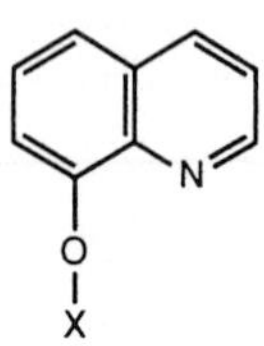

17 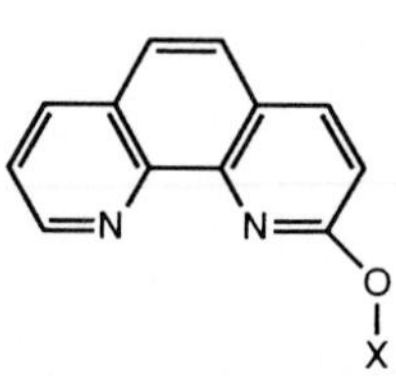   **18**

In **17**, X may be PO_3^{2-} [56] (but $COCH_3$,[57] SO_3 [58] and other groups have also been examined by this means). In the type of structure shown in **18** we have already encountered the 2-nitrile hydrolyses. With $X = PO_3^{2-}$ in **18**, divalent metal ions show a pronounced catalysis of the hydrolysis of the dianionic species. The metal is strongly chelated to the phenanthroline but in the product it is unlikely that the O^- is coordinated since a four-membered ring would result (see Sec. 6.8). The monoanionic form ($X = PO_3H^-$) is the reactive species (Prob. 3). Reaction of the dianion in the absence of metal ion cannot be observed and with Cu^{2+}, for example, accelerating effects of $>10^8$ are estimated.[59]

6.3.5 Other Groups

There have been a number of isolated studies of metal-ion catalyzed nucleophilic reactions of other groupings.[1,60] Particularly interesting is the induced nucleophilic attack on olefins. Hydration is normally very sluggish. Enzymes can speed up such reactions. Aconitase, an iron-containing enzyme, catalyzes the isomerization of citric acid to isocitric acid, through the intermediacy of *cis*-aconitic acid. A possible mechanism has been suggested based on the following Co(III) model chemistry.[61] Rapid cyclization of the maleate ester produces ΛR and ΛS chelated malate half ester:

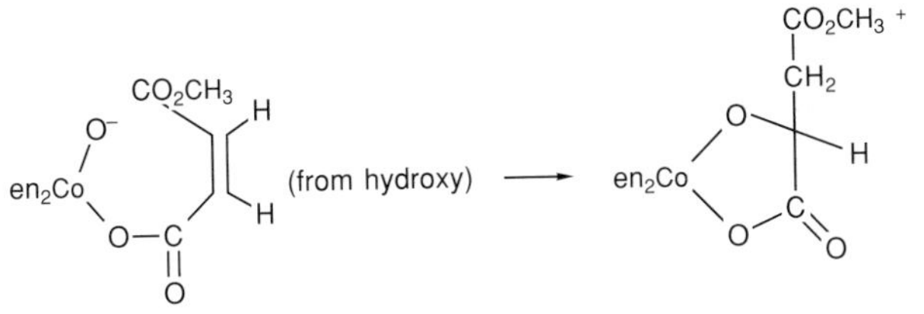

$$(6.29)$$

as well as the fumarato isomer of the reactant (which ring closes very much slower than the maleate isomer).

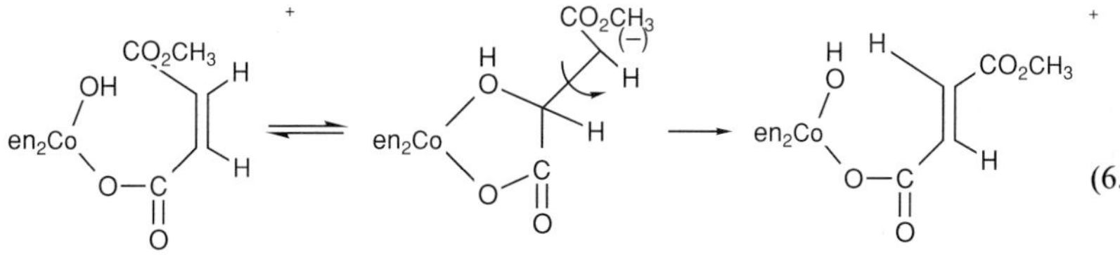

$$(6.30)$$

The process is 10^7 fold faster than hydration of uncoordinated half-ester. More details and a full kinetic scheme are provided in Ref. 61.

6.4 The Acidity of Coordinated Ligands

An important effect of the metal ion lies in its ability to enhance the acidity of certain coordinated ligands. Any ligand that in the free state can release a proton can also do so when coordinated to a metal. The positive charge originally associated with the central metal is dissipated over the whole complex, and the resultant neutralization of negative charge at the coordinated ligand center will result in the center's being more acidic. Some examples of this effect, with comparisons of the free and coordinated ligand, are shown in Table 6.5. A comprehensive list of cobalt(III) complexes is given in Ref. 1 and of many mononuclear and dinuclear metal complexes in Refs. 62–65.

Table 6.5 Values of pK_a for Some Free and Coordinated Ligands[a]

Ligand or Complex	pK_a^b
H_2O	15.8
$Cr(H_2O)_6^{3+}$	4.3
cis-$Cr(en)_2(H_2O)_2^{3+}$	4.8 (7.4)[c]
$Ir(H_2O)_6^{3+}$	4.4 (5.2)[c]
$Co(NH_3)_5H_2O^{3+}$	5.8
NH_3	>16
$Ru(NH_3)_6^{3+}$	13.1
$Au(en)_2^{3+}$	6.5
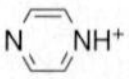 $N\bigcirc NH^+$ (pzH$^+$)	0.6
$(NH_3)_5Ru^{II}(pzH)^{3+}$	2.9
$(NH_3)_5Ru^{III}(pzH)^{4+}$	-0.8
$HC_2O_4^-$	4.2
$Co(NH_3)_5C_2O_4H^{2+}$	2.2
HCO_3^-	10.3
$Co(NH_3)_5OCO_2H^{2+}$	6.4
$(NH_2)_2CO$	≈ 14
$Co(NH_3)_5OC(NH_2)_2^{3+}$	13.2

[a] From Refs. 1, 62–65   [b] $pK_a = -\log [\text{base}] [H^+] [\text{acid}]^{-1}$ [c] pK_a for second ionization

Generally, the higher the positive charge, the greater the enhancement of the ionization constant. The increase in K_a^M compared with K_a^H, (6.32) vs (6.31),

$$HA \; \rightleftharpoons \; H^+ + A^- \qquad\qquad k_1^H, \, k_{-1}^H, \, K_a^H (= k_1^H/k_{-1}^H) \qquad (6.31)$$

$$M(HA)^{n+} \; \rightleftharpoons \; H^+ + M(A)^{(n-1)+} \qquad k_1^M, \, k_{-1}^M, \, k_a^M (= k_1^M/k_{-1}^M) \qquad (6.32)$$

which may amount to as much as 13 orders of magnitude, resides in different values for k_1^M compared with k_1^H since k_{-1} and k_{-1}^M values are approximately constant and diffusion-controlled for most proton-base reactions.[63,65,66] (See Prob. 8)

One of the very few exceptions to the rule that the acidity of the complexed ligand exceeds that of the free ligands involves the Ru(II) complexes shown in Table 6.5. It is believed that back bonding from the filled t_{2g} orbitals of Ru(II) to unoccupied π-antibonding orbitals of the ligands more than compensates for the usual electrostatic effects of the metal that makes the nitrogen less basic. This π-bonding is less likely with the Ru(III) complex and its pK_a is lower than that for the protonated pyrazine (see also Sec. 6.3.3. for the effects of Ru(II) and Ru(III) on hydrolysis of nitriles).[67,68]

6.4.1 Coordinated Water

The marked increase in the acidity of water when it becomes metal-coordinated, as shown in Table 6.5, has very important ramifications. Many metal complexes will be involved in an aqua-hydroxo equilibrium in the common pH region of 3 to 11. Since the hydroxo form often has a different reactivity than its acidic partner, there will be marked effects of pH on the rates of water exchange, substitution, redox reactions and isomeric change in which aqua complexes are involved. We have seen in the previous sections that certain metal complexes donate OH^- in conditions (neutral pH) where the concentration of free OH^- is small. It has been shown that the $M-OH$ species is a good nucleophile with a number of substrates which are either very electrophilic such as CO_2 or SO_2 or when a very good leaving group is involved as with a variety of activated carbonyl substrates. [1,31,69] The Brønsted plots of $\log k_{MOH}$ vs the pK_a of $M-OH_2$ are linear and the β coefficient (Sec. 2.5.3) is 0.2–0.4. An example is shown for the promoted hydrolysis of 2,4-dinitrophenyl acetate, dnpa, e.g.

$$Co(NH_3)_5OH^{2+} + MeCO{-}\overset{\parallel}{\underset{O}{}}\!\!\!\bigcirc\!\!-NO_2 \longrightarrow Co(NH_3)_5OCMe^{2+} + HO{-}\bigcirc\!\!-NO_2 \tag{6.33}$$

The first-order loss of dnpa (k_{obs}) is given by

$$k_{obs} = k_{H_2O} + k_{OH}[OH^-] + k_{CoOH}[CoOH] \tag{6.34}$$

with k_{H_2O} and k_{OH} independently measured ($6.2 \times 10^{-6} s^{-1}$ and $37 M^{-1}s^{-1}$ respectively). For a number of labile and inert metal hydroxy species there is a good linear correlation ($\beta = 0.33$) extending over a range of 10^{10} in nucleophilic basicity (Fig. 6.5). [69] Although the second-order rate constant for hydrolysis of dnpa by $Zn(nta)OH^{2-}$ ($0.3 M^{-1}s^{-1}$) is about 10^2 less than for that by OH^- ($37 M^{-1}s^{-1}$), it is obvious that at pH ≈ 7, even mM concentrations of the Zn(II) complex will compete favorably with the OH^--promoted hydrolysis. Since the Brønsted slope is shallow, the same situation will apply for poorer nucleophiles also e.g. $Cr(NH_3)_5OH^{2+}$ ($k = 9 \times 10^{-3}M^{-1}s^{-1}$). The metal ion thus provides an efficient nucleophile at neutral pH. This is of paramount importance for biological reactions. This effect is considerably enhanced when the coordinated nucleophile is forced into a juxtaposition relative to the substrate (see previous sections). $M-OH$ reactivity is more sensitive to the substrate character than is OH^- reactivity. This is demonstrated by a plot of $\log k_{MOH}$ vs $\log k_{OH}$ for a number of substrates being beautifully linear with a slope 2.0. [69]

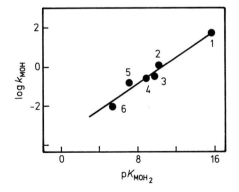

Fig. 6.5 Brønsted plot for the MOH (including OH^-) promoted hydrolysis of dnpa at 25°C and $\mu = 1.0$ M $NaClO_4$. The points are for OH^- (1), $Co(CN)_5OH^{3-}$ (2), $Zn(nta)OH^{2-}$ (3), $Cu(nta)OH^{2-}$ (4), $Co(NH_3)_4(NO_2)OH^+$ (5), and $Cr(NH_3)_5OH^{2+}$ (6). [69]

6.4.2 Coordinated Ammonia and Amines

The enhanced ionization effect with the coordinated-NH_2 entity is sometimes much less marked than with coordinated $-OH_2$. It plays an important role in the conjugate base mechanism for hydrolysis (Sec. 4.3.4(a)). Aqueous alkaline solutions of $Ru(NH_3)_6^{3+}$ are deep yellow due to $Ru(NH_3)_5NH_2^{2+}$. The deprotonated species is implicated in a number of reactions, base-catalyzed proton exchanges, and with NO to give $Ru(NH_3)_5N_2^{2+}$, for example. [70] When the coordinated N in amines is asymmetric the relation between proton exchange and inversion can be explored (Sec. 7.9). The $-NH$ ionization of the ligand in $Ru(sar)^{3+}$ **19** leads to interesting autoxidation behavior. [71] The acidity is enormously enhanced ($pK_a = 6.3$) compared even with $Ru(NH_3)_6^{3+}$. The rate constants for ionization of coordinated $-NH$ groups have been determined by nmr exchange methods. If the acidic group is part of a ligand but not involved directly in coordination its acidity may be modified less than in some of the examples considered above. This has been demonstrated with a number of polyamine complexes in which because of the geometry of the ligand or because of a limited number of available coordination sites on the metal, there are uncoordinated NH_2 groups. [72] Generally, the basicity is decreased by some 2–3 pK units. Thus the pK_a of $NH_2(CH_2)_2NH_3^+$ is 10.5 and that of $Pt(en)(NH_2(CH_2)_2NH_3)Cl^{2+}$ is 8.0.

19

6.4.3 Other Coordinated Ligands

Table 6.5 indicates that for still different types of ligands than H_2O and NH_3 the pK_a is modified by attachment to the metal. This will play a role in the reactivity of the resulting complex. We observe therefore that the reduction of $Co(NH_3)_5C_2O_4H^{2+}$ in acid by Fe^{2+} and V^{2+} has an inverse $[H^+]$ term in the rate law, whereas the same reductions of $Co(NH_3)_4C_2O_4^+$, which is aprotic, are pH-independent. A number of metal ions have the ability to promote proton ionization from a coordinated amide or peptide group. The observation of relatively slow rates associated with the protonation supports the idea of bond rearrangement from $M-O$ to $M-N$ accompanying the ionization, (6.19). [73] In the structural grouping $X=Y-ZH$, H tends to be acidic and once again this is enhanced by involvement of any of the centers, X, Y or Z with a metal. The phenomenon shows up well in the metal complexes of amino acids and amino polycarboxylates. The increased chemical reactivity at the α-carbon atom is manifested in the following three ways: [74]

(a) base catalyzed aldol-type condensations[75,76]

Glycine complexes undergo aldol reactions with aldehydes yielding free β-hydroxy-α-amino acids after cleavage of the product. Although the NH_2 of amino acids is protected by metal coordination (Sec. 6.6) nevertheless it does react with aldehydes to form N-hydroxymethyl derivatives. This, together with the enhanced acidity in the neighboring CH_2 group, leads to formation of an oxazolidine complex

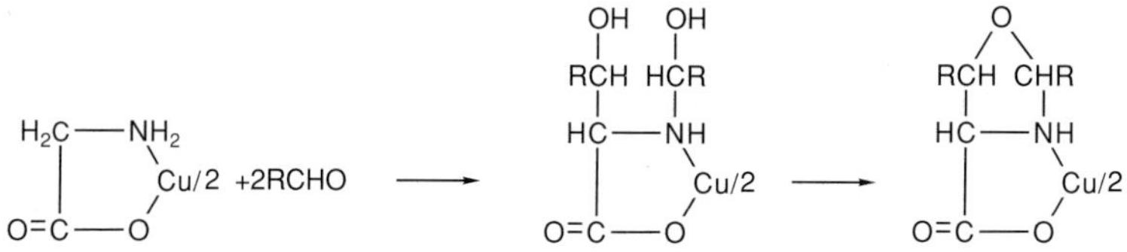

$$(6.35)$$

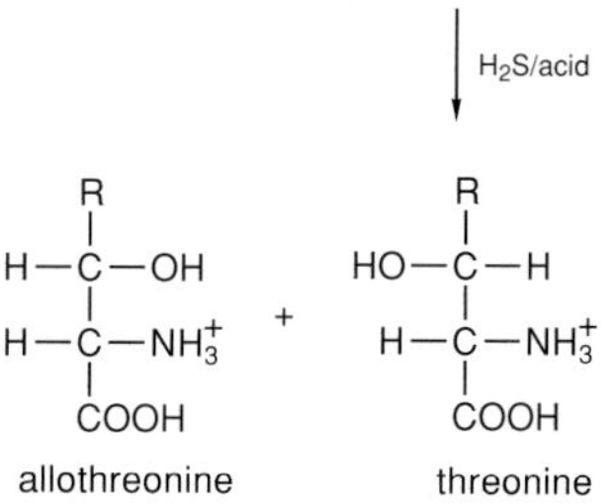

$$(6.36)$$

allothreonine   threonine

Decomposition of the copper complex with H_2S in acid leads to allothreonine and (in about 2-fold excess) threonine. Following the pioneering work of Akabori *et al.*,[75] this reaction has been extended to a variety of metals and aldehydes and to complexes such as Δ- and Λ-Co(en)$_2$gly^{2+} with intriguing stereoselectivity behavior (optically-active allothreonine and threonine products).[76]

(b) enhanced C—H exchange with basic D_2O.[77-79]

In Co(edta)$^-$, **20**, for example, out-of-plane hydrogens exchange more readily than those in the in-plane rings.[77] The latter in fact take place only via the interchange of in-plane and out-of-plane rings.[78] Since this is a mechanism for racemization, the exchange and racemization rates are similar. The distinctive nmr for the different CH_2 hydrogens in Co(edta)$^-$ is invaluable for diagnosis and monitoring purposes.

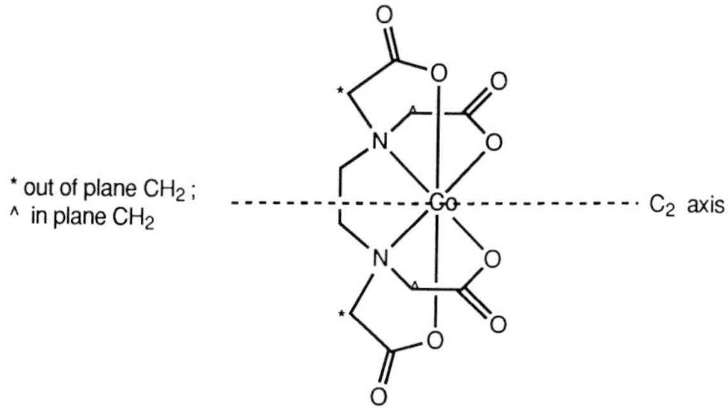

* out of plane CH_2;
^ in plane CH_2 ------- Co ------- C_2 axis

20

(c) enhanced racemization rates of chelated optically-active amino acids. [74, 80]

The racemization rate constant for $L-$Ala coordinated to Co(III) is much larger than for free $L-$Ala. In addition, the larger the positive charge on the complex, the faster is the rate of racemization. These observations give credance to the idea that racemization (and exchange) takes place by an initial abstraction of an α-H by base, resulting in a carbanion. [77]

6.5 Electrophilic Substitution in Metal Complexes

When a metal atom donates electron density to a bound ligand, usually by means of π-back bonding, electrophilic substitution reactions may be promoted. This is observed then usually with metals in low oxidation states and is therefore prevalent with organometallic complexes [2, 81] and less with those of the Werner-type, where the metals are usually in higher oxidation states. Nevertheless there have been detailed studies of electrophilic substitution in metal complexes of β-diketones, [82] 8-hydroxyquinolines [83] and porphyrins. [84] Usually the detailed course of the reaction is unaffected. It is often slower in the metal complexes than in the free ligand but more rapid than in the protonated form.

The apparently first kinetic study of a metal-assisted electrophilic substitution in a Co(III) complex is recent. [85] The bromination of $Co(NH_3)_5imidH^{3+}$ is complicated by the presence of different bromine species in solution (Br_2, HOBr and Br_3^-). In addition, successive brominations of the coordinated imidazole occur. Rate data can be interpreted in terms of reaction of the conjugate base of the Co(III) complex with Br_2, and a suggested mechanism for the first steps is ($R_0 = Co(NH_3)_5^{3+}$)

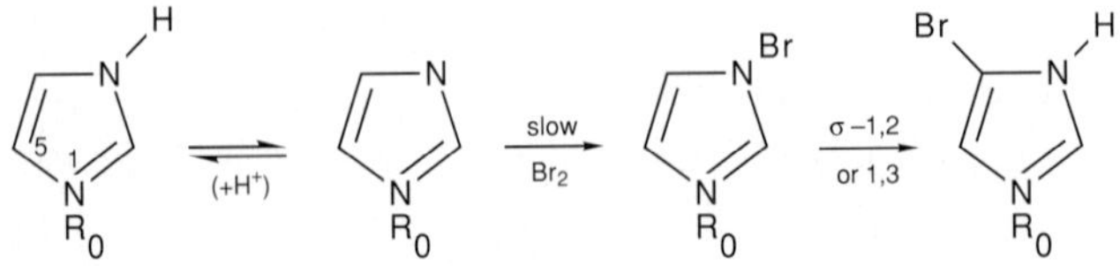

$$(6.37)$$

The first observed product is however $Co(NH_3)_5(4,5Br_2imidH)^{3+}$. The rate constant $(6 \times 10^9\ M^{-1}s^{-1})$ is $> 10^3$ fold larger than for reaction of $imidH^+$ ($1.1 \times 10^6\ M^{-1}s^{-1}$). The initial site of attack is uncertain in both instances. [85]

6.6 Masking Effects

The masking of the normal reactions of simple ligands, such as the nitro, cyano, and ammonia groups, by coordination to a metal is a phenomenon encountered early by a chemist. One of the first examples of masking in a chelate complex was reported, significantly, in biological journals. [86] It involves the protection by copper ion of the α-amino group in ornithine and lysine:

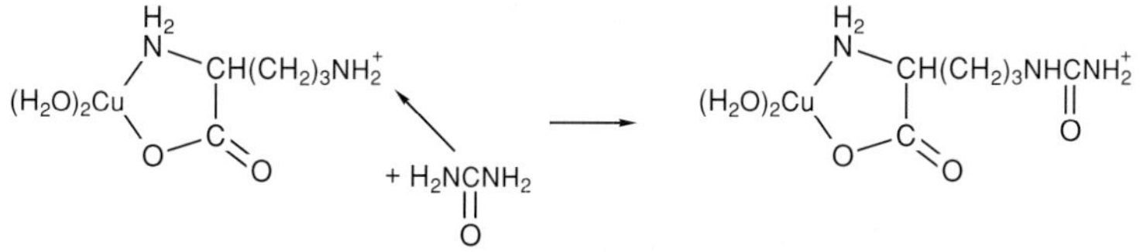

$$(6.38)$$

Partial protection of the $-NH_2$ group by copper ion results in a fiftyfold decrease in the rate of the ring-closure reaction:

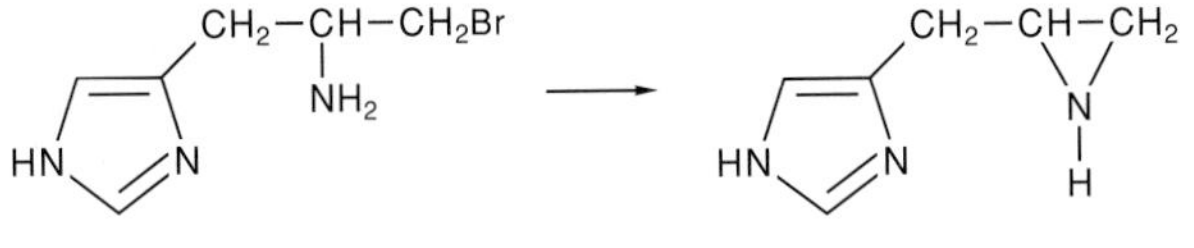

$$(6.39)$$

in the presence of the metal ion compared with the free ligand. However, Cu(II) binds strongly to the product and if significant amounts of the free metal ion are removed in this way, the initial reaction rate increases sharply.[87] An opposite effect is noted in the metal-ion-catalyzed hydrolysis of amino acid esters. Removal of the metal ion as it complexes with the amino acid product causes *deceleration* of the reaction.[88]

6.7 Disturbance of Reaction Stoichiometry

The intervention of a metal ion in the stoichiometry of a reaction has been illustrated several times previously. Reaction is forced to completion in ester hydrolysis since the carboxylate grouping forms a more stable complex than the ester moiety does. A similar driving force underlies the formation of macrocycles and the completion of transamination by formation of the metal-Schiff base complex.[89] The latter is particularly relevant in dilute solution and at low pH. For example, the extent of aldimine formation between pyridoxal and alanine is undetectable at the physiological pH but occurs to the extent of $\approx 10\%$ in the presence of zinc ions.[89]

6.8 Molecular Strain Alterations

Metal complexing can subject a ligand to severe internal strain or, alternatively, it can relieve strain; or it can freeze the conformation of the coordinated ligand. These modifications often lead to enhanced reactivity by reducing the energy difference between ground and transition state. The subjection of the cyano group to strain when 2-cyanophenanthroline is coordinated to a metal ion, and its resultant accelerated hydrolysis, has been referred to previously (Sec.

6.3.3). The rates of a number of different types of reactions of 2,2'-bipyridyl derivatives are markedly enhanced by metal chelation.[90] The elimination reaction (6.40) in wet dmso

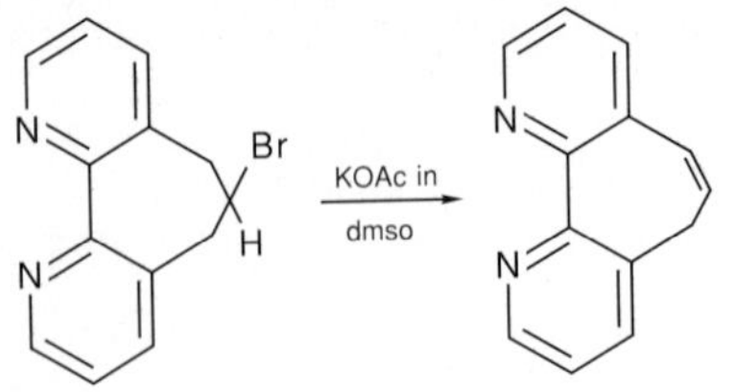

$$\text{(6.40)}$$

is very slow with a second-order rate constant (reaction with KOAc) of $3.4 \times 10^{-2}\,M^{-1}s^{-1}$ at 80°C. Smooth elimination occurs with the Pd(OAc)$_2$ complex of the bipyridyl derivative ($>3.0\,M^{-1}s^{-1}$ at 0°C). A favored explanation for the enhancement involves distortion of the ligand by coordination with Pd to produce a conformation which seems elimination prone (**21**), M = Pd. The rates of racemization of asymmetric bipyridyl crown ethers **22** are also markedly enhanced ($\approx 10^{8}$ fold) when complexed with Pd. There again the metal may force in-plane distortions of the bipyridyl nucleus in such a way that substituent bending (required for racemization) is much reduced with a concommitant lower $\Delta G^{\ddagger}$.[90]

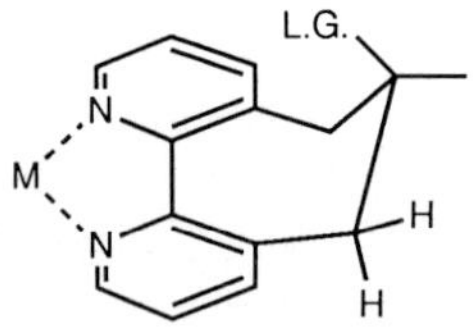

21 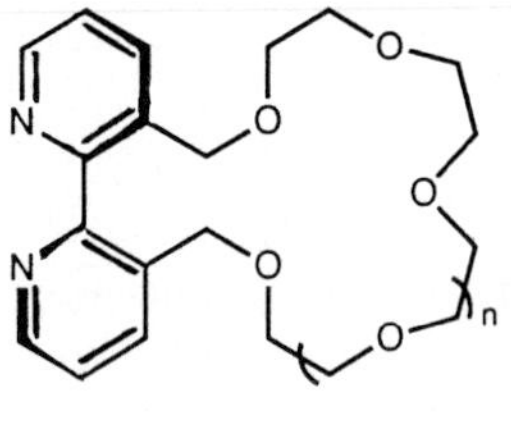   22

   A number of complexes, mainly of Ag(I) but also Rh(I) and Pd(II), can catalyze the rearrangement and degradation of a wide variety of highly strained polycyclics, which may not react in the absence of catalysts. The products vary with the catalyst, the concentration of which features in the rate law. Again, the relief of ring strain which is a driving force for the reaction, appears aided by metal complexing.[91-93]

6.9 Function of the Ligand

The principal upset of the properties of the ligand is due to the metal. However effects of one ligand on another are not unimportant. A ligand may have a strong effect on the reaction at a *trans* position in substitution (Sec. 4.10.1) and redox (Sec. 5.6.(d)) reactions. Ligands may modify reactivity patterns, such as in the smoother catalyzed decarboxylation of dimethyloxaloacetic acid by aromatic compared with aliphatic amine complexes.[94] Complex ions containing ligands additional to water have a decided advantage over the metal ion alone for the metal complex can function at a pH at which metal ion might precipitate. It has been pointed out that if the "catalase" ability of ferric ion in acid were extrapolated to pH $\approx$ 10 (where obviously it would precipitate) it would be almost as effective a catalyst as the enzyme itself and much superior to other iron(III) complexes.[95]

6.10 Conclusions

Enzymes are much more effective as catalysts than any man-devised model. The 10^5–10^7 enhanced factor observed in the hydrolysis of esters, amides etc. in specially constructed Co(III) complexes is beginning to approach that sometimes observed with enzymes (10^{11}). The neighboring group and nucleophilic effects which act in concert with Co(III) are being augmented by other factors with enzymes, such as strain induced in the substrate by the enzyme in which the protein structure itself plays a key role. Thus, the principles being developed by using models (particularly, it seems, of Co(III)) are clearly applicable to metalloenzyme function.[96] Probably the most important single effect that metal centers have in metalloenzymes is to be a focal point for reactants, often stereospecifically, although it appears that only one substrate is ever coordinated at a time. The metal ion acts as a Lewis acid but can exist in much higher concentrations than the proton at the physiologically important neutral pH.

Many of the properties of metal ions in aiding or discouraging reactions are beautifully illustrated in the work on the chemical synthesis of corrins. Thus, it is appropriate to conclude this chapter by a relevant quotation from Eschenmoser:[14]

The role of transition metals in the chemical synthesis of corrins is more than just adding a touch of "inorganic elegance" to organic synthesis-"leave elegance to tailors and cobblers", a physicist once said — it is a vital role in the sense that perhaps no synthetic corrin would as yet exist without recourse to metal templates. Beside the purely topological function of *arranging the proximity* of reaction centres, metal ions have served this purpose as follows:

(i) by *stabilizing* labile organic intermediates and thereby facilitating their isolation and characterization,

(ii) by *activating* organic ligands electronically for base-catalyzed processes,

(iii) by subjecting organic ligands to heavy *steric strain* so that they perform strain-releasing reactions which they would otherwise certainly never undergo,

(iv) by *protecting* organic coordination sites against the detrimental attack of aggressive alkylation reagents,

(v) and, last but not least-by converting the organic chemists involved in this work to genuine admirers of the depth potentials and wonders of *transition metal chemistry.*

References

1. N. E. Dixon and A. M. Sargeson, in Zinc Enzymes, T. Spiro, ed. Wiley, NY, 1983, Chap. 7.
2. C. J. Hipp and D. H. Busch in B32, Chap. 3; D. H. Busch, Acc. Chem. Res. **11**, 392 (1978).
3. D. S. C. Black, Reactions of Coordinated Ligands in B35, Vol. 1, Chap. 7.4.
4. R. W. Hay, Lewis Acid Catalysis and the Reactions of Coordinated Ligands in B35, Vol. 6, Chap. 61.4.
5. W. P. Jencks, Catalysis in Chemistry and Enzymology, Mc.Graw-Hill, NY, 1969, p. 8.
6. G. Albertin, E. Bordignon, A. A. Orio, B. Pavoni and H. B. Gray, Inorg. Chem. **18**, 1451 (1979).

7. M. B. Stevenson, R. D. Mast, and A. G. Sykes, J. Chem. Soc. A. 937 (1969).

8. J. Springborg, Adv. Inorg. Chem. 32, 56 (1988); L. Mønsted and O. Mønsted, Coordn. Chem. Revs. **94**, 109 (1989).

9. P. Hendry and A. M. Sargeson, Inorg. Chem. **25**, 865 (1986). The $Co-NH_2$ entity is generated in a rapid preequilibrium involving OH^- attack on the $Co-NH_3$ bond.

10. H. Ogino, J. Coordn. Chem. **15**, 187 (1987); R. Bhula, P. Osvath and D. C. Weatherburn, Coordn. Chem. Revs. **91**, 89 (1988).

11. G. A. Melson, ed. Coordination Chemistry of Macrocyclic Compounds, Plenum, NY 1979 contains several chapters in which the template effect is utilised; J. F. Lindoy, The Chemistry of Macrocyclic Ligand Complexes, Cambridge 1989.

12. G. Ercolani, L. Mandolini and B. Masci, J. Amer. Chem. Soc. 105, 6146 (1983).

13. C. O. D.-Buchecker, J.-P. Sauvage and J.-M. Kern, J. Amer. Chem. Soc. **106**, 3043 (1984); J. C. Chambron, C. D.-Buchecker, C. Hemmert, A. K. Khemiss, D. Mitchell, J. P. Sauvage and J. Weiss, Pure Appl. Chem. **62**, 1027 (1990). J. P. Sauvage, Acc. Chem. Res. **23**, 319 (1990).

14. A. Eschenmoser, Pure Appl. Chem. **20**, 93 (1969); Chem. Soc. Revs. **5**, 377 (1976); R. B. Woodward, Pure Appl. Chem., **33**, 145 (1973).

15. W.-S. Hwang, I. B. Joedicke and J. T. Yoke, Inorg. Chem. **19**, 3225 (1980).

16. A. Sen and J. Halpern, J. Amer. Chem. Soc. **99**, 8337 (1977).

17. G. Read and M. Urgelles, J. Chem. Soc. Dalton Trans. 1383 (1986).

18. H. Sigel, Angew. Chem. Int. Ed. Engl. **8**, 167 (1969).

19. J. E. Frew and P. Jones, Adv. Inorg. Bioinorg. Mechs. **3**, 175 (1984).

20. Oxygen binds to iron in cytochrome P-450 and this allows the introduction of O into a $C-H$ bond in substrate near the iron site, J. T. Groves, J. Chem. Educ. **62**, 928 (1985).

21. Histidine-64 plays an important role in accepting H^+ from the $Zn-H_2O$ moiety in carbonic anhydrase which catalyses the CO_2-H_2O reaction, D. N. Silverman and S. Lindskog, Acc. Chem. Res. **21**, 30 (1988).

22. R. Breslow and M. Schmir, J. Amer. Chem. Soc. **93**, 4960 (1971).

23. T. C. Bruice and A. Turner, J. Amer. Chem. Soc. **92**, 3422 (1970).

24. G. J. Lloyd and B. S. Cooperman, J. Amer. Chem. Soc. **93**, 4883 (1971). See also D. S. Sigman, G. N. Wahl and D. J. Creighton, Biochemistry, **11**, 2236 (1972), who report zinc-ion-catalyzed phosphorylation of 1,10-phenanthroline-2-carbinol by ATP.

25. D. Hopgood and D. L. Leussing, J. Amer. Chem. Soc. **91**, 3740 (1969).

26. A. E. Martell, Acc. Chem. Res. **22**, 115 (1989).

27. H. Kroll, J. Amer. Chem. Soc. **74**, 2036 (1952).

28. R. W. Hay and P. J. Morris, J. Chem. Soc. Chem. Communs. 732 (1968); R. W. Hay and L. J. Porter, J. Chem. Soc. A, 127 (1969).

29. H. L. Conley, Jr. and R. B. Martin, J. Phys. Chem. **70**, 2914 (1965).

30. D. Tschudin, A. Riesen and T. A. Kaden, Helv. Chim. Acta **72**, 131 (1989).

31. P. A. Sutton and D. A. Buckingham, Acc. Chem. Res. **20**, 357 (1987).

32. J. Chin and M. Banaszczyk, J. Amer. Chem. Soc. **111**, 2724 (1989).

33. M. D. Alexander and D. H. Busch, J. Amer. Chem. Soc. **88**, 1130 (1966).

34. D. A. Buckingham, D. M. Foster and A. M. Sargeson, J. Amer. Chem. Soc. **90**, 6032 (1968); **91**, 4102 (1969); **92**, 5701 (1970).

35. E. Baraniak, D. A. Buckingham, C. R. Clark, B. H. Moynihan and A. M. Sargeson, Inorg. Chem. **25**, 3466 (1986).

36. An alternative, kinetically indistinguishable, scheme replaces k_3 in (6.18) by a rapid preequilibrium followed by a slow step (k_3) instead of a fast step (k_4).

37. H. Sigel and R. B. Martin, Chem. Revs. **82**, 385 (1982).

38. C. J. Boreham, D. A. Buckingham and F. R. Keene, J. Amer. Chem. Soc. **101**, 1409 (1979); Inorg. Chem. **18**, 28 (1979).

39. D. A. Buckingham, F. R. Keene and A. M. Sargeson, J. Amer. Chem. Soc. **96**, 4981 (1974).

40. L. M. Sayre, J. Amer. Chem. Soc. **108**, 1632 (1986).

41. J. T. Groves and J. R. Olson, Inorg. Chem. **24**, 2715 (1985) and previous references. Metal coordination forces the boat-chair conformation in the bicyclic azalactam portion of the tridentate ligand, alleviating severe steric interactions present in the chair-chair conformation of the free ligand.

42. B. N. Storhoff and H. C. Lewis, Jr. Coordn. Chem. Revs. **23**, 1 (1977).

43. N. G. Granik, Russ. Chem Revs. (Engl. Trans.) **52**, 377 (1983).

44. R. Breslow, R. Fairweather and J. Keana, J. Amer. Chem. Soc. **89**, 2135 (1967).

45. D. A. Buckingham, P. Morris, A. M. Sargeson and A. Zanella, Inorg. Chem. **16**, 1910 (1977).

46. N. J. Curtis, K. S. Hagen and A. M. Sargeson, J. Chem. Soc. Chem. Communs. 1571 (1984).

47. R. L. De La Vega, W. R. Ellis, Jr. and W. L. Purcell, Inorg. Chim. Acta **68**, 97 (1983).

48. W. R. Ellis, Jr. and W. L. Purcell, Inorg. Chem. **21**, 834 (1982).

49. J. E. Coleman and P. Gettins in Zinc Enzymes, T. Spiro, ed. Wiley, NY, 1983, p. 153.

50. G. P. Haight, Coordn. Chem. Revs. **79**, 293 (1987).

51. J. Chin, M. Banaszczyk, V. Jubian and X. Zou, J. Amer. Chem. Soc. **111**, 186 (1989) and references therein.

52. D. R. Jones, L. F. Lindoy and A. M. Sargeson, J. Amer. Chem. Soc. **105**, 7327 (1983).

53. J. Chin and X. Zou, J. Amer. Chem. Soc. **110**, 223 (1988).

54. P. Hendry and A. M. Sargeson, J. Amer. Chem. Soc. **111**, 2521 (1989).

55. J. MacB. Harrowfield, D. R. Jones, L. F. Lindoy and A. M. Sargeson, J. Amer. Chem. Soc. **102**, 7733 (1980).

56. Y. Murakami and J. Sunamoto, Bull Chem. Soc. Japan **44**, 1827 (1971).

57. R. W. Hay and C. R. Clark, J. Chem. Soc. Dalton Trans. 1993 (1977).

58. R. W. Hay and J. A. G. Edmonds, J. Chem. Soc. Chem. Communs. 969 (1967).

59. T. H. Fife and M. P. Pujari, J. Amer. Chem. Soc. **110**, 7790 (1988).

60. B 41.

61. L. R. Gahan, J. M. Harrowfield, A. J. Hearlt, L. F. Lindoy, P. O. Whimp and A. M. Sargeson, J. Amer. Chem. Soc. **107**, 6231 (1985).

62. D. A. House, Coordn. Chem. Revs. **23**, 223 (1977).

63. R. van Eldik, Adv. Inorg. Bioinorg. Mechs. **3**, 275 (1984).

64. J. Springborg, Adv. Inorg. Chem. **32**, 55 (1988) pps. 111–113.

65. B 27, B 28.

66. M. Eigen, W. Kruse, G. Maass, and L. DeMaeyer, Prog. React. Kinetics **2**, 287 (1964).

67. P. Ford, De F. P. Rudd, R. Gaunder and H. Taube, J. Amer. Chem. Soc. **90**, 1187 (1968).

68. C. R. Johnson and R. E. Shepherd, Inorg. Chem. **22**, 2439 (1983); H. E. Toma and E. Stadler, Inorg. Chem. **24**, 3085 (1985).

69. D. A. Buckingham and C. R. Clark, Aust. J. Chem. **35**, 431 (1982).

70. D. Waysbort and G. Navon, Inorg. Chem. **18**, 9 (1979).

71. P. Bernhard, A. M. Sargeson and F. C. Anson, Inorg. Chem. **27**, 2754 (1988).

72. E. G.-España, M. Micheloni, P. Paoletti and A. Bianchi, Inorg. Chem. **25**, 1435 (1986).

73. D. W. Margerum in Mechanistic Aspects of Inorganic Chemistry, D. B. Rorabacher and J. F. Endicott, eds. ACS Symposium Series, 198, 1982, Chapter 1; C. J. Hawkins and M. T. Kelso, Inorg. Chem. **21**, 3681 (1982).

74. G. S. Reddy and G. G. Smith, Inorg. Chim. Acta **96**, 189 (1985) and references therein.

75. S. Akabori, K. Okawa and M. Sato, Bull. Chem. Soc. Japan **29**, 608 (1956).

76. M. G. Weller and W. Beck, Inorg. Chim. Acta **57**, 107 (1982) and references therein.

77. D.H. Williams and D. H. Busch, J. Amer. Chem. Soc. **87**, 4644 (1965).

78. P. R. Norman and D. A. Phipps, Inorg. Chim. Acta **24**, L 35 (1977) and references therein.

79. A. Miyanaga, U. Sakaguchi, Y. Morimoto, Y. Kushi and H. Yoneda, Inorg. Chem. **21**, 1387 (1982).

80. G. G. Smith, A. Khatib and G. S. Reddy, J. Amer. Chem. Soc. **105**, 293 (1983).

81. C. Elschenbroich and A. Salzer, Organometallics, VCH, Weinheim, 1989.
82. K. C. Joshi and V. N. Pathak, Coordn. Chem. Revs. **22**, 37 (1977).
83. R. J. Kline and J. G. Wardeska, Inorg. Chem. **8**, 2153 (1969).
84. J.-H. Fuhrhop, Irreversible Reactions at the Porphyrin Periphery, in Porphyrins and Metallopor-phyrins, K. M. Smith, ed. Elsevier, Amsterdam, 1975, Chap. 15.
85. a. G. Blackman, D. A. Buckingham, C. R. Clark and S. Kulkarni, Aust. J. Chem. **39**, 1465 (1986).
86. A. C. Kurtz, J. Biol. Chem. **122**, 477 (1937); A. Neuberger and F. Sanger, Biochem. J. 37, 515 (1943); 38, 125 (1944).
87. D. A. Usher, J. Amer. Chem. Soc. **90**, 367 (1968).
88. M. L. Bender and B. W. Turnquest, J. Amer. Chem. Soc. **79**, 1889 (1957).
89. D. A. Gansow and R. H. Holm, J. Amer. Chem. Soc. **91**, 573 (1969).
90. J. Rebek, Jr., T. Costello and R. Wattley, J. Amer. Chem. Soc. **107**, 7487 (1985).
91. P. G. Gassman and T. J. Atkins, J. Amer. Chem. Soc. **93**, 1042, 4597 (1971).
92. L. A. Paquette, Acc. Chem. Res. **4**, 280 (1971).
93. Ref. 2 p. 320; G. W. Griffin and A. P. Marchand, Chem. Revs. **89**, 997 (1989); A. P. Marchand, Chem. Revs. **89**, 1011 (1989).
94. J. V. Rund and K. G. Claus, Inorg. Chem. **7**, 860 (1968).
95. S. B. Brown, P. Jones and A. Suggett in Inorganic Reaction Mechanisms, Part 1, J. O. Edwards, ed Interscience, NY, 1970, p. 170.
96. A. S. Mildvan, in The Enzymes, Volume 2, 3rd Edit., P. D. Boyer, ed. Academic, NY, 1970. Examples are cited where coordination compounds and enzymes show common effects.

Selected Bibliography

B41. P. S. Braterman, ed. Reactions of Coordinated Ligands Vol. 1, Plenum 1986. Comprehensive but mainly descriptive; Vol. 2, Plenum, 1989 covers coordinated CO_2, N_2, nitrosyls, O- and N-bound, and S- and P-bound ligands.
B42. H. Dugas and C. Penney, Bioorganic Chemistry, Springer-Verlag 1981. Chap. 6.

Problems

1. Suggest the origin of the rate law for (internal) amidolysis of the coordinated ester in $Co(NH_3)_5NH_2CH_2CO_2R^{3+}$ that contains a second-order dependence on $[OH^-]$.

2. The hydrolysis of **I**

(I)

to the corresponding amide is first order in complex and has a rate/pH dependence with a plateu at high pH corresponding to:

$$\text{Rate} = \frac{a[\mathbf{I}][OH^-]}{1 + b[OH^-]}$$

Suggest a mechanism. What would you anticipate would be the effect of SCN⁻ on the hydrolysis rate? The hydrolysis of the corresponding dinitrile (substituted on adjacent N's) gives the monoamide after a few seconds. No bisamide is formed after a longer time. Why?

T. A. Kaden, Pure Appl. Chem. **60**, 1117 (1988).

3. The plot is shown of k_{obsd} vs pH for hydrolysis of **II**

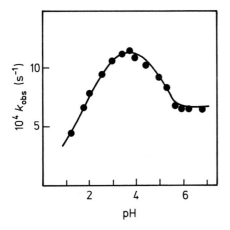

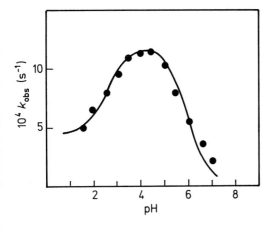

Problem 3. Reproduced with permission from T. H. Fife and M. P. Pujari, J. Amer. Chem. Soc. **110**, 7790 (1988) © (1988) American Chemical Society.

in the presence of 2×10^{-3} M Cu^{2+} at 30°C, $\mu = 0.1$ M with KCl. [Ester] $= 2.6 \times 10^{-5}$ M and a saturating effect of Cu^{2+} on the rate has been reached. The pK's of free ligand are 2.9 and 6.0 in the pH range 2–11 and the pH profile for hydrolysis of the free ligand at 85°C, $\mu = 0.1$ M (KCl) is shown in the lower figure. Rationalize these behaviors. T. H. Fife and M. P. Pujari, J. Amer. Chem. Soc. **110**, 7790 (1988).

4. The difference in the two pK's (4.8 and 7.4) for mononuclear complexes such as *cis*-$Cr(en)_2(H_2O)_2^{3+}$ is approx. 3 (see Table 6.5). The difference is much larger for the binuclear complex Δ, Λ-$(H_2O)(en)_2Cr(OH)Cr(en)_2(H_2O)^{5+}$, p$K$'s 0.5 and 7.9 (i.e. the aquahydroxo cation is stabilized compared with the extreme forms). Give a reasonable explanation for this larger difference.

5. The rate constant for ionization of aqua species can be easily estimated from a knowledge of the pK of the coordinated water. It can then be verified whether H exchange between the complex species and water is controlled by this ionization or by dissociation of water from the complex. Calculate which path is responsible for the H exchange of VO_{aq}^{2+} ($k = 7.7 \times 10^3$ s^{-1}) and Ni_{aq}^{2+} ($k = 3 \times 10^4$ s^{-1}) with H_2O, determined by nmr line-broadening techniques.

6. The ^{1}H nmr spectrum of $Os(bpy)_3^{2+}$ in [D_6]dmso is shown in Fig. (a) If to the solution is added NaOD/D_2O, after about 15 min. at 43°C the signal at 8.6 ppm is lost (b); The rate law is determined in dmso-H_2O mixtures and k decreases as the molar fraction of water in dmso increases

$$-d[8.6 \text{ ppm signal}]/dt = k[Os(bpy)_3^{2+}][\text{NaOD}]$$

Further loss of the other signals occurs in three much slower steps. The uv/vis absorption spectra of the complex remains unchanged and no resonances due to the free bpy are observed during this time. The nmr of free bpy does not undergo any changes in the same conditions. Suggest what may be happening and any implications of the results.
O. Wernberg, J. Chem. Soc. Dalton Trans. 1993 (1986); E.C. Constable and K. R. Seddon, J. Chem. Soc. Chem. Communs. 34 (1982).

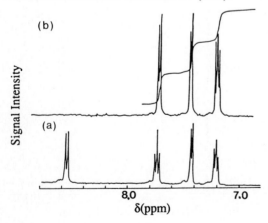

7. The ^{1}H nmr spectrum of *mer*-Co(gly)$_3$ (**III**) in D$_2$O exhibits two resonances at δ 3.68 and 3.42 due to methylene hydrogens with an intensity ratio of 2:1 when measured at 60 MHz. The low field peak is split at 100 MHz into two marginally resolved peaks. The rate constants for exchange of α-H's of *mer*-Co(gly)$_3$ are one at $1.6 \times 10^{-5} s^{-1}$ (δ 3.42) and two at $k = 4.1 \times 10^{-6} s^{-1}$ ($\delta = 3.68$) at pD 10.77 and 25°C. Explain.
A. Miyanaga, U. Sakaguchi, Y. Morimoto, Y. Kushi and H. Yoneda, Inorg. Chem. **21**, 1387 (1982).

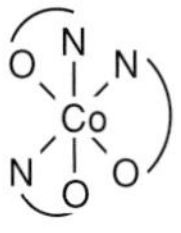

(III)

8. The rate constant for the reactions of H$^+$ with many bases, and of OH$^-$ with many acids, are between 10^9 and 10^{11} M^{-1}s^{-1}. Suggest why the reaction of Co(dmg)$_2$(CN)$_2^-$ (**IV**)X $=$ CN$^-$ with base is much slower, $k = 1.9 \times 10^5$ M^{-1}s^{-1}, and that of Co(dmg)$_2$(NH$_3$)$_2^+$ (**IV**)X $=$ NH$_3$ with base is a little higher, $k = 1.3 \times 10^6$ M^{-1}s^{-1}

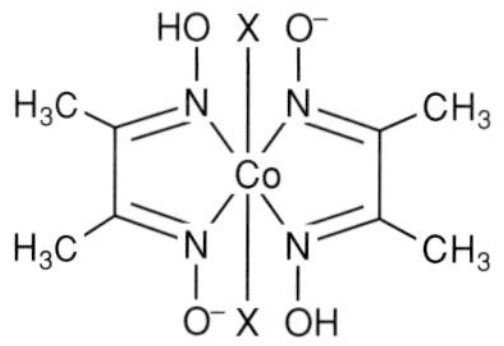

(IV)

J. P. Birk, P. B. Chock and J. Halpern, J. Amer. Chem. Soc. **90**, 6959 (1968).

9. Give plausible explanations for the following:
a. The hydrolysis of **V** when R $=$ C$_2$H$_5$ is 10^4 faster than when R $=$ C(CH$_3$)$_3$. Both give the chelated glycine

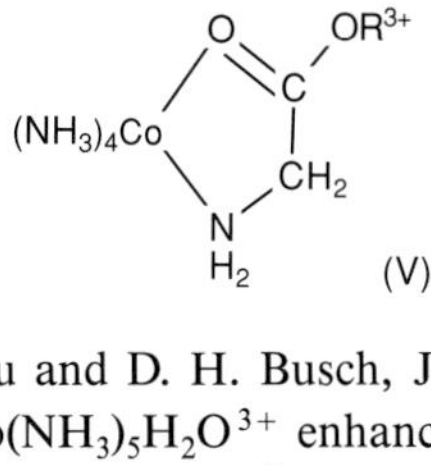

(V)

Y. Wu and D. H. Busch, J. Amer. Chem. Soc. **92**, 3326 (1970).
b. Co(NH$_3$)$_5$H$_2$O^{3+} enhances the decomposition of urea, and the result is the formation of Co(NH$_3$)$_5$NCO^{2+} and NH$_3$. No cyanate complex results, however, from using ((CH$_3$)$_2$N)$_2$C$=$O with the cobalt(III) complex.
R. J. Balahura and R. B. Jordan, Inorg. Chem. **9**, 1567 (1970).
c. The effect of acid on the hydrolysis of Co(NH$_3$)$_5$C$_2$O$_4^+$ is very small even though the coordinated oxalate is basic.
C. Andrade and H. Taube, Inorg. Chem. **5**, 1087 (1966).

d. The acid-base reaction

$$Co(CN)_5H^{3-} + OH^- \rightarrow Co(CN)_5^{4-} + H_2O$$

has a very low rate constant (0.1 $M^{-1}s^{-1}$ at 20°C).
G. D. Venerable II and J. Halpern, J. Amer. Chem. Soc. **93**, 2176 (1971).
e. Metal ion catalyzes the reversible tautomerization of the enol form of acetylacetone, while proton does not.
J. E. Meany, J. Phys. Chem. **73**, 3421 (1969).

10. The oxidation of $Co(NH_3)_5C_2O_4^+$ by Ce(IV) leads to Co(II) ion:

$$Co(NH_3)_5C_2O_4^+ + Ce^{IV} + 5H^+ \rightarrow Ce^{III} + 5NH_4^+ + Co^{2+} + 2CO_2$$

whereas oxidation by Cl_2 preserves Co(III) as a complex:

$$Co(NH_3)_5C_2O_4^+ + Cl_2 \rightarrow Co(NH_3)_5H_2O^{3+} + 2Cl^- + 2CO_2$$

Explain this behavior.
P. Saffir and H. Taube, J. Amer. Chem. Soc. **82**, 13 (1960).

Chapter 7

Isomerism and Stereochemical Change

If two or more substances have the same empirical formula but a different arrangement of atoms in the molecule, they are said to be isomeric. The types of isomerism and isomeric change with which we are concerned in this chapter are contained in Table 7.1.

Table 7.1 Classifications of Isomerism

Type of isomerism	Basis of isomerism	Example of associated isomeric change
Conformational	Different dispositions of ligand around central atom.	$\delta \rightleftharpoons \lambda$ (see Fig. 7.1)
Configurational	Different geometries of central atom.	Tetrahedral $\rightleftharpoons$ planar
Spin	Low and high spin states.	Fe(II), $^1A \rightleftharpoons {}^5T$
Linkage	Different donor centers in identical ligand.	$M-ONO \rightleftharpoons M-NO_2$
Geometrical	Different spatial distribution of atoms or groups around central atom.	*Cis* $\rightleftharpoons$ *trans* in square planar and octahedral complexes.
Optical	Nonsuperimposable mirror images.	In *R, S* tetrahedral and $\Delta \rightleftharpoons \Lambda$ octahedral complexes; chiral centers in coordinated ligands.

The ability of certain metal complexes to exist in stereoisomeric forms, and particularly to interconvert, adds another dimension to the study of the mechanisms of their reactions. There are two aspects from which the phenomenon of stereochemical change may be regarded.

(a) The examination of the stereochemical course of a reaction may allow some reasonable deductions about the mechanism of the reaction and the structure of any activated complexes or intermediates. Such information has already been used in the preceding chapters of this book. The application has been mainly to inert complexes because of the difficulty in preparing isomeric forms of labile complexes, a hurdle that the use of nmr detection and monitoring of systems *in situ* is eliminating.[1,2] (See Secs. 7.6 and 7.7).

(b) The study of the isomerism *per se*. Rearrangement may occur via intermolecular substitutive processes or, what appears to be more often the case, via intramolecular mechanisms. The latter are processes we have hitherto not much encountered, but the mobility of unidentate and bidentate ligands in rotation or twisting around the metal-ligand bond often allows for easier paths for isomerization than complete ligand breakage.

R and S Designations and Definitions

The most widely used method for specifying the configuration about a chiral center uses the R/S designations. Groups are assigned priorities based on the atomic weight of the atom adjacent to the chiral center i.e. decreasing priorities for $O > C > H$; $CO > CH$ and so on. The chiral center is then viewed with the group of lowest priority directed away from the observer. If with the three remaining groups directed towards the observer, the movement from the group of highest priority to that of second highest priority is clockwise then the configuration is R — if anticlockwise, S.

OH   OH

H_3C CO_2H HO_2C CH_3

R S

The two molecules are *enantiomers*. If two of four groups attached to the C are identical as in $CH_3CH_2CO_2H$, the 2 H's are *enantiotopic* and are attached to a *prochiral* center. Replacement of one of the H's by OH obviously leads to enantiomers. If there is already one chiral center in the molecule, as in (R)-malic acid

OH H_R

$HO_2C - C - C - CH_3$

H H_S

R

the H's are termed *diastereotopic* and replacement of H_R (say by D) leads to the RR *diastereoisomer. The H_R is termed a pro-R* ligand. The two hydrogens H_R and H_S are of course identical.

7.1 Conformational Isomerism

Five- and six-membered rings formed by coordination of diamines with a metal ion have the stereochemical characteristics of cyclopentane and cyclohexane. The ethylenediamine complexes have puckered rings and the trimethylenediamine complexes have chair conformations.[3] The methylene protons are nonequivalent in these nonplanar conformations, taking on the character of equatorial and axial substituents. They are made equivalent as the result of *rapid* conformational inversion at room temperature, just as in the alicyclic compounds (Fig. 7.1). This has been observed in nmr studies of planar and octahedral complexes of ethylenediamine-type ligands with a number of metals.[3-5]

Rate parameters for only a few substituted ethylenediamine chelates have been reported and, until recently, none for ethylenediamine chelates because the chemical shifts between the two forms in diamagnetic complexes (on which the analysis depends) are small and because of the extreme lability of the system.[3-5] For the interconversion of the δ and λ forms of

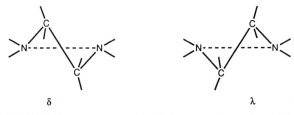

δ λ

Fig. 7.1 A simple representation of the two conformations of the five membered puckered ring in metal complexes of ethylenediamine and derivatives. Rings are viewed along the plane containing the two nitrogens and the metal. The $C-C$ bond is nearer the observer than the two nitrogens. It is skewed down to the right for a δ conformation and to the left for a λ conformation.

$Fe(CN)_4(1,2\text{-diamine})^-$, (7.1) represented in Fig. 7.2, the activation parameters are shown in Table 7.2.[5]

$$\delta \underset{}{\overset{k}{\rightleftharpoons}} \lambda \qquad\qquad (7.1)$$

Table 7.2 Activation Parameters for the $\delta \rightleftharpoons \lambda$ Conformational Interconversion of $Fe(CN)_4(1,2\text{-diamine})^-$ in $CD_3OD-DCl$ at 25 °C

Diamine	k s^{-1}	$\Delta H^{\ddagger}$ $kJ\ mol^{-1}$	$\Delta S^{\ddagger}$ $J\ K^{-1}\ mol^{-1}$
en	3×10^8	25	0
meso-bn (2 R 3 S)	2×10^7	30	-3
cis-chxn (1 R 2 S)	8×10^4	43	-8

The interconversion slows down and $\Delta H^{\ddagger}$ increases as the bulkiness of the carbon substitution increases. It is suggested that the interconversion proceeds through an envelope conformation.[3-5]

$(NC)_4Fe$ H²   H¹   H¹ $(NC)_4Fe$ H¹   H²   H²   H¹

δ λ

Fig. 7.2 The $\delta \rightleftharpoons \lambda$ equilibrium for $Fe(CN)_4(1,2\text{-diamine})^-$. This shows an alternative representation to that in Fig. 7.1 for the two conformations of the puckered ring.[5]

7.2 Configurational Isomerism

Configurational isomerism arises with complexes that are identical in all respects except for the stereochemistry of the central atom. The isomeric forms are also, incidentally, associated with different electronic configurations for the central metal. The thermodynamic aspects have been much more pursued than the dynamic features, which mainly concern (Ni(II).[6]

7.2.1 Planar, Tetrahedral

The lability inherent in the planar, tetrahedral equilibria which nearly all involve Ni(II) requires that nmr line broadening[2,7] or photochemical perturbation[8] methods be used for their kinetic resolution. First-order interconversion rate constants for

$$Ni(RR_1R_2P)_2X_2 \rightleftharpoons Ni(RR_1R_2P)_2X_2 \qquad (7.2)$$
$$\text{(square-planar)} \qquad \text{(tetrahedral)}$$

in $CDCl_3$ and CD_2Cl_2 are around 10^5–$10^6\,s^{-1}$ at 25 °C, Ref. 7. The complex $Ni(dpp)Cl_2$ (dpp $= (C_6H_5)_2P(CH_2)_3P(C_6H_5)_2)$ in CH_3CN at room temperature exists as a mixture of comparable amounts of planar and tetrahedral forms. Irradiation of millimolar solutions at 1060 nm

$$\text{planar } Ni(dpp)Cl_2 \rightleftharpoons \text{tetrahedral } Ni(dpp)Cl_2 \qquad k_1, k_{-1}, K_1 \qquad (7.3)$$

leads to absorbance by, and depletion of, the tetrahedral species with the formation of an excited distorted planar isomer within the time of the laser pulse. The reestablishment of the original equilibrium in the dark following the pulse can be monitored at 470 nm (where there is a decrease in absorbance due to planar loss) or 380 nm (an increase due to tetrahedral gain). The observed relaxation time (0.95 μs at 24 °C) means that $(k_1 + k_{-1})$ in (7.3) is $1.05 \times 10^6\,s^{-1}$ and since $K_1 = k_1/k_{-1} = 0.75$, k_1 and k_{-1} are $4.5 \times 10^5\,s^{-1}$ and $6.0 \times 10^5\,s^{-1}$, respectively.[8]

7.2.2 Planar, Square Pyramidal

The basis of this rearrangement, as well as those in Secs. 7.2.3 and 7.2.4, is a change in the coordination number of the metal.

The kinetics associated with the equilibrium:

$$Ni(aex)^{2+} + H_2O \rightleftharpoons Ni(aex)(H_2O)^{2+} \qquad k_1, k_{-1} \qquad (7.4)$$
$$\mathbf{1}$$

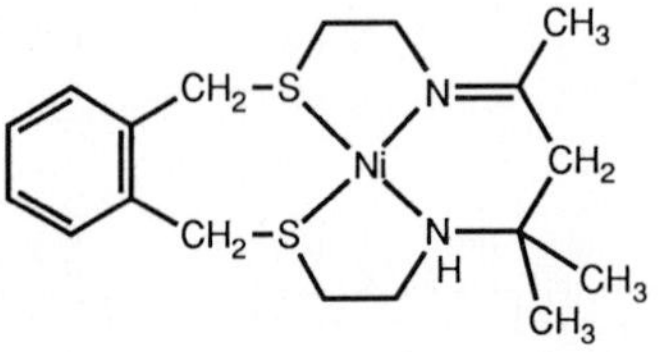

1

have been measured in a similar manner to those of (7.3), using a photochemically-induced concentration jump. The values of k_1 and k_{-1} are $4.5 \times 10^6\,s^{-1}$ and $1.5 \times 10^7\,s^{-1}$ respectively at 20 °C in water.[9] Another example is mentioned in Sec. 3.4.3.

7.2.3 Planar, Octahedral

This equilibrium can be represented in general terms as the two steps (7.5) and (7.6),

$$ML + S \rightleftharpoons ML(S) \qquad k_1, k_{-1} \tag{7.5}$$

$$ML(S) + S \rightleftharpoons ML(S)_2 \qquad k_2, k_{-2} \tag{7.6}$$

where S is a unidentate ligand (often solvent) and L is a tetradentate ligand or two bidentate ligands. The equilibria are invariably labile (see however Ref. 10) and since ML(S) is usually in small concentration compared with the other species, the perturbation approach (by T-jump, [11,12] ultrasonic absorption [13,14] or photolytic perturbation [15]) gives only one relaxation and therefore an incomplete description of the system (Sec. 1.8.2). The study of the exchange of ML(S)$_2$ with S (by nmr line broadening) defines the second step (7.6), and, combined with the relaxation data, allows the designation of the rds. The study of the equilibria (7.7) illustrates this approach.

$$\text{Ni([12]aneN}_4)^{2+} + 2\,H_2O \underset{k_{-1}}{\overset{k_1}{\rightleftharpoons}} \text{Ni([12]aneN}_4)(H_2O)^{2+} + H_2O$$
(low spin square planar)

$$\underset{k_{-2}}{\overset{k_2}{\rightleftharpoons}} \text{Ni([12]aneN}_4)(H_2O)_2^{2+} \tag{7.7}$$
(high-spin *cis*-octahedral)

Exchange studies yield $k_{-2} = 4.2 \times 10^7 \text{ s}^{-1}$, $\Delta H^{\ddagger}_{-2} = 33 \text{ kJ mol}^{-1}$ and $\Delta S^{\ddagger}_{-2} = 5 \text{ J K}^{-1}\text{mol}^{-1}$. The values $k_1 = 5.8 \times 10^3 \text{ M}^{-1}\text{s}^{-1}$, $\Delta H^{\ddagger}_1 = 43 \text{ kJ mol}^{-1}$ and $\Delta S^{\ddagger}_1 = -28 \text{ J K}^{-1}\text{mol}^{-1}$ and $k_{-1}/k_2 = 0.016$ result from the T-jump work (all data at 25 °C in 3.0 M LiClO$_4$). [11] It is apparent then that the first step is rate determining in the establishment of the overall equilibrium. This presumably arises because the *trans, cis* rearrangement of the macrocyclic ligand occurs at this stage. Two possible free energy profiles are illustrated in Fig. 7.3. [6]

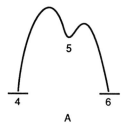

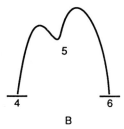

	5				5	
4		6		4		6
	A				B	

Fig. 7.3 Free energy profiles for planar-octahedral equilibria. In A, ligand or solvent exchange is more rapid than the planar, octahedral equilibrium as established in (7.7). In B, the formation and dissociation of the octahedral complex is rate-determining and the interconversion of the planar and five-coordinated species is more rapid. [6,12]

A related equilibrium involving the conversion of *cis*-Ni([13]aneN$_4$)(H$_2$O)$_2^{2+}$ to β-*trans* Ni([13]aneN$_4$)$^{2+}$ is believed to involve a planar isomer, (α-*trans*-Ni([13]aneN$_4$)$^{2+}$), as an intermediate:

$$\text{cis-Ni([13]aneN}_4)(H_2O)_2^{2+} \rightleftharpoons \alpha\text{-trans-Ni([13]aneN}_4)^{2+} \rightleftharpoons \beta\text{-trans-Ni([13]aneN}_4)^{2+} \tag{7.8}$$

or a "dead-end" species (Sec. 1.6.4):

$$\alpha\text{-}trans\text{-}Ni([13]aneN_4)^{2+} \rightleftharpoons cis\text{-}Ni([13]aneN_4)(H_2O)_2^{2+} \rightleftharpoons \beta\text{-}trans\text{-}Ni([13]aneN_4)^{2+} \quad (7.9)$$

The α- and β-planar forms are configurational isomers and their interconversion require inversions at two nitrogens in the macrocycle ring (Sec. 7.9).[10] A number of systems of the general type (7.5), (7.6) have been described[12-15] including those in which S may be an exogenous ligand, or may represent an endogenous donor atom of a multidentate ligand:[13]

$$Ni(N,O\text{-salpip})_2 \rightarrow Ni(N,N,O\text{-salpip})_2 \quad (7.10)$$
$$\textit{trans}\text{-O-planar} \qquad \textit{cis}\text{-meridional}$$

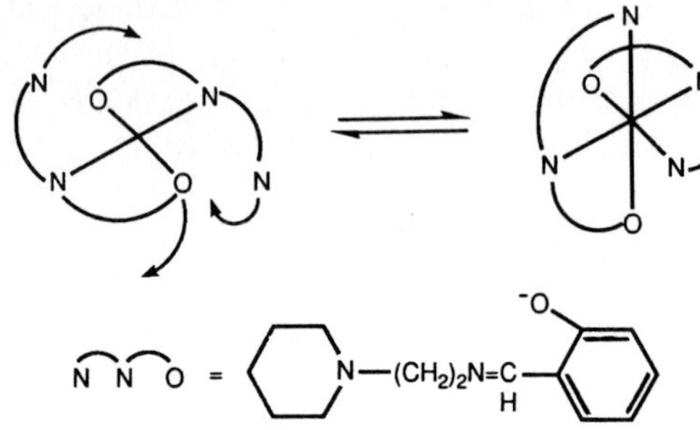

(planar when tridentate and must bind meridonal)

7.2.4 Tetrahedral, Octahedral

Equilibria between tetrahedral and octahedral cobalt(II) complexes in nonaqueous solution are well characterized. Kinetic data are sparse but those available from T-jump experiments in pyridine solution are interpreted in terms of (X = Cl and Br):[16]

$$Co(py)_4X_2 \rightleftharpoons Co(py)_3X_2 + py \quad \text{rapid} \quad (7.11)$$

$$Co(py)_3X_2 \rightleftharpoons Co(py)_2X_2 + py \quad (7.12)$$

Stereochemical nonrigidity is ubiquitous with seven and higher coordinated complexes because the geometries associated with them are easily interconverted by relatively small atomic displacements.[2,17,18] Intramolecular rearrangements are complex but their understanding is helped considerably, but not solved, by nmr techniques.[2]

Magnetic susceptibilities of solutions — These are useful parameters for determining equilibrium constants for reactions involving spin changes. The Evans nmr method utilizes the observed shift in the resonance line (say of a proton of t-BuOH or hexamethyldisiloxane) in solution when a paramagnetic substance is added. The paramagnetic shift Δf is related to the magnetic moment (μ, of the solution at TK by the approximate expression

$$\mu \approx 0.49(\Delta f T/f M)^{1/2}$$

where f is the oscillator frequency (in MHz) and M is the molarity of the substance (D. F. Evans and T. A. James, J. Chem. Soc. Dalton Trans. 723 (1979)).

7.3 Spin Equilibria in Octahedral Complexes

High-spin and low-spin electronic states exist in thermal equilibrium for a few octahedral Fe(II), Fe(III), Co(II), Co(III), and Mn(III) complexes.[6,19,20] Some differences in bond distances accompany the spin changes. The extreme lability of the systems necessitates the most rapid relaxation or (increasingly popular) photoperturbation techniques (Chap. 3) for the examination of their dynamics. Data for two of a number[6] of equilibria which have been measured are shown in Table 7.3. For Co(II), rate constants for the interconversion of the spin states are estimated as approximately 10^{10} s^{-1}, (Ref. 6) so that in general the order of decreasing lability is Co(II) > Fe(III) > Fe(II). The first reported dynamics were for the spin equilibrium in an iron(II) complex using Raman laser T-jump.[21] Since then the Fe(II) systems have been the more extensively examined.[22-26] Iron(II)-ligand bond distances are anywhere up to 0.20 Å shorter in the low spin than in the high spin isomer and the accompanying volume increase in going to the high spin form is 8–10 cm^3 mol^{-1} although this is smaller than would be expected.[22] The pronounced differences in the activation parameters for the spin change for the Fe(II)[23,24,26] and Fe(III)[27] systems in different solvents suggest that solvation effects are important, although the situation is complicated.[26] The values for $\Delta V^{\ddagger}$ in either direction indicate that the transition state lies well along the reaction coordinate between the low and high-spin states.

Table 7.3 Dynamics of Spin-Crossover (Low Spin $\rightleftharpoons$ High Spin; k_1, k_{-1}) in Iron(II) and (III) Complexes in Solution at 25°C

Complex	k_1 s^{-1}	$\Delta H_1^{\ddagger}$ kJ mol^{-1}	$\Delta S_1^{\ddagger}$ J K^{-1}mol^{-1}	$\Delta V_1^{\ddagger}$ cm^3 mol^{-1}	Refs.
Fe(II)L$_3$ ($t_{2g}^6 \rightleftharpoons t_{2g}^4 e_g^2$)	5×10^6	31	−9	5.5(5)[24]	
					23 (photoperturbation), see also
L =	k_{-1} 1.0 × 10^7	$\Delta H_{-1}^{\ddagger}$ 15	$\Delta S_{-1}^{\ddagger}$ −58	$\Delta V_{-1}^{\ddagger}$ −5(−3)[24]	24 (photoperturbation) and 25 (Raman T-jump)
(Me$_2$CO)					
Fe(III)L	k_1 s^{-1}	$\Delta H_1^{\ddagger}$ kJ mol^{-1}	$\Delta S_1^{\ddagger}$ J K^{-1} mol^{-1}		28 (ultrasonics), see
($t_{2g}^5 \rightleftharpoons t_{2g}^3 e_g^2$)	6.1 × 10^7	31	9		also 27 (photopertur- at 200–255 K in
L = Sal$_2$trien (H$_2$O)	k_{-1} 1.3 × 10^8	$\Delta H_{-1}^{\ddagger}$ 10	$\Delta S_{-1}^{\ddagger}$ −57		CH$_3$OH (extrapolated $\tau = 5 \pm 1$ ns at 25°C).

The study of the dynamics of spin-state changes is important for the understanding of the kinetics of bimolecular electron transfer reactions[6,29] and racemization and isomerization processes (Sec. 7.5.1).[6,22] Low spin − high spin equilibria, often attended by changes in coordination numbers, are observed in some porphyrins and heme proteins,[30] although their biological significance is, as yet, uncertain.

7.4 Linkage Isomerism

There are a large number of unidentate ligands that can coordinate through only one donor center but that contain more than one donor site. Many linkage isomers based on the framework $M(NH_3)_5L^{n+}$ M = Co(III), Ru(II) or Ru(III), have been prepared. A list is collected in Refs. 31–33. The classic pair of linkage isomers is $Co(NH_3)_5ONO^{2+}$ and $Co(NH_3)_5NO_2^{2+}$ which have been extensively investigated, with new features continually appearing (see below). [34]

The preparation of linkage isomers relies mainly on kinetically-controlled processes. Thus the reaction of $Co(NH_3)_5H_2O^{3+}$ with SCN^- ion gives both the stable N-bonded, $Co(NH_3)_5NCS^{2+}$ ($\approx 75\%$) and the unstable S-bonded, $Co(NH_3)_5SCN^{2+}$ ions. The two isomers can be separated by ion exchange chromatography. [35] If the Co(III)-O bond is not severed when H_2O is replaced in $Co(NH_3)_5H_2O^{3+}$ by another ligand an *unstable* O-bound linkage isomer may result, as for example, in the production of the nitrito isomer in acid solution [34, 36]

$$Co(NH_3)_5H_2O^{3+} \rightleftharpoons Co(NH_3)_5OH^{2+} + H^+ \tag{7.13}$$

$$2\,HNO_2 \rightleftharpoons N_2O_3 + H_2O \tag{7.14}$$

$$Co(NH_3)_5OH^{2+} + N_2O_3 \rightarrow Co(NH_3)_5ONO^{2+} + HNO_2 \tag{7.15}$$

$$d\,[Co(NH_3)_5ONO^{2+}]/dt = k\,[Co[NH_3)_5OH^{2+}]\,[HNO_2]^2 \tag{7.16}$$

The base hydrolysis of $Co(NH_3)_5X^{n+}$ generates the species $Co(NH_3)_4(NH_2)^{2+}$ and either donor site of $S_2O_3^{2-}$, NO_2^- or SCN^- (when present in solution) can attack the transient to generate both linkage isomers (Sec. 2.2.1 (b)).

One important approach to preparing linkage isomers uses the inner-sphere redox concept, combined with the fact that a ligand capable of forming linkage isomers is usually a good bridging group. Thus, the reaction of Cr^{2+} with $FeNCS^{2+}$ produces by remote attack (Sec. 5.6 (c)) green $CrSCN^{2+}$ which slowly rearranges to the stable purple $CrNCS^{2+}$. [37] Most Co(III) complexes with the sulfito, sulfinato and sulfenato groups contain Co-S bonding. Photochemical irradiation may effect isomerization, e. g. [38]

$$Co(en)_2(SO_2CH_2CH_2NH_2)^{2+} \underset{dark}{\overset{h\nu}{\rightleftharpoons}} Co(en)_2(OS(O)CH_2CH_2NH_2)^{2+} \tag{7.17}$$

The reversion to the S-bonded chelate in the dark is very slow $k = 1.5 \times 10^{-5}\,s^{-1}$ (59.9°C). [39]

The characterization of linkage isomers is usually accomplished by ir or nmr methods. Fortunately there are often strong spectral differences in the linkage isomers. For example, in $Co(NH_3)_5L^{2+}$ L = $-ONO$, $\varepsilon = 1.5 \times 10^2$ $M^{-1}cm^{-1}$ and L = $-NO_2$, $\varepsilon = 1.7 \times 10^3$ $M^{-1}cm^{-1}$ at 325 nm; Co-S bonding in the sulfur-bound $S_2O_3^{2-}$ or SCN^- complexes is characterized by a strong band at 289 nm ($\varepsilon = 1.5 \times 10^4$ $M^{-1}cm^{-1}$). [40] Kinetic studies of linkage isomerism invariably uses these uv-vis absorbance differences.

7.4.1 Rearrangement Studies

There are basically two types of mechanisms which can explain rearrangement of linkage isomers. In the intermolecular mechanism, the ligand giving rise to linkage isomerism $(L-L_1$, with two donor centers) dissociates from and recombines with the metal M:

$$M(L - L_1) \rightleftharpoons M + (L - L_1) \rightleftharpoons M(L_1 - L) \qquad (7.18)$$

In the intramolecular mechanism, the ligand $L-L_1$ rearranges its mode of coordination in $M(L - L_1)$ without severence from the metal M. Intramolecular rearrangements dominate the field.

The isomerizations of the nitrito complexes $M(NH_3)_5ONO^{n+}$ to the corresponding nitro compounds $M(NH_3)_5NO_2^{n+}$, $M = Co(III)$, $Rh(III)$, $Ir(III)$ and $Pt(IV)$, are complete in solution, and have similar first-order rate constants for rearrangement. This strongly suggests that M–O bond cleavage does not occur during the isomerization. This is substantiated by ^{18}O[41] or ^{17}O[42] studies of the Co(III) complex. A suggested rearrangement mechanism is [41,42]

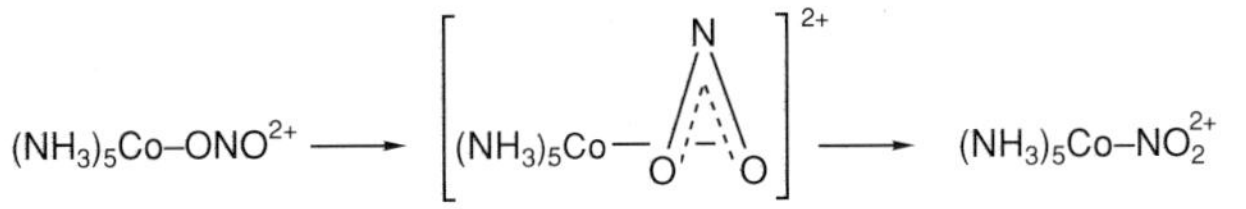

$$(7.19)$$

The tight π-bonded intermediate is consistent with a negative volume of activation.[43] It also explains the (unanticipated) O-to-O exchange in the nitrito ligand at a rate comparable to the O-to-N isomerization (Sec. 2.2.2). $k_1/k_2 = 1.2$[42]

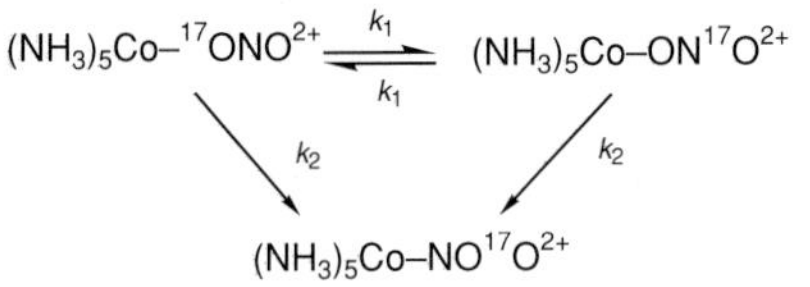

$$(7.20)$$

The rearrangements of $(NH_3)_5M(ONO)^{2+}$, $M = Co$, Rh and Ir, are all base catalyzed (Sec. 1.10.1)

$$k = k_1 + k_2[OH^-] \qquad (7.21)$$

For the spontaneous change examined in 16 solvents, k_1 spans 2 orders of magnitude, but $\Delta V^{\ddagger}$ varies only from $- 3.5$ to -7.0 cm^{3}mol^{-1}. The plot of $\Delta H^{\ddagger}$ against $\Delta S^{\ddagger}$ is linear, so that an intramolecular mechanism in all solvents is favored.[44] For the k_2 path, $\Delta V^{\ddagger}$ is much more positive, $+27$ cm^{3}mol^{-1} in water, consistent with a conjugate base pre-equilibrium, with most of $\Delta V^{\ddagger}$ residing in the release of electrostricted water (from OH^-) in the pre-equilibrium. An intramolecular mechanism is favored.[45] Other linkage isomerizations which have been examined are also intramolecular.[31,32] This is shown by one of several observations – isotopic labelling experiments, the values of the activation parameters[34,46(a)] or by a more rapid rate of rearrangement than dissociation (solvolysis). Table 7.4 contains some novel examples of linkage isomerization.[47-51] An unusual type of linkage isomerism is shown by some cryptand complexes. In the *inclusive* form the solvent shell of the metal ion is completely

Table 7.4 Some Recent Examples of Linkage Isomerism

System		Comments	Ref.
		Unusual example of bidentate ligand involved in linkage isomerism. Forms characterized by ^{31}P nmr and x-ray diffraction. $\beta\gamma$ isomer favored at 40°C pH 6.5. No kinetic data on rearrangement.	47
		N-1-bonded prepared by novel method (Sec. 6.3.3). Isomerises to N-2-bonded, $k = 1.9 \times 10^{-6} s^{-1}$ at 25°C and $\mu = 1.0$ M ($R = CH_3$) Rearrangements are acid and base catalyzed. Intramolecular (no hydrolysis observed). See also Ref. 49.	48
$(NH_3)_5Os-NH_2^{2+}$ N–bound	 π–bound	Linkage isomerism from N-bound to π-bound aniline to Os. $k(N \to \pi) = 2.3 \times 10^{-5} s^{-1}$ and $k(\pi \to N) = 4.6 \times 10^{-5} s^{-1}$ in acetone-d_6 at 20°C. Intramolecular (no solvolysis is observed).	50
		Alterdentate ligands which offer two *equivalent* sites for metal binding.. Intramolecular site transfer (k) by epr line broadening (Sec. 3.9.8). Rate dependent on closed-shell metal ion M, with a range of $k = 1.3 \times 10^4 s^{-1}$ (Mg^{2+}) to $4 \times 10^6 s^{-1}$ (Cd^{2+}) in dmf at 25°C	51

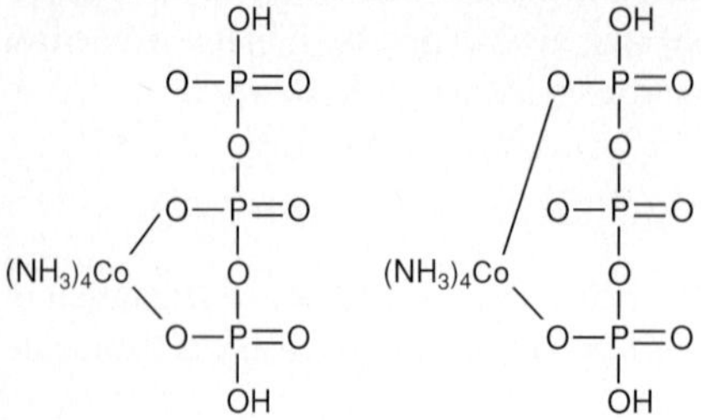

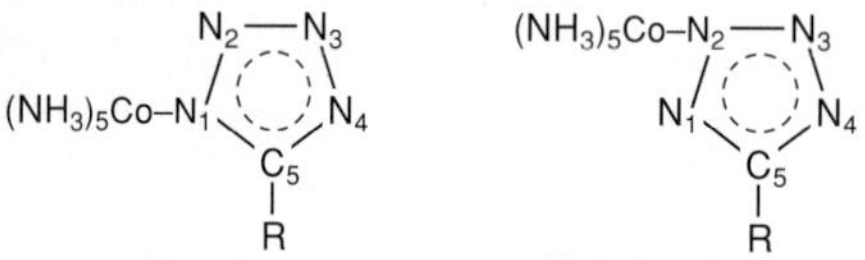

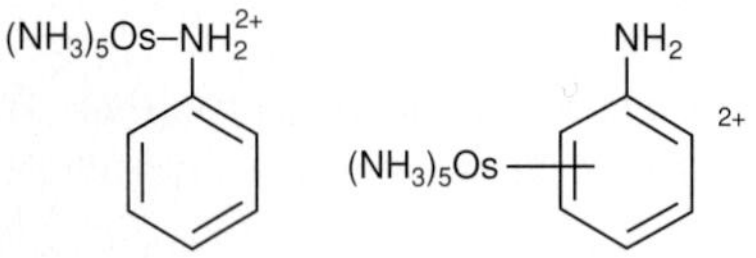

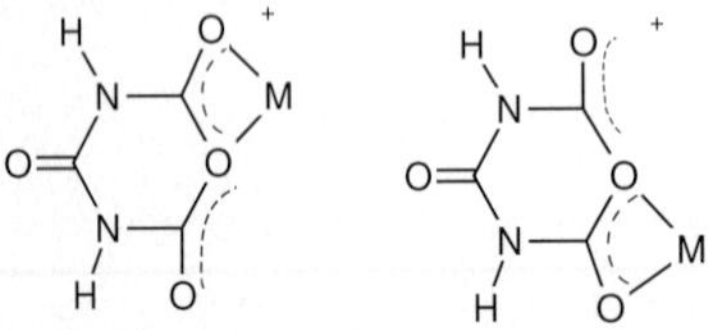

(complex of radical anion of alloxan, prepared by electrolytic reduction in dmf in esr cavity)

replaced by the cryptand donor atoms. The coordination shell of the metal ion in the *exclusive* isomer consists of both cryptand and other (unidentate) ligands. The equilibrium (compare Structure **15**, Chap. 2):

$$Cs^+ + 2_02_02_0 \rightleftharpoons Cs^+ 2_02_02_0 \rightleftharpoons Cs^+ 2_02_02_0 \qquad (7.22)$$
$$\qquad\qquad\qquad\quad \text{exclusive} \qquad \text{inclusive}$$

has been examined by ^{133}Cs nmr in nonaqueous solution.[52]

In the solid state, the remarkable complex $Pt(Me_2SO)_4(CF_3SO_3)_2$ has two O- and two S-bonded ligands attached in a *cis*-Pt(II) cation configuration, **2**. This structure is retained in solution.[53] The exchange of Me_2SO with $Pt(OSMe_2)_2(S(O)Me_2)_2^{2+}$ in CD_3NO_2 is necessarily complicated, but the details have been resolved using ^1H nmr. Kinetic parameters have been obtained. The exchange of the O-bonded dmso is about 10^8 times faster than that of water exchange on $Pt(H_2O)_4^{2+}$, showing the powerful trans effect of S-bonded dmso.[53]

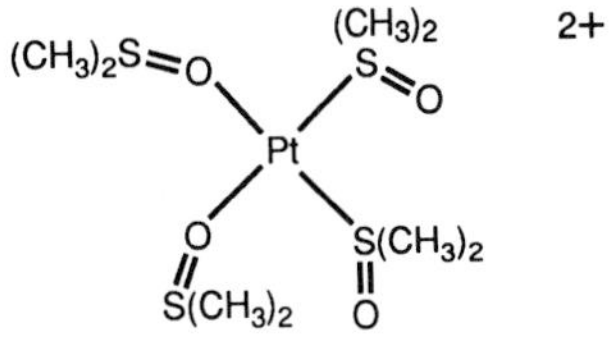

2

7.5 Geometrical and Optical Isomerism

This is by far the most studied of the isomerisms, with those of square planar and octahedral complexes receiving the major attention. As we shall see, developments in the preparation and separation of isomeric forms and the clever use of nmr techniques have played a major part in the increased understanding of the field.

7.6 Octahedral Complexes

There are a variety of combinations of unidentate ligand and/or multidentate ligands placed around an octahedral frame which can lead to geometrical and/or optical isomeric forms.[2,54,55] The most studied types are undoubtedly $M(AA)_3$, $M(AA_1)_3$ and $M(AA)_2XY$, where AA and AA_1 are symmetrical and unsymmetrical bidentate ligand respectively, and X and Y are unidentate ligands. Subsequent discussion will show that geometrical isomerization and racemization are obviously interrelated and it is therefore sensible not to separate their discussion. Indeed, examination of *both* processes for specific complexes can be more revealing than their separate characterization alone.[2,54] Most studies refer to behavior in solution, there having been a limited number of investigations of rearrangements in the solid state. In the past decade, attempts have been made to elucidate mechanisms of rearrangement by using $\Delta V^{\ddagger}$ values, with variable success.[46(b), 56]

7.6.1 Complexes of the Type $M(AA)_3$

The two optical forms designated Λ and Δ are shown in Fig. 7.4. Since the bidentate ligand is symmetrical, geometrical isomerism cannot arise.

(a) Λ-configuration

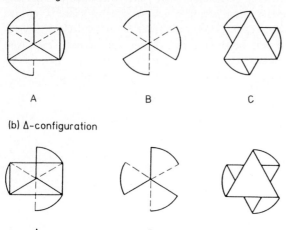

A   B   C

(b) Δ-configuration

A   B   C

Fig. 7.4 Equivalent representations of the complex M(AA)$_3$. The usual octahedral representation is shown in A. In B, viewing is down the C_3 axis; three metal-ligand bonds point towards the viewer, and three point away. In C, the opposite faces of the octahedron are shown when viewing is down the C_3 axis. Representation B will be often used to illustrate configurational changes.

In (a) the Λ configuration is shown and in (b) the mirror-image, Δ configuration, is represented. The designations Λ and Δ were previously termed d and l, or D and L forms (not necessarily respectively).

Stereochemical rearrangement of the two isomers can occur by intermolecular and intramolecular mechanisms. These take the following form:

(a) Complete dissociation of one ligand,

$$M(AA)_3 \rightleftharpoons M(AA)_2 + AA \tag{7.23}$$

This mechanism is supported by identical dissociation and racemization rate constants. This further implies either that the *bis* species M(AA)$_2$ is racemic as formed, or that it may racemize (by a *cis-trans* change, or by a dissociative or intramolecular path) more rapidly than it re-forms *tris* in the dynamic equilibrium (7.23). Identical activation parameters for the dissociation (to the *bis* species) and racemization in aqueous acid[57] (Table 7.5) and other solvents[58] of Ni(phen)$_3^{2+}$ and Ni(bpy)$_3^{2+}$ indicate that these ions racemize by an intermolecular mechanism. This is the only such example for an M(phen)$_3^{n+}$ or M(bpy)$_3^{n+}$ species (see Table 7.5)[57-62] although recently it has been observed that Fe(bps)$_3^{4-}$ (bps is the disulfonated phenanthroline ligand shown in **13**, Chap. 1) but not Fe(phen)$_3^{2+}$ also racemizes predominantly by a dissociative mechanism in water.[59] For the other *tris*-phenanthroline complexes (and for Fe(bps)$_3^{4-}$ in MeOH rich, MeOH/H$_2$O mixtures[59]) an intramolecular mechanism pertains since the racemization rate constant is larger than that for complete dissociation of one ligand, Table 7.5.

There are basically two types of intramolecular rearrangements which are not easy to distiguish experimentally. In one type, intramolecular twisting occurs without any metal-donor atom bond rupture. In the other, rupture of one metal-ligand bond of the chelate occurs to give a five-coordinated intermediate.[2, 46(b), 56]

Table 7.5 Activation Parameters for Racemization and Ligand Dissociation of $M(AA)_3^{n+}$ Ions at 25 °C

Ion		k s^{-1}	$\Delta H^{\ddagger}$ kJ mol^{-1}	$\Delta S^{\ddagger}$ J K^{-1} mol^{-1}	$\Delta V^{\ddagger}$ cm^3 mol^{-1}	Ref.
$Fe(phen)_3^{2+}$	rac. (0.01 M HCl)	6×10^{-4}	118	$+89$	$+16$	57,60
	dissoc. (1 M HCl)	7×10^{-5}	135	$+117$	$+15$	
$Fe(bps)_3^{4-}$	rac. (H$_2$O)	2.7×10^{-5}	130	$+90$	$-$	59
	dissoc. (H$_2$O)	2.8×10^{-5}	130	$+90$	$-$	59
$Co(phen)_3^{2+}$	rac. (H$_2$O)	6.9	31	-125	$-$	61
	dissoc. (H$_2$O)	0.16	85	$+21$	$-$	61
$Ni(phen)_3^{2+}$	rac.[a] (1 M HCl)	1.7×10^{-4}	105	$+12$	-2	57,58
	dissoc.[a] (1 M HCl)	1.6×10^{-4}	102	$+3$	-1	
$Cr(phen)_3^{3+}$	rac.[b] (0.05 M HCl)	6.7×10^{-5}	94	-56	$+3$	62
	dissoc.[b]	very slow	$-$	$-$	$-$	
$Cr(C_2O_4)_3^{3-}$	rac. (0.05 M HCl)	1.6×10^{-3}	66	-76	-16	62
	dissoc.	c	$-$	$-$	$-$	

[a] At 45.0 °C   [b] At 75 °C   [c] Does not exchange with $^{14}C_2O_4^{2-}$ during racemization.

(b) Twist Mechanism — This mechanism can be considered as arising from the twisting of opposite faces of an octahedron through 60 degrees to form a roughly trigonal prismatic state (Fig. 7.5). Further twisting through another 60 degrees leads to inversion. There are two types of transition state depending on the pair of trigonal faces chosen for the rotation operation, that is, depending on the axis about which rotation occurs, the C_3 axis (a) or an "imaginary" C_3 axis (b). The form (a) produces a transition state that has a C_3 or pseudo C_3 axis and is termed a trigonal twist.[63] A "rhomboid (rhombic) twist,"[64] pictured in (b), produces a transition state with no such elements of symmetry. Both twists lead to optical inversion. The two mechanisms are very similar.[62,65] The Ray-Dutt twist is a rigid-ring process in which donor-metal-donor angles and chelate internal ring angles remain constant throughout the inversion. The Bailar twist involves a flexing of rings and changes in all of the internal chelate ring angles can occur. Because of these difference, the types of chelate complexes, for which one of the twists is more likely, can be assessed.[65]

(c) One-Ended Dissociation Mechanism — This intramolecular mechanism can proceed through a bipyramidal or square pyramidal intermediate (Fig. 7.6).[62] It is usually easy to distinguish between an intermolecular (a) and either of the intramolecular mechanisms (b) or (c) by comparing the racemization and dissociation rate constants.

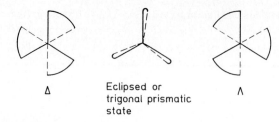

(a) Trigonal or Bailar twist

Δ   Eclipsed or trigonal prismatic state   Λ

(b) Rhomboid or Ray-Dutt twist

Δ   Λ

Fig. 7.5 (a) The trigonal, or Bailar, twist, in which opposite faces of the octahedron are twisted around a C_3 axis, through an angle of 60° to form the eclipsed transition state. Further twisting of 60° leads to an inverted configuration. (b) The rhomboid, or Ray-Dutt twist. Opposite faces of an octahedron are twisted but about a pseudo C_3 axis. There are eight C_3 axes (normal to the eight trigonal faces of the octahedron). When three bidentate ligands span the octahedron, there are then only two C_3 axes and six pseudo, or "imaginary", C_3 axes. Rotation around one of the two C_3 axes constitutes a trigonal twist. Rotation around one of the six pseudo axes constitutes a rhomboid twist. It can be seen from Fig. 7.5 that viewed down the C_3 axis, one corner of the face is always joined by the ligand to a corner of the *other* face. Viewed down the "imaginary" axis (Fig. 7.5b), only one (of the three) corners is seen joined from one face to the other.

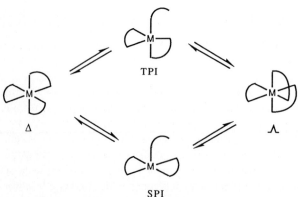

TPI

Δ   Λ

SPI

Fig. 7.6 One-ended dissociation mechanisms for the racemization of M(AA)$_3$ [from Ref. 62].

Two ^{1}H nmr signals are observed at lowered temperature ($-20°C$) for the two non-equivalent N,N-dimethyl substituents in Co(ompa)$_3^{2+}$ which are indicated as N(1)Me$_2$ and N(2)Me$_2$ in Structure **3**.[66] When the temperature is raised the signals are modified as the interchange N(1)Me$_2$ ⇌ N(2)Me$_2$ becomes increasingly rapid (slow exchange region from $-21°C$ to $+3°C$ and fast exchange region from $44°C$ to $70°C$ with an intermediate coalescence temperature). This interchange is brought about by the inversion Δ ⇌ Λ about the chiral cobalt, the dynamics of which are therefore indirectly measured. The inversion kinetic parameters are shown in Table 7.6.[67] Only at higher temperatures ($60-100°C$) does the exchange of Co(ompa)$_3^{2+}$ with free ompa become fast enough to be examined, also by nmr line broadening methods. This behavior shows in a striking manner the faster inversion than dissociation rate. The approach also illustrates the power of the nmr method for measuring inversion rates particularly in paramagnetic complexes,[67] without resorting to tedious resolution techniques. The latter is not a viable option with labile systems anyway.

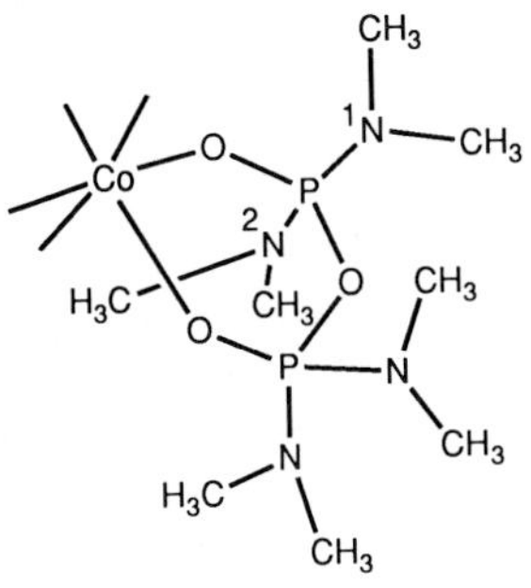

3

Table 7.6 Inversion of $Co(ompa)_3^{2+}$ and Exchange with ompa in CD_2Cl_2/CD_3NO_2 mixture at 25°C

Reaction	k	$\Delta H^{\ddagger}$ kJ mol^{-1}	$\Delta S^{\ddagger}$ J K^{-1} mol^{-1}
inversion	$1.6 \times 10^4 \, s^{-1}$	56	25
exchange	$3.8 \, M^{-1} s^{-1}$	54	-51

It is difficult, in general, to assess the relative importance of the twist and dissociation mode of intramolecular rearrangement. With $Co(ompa)_3^{2+}$, one-ended dissociation is favored since there is a general parallelism between optical inversion and ligand exchange rates for complexes of this type with a variety of metals. A common five-coordinated intermediate is postulated for both processes.[67,68] On the other hand, one-ended dissociation is not easily envisaged with $M(phen)_3^{n+}$ complexes since the phenanthroline ligand is rigid. The more rapid racemization of $Fe(phen)_3^{2+}$, $Co(phen)_3^{2+}$ and $Cr(phen)_3^{2+}$ compared with ligand dissociation (Table 7.5) strongly suggests that here a purely twist mechanism is operative. Nowhere has the difficulty in distinguishing between the intramolecular mechanisms for racemizations been more evident than with chelated complexes containing oxalate groups particularly of the type $M(C_2O_4)_3^{3-}$. Racemization is invariably faster than the complete loss by aquation of one oxalate group, so that an intermolecular mechanism can be ruled out.[69,70] With these complexes, we have an additional probe that should in principle help us to distinguish between the intramolecular paths (b) and (c). This is the examination of the exchange of ^{18}O between H_2O and the oxygens of the coordinated oxalate (Sec. 1.9). Exchange of the outer oxygens of $Rh(C_2O_4)_3^{3-}$ (Table 7.7)[70] and $Co(C_2O_4)_3^{3-}$ is much faster than racemization. Rate constants for inner-oxygen exchange and racemization are similar, suggesting a common intermediate for both processes involving one-ended dissociation. A similar conclusion has been reached for the racemization of $Cr(C_2O_4)_3^{3-}$, although the argument is circuitous.[62]

Table 7.7 Rate Parameters for Acid-Catalyzed Oxygen Exchange and Racemization of $Rh(C_2O_4)_3^{3-}$ at 56 °C (Ref. 70)

	$10^5 \times k\,(exch)^a$ s^{-1}	$\Delta H^{\ddagger}\,(exch)$ kJ mol^{-1}	$10^5 \times k\,(rac)^a$ s^{-1}	$\Delta H^{\ddagger}\,(rac)$ kJ mol^{-1}
Outer O	148	71		
Inner O	8.4	99		
Racemization			4.2	98

^{a} Value of k in $k\,[H^+]$ term in rate law

It has not been easy to use $\Delta V^{\ddagger}$ values to support a racemization mechanism, nor indeed to rationalize $\Delta V^{\ddagger}$ in terms of known mechanisms. For racemization, ΔV° is zero and to a first approximation $\Delta V^{\ddagger}$ might also be expected to be near zero for a twist mechanism. For both $Cr(phen)_3^{3+}$ and $Ni(phen)_3^{2+}$ however $\Delta V^{\ddagger}$ values are near zero, although the mechanisms are undoubtedly intra- and inter-molecular respectively. A large negative value for $\Delta V^{\ddagger}$ for $Cr(C_2O_4)_3^{3-}$ can however be rationalized in terms of a one-ended dissociative mechanism in which solvent electrostriction of the newly developed charge center $(-CO_2^-)$ dominates in the overall volume of activation. It is probably true that for strong-field d^6 complexes, a spin change occurs along the reaction coordinate for isomerization, racemization and substitution processes.[71] Thus the large positive $\Delta V^{\ddagger}$ value for racemization of $Fe(phen)_3^{2+}$ may arise from a contribution from the preequilibrium low-spin to high-spin excitation (Sec. 7.3). The small residual component of $\Delta V^{\ddagger}$ is consistent with the proposed twist mechanism.[65] See also Ref. 72. It is apparent that rationalization in hindsight abounds, which is understandable in these relatively early days of the subject.

7.6.2 Complexes of the type M(AA$_1$)$_3$

One valuable, although not easy, approach for distinguishing the various types of intramolecular change is by using chelates of the kind $M(AA_1)_3$, where AA_1 represents an unsymmetrical bidentate ligand.[62] Geometrical[73] as well as optical isomers now exist (see Fig. 7.7), so that there are a number of isomeric changes that can be studied. These are thoroughly discussed in Ref. 2. Different results, dependent on the modes of intramolecular rearrangement, can be expected. This idea is illustrated by considering the likely consequences when the *cis*-Δ form rearranges by any of five distinct routes (a) to (e).[74] Figure 7.8 shows the product(s) and the ratio of rate constants that would result for the various changes. The twist mechanism (a) leads to inversion without isomerization. The rhombic mechanism (b) leads to inversion and isomerization. Bond rupture via a trigonal-bipyramidal intermediate (TBI) with the dangling ligand in an axial (c) or equatorial position (d) causes isomerization with or without inversion, respectively. Rearrangement via a square pyramidal intermediate (SPI), with the energetically most likely structure shown (e), leads to equivalent amounts of all forms.

The observed ratios of rate constants for the rearrangement of *cis*-Δ-Co(bzac)$_3$, **4** immediately rules out mechanisms (a), (d), and (e) as *primary paths*. (Fig. 7.8)[74] The very similar activation parameters for isomerization and inversion of *cis*- and *trans*-Co(bzac)$_3$

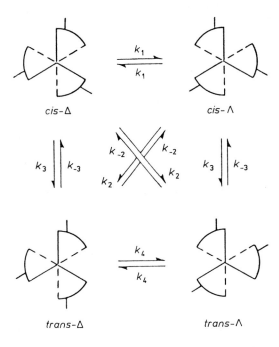

Fig. 7.7 Geometrical and optical isomers of chelates of the type $M(AA_1)_3$ and their interrelationships. The ligand AA_1 is represented as shown.

support a common mechanism for the two processes. This cannot be a completely dissociative one since ligand breakage is some 300 times slower than rearrangement at 96 °C. This leaves, as the favored paths, either a twisting mechanism with 80% rhombic (b) and 20% trigonal (a) character in the activated complex, or a bond-rupture mechanism with 80% axial TBI (c) and 20% SPI (e) contributions. The bond rupture is favored by a high $\Delta H^{\ddagger}$ value (138 kJ mol^{-1} compared with 200 kJ mol^{-1} for complete rupture of one bzac ligand). Inversion must occur at least partly by Co–O bond rupture because inversion is accompanied by linkage isomerism, the latter ingeniously shown by using the deuterated complex:[74]

4

(7.24)

Linkage isomerization cannot occur by a twist mechanism.

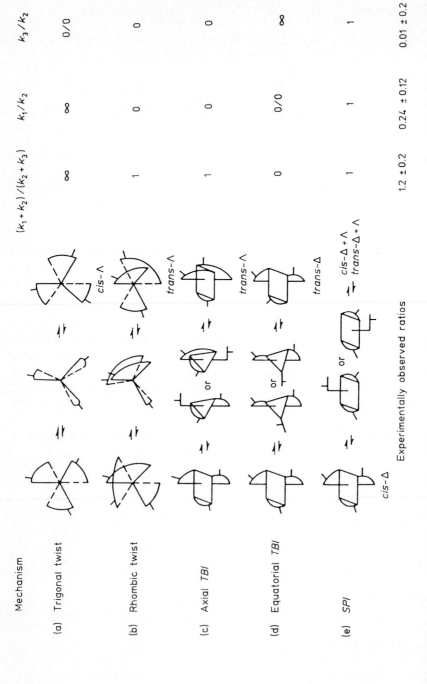

Fig. 7.8 Ratios of rate constants (see Fig. 7.7) anticipated for various intramolecular mechanisms, and values found experimentally for rearrangement of *cis*-Δ-Co(bzac)$_3$ in C$_6$H$_5$Cl at 96°C, Ref. 74.

7.6.3 Complexes of the Type M(L₁)(L₂)

In this group, L_1 and L_2 represent symmetrical terdentate ligands, which may be identical. The main product from the reaction of $[Co(NH_3)_5Cl]Cl_2$ with dien is the *s-fac* $Co(dien)_2^{3+}$ isomer **5a**. On heating solutions of this isomer for long times, small quantities of the other isomeric forms, *u-fac* and *mer*, result, **5b**, and **c** ($R_1 = R_2 = H$). Separation of the three forms is carried out by using Sephadex cation exchange columns. Characterization of the isomers is possible by ^{13}C nmr which together with absorption spectroscopy, is used to study the interconversions.

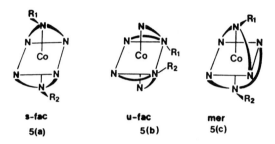

s-fac	u-fac	mer
5(a)	5(b)	5(c)

These occur by intramolecular twist (b) or bond-rupture (c) mechanisms, (See 7.6.1) consult Ref. 75 for full details. Subsequent to this early work, the rearrangements of $Co(ida)_2^-$, $Co(ida)(dien)^+$ and Co(III) complexes of *N*-methyl derivatives of ida and dien have been quantitatively examined, by batch analyses.[76,77] For $Co(ida)(dien)^+$, the kinetics of the sequence:

$$u\text{-}fac \underset{8.0 \times 10^{-6}}{\overset{6.2 \times 10^{-6}}{\rightleftharpoons}} mer \underset{1.0 \times 10^{-6}}{\overset{3.2 \times 10^{-6}}{\rightleftharpoons}} s\text{-}fac \tag{7.25}$$

have been determined with the results shown (40°C, pH 10.0, $\mu = 0.1$ M; all rate constants s^{-1}). The rate constant for the decrease of the circular dichroism of $(+)_{583}$ *u-fac* is $24 \times 10^{-6}s^{-1}$, which being larger than the sum of the rate constants of *u-fac* → *mer* and *u-fac* → *s-fac* (for which k is only $0.5 \times 10^{-6}s^{-1}$), indicates that racemization of *u-fac* occurs without geometrical rearrangement. The *mer* ⇌ *s-fac* change is most easily envisaged in terms of bond rupture, whereas the *u-fac* ⇌ *s-fac* interconversion involves gross changes and both types of racemization mechanisms are likely.[77] All interconversions are base catalyzed:

$$V_{isom} = k[\text{complex}][OH^-] \tag{7.26}$$

and a mechanism similar to that discussed in Sec. 7.9 is suggested.[76,77]

7.6.4 Complexes of the Type M(AA)₂X₂ and M(AA)₂XY

Mechanisms leading to geometrical isomerization in complexes of this type resemble those already discussed with $M(AA_1)_3$ and $M(L_1)(L_2)$, Secs. 7.6.2 and 7.6.3. As well as twist mechanisms, dissociation of either the unidentate ligand or one-ended dissociation of the chelate or even an associative mechanism, must be considered, Fig. 7.9.[56] In general, total dissociation of one AA grouping, as a contributing path, can be discounted. Usually the rates of geometrical isomerization and racemization of complexes of this type are similar to the rate of aquation. This means that rearrangement is often accompanied by a net chemical reaction.

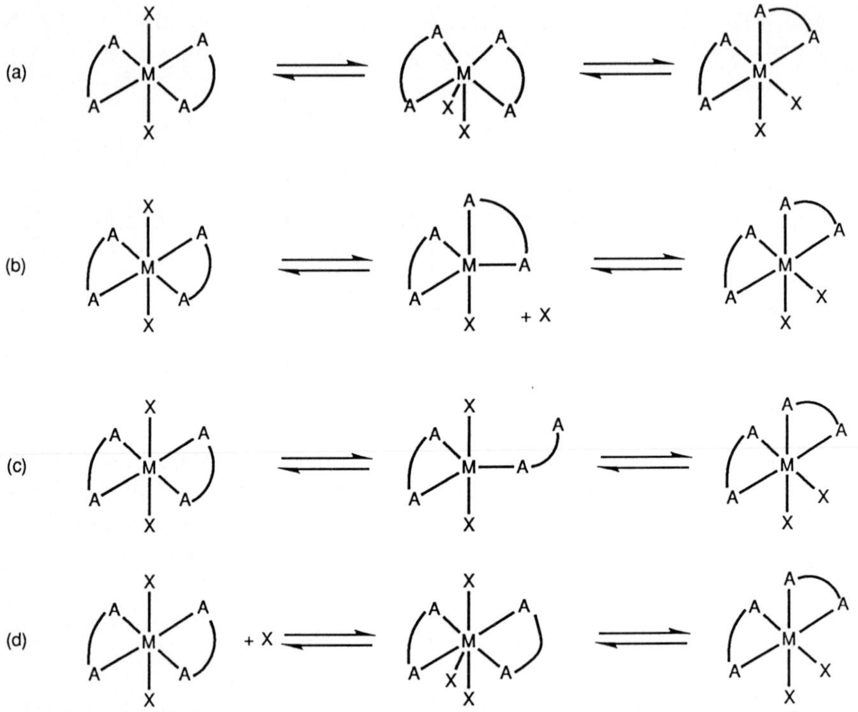

Fig. 7.9 Mechanisms which can interconvert *trans, cis* isomers of the type $M(AA)_2X_2$ include (a) twist, (b) dissociation of a unidentate ligand, (c) one-ended dissociation of one AA ligand and (d) associative attack by X.[56]

For a number of optically active ions of the type *cis*-$M(AA)_2XY^+$, where M = Co and Cr, there is an initial optical rotation change (mutarotation) that is similar in rate to that of acid hydrolysis, for example,

$$cis\text{-}\Lambda\text{-}(+)\text{-}M(AA)_2XY^+ + H_2O \rightarrow cis\text{-}\Lambda\text{-}(+)\text{-}M(AA)_2(H_2O)X^{2+} + Y^- \qquad (7.27)$$

The resultant aqua ion then racemizes to zero rotation more slowly and *without* loss of X^-. In many respects then, the aqua ion is the most suitable one to examine for the relationship of isomerization, racemization, and substitution (using water exchange).[78] For the interconversion

$$cis\text{-}Co(en)_2(H_2O)Cl^{2+} \rightleftharpoons trans\text{-}Co(en)_2(H_2O)Cl^{2+} \qquad (7.28)$$

the associated volumes of activation are 8.0 and 5.1 $cm^3 mol^{-1}$ for the forward and reverse directions.[79] The volume of the activated complex, $V^{\ddagger}$, is therefore some 5–8 $cm^3 mol^{-1}$

larger than either the *cis*- or *trans*-isomer. This appears to rule out a twist mechanism. Bond cleavage of the $Co-H_2O$ bond is believed to occur during the rearrangement and since the exchange

$$trans\text{-}Co(en)_2(H_2{}^*O)_2^{3+} + 2\,H_2O \rightleftharpoons trans\text{-}Co(en)_2(H_2O)_2^{3+} + 2\,H_2{}^*O \qquad (7.29)$$

is stereoretentive and has an associated $\Delta V^{\ddagger} = 5.9$ cm^{3}mol^{-1}, an I_d mechanism, with a tetragonal pyramid transition state, is suggested for both exchange (7.29) and *cis* $\rightleftharpoons$ *trans* rearrangement (7.28).[79, 80] A similar mechanism is suggested for a number of complexes of the type $Co(en)_2(H_2O)X^{n+}$ Ref. 79. To illustrate the complexity of the field consider however the volumes of activation for the equilibration of *cis*- and *trans*- $Co(en)_2(H_2O)_2^{3+}$. The values of $\Delta V^{\ddagger}$ are 15.3 cm^{3}mol^{-1} and 14.3 cm^{3}mol^{-1} in the two directions. They suggest a D mechanism and a trigonal bipyramidal transition state![81] A final (encouraging) comparison of k(cis $\rightarrow$ trans) and k(rac) in Table 7.8[78] shows that for X = OH$^-$, Cl$^-$, Br$^-$, N$_3^-$ and probably SCN$^-$ and H$_2$O (but not NH$_3$) in $Co(en)_2(H_2O)X^{n+}$, racemization must arise predominantly because of the *cis* $\rightarrow$ *trans* conversion, rather than through the operation of a distinct 2-(+)-*cis* $\rightarrow$ (+)-*cis* + (−)-*cis* process.

Table 7.8 Rate Constants (k, s^{-1}) for the Isomerization, Racemization, and Water Exchange of Complexes of the Type $Co(en)_2(H_2O)X^{n+}$ at 25 °C Ref. 78.

X	$10^5 \times k$ (cis $\rightarrow$ trans)	$10^5 \times k$ (trans $\rightarrow$ cis)	$10^5 \times k$(rac)	$10^5 \times k$(exch)
OH$^-$	220	296	220	160
Br$^-$	5.1	15.3	4.7	...
Cl$^-$	2.0	5.4	2.0	...
N$_3^-$	5.2	11.1	4.3	...
NCS$^-$	0.014	0.071	0.022	0.13
H$_2$O	≈ 0.012	0.68	≈ 0.015	1.0
NH$_3$	<0.0001	0.002	0.003	0.10

A number of complexes of the type $M(AA)_2X_2$ are neutral and soluble in organic solvents. They are stereochemically non-rigid on an nmr time scale, that is they display nmr line broadening or collapse in accessible temperature ranges.[2] Concommitant solvolysis problems can usually be avoided. The behavior shown is typified by the complexes $Ti(\beta\text{-diketone})_2X_2$. The *cis*-configuration is invariably the stable one in solution, e.g. **6**

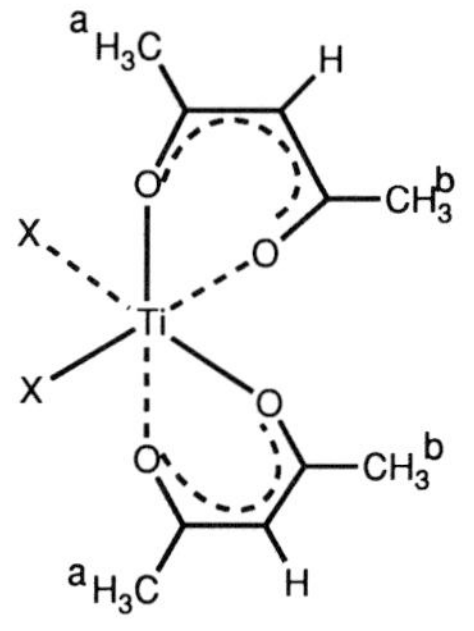

6

Inversion of configuration results in the broadening and collapse of the two distinct nmr signals of the diastereotopic geminal CH_3 groups in X (= $OCH(CH_3)_2$ in **6**). Exchange of CH_3 groups between the two inequivalent sites a and b in acac in **6** is probed by the usual changes in nmr signals (Sec. 3.9.6). The activation parameters associated with these two processes are essentially identical. Since these rates decrease markedly as the R group in X (= OR) increases in size, a common twist mechanism for inversion and exchange is strongly supported.[82] The nmr approach is vital for even a glimmering of understanding of these mechanisms, particularly with the unsymmetrical bidentate ligands (Sec. 7.6.2).[2]

7.6.5 Isomeric Forms As Biological Probes

Although outside the main thrust of the book, attention should be drawn to the important use that octahedral complexes are receiving as molecular probes. Metal complexes can sometimes recognize specific sites on biological material.[83, 84] The array of isomeric forms possible with some metal complexes (previous section) adds a dimension to their use as probes.

The racemic Fe(III) complex containing a hexadentate ligand, Fe(5-Br-ehpg)$^-$, can exist in *racemic* (*RR* and *SS*) and *meso* (*RS*) forms (Structures **7a** and **7b**)[85] which have different solubilities in methanol.

R,R-[Fe(5-Br-ehpg)$^-$]

7(a)

meso-[Fe(5-Br-ehpg)$^-$]

7(b)

The *racemic* isomers have much higher binding affinities than the *meso* form for the bilirubin site on human serum albumin. Bilirubin is the heme breakdown product which binds to albumin and is transported to the liver. The resemblance of the shape of the *RR* form to that of the extended form of bilirubin is striking. (Figure 7.10[85]). This illustrates the potential for using metal complexes for probing macromolecules such as DNA.[84] Chiral metal complexes, particularly of the tris(phenanthroline) type can recognize and react at discrete sites along the DNA strand. They are thus sensitive probes for local DNA helical structures. The binding of the complex may be either through noncovalent insertion between base pairs or by covalent attachment.[84] Ru(II) complexes are particularly useful for these studies since they have intense absorbancies, and are inert to substitution and racemization. The overall affinity increases as the matching of the shape of the complex with the DNA site improves. Strong binding increases the tendency for enantioselectivity.[84] Finally, the β, γ or α, β, δ bound Cr[III]ATP and Co[III]ATP diasteroisomers (see Table 7.4) have been used effectively as chiral probes and inhibitors. Usually only one isomer shows activity, whereas the other isomer may bind, but nonproductively. Thus it is possible to deduce substrate specificity and mechanisms of action of enzymes for which MgATP or similar molecules are substrates.[86]

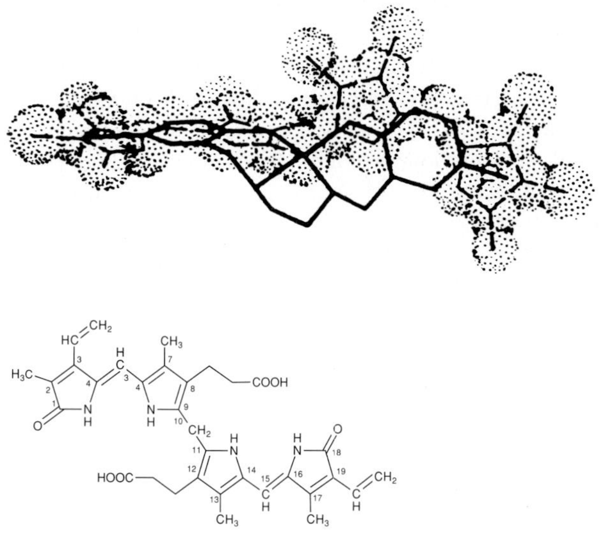

bilirubin IX

Fig. 7.10 Matching of *RR* from **7**a (thick lines) with the extended form of bilirubin IX (dashed lines). The extended conformation of bilirubin is one of several possibilities and it is uncertain which of these it adopts when binding to albumin. The iron atom of **7**a is placed at the central methylene of bilirubin. A number of the peripheral groups have been omitted from the structures for clarity.[85] Reproduced with permission from R. B. Lauffer, A. C. Vincent, S. Padmanabhan and T. I. Meade, J. Amer. Chem. Soc. **109**, 2216 (1987). © (1987) American Chemical Society.

7.7 Four-Coordinated Complexes

The two shapes associated with four-coordination are tetrahedral and square-planar. In the former only optical activity and in the latter only geometrical isomerism is generally encountered in the appropriate molecules.[87]

7.7.1 Optical Isomerism in Tetrahedral Complexes

The preparation, and even more the resolution, of an asymmetric tetrahedral center in Werner-type complexes has been thwarted by the configurational instability of tetrahedral complexes. However the use of ligands of the strongly σ, π bonding type imposes stability and the forma-

tion of optically active tetrahedral complexes of a large number of metals have been reported.[88] Some enantiomeric forms are configurationally stable, e.g. $Mn(CO)(NO)(cp)Ph_3P$
does not racemize in tetrahydrofuran solution for several weeks. The optical forms of
$Mn(COOCH_3)(NO)(cp)Ph_3P$ racemize slowly in benzene solution and the large enthalpy of
activation (128 kJmol^{-1}) suggests a dissociative mechanism for racemization. This is strongly
supported by similar exchange rates with Ph_3P and by the observation that racemization is
slowed in the presence of Ph_3P.[88]

The stereoisomerism in some asymmetric tetrahedral complexes of the type $M(AA_1)_2$ has
been characterized by nmr.[2,88] Tetrahedral complexes of the type $M(AA_1)_2$ are chiral, but
their lability requires nmr methods for measurement of the rates of interconversion of the
enantiomers. The pioneering work of Holm and his collaborators[2] has been consolidated by
Minkin and his Russian group[89] and Zn, Cd and, more recently, Ni chelates of the type
$M(AA_1)_2$ have been extensively studied.

The presence of the prochiral i-Pr(R_3) and $C_6H_5CH_2(R_1)$ in 8[89] allows the investigation of
$\Delta \rightleftharpoons \Lambda$ in CDCl$_3$ by nmr (compare the observations with 3). It is found that the bulkier the
R_3 substituent, the higher the coalescence temperature, and thus the higher the energy barrier
for inversion. This is believed to arise from steric hindrance effects in forming the intermediate
planar form, i.e. (viewed down C_2 axis)[2,89].

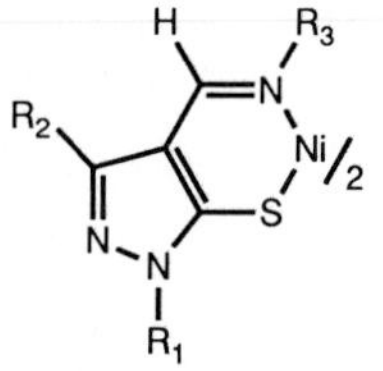

8

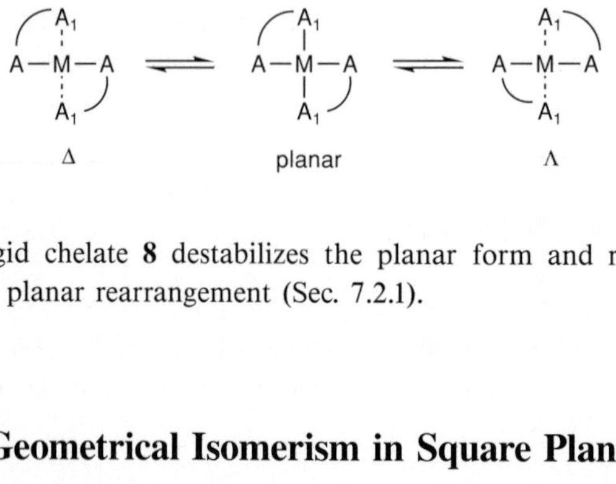

(7.30)

The rigid chelate **8** destabilizes the planar form and reduces the lability of the tetrahedral $\rightleftharpoons$ planar rearrangement (Sec. 7.2.1).

7.7.2 Geometrical Isomerism in Square Planar Complexes

Most of the complexes studied are of the type PtL_2X_2 and PtL_2XR where L is a neutral
ligand (often a substituted phosphine) and R and X are alkyl and halide groups respectively.
Usually the *cis*- and *trans*-isomers are separately stable in solution, but equilibrate on the addition of free ligand L (or L_1). Two mechanisms have found general support for explaining the
catalyzed isomerizations.[90]

(a) the consecutive displacement mechanism

$$X-\underset{\underset{L}{|}}{\overset{\overset{X}{|}}{Pt}}-L \quad \underset{-L}{\rightleftharpoons} \quad X-\underset{\underset{L}{|}}{\overset{\overset{X}{|}}{Pt}}\cdots\overset{L}{\underset{L}{\diagdown}} \tag{7.31}$$

A

$$\left[\underset{\underset{L}{|}}{\overset{\overset{X}{|}}{L-Pt}}-L\right]^{+} X^{-}$$

$$X-\underset{\underset{L}{|}}{\overset{\overset{L}{|}}{Pt}}-X \quad \underset{L}{\overset{-L}{\rightleftharpoons}} \quad L-\underset{\underset{L}{|}}{\overset{\overset{L}{|}}{Pt}}\cdots\overset{X}{\underset{X}{\diagdown}} \tag{7.32}$$

B

In the first step of scheme (7.31–7.32), L replaces Y from the *cis* isomer to form an ionic intermediate $[PtL_3X]^+X^-$. In the second step, X^- replaces L to produce the *trans*-isomer (or revert to the *cis*-form). The two steps are required to get around the fact that substitution at Pt(II) is stereospecific (Sec. 4.6). It appears to operate in a variety of conditions but may be more complicated than indicated in Scheme (7.31) and (7.32).[90]

(b) the pseudorotation mechanism

The five-coordinated species A may pseudorotate to form B directly without the intermediacy of the ionic species (Schemes (7.31) and (7.32)).

Pseudorotation is well established in 5-coordinated species involving the main group elements and is best described by the Berry mechanism which interconverts two trigonal bipyramids via a square pyramid. Its operation here is difficult to reconcile with the highly stereospecific nature of substitution in Pt(II). Nevertheless, the mechanism has had substantial support. It may very well be[90] that (a) is favored by polar solvents and that (b) is prevalent in nonpolar media. The associated reaction profiles are shown in Fig. 7.11.

Spontaneous isomerization is difficult to validate since traces of ligand are almost impossible to avoid. Photochemically-induced rearrangements may go *via* a tetrahedral intermediate (Sec. 7.2.1), although there is no evidence for this.

(c) the dissociative mechanism

Perhaps the most controversial suggestion is that spontaneous isomerizations may proceed through a dissociative mechanism, invoking two three-coordinated intermediates (probably T-shaped). The labile intermediates may be "cis-like" (A) or "trans-like" (B) in structure. Applied to the monoorganoplatinum(II) complexes, PtL_2XR, the scheme is[91]

$$\underset{L}{\overset{L}{\diagdown}}\underset{R}{\overset{X}{\diagup}}Pt \quad \rightleftharpoons \quad \underset{L}{\overset{L}{\diagdown}}\underset{R}{\diagup}Pt^{+} \quad +X^{-}$$

(A)

$$\tag{7.33}$$

$$\underset{R}{\overset{L}{\diagdown}}\underset{L}{\overset{X}{\diagup}}Pt \quad \underset{X^{-}}{\longleftarrow} \quad \underset{R}{\overset{L}{\diagdown}}\underset{L}{\diagup}Pt^{+}$$

(B)

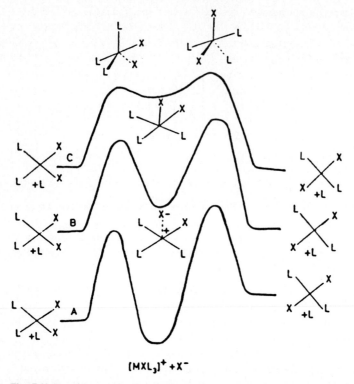

$$(MXL_3)^+ + X^-$$

Fig. 7.11 Reaction profiles for L-catalyzed isomerization of *cis*-to-*trans* ML_2X_2. In (A) an ionic intermediate is favored by a polar solvent. In (B) ion-pair formation arises with a less polar solvent. In (C) a non-polar solvent promotes a 5-coordinated intermediate. In (C), pseudo-rotation occurs. Based on D. G. Cooper and J. Powell, J. Amer. Chem. Soc. **95**, 1102 (1973); see also Ref. 90. Reproduced with permission from D. G. Cooper and J. Powell, J. Amer. Chem. Soc. **95**, 1102 (1973). © (1973) American Chemical Society.

The idea of a 3-coordinated intermediate, although unusual, has been supported in other work.[90,91] Not surprisingly, the mechanism has been challenged. It is suggested that solvent (S) replaces X from *cis*-PtL_2XR in a pre-equilibrium to produce *cis*-PtL_2RS^+ which undergoes slower rearrangement to *trans*-PtL_2RS^+ (7.34). The latter is converted by X^- to *trans*-PtL_2XR. It is unclear how the *cis, trans* rearrangement occurs.[46(c),92] Some features of both (b) and (c) may apply, particularly in a strong dissociating medium.[91]

$$\underset{L}{\overset{L}{}}\!\!\!\!\!\!\underset{R}{\overset{S}{\text{Pt}}} \quad \longrightarrow \quad \underset{R}{\overset{L}{}}\!\!\!\!\!\!\underset{L}{\overset{S}{\text{Pt}}} \tag{7.34}$$

Finally, the relationship of structural isomerization to other processes is illustrated in the behavior of $PtH_2(Me_3P)_2$. This exists as an equilibrium mixture of *trans*- and *cis*-isomers. For the equilibrium

$$\textit{trans-}PtH_2(Me_3P)_2 \;\rightleftharpoons\; \textit{cis-}PtH_2(Me_3P)_2 \quad k_1, k_{-1} \tag{7.35}$$

$k_1 = 0.079$ s^{-1} and $k_{-1} = 0.027$ s^{-1} at $-60\,^{\circ}$C in acetone-d_6. The isomerization takes place either by the pseudorotation (b) pathway or by the dissociation (c) mechanism. At temperatures from $-35\,^{\circ}$C to $+35\,^{\circ}$C nmr shows exchanges of the *cis*- and *trans*-isomers (with traces of PMe$_3$) which are more rapid than the isomerization. The Pr-H bonds remain intact. The faster rates of phosphine exchange from the *cis* isomer suggest a dissociative exchange process arising from a strong labilizing *trans* hydride ligand. Above $10\,^{\circ}$C, reductive elimination of H$_2$ occurs slowly. This is monitored by ir and a mechanism (7.36) is suggested.

$$\text{trans-PtH}_2(\text{Me}_3\text{P})_2 \underset{}{\overset{\text{fast}}{\rightleftharpoons}} \text{cis-PtH}_2(\text{Me}_3\text{P})_2 \longrightarrow \text{Pt}(\text{Me}_3\text{P})_4 + \text{Pt}_x + \text{H}_2 \qquad (7.36)$$

There is an inverse kinetic isotope effect, $k_H/k_D = 0.45 \pm 0.1$ (for decomposition of PtH$_2$(Me$_3$P)$_2$) compared with PtD$_2$(Me$_3$P)$_2$) in THF at $21\,^{\circ}$C. This supports the previous prediction of nearly complete H$-$H formation in a "late" transition state.

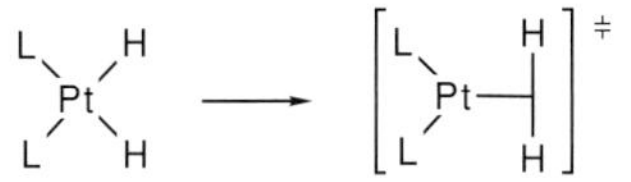

$$(7.37)$$

Thus, geometrical isomerization, phosphine ligand exchange and reductive elimination of H$_2$ all proceed by independent kinetic processes.[93]

7.8 Five-, Seven- and Eight-Coordinated Complexes

Werner-type complexes with these coordination numbers have been characterized. However a large majority of the complexes showing these coordination numbers are organometallic in nature and generally outside the scope of this book. Examples are shown in Structures **9–11**, and discussion of the associated rearrangements will be necessarily brief.

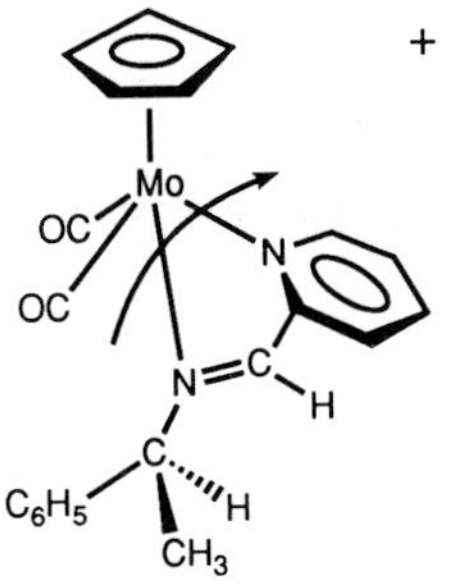

9

A large number of optically active square pyramidal organometallic complexes have been described.[88] That shown in **9** is one of a pair of diastereoisomers that can be separated by fractional crystallization into (+) and (−) rotating components. It is optically stable in acetone at room temperature. At higher temperatures a single first-order epimerization[94] is observed and an equilibrium involving the two diastereoisomers of **9** is established, possibly by intramolecular rotation of the chelate by 180° with respect to the rest of the molecule. The fast interconversion of square-pyramidal complexes of the type $M(cp)(CO)_2LL_1$ are better studied by dynamic nmr spectroscopy rather than by polarimetry of the resolved species.[88]

The complexity of rearrangements in molecules with the higher coordination numbers 7 and 8 is awesome! Seven-coordinated complexes such as $Ti(Me_2dtc)_3Cl$ are stereochemically nonrigid in solution, even at lowered temperatures, because of the easy interconversion of the shapes associated with this coordination number. The complex $Ti(Me_2dtc)_3(cp)$ **10** on the other hand has a pentagonal bipyramidal structure (depicted in **10**, cp occupying one axial position) and metal centered rearrangements are slow on an nmr time scale even at room temperature. A double facial twist mechanism is postulated for the exchange of the equatorial ligands and the unique Me_2dtc ligand.[95]

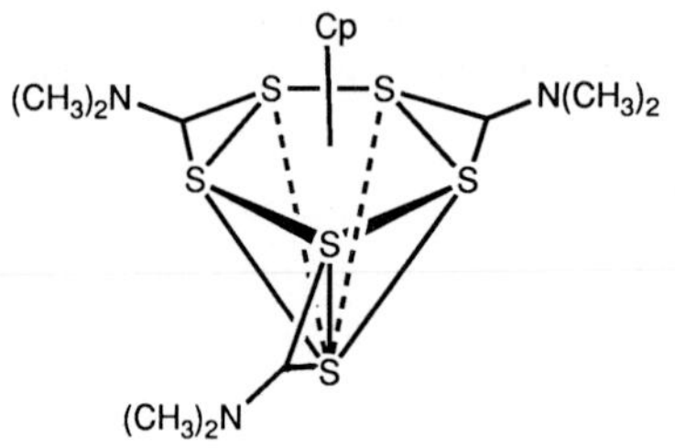

10

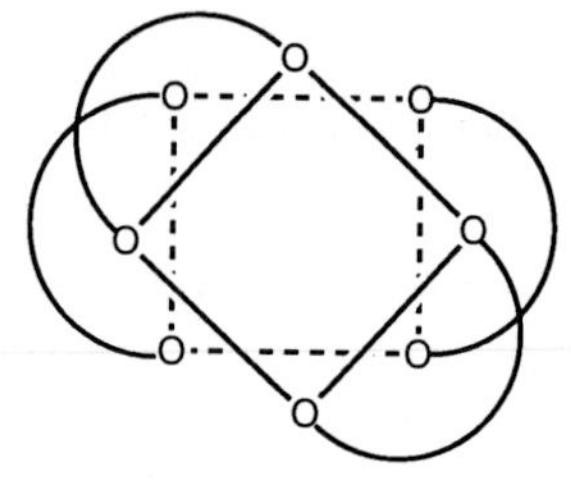

11

$$\left(\bigcirc\hspace{-0.3em}\bigcirc = \beta\text{-diketones} \right)$$

The ^1H nmr spectrum of $Zr(Hf)(acac)_4$ in $CHClF_2$ shows two Me proton resonances of equal intensity in the slow exchange region only at lowered temperature ($-170\,°C!$). This observation is consistent with a square antiprism geometry **11**. At $-145\,°C$ coalescence of the signals occurs arising from complicated intramolecular rearrangements.[96]

7.9 Inversion and Proton Exchange at Asymmetric Nitrogen

Coordination of ammonia or a substituted ammonia to a metal ion alters markedly the $N-H$ dissociation rate (see Sec. 6.4.2). Since also proton dissociation of complexed ammines is base-catalyzed, then exchange can be made quite slow in an acid medium. Thus, in a coordinated system of the type **12**, containing an asymmetric nitrogen atom (and this is the only potential source of optical activity), there is every chance for a successful resolution in acid conditions, since inversion is expected *only* after deprotonation. It was not until 1966 that this was successfully performed, however, using the complex ion **12**.[97] A number of Co(III)[32], Pt(II)[98] and Pt(IV)[99] complexes containing sarcosine or secondary amines have been resolved and their racemizations studied.[32] Asymmetric nitrogen centers appear confined to d^6 and d^8

systems. The inversion at the N center (as well as the H exchange) has also been probed for Pd(II)[100] and Pt(II)[101] complexes, by using dynamic nmr without the necessity of resolving the complexes.

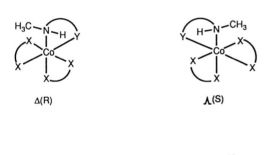

12

Molecules which contain a chiral cobalt as well as an asymmetric nitrogen exist in four possible optical isomeric forms. These are represented for $Co(sar)(hbg)_2^{2+}$, hbg = $NH_2C(=NH)NHC(=NH)NH_2$ in Fig. 7.12. All four optically-active isomers have been isolated and characterized by cd, nmr and vis/uv absorption spectroscopy. The kinetics of

Δ(R)   Λ(S)

Δ(S)   Λ (R)

Fig. 7.12 The four possible isomers of $Co(XX)_2(NY)^{n+}$. XX may represent $C_2O_4^{2-}$ or $NH_2C(=NH)NHC(=NH)NH_2$ with NY being $CH_3NH(CH_2)_2NH_2$ or $CH_3NHCH_2CO_2^-$.

racemization for the chiral cobalt centre $(\Delta \rightleftharpoons \Lambda)$ is faster than for the asymmetric nitrogen $(R \rightleftharpoons S)$.[102] (Table 7.9) It is however not necessary to resolve these complexes in order to study the inversion behavior. For example, the complex $Co(C_2O_4)_2(Me_3en)^-$ could be separated by column chromatography under acid conditions, into the racemic pairs $\Lambda(S) \Delta(R)$ and $\Lambda(R) \Delta(S)$. Either racemic pair will revert in solution into a mixture

$$\Lambda(S) \; \Delta(R) \; \rightleftharpoons \; \Lambda(R) \; \Delta(S) \quad k_1, k_{-1}, K_1 \qquad (7.38)$$

the dynamics of which may be measured by periodic withdrawal of a sample, separation of the pairs by a HPLC method and analysis of the chromatograms by area measurement (Fig. 7.13).[103] The first-order epimerization[94] rate constant, k_{ep}, is equal to the sum of k_1 and k_{-1}, and the ratio of the signals at equilibrium yields $K_1 = k_1/k_{-1}$.[103, 104]

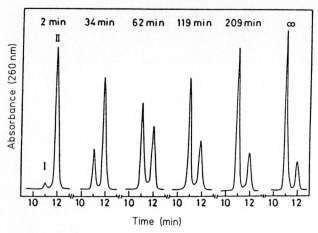

Fig. 7.13 Variation in the elution curve during the epimerization of $\Delta(R)\Lambda(S)$-Co$(C_2O_4)_2$(Me$_3$en)$^-$ at pH 6.0 and 34.0 °C. I = $\Delta(S)\Lambda(R)$ isomer; II = $\Delta(R)\Lambda(S)$ isomer. [104] Reproduced with permission from M. Kojima, T. Hibono, Y. Ouyang and J. Fujita, Inorg. Chim. Acta **117**, 1 (1986).

For all systems studied so far it is found that the rates of racemization R and proton exchange E (measured either by D$_2$O exchange (Sec. 3.9.5) or by nmr line coalescence methods (Sec. 3.9.6)) are both first-order in [OH$^-$].

$$V_{R,E} = k_{R,E}[\text{complex}]\,[\text{OH}^-] \tag{7.39}$$

Since proton exchange is usually measured by nmr methods in D$_2$O (Secs. 3.9.5 and 3.9.6), the more appropriate rate law is

$$V_{R,E} = k_{R,E}[\text{complex}]\,[\text{OD}^-] \tag{7.40}$$

where [OD$^-$] is estimated from pD = pH + 0.4 and $K_{D_2O} = 0.195\,K_w$. Racemization and exchange data are shown in Table 7.9 for a selected number of cobalt(III) systems. [97, 99, 102–105] Since $k_E \gg k_R$, the loss of a proton from the N$-$H group must rarely lead to racemization. The results can be accommodated by a scheme (M$_1$ contains the metal center):

$$\begin{array}{c} M \\ \diagdown \\ R_1\!-\!\overset{|}{N}\!:\!H + OH^- \\ \diagup \\ R_2 \\ R \end{array} \quad\rightleftharpoons\quad \begin{array}{c} M \\ \diagdown \\ R_1\!-\!\overset{|}{N}\!: + H_2O \\ \diagup \\ R_2 \\ R \end{array} \qquad k_1, k_{-1} \tag{7.41}$$

$$\begin{array}{c} M \\ \diagdown \\ R_1\!-\!\overset{|}{N}\!: \\ \diagup \\ R_2 \\ R \end{array} \quad\rightleftharpoons\quad \begin{array}{c} M \\ \diagup \\ :\!\overset{|}{N}\!-\!R_1 \\ \diagdown \\ R_2 \\ S \end{array} \qquad k_2, k_2 \tag{7.42}$$

$$\begin{array}{c} M \\ \diagup \\ :\!\overset{|}{N}\!-\!R_1 + H_2O \\ \diagdown \\ R_2 \\ S \end{array} \quad\rightleftharpoons\quad \begin{array}{c} M \\ \diagup \\ H\!:\!\overset{|}{N}\!-\!R_1 + OH^- \\ \diagdown \\ R_2 \\ S \end{array} \qquad k_{-1}, k_1 \tag{7.43}$$

Table 7.9 Rate Parameters (k_E, k_R, $M^{-1}s^{-1}$) for Secondary $N-H$ Hydrogen Exchange and Racemization in Metal Complexes

Complex	Temp.	k_E^a	k_R	k_E/k_R	Ref.
$Co(NH_3)_4(Meen)^{3+}$	34.3	3.0×10^7	2.5×10^2	1.2×10^5	99
$Co(C_2O_4)_2(Meen)^-$	34.0	1.8×10^5	5.7×10^{2b}	3.2×10^2	103
$Co(acac)_2(Meen)^+$	34.0	9×10^3	7.0×10^{-2b}	1.3×10^5	103
$Co(acac)_2(N-Phen)^+$	34.0	4×10^7	1.4×10^{5b}	2.9×10^2	104
$Co(NH_3)_4(sar)^{2+}$	33.3	1.2×10^8	3.3×10^4	3.6×10^3	97
$Co(sar)(hbg)_2^{2+}$	39.8	3.6×10^3	21.5^c	1.7×10^2	102
trans-$Co(dien)_2^{3+}$	35.0	1.0×10^8	2.4×10^2	4.2×10^5	105
$Pt(NH_3)_2(Meen)^{2+}$	25.0	6.6×10^4	3.2×10^2	2.1×10^2	98
$Pt(en)(Meen)Cl_2^{2+}$	34.3	2×10^{10}	6.5×10^5	3.1×10^4	99

[a] Deuteration rate constant in D_2O.   [b] for k_1 in Eqn. (7.41).   [c] For the $\Lambda \to \Delta$ conversion, $k = 1.5 \times 10^2 M^{-1}s^{-1}$ at 35.0°C

Step (7.41) leads to hydrogen exchange, $k_1 = k_E$. Only step (7.42) leads to racemization and is considered pH-independent. Most of the time the amide complex becomes reprotonated with retention of configuration. If a steady-state concentration for the amide form is assumed,

$$k_R = \frac{2k_1 k_2}{k_{-1} + k_2} = \frac{2k_1 k_2}{k_{-1}} = \frac{2k_2 K_a}{K_w} \tag{7.44}$$

with K_a, the acid dissociation constant for the $N-H$ proton. The factor 2 enters into (7.44) because the racemization rate constant is twice the inversion rate constant (k_2). A more involved treatment is discussed in Ref. 32. Attention has been focused on the factors affecting the values of k_R, k_E and their ratio. The values of k_E/k_R are in the range 1.7×10^2 to 4.2×10^5 (Table 7.9). The reasons for these variations are not well understood [103] but are undoubtedly linked to the stability of the racemized amido intermediate. [32, 98, 100, 103] In general, enthalpies of activation for exchange and racemization are in the range 50–60 kJ mol^{-1} and 80–100 kJ mol^{-1} respectively. The replacement of NH_3 by acac or hbg (both producing a planar chelate ring with π-electrons) reduces both exchange and racemization rates by 10^3–10^4. The similarity of $Co(dien)_2^{3+}$ and $Co(NH_3)_4(Meen)^{3+}$ in behavior indicates that the coupling of chelate rings across the NH center has little effect. [105]

The asymmetry at certain nitrogen centers has been cleverly exploited to reveal features of base hydrolysis. The complex $Co(trenen)Cl^{2+}$, represented in **13**, contains an asymmetric N *trans* to a chloride group and it has been resolved. It is found that the (+) chloro to ($\pm$)

13

hydroxo transformation is unimportant during the base hydrolysis. It is concluded that the stereochemically labelled nitrogen could not have gone planar in the course of base hydrolysis. [106] It is uncertain whether it is the site of the labilising amido group even though it is the most labile $N-H$ moiety (Sec. 4.3.5). A similar approach used a cobalt(III) complex where the asymmetric nitrogen is *cis* to the departing chloride, and probably the site of amido formation. Now, base hydrolysis is accompanied by complete racemization and this is reasonably explainable in terms of a symmetrical five-coordinate intermediate with planar nitrogen. [107]

Inversion at the N center is coupled to conformational changes in a chelate ring. The kinetics of inversion at asymmetric N centers in complexes of tetraaza linear or macrocyclic ligands have received scant attention. There are five configurational isomers of the planar complex $Ni([14]aneN_4)^{2+}$, Sec. 3.1.1. The interconversions between such structures are base catalyzed with second-order rate constants covering a small range from 1.2×10^2 to 2.4×10^3 $M^{-1}s^{-1}$ Refs. 108–110.

The blue and (stable) red isomers of $Cu(tet\,a)^{2+}$ **14** and **15** differ in the configuration of a single asymmetric N center. [111] The blue-to-red conversion entails, as well as the single

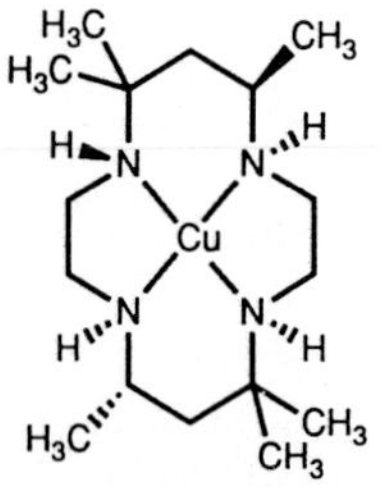

blue complex
14

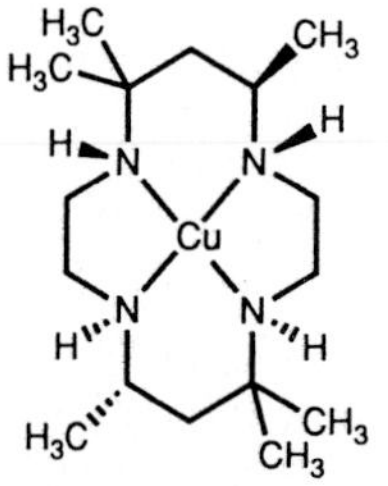

red complex
15

nitrogen inversion, the favorable inversion of two adjacent unstable chelate rings. These are an eclipsed to gauche five-membered ring and a skew-boat to chair six-membered ring. The conversion is both acid catalyzed (to rupture $Cu-N$ bonding) and base catalyzed (to produce the deprotonated species). The kinetics of the conversion at pH >10 requires stopped-flow monitoring. For the major path,

$$-d\,[Cu(tet\,a)\,blue^{2+}]_{total}/dt = \frac{k_1 K_{OH}[OH^-]\,[Cu(tet\,a)\,blue^{2+}]_{total}}{1 + K_{OH}[OH^-]} \qquad (7.45)$$

consistent with a scheme

$$Cu(tet\,a)\,blue^{2+} + OH^- \rightleftharpoons Cu(tet\,a)(OH)\,blue^+ \qquad K_{OH},\ fast \qquad (7.46)$$

$$Cu(tet\,a)(OH)blue^+ \rightarrow Cu(tet\,a)red^{2+} + OH^- \qquad k_1,\ slow \qquad (7.47)$$

Inversion occurs through a copper-hydroxy species, $K_{OH} = 51.6\ M^{-1}$, $k_1 = 5.5\ s^{-1}$ in $\mu = 5.0\ M$ at 25°C Ref. 112. A similar kinetic pattern is observed for the blue-to-red change with

Cu(tet b)$^{2+}$ but the reactions are much slower since two inverting nitrogens and less favorable ring conformational changes are involved.[113]

Finally, other chiral donor atoms, P,[114] As (see Problem 9) and S[115] in metal complexes have been less systematically studied than nitrogen.[32] Inversion rates in metal ion-thioether complexes have been measured by nmr, but the species are too labile to allow a successful resolution.[32, 115]

References

1. C. J. Hawkins, Absolute Configuration of Metal Complexes, Wiley-Interscience, NY, 1971, Chap. 1.
2. L. H. Pignolet and G. N. La Mar, Dynamics of Intramolecular Rearrangements in NMR of Paramagnetic Molecules G. N. La Mar, W. G. Horrocks, Jr., and R. H. Holm, eds. Academic, NY, 1974, Chap. 8.; R. H. Holm, Stereochemically Nonrigid Metal Chelate Complexes in Dynamic Nuclear Magnetic Resonance Spectroscopy, L. M. Jackman and F. A. Cotton, eds. Academic, NY, 1974 Chap. 9.
3. C. J. Hawkins and J. A. Palmer, Coordn. Chem. Revs. **44**, 1 (1982).
4. C. J. Hawkins, R. M. Peachey and C. L. Szoredi, Aust. J. Chem. **31**, 973 (1978) and references therein.
5. Y. Kuroda, N. Tanaka, M. Goto and T. Sakai, Inorg. Chem. **28**, 997 (1989) and references therein.
6. J. K. Beattie, Adv. Inorg. Chem. **32**, 1 (1988).
7. L. H. Pignolet, W. D. Horrocks, Jr. and R. H. Holm, J. Amer. Chem. Soc. **92**, 1855 (1970); G. N. La Mar and E. O. Sherman, J. Amer. Chem. Soc. **92**, 2691 (1970).
8. J. L. McGarvey and J. Wilson, J. Amer. Chem. Soc. **97**, 2531 (1975).
9. L. Campbell, J. J. McGarvey and N. G. Samman, Inorg. Chem. **17**, 3378 (1978).
10. A. M. Martin, K. J. Grant and E. J. Billo, Inorg. Chem. **25**, 4904 (1986).
11. J. H. Coates, D. A. Hadi, S. F. Lincoln, H. W. Dodgen and J. P. Hunt, Inorg. Chem. **20**, 707 (1981).
12. Microwave T-jump and ^{14}N-nmr studies rule out 4 → 5 coordinated interconversion as the rds in the system in C$_6$H$_5$Cl: Ni(bbh) + 2 L ⇌ Ni(bbh)L + L ⇌ Ni(bbh)L$_2$ (bbh is a tetradentate ligand and L is a substituted pyridine), M. Cusamano, J. Chem. Soc. Dalton Trans. 2133, 2137 (1976); J. Sachinidis and M. W. Grant, J. Chem. Soc. Chem. Communs. 157 (1978).
13. A. F. Godfrey and J. K. Beattie, Inorg. Chem. **22**, 3794 (1983).
14. J. K. Beattie, M. T. Kelso, W. E. Moody and P. A. Tregloan, Inorg. Chem. **24**, 415 (1985).
15. K. J. Ivin, R. Jamison and J. J. McGarvey, J. Amer. Chem. Soc. **94**, 1763 (1972).
16. R. D. Farina and J. H. Swinehart, Inorg. Chem. **11**, 645 (1972).
17. M. G. B. Drew, Prog. Inorg. Chem. **23**, 67 (1977).
18. D. L. Kepert, Inorganic Stereochemistry, Springer Verlag, Berlin 1982.
19. P. Gutlich, Struct. Bonding Berlin **44**, 83 (1981).
20. R. L. Martin and A. H. White, Trans. Metal Chem. **5**, 113 (1969).
21. J. K. Beattie, N. Sutin, D. H. Turner and G. W. Flynn, J. Amer. Chem. Soc. **95**, 2052 (1973).
22. H. Toftlund, Coordn. Chem. Revs. **94**, 67 (1989).
23. J. McGarvey and I. Lawthers, J. Chem. Soc. Chem. Communs. 906 (1982); J. J. McGarvey, I. Lawthers, K. Heremans and H. Toftlund, J. Chem. Soc. Chem. Communs. 1575 (1984).
24. J. DiBenedetto, V. Arkle, H. A. Goodwin and P. C. Ford, Inorg. Chem. **24**, 455 (1985).
25. J. K. Dose, M. A. Hoselton, N. Sutin, M. F. Tweedle and L. J. Wilson, J. Amer. Chem. Soc. **100**, 1141 (1978); K. A. Reeder, E. V. Dose and L. J. Wilson, Inorg. Chem. **17**, 1071 (1978).
26. A. J. Conti, C. L. Xie and D. N. Hendrickson, J. Amer. Chem. Soc. **111**, 1171 (1989).
27. I. Lawthers and J. J. McGarvey, J. Amer. Chem. Soc. **106**, 4280 (1984).
28. R. A. Binstead, J. K. Beattie, T. G. Dewey and D. H. Turner, J. Amer. Chem. Soc. **102**, 6442 (1980).

29. N. Sutin, Prog. Inorg. Chem. **30**, 441 (1983) and references therein.

30. J. H. Dawson and K. S. Eble, Adv. Inorg. Bioinorg. Mechs. **4**, 1 (1986); T. L. Poulos, Adv. Inorg. Bioinorg.Mechs **7**, 1 (1988).

31. D. P. Fairlie and H. Taube, Inorg. Chem. **24**, 3199 (1985).

32. W. G. Jackson and A. M. Sargeson in Rearrangements in Ground and Excited States, P. de Mayo, ed., Academic, NY, 1980 Vol. 2, p. 273.

33. W. G. Jackson and S. S. Jurisson, Inorg. Chem. **26**, 1060 (1987).

34. R. van Eldik, Adv. Inorg. Bioinorg. Mechs. **3**, 275 (1984).

35. W. G. Jackson, S. S. Jurrisson and B. C. McGregor, Inorg. Chem. **24**, 1788 (1985).

36. R. G. Pearson, P. M. Henry, J. G. Bergmann and F. Basolo, J. Amer. Chem. Soc. **76**, 5920 (1954).

37. A. Haim and N. Sutin, J. Amer. Chem. Soc. **88**, 434 (1966).

38. V. Houlding, H. Maecke and A. W. Adamson, Inorg. Chem. **20**, 4279 (1981).

39. W. Weber, H. Maecke and R. van Eldik, Inorg. Chem. **25**, 3093 (1986).

40. W. G. Jackson, D. P. Fairlie and M. L. Randall, Inorg. Chim. Acta **70**, 197 (1983).

41. R. K. Murmann and H. Taube, J. Amer. Chem. Soc. **78**, 4886 (1956).

42. W. G. Jackson, G. A. Lawrance, P. A. Lay and A. M. Sargeson, J. Chem. Soc. Chem. Communs. 70 (1982).

43. M. Mares, D. A. Palmer and H. Kelm, Inorg. Chim. Acta **27**, 153 (1978).

44. W. G. Jackson, G. A. Lawrance, P. A. Lay and A. M. Sargeson, Aust. J. Chem. **35**, 1561 (1982).

45. W. G. Jackson, G. A. Lawrance, P. A. Lay and A. M. Sargeson, Inorg. Chem. **19**, 904 (1980).

46. B17 (a) p. 205 (b) p.195 (c) p.250

47. J. Reibenspies and R. D. Cornelius, Inorg. Chem. **23**, 1563 (1984).

48. W. L. Purcell, Inorg. Chem. **22**, 1205 (1983).

49. M. F. Hoq, C. R. Johnson, S. Paden and R. E. Shepherd, Inorg. Chem. **22**, 2693 (1983).

50. W. D. Harman and H. Taube, J. Amer. Chem. Soc. **110**, 5403 (1988).

51. C. Daul, E. Diess, J.-N. Gex, D. Perret, D. Schaller and A. von Zelewsky, J. Amer. Chem. Soc. **105**, 7556 (1983).

52. E. Kauffmann, J. L Dye, J.-M. Lehn and A. I. Popov, J. Amer. Chem. Soc. **102**, 2274 (1980).

53. Y. Ducommun, L. Helm, A. E. Merbach, B. Hellquist and L. I. Elding, Inorg. Chem. **28**, 377 (1989).

54. N. Serpone and D. G. Bickley, Prog. Inorg. Chem. **17**, 392 (1972).

55. B43.

56. G. A. Lawrance and D. R. Stranks, Acc. Chem. Res. **12**, 403 (1979).

57. G. A. Lawrance and D. R. Stranks, Inorg. Chem. **17**, 1804 (1978).

58. R. G. Wilkins and M. J. G. Williams, J. Chem. Soc. 1763 (1957).

59. A. Yamagishi, Inorg. Chem. **25**, 55 (1986).

60. F. Basolo, J. C. Hayes and H. M. Neumann, J. Amer. Chem. Soc. **76**, 3807 (1954).

61. E. L. Blinn and R. G. Wilkins, Inorg. Chem. **15**, 2952 (1976).

62. G. A. Lawrance and D. R. Stranks, Inorg. Chem. **16**, 929 (1977) and references therein.

63. J. C. Bailey, Jr., J. Inorg. Nucl. Chem., **8**, 165 (1958); W. G. Gehman, Ph. D. thesis, Penn. State, 1954; L. Seiden, Ph. D. thesis, Northwestern, 1957.

64. P. C. Ray and N. K. Dutt, J. Indian Chem. Soc. **20**, 81 (1943).

65. A. Rodger and B. F. G. Johnson, Inorg. Chem. **27**, 3061 (1988).

66. These are termed diastereotopic centers,[2] see Inset, p. 334.

67. P. R. Rubini, Z. Poaty, J.-C. Boubel, L. Rodehuser and J.-J. Delpuech, Inorg. Chem. **22**, 1295 (1983).

68. P. R. Rubini, L. Rodehuser and J. J. Delpuech, Inorg. Chem. **18**, 2962 (1979).

69. This was demonstrated for $Co(C_2O_4)_3^{3-}$ and $Cr(C_2O_4)_3^{3-}$ in one of the earliest radioisotopic exchange experiments. Exchange of these ions with ^{11}C-labeled $C_2O_4^{2-}$ is very much slower than racemization; F. A. Long, J. Amer. Chem. Soc. **61**, 571 (1939); 63, 1353 (1941).

70. L. Damrauer and R. M. Milburn, J. Amer. Chem. Soc. **90**, 3884 (1968); **93**, 6481 (1971).

71. L. G. Vanquickenborne and K. Pierloot, Inorg. Chem. **20**, 3673 (1981).

72. G. A. Lawrance, M. J. O'Connor, S. Suvachittanont, D. R. Stranks and P. A. Tregloan, Inorg. Chem. **19**, 3443 (1980).

73 The enhanced symmetry of the *cis* compared with the *trans* isomer allows their easy characterization by early elegant nmr experiments; R. C. Fay and T. S. Piper, J. Amer. Chem. Soc. **84**, 2303 (1962); **85**, 500 (1963).

74. A. Y. Girgis and R. C. Fay, J. Amer. Chem. Soc. **92**, 7061 (1970).

75. G. H. Searle, F. R. Keene and S. F. Lincoln, Inorg. Chem. **17**, 2362 (1978).

76. H. Kawaguchi, T. Ama and T. Yasui, Bull. Chem. Soc. Japan **57**, 2422 (1984).

77. H. Kawaguchi, M. Shimizu, T. Ama and T. Yasui, Bull Chem. Soc. Japan **62**, 753 (1989).

78. M. L. Tobe, in Studies on Chemical Structure and Reactivity, J. H. Ridd, ed. Methuen, London, 1966; W. G. Jackson and A. M. Sargeson, Inorg. Chem. **17**, 1348 (1978); W. G. Jackson, Inorg. Chim. Acta **126**, 147 (1987).

79. Y. Kitamura, S. Nariyuki and K. Yoshitani, Inorg. Chem. **24**, 3021 (1985).

80. S. B. Tong, H. R. Krouse and T. W. Swaddle, Inorg. Chem. **15**, 2643 (1976).

81. D. R. Stranks and N. Vanderhoek, Inorg. Chem. **15**, 2639 (1976).

82. R. C. Fay and A. F. Lindmark, J. Amer. Chem. Soc. **105**, 2118 (1983).

83. S. E. Sherman, D. Gibson, A. H.-J. Wang and S. J. Lippard, Science, Washington, DC **230**, 412 (1985); P. B. Dervan, Science, Washington, DC **232**, 465 (1986).

84. J. K. Barton, Science Washington, DC **233**, 727 (1986); A. M. Pyle, J. P. Rehmann, R. Meshoyrer, C. V. Kumar, N. J. Turro and J. K. Barton, J. Amer. Chem. Soc. **111**, 3051 (1989).

85. R. B. Lauffer, A. C. Vincent, S. Padmanabhan and T. J. Meade, J. Amer. Chem. Soc. **109**, 2216 (1987) and references therein.

86. W. W. Cleland, Methods in Enzym. **87**, 159 (1982).

87. B43.

88. H. Brunner, Adv. Organomet. Chem. **18**, 151 (1980).

89. L. E. Nivorozhkin, A. L. Nivrorozhkin, M. S. Korobov, L. E. Konstantinovsky and V. I. Minkin, Polyhedron, **4**, 1701 (1985) and previous work cited. Inversion at the metal center on the ^{1}H nmr time scale results in coalescence of signals belonging to the i-Pr or $C_6H_5CH_2$ protons.

90. G. K. Anderson and R. J. Cross, Chem. Soc. Revs. **9**, 185 (1980) for a comprehensive and incisive review of isomerization mechanisms of square-planar complexes.

91. G. Alibrandi, D. Minniti, L. M. Scolaro and R. Romeo, Inorg. Chem. **27**, 318 (1988) and previous work.

92. R. van Eldik, D. A. Palmer, H. Kelm and W. J. Louw, Inorg. Chem. **19**, 3551 (1980).

93. D. L. Packett and W. C. Trogler, Inorg. Chem. **27**, 1768 (1988).

94. Whereas *racemization* is the complete loss of optical activity with time, *epimerization* is the reversible interconversion of diastereoisomers to an equilibrium mixture which is not necessarily optically inactive. Diastereoisomers arise from the combination of the two chiral centers in **9**, namely the metal centered, *R* and *S*, and the resolved (*S*) optically active ligand center. The diastereoisomers (*RS*) and (*SS*) differ in their properties.

95. R. C. Fay, J. R. Weir and A. H. Bruder, Inorg. Chem. **23**, 1079 (1984).

96. R. C. Fay and J. K. Howie, J. Amer. Chem. Soc. **101**, 1115 (1979).

97. B. Halpern, A. M. Sargeson and K. R. Turnball, J. Amer. Chem. Soc., **88**, 4630 (1966).

98. J. B. Goddard and F. Basolo, Inorg. Chem. **8**, 2223 (1969).

99. D. A. Buckingham. L. G. Marzilli and A. M. Sargeson, J. Amer. Chem. Soc. **89**, 825 (1967); **91**, 5227 (1969).

100. T. P. Pitner and R. B. Martin, J. Amer. Chem. Soc., **93**, 4400 (1971).

101. P. Haake and P.C. Turley, J. Amer. Chem. Soc. **90**, 2293 (1968); L. E. Erickson, A. J. Dappen and J. C. Uhlenhopp, J. Amer. Chem. Soc. **91**, 2510 (1969); L. E. Erickson, H. L. Fritz, R. J. May and D. A. Wright, J. Amer. Chem. Sec. **91**, 2513 (1969).

102. H. Kawaguchi, M. Matsuki, T. Ama and Y. Yasui, Bull. Chem. Soc. Japan **59**, 31 (1986).

103. G. Ma, T. Hibino, M. Kojima and J. Fujita, Bull. Chem. Soc. Japan **62**, 1053 (1989).
104. M. Kojima, T. Hibino, Y. Ouyang and J. Fujita, Inorg. Chim. Acta **117**, 1 (1986).
105. G. H. Searle and F. R. Keene, Inorg. Chem. **11**, 1006 (1972).
106. D. A. Buckingham, P. A. Marzilli and A. M. Sargeson, Inorg. Chem. **8**, 1595 (1969).
107. P. Comba and W. Marty, Helv. Chim. Acta **63**, 693 (1980). See also A. A. Watson, M. R. Prinsep and D. A. House, Inorg. Chim. Acta **115**, 95 (1986).
108. R. W. Hay and M. Akbar Ali, Inorg. Chim. Acta **103**, 23 (1985). *cis*-Ni([13]aneN$_4$)(H$_2$O)$_2^{2+}$ change to Ni([13]aneN$_4$)$^{2+}$ ($+ - - +$), spectrally.
109. A. M. Martin, K. J. Grant and E. J. Billo, Inorg. Chem. **25**, 4904 (1986). Probably *trans* V ($+ - + +$) to *trans* II ($+ - - +$) Ni([13]aneN$_4$)$^{2+}$ from an analysis of the absorbance changes. Inversion at one N center is involved.
110. P. J. Connolly and E. J. Billo, Inorg. Chem. **26**, 3224 (1987). *Trans* III $\rightleftharpoons$ *trans* I Ni([14]aneN$_4$)$^{2+}$ by ^{1}H nmr. Inversion at two N centers must occur.
111. The isomers correspond to the *trans*-II (blue) and *trans*-III (red) isomers shown in Sec. 3.1.1. There are two additional asymmetric (carbon) atoms in the macrocycle.
112. C.-S. Lee and C.-S. Chung, Inorg. Chem. **23**, 639 (1984).
113. B.-F. Liang, D. W. Margerum and C.-S. Chung, Inorg. Chem. **18**, 2001 (1979); C.-S. Lee, G. T. Wang and C.-S. Chung, J. Chem. Soc. Dalton Trans. 109 (1984); tet b differs from tet a in the configuration at one asymmetric carbon (p. 00).
114. P. H. Leung, J. W. L. Martin and S. B. Wild, Inorg. Chem. **25**, 3396 (1986).
115. W. G. Jackson and A. M. Sargeson, Inorg. Chem., **17**, 2165 (1978).

General Bibliography

B43. B. B. Douglas, D. H. McDaniel and J. J. Alexander, Concepts and Models of Inorganic Chemistry, 2nd. Edit., Wiley, 1983, Chap. 8.
B44. W. G. Jackson and A. M. Sargeson, Rearrangement in Coordination Complexes in Rearrangements in Ground and Excited State, P. de Mayo, ed., Academic, NY, 1980, Vol. 2, p. 273.
B45. J. MacB. Harrowfield and S. B. Wild, Isomerism in Coordination Chemistry B35, Chap. 5.

Problems

1. The equilibrium

$$\text{NiL}^{2+} + 2\,\text{H}_2\text{O} \rightleftharpoons \text{NiL(H}_2\text{O)}_2^{2+}$$

is slightly displaced to the right on application of pressure. What is the sign of ΔV? Explain its small value. (L are 5 different macrocycles).
Y. Kitamura, T. Ito and M. Kato, Inorg. Chem. **23**, 3836 (1984).

2. The relaxation time τ for the Fe(II) entry in Table 7.3 was measured in the pressure range 0.1–300 MPa

$$\left(\frac{d\ln\tau^{-1}}{dP}\right)_T = \frac{\Delta V^{\ddagger}}{RT}$$

$$\Delta V^{\ddagger} = 0$$

ΔV (difference in partial molar volumes of the two spin states) $= 8.1 \text{ cm}^3\text{mol}^{-1}$

$$K = 0.56 \text{ for [hs]/[ls]}$$

Calculate $\Delta V^{\ddagger}_1$ and $\Delta V^{\ddagger}_{-1}$ for ls $\underset{k_{-1}}{\overset{k_1}{\rightleftharpoons}}$ hs

C. Creutz, M. Chau, T. L. Netzel, M. Okumura and N. Sutin, J. Amer. Chem. Soc. **102**, 1309 (1980); J. DiBenedetto, V. Arkle, H. A. Goodwin and P. C. Ford, Inorg. Chem. **24**, 455 (1985).

3. Temperature-jump relaxation kinetics of the P-450$_{cam}$ spin equilibrium have been measured.

(a) How does the high spin (hs), low spin (ls) equilibrium in ferric P-450$_{cam}$ arise? For what other transition metal ion species has such a spin-state equilibrium been observed? What other methods are available for studying the kinetics of the spin-state equilibrium?

(b) The suggested mechanism for the P-450/substrate (S) interaction is

$$\text{P-450}_{ls} + \text{S} \underset{}{\overset{K_1}{\rightleftharpoons}} \text{P-450}_{ls}\cdot\text{S} \underset{k_{-2}}{\overset{k_2}{\rightleftharpoons}} \text{P-450}_{hs}\cdot\text{S}$$

How many relaxations should be observed? Why is only one observed? Can you derive

$$k_{obsd} = \frac{K_1 k_2([\text{P-450}]_{free} + [\text{S}]_{free})}{1 + K_1([\text{P-450}]_{free} + [\text{S}]_{free})} + k_{-2}$$

for the single observed relaxation rate constant k_{obsd}?

(c) Show that

$$K_d^{-1} = K_1(1 + (k_2/k_{-2}))$$

(d) Suggest the significance of the fact that k_{-2} increases as the percentage of low spin increases.

M. T. Fischer and S. G. Sligar, Biochemistry **26**, 4797 (1987).

4. (a) Suggest how you would prepare $\text{Co(NH}_3)_5{}^{17}\text{ONO}^{2+}$ and $\text{Co(NH}_3)_5\text{ON}^{17}\text{O}^{2+}$ for the experiments which result in (7.20).

(b) $\text{Co(NH}_3)_5\text{NO}_2^{2+}$ reacts rapidly and completely in neat anhydrous trifluoromethanesulfonic acid to produce $\text{Co(NH}_3)_5\text{H}_2\text{O}^{3+}$. Suggest what experiments you would carry out to lend support to an acid-catalyzed nitro-to-nitrito rearrangement as part of the reaction.

W. G. Jackson, Inorg. Chem. **26**, 3857 (1987).

5. Suggest how linkage isomerism might arise in the following complexes

$$Os(NH_3)_5[(CH_3)_2CO]^{2+}$$

W. D. Harman, M. Sekine and H. Taube, J. Amer. Chem. Soc. **110**, 2439 (1988).

$$Ru(NH_3)_5(asc)^+ \quad asc = ascorbate\ anion$$

D. M. Bryan, S. D. Pell, R. Kumar, M. J. Clarke, V. Rodriguez, M. Sherban and J. Charkoudian, J. Amer. Chem. Soc. **110**, 1498 (1988).

$$Co(NH_3)_5(4\text{-Meimid})^{3+} \quad 4\text{-Meimid} = 4\text{-methylimidazole}$$

M. F. Hoq, C. R. Johnson, S. Paden and R. E. Shepherd, Inorg. Chem. **22**, 2693 (1983). How might you monitor the rearrangements?

6. The acid hydrolysis of $Ru^{III}(NH_3)_5(NH_2CH_2CO_2Et)^{3+}$ is first order in Ru(III) and the dependence of the rate constant k_{obs} is given by

$$k_{obs} = \frac{a[H^+]}{b + c[H^+]}$$

At $\mu = 1.0$ M and 25.0°C, $b/a = 116$ Ms and $c/a = 880$ s from a k_{obs}^{-1} vs $[H^+]^{-1}$ plot. The acid-independent rate constant for the rearrangement of $Ru(NH_3)_5NH_2CH_2CO_2H^{3+}$ to $Ru(NH_3)_5O_2CCH_2NH_3^{3+}$ is $2 \times 10^{-3}s^{-1}$. Suggest a mechanism for the acid hydrolysis. A. Yeh and H. Taube, J. Amer. Chem. Soc. **102**, 4725 (1980).

7. Interpret the changing nmr signals when $\Delta(R)$-Co(sar)(hbg)$_2^{2+}$ is added to D_2O, pD = 7.0 at 39.6°C (Table 7.9 is helpful!).
   H. Kawaguchi, M. Matsuki, T. Ama and T. Yasui, Bull. Chem. Soc. Japan **59**, 31 (1986).

8. *Fac* and *mer* isomers of $IrH_3(CO)(Ph_3P)_2$ were characterized by ^1H and ^{31}P nmr spectra. Their rates of interconversion in CH_2Cl_2 at 25°C

$$mer \underset{k_{-1}}{\overset{k_1}{\rightleftharpoons}} fac$$

were measured at 1780 cm^{-1} (infrared band due to *mer* isomer) in the presence of free H_2, to suppress a side reaction with CH_2Cl_2. The transformations *mer* $\rightleftharpoons$ *fac* and *fac* $\rightleftharpoons$ *mer* are first order in Ir complex and the values of k_1 and k_{-1} are independent of $[H_2]$.

$$IrH_3(CO)(Ph_3P)_2 + Ph_3P \overset{k_2}{\longrightarrow} IrH(CO)(Ph_3P)_3 + H_2$$

In addition the *fac* and *mer* isomers undergo reaction with Ph_3P. Using excess Ph_3P to prevent any back reaction, the rates of reaction were followed at the 330 nm band of the product. With the same conditions,

$k_1 \quad\;\; = 2.0 \times 10^{-4}\mathrm{s}^{-1}$

$k_{-1} \quad = 1.3 \times 10^{-4}\mathrm{s}^{-1}$

$k_2(fac) \; = 2.3 \times 10^{-4}\mathrm{s}^{-1}$

$k_2(mer) = 2.2 \times 10^{-4}\mathrm{s}^{-1}$

Both k_2 values were independent of [Ph$_3$P]. Suggest a mechanism for the isomerization and PPh$_3$ substitution reactions.

J. F. Harrod and W. J. Yorke, Inorg. Chem. **20**, 1156 (1981).

9. Draw the *meso* and *racemic* forms of the tetradentate arsine

$$(CH_3)_2As(CH_2)_3As(C_6H_5)CH_2CH_2As(C_6H_5)(CH_2)_3As(CH_3)_2$$

Then draw the five possible isomers that may be formed when the *meso* and *racemic* ligands are coordinated to an octahedral metal.

B. Bosnich, W. R. Kneen and A. T. Phillips, Inorg. Chem. **8**, 2567 (1969); B. Bosnich, W. G. Jackson and S. B. Wild, J. Amer. Chem. Soc. **95**, 8269 (1973).

10. Estimate the nitrogen inversion rate constants in the complexes Co(NH$_3$)$_4$(sar)$^{2+}$, Pt(NH$_3$)$_2$(Meen)$^{2+}$ and Pt(en)(Meen)Cl$_2^{2+}$ using (7.44) and Table 7.9. Comment on their values in Pt(II), Co(III) and Pt(IV).

T. P. Pitner and R. B. Martin, J. Amer. Chem. Soc. **93**, 4400 (1971); W. G. Jackson and A. M. Sargeson in Rearrangements in Ground and Excited States, P. de Mayo, ed. Academic, NY, 1980, Vol. 2, p. 277.

Chapter 8
A Survey of the Transition Elements

In this final chapter the salient features of the transition elements are surveyed. This affords an opportunity to assemble some important mechanistic chemistry discussed in the previous chapters, thus furnishing an index. More important, each element is reviewed with key references mainly to recent literature, which gives access to the older literature. References in previous chapters give fuller tables of data.

Titanium(III)

The water exchange of $Ti(H_2O)_6^{3+}$ has been examined using a $CF_3SO_3^-$ counteranion, ^{17}O nmr and the full Swift and Connick equation (3.56).[1] The plot of $\ln(k^P/k^o)$ against pressure shows slight curvature. The value of $\Delta V^{\ddagger}$ is independent of temperature and is the largest negative value for an hexaaqua ion (Table 4.5). It is close to that expected for an A or nearly A mechanism as would be anticipated for a t_{2g}^1 configuration.[2] The rate constants for reaction of $Ti(H_2O)_6^{3+}$ with a number of anions measured by temperature-jump are shown in Table 8.1.[1,3] There is a nearly 10^3 variation in the second-order formation rate constants which increase in value with increased basicity of entering ligand.

Table 8.1 Formation Rate Constants k $(M^{-1}s^{-1})$ for the Reaction of $Ti(H_2O)_6^{3+}$ with Various Ligands in Water at 15°C Refs. 1, 3

Ligand	k	pK_a
$ClCH_2COOH$	7×10^2	–
CH_3COOH	1×10^3	–
H_2O	8.6×10^3 [a]	-1.74
$ClCH_2CO_2^-$	2.1×10^5	2.46
$CH_3CO_2^-$	1.8×10^6	4.47

[a] This rate constant at 12°C corresponds to the exchange of one of the six coordinated water molecules in $Ti(H_2O)_6^{3+}$ and is converted to second-order units i.e. $6k$ $(s^{-1})/55.5$ (M)

This is also suggestive of associative character for ligation.[3]

Titanium(III) is a strongly reducing d^1 ion. The kinetics of reduction by Ti(III) are dominated by an inverse $[H^+]$ dependence. With a large variety of oxidants, including O_2, the reactive species $TiOH^{2+}$ shows outer-sphere Marcus behavior[4-6] (Prob. 1). Groups which can provide a good "lead-in" substituent can promote an inner-sphere redox process. A binuclear intermediate is sometimes detected[7-9] e.g. with $Co(NH_3)_5ntaH^+$ using rapid scan spectrophotometry.[8] Substitution on Ti(III) can sometimes be rate limiting in redox reactions.[10]

Titanium(IV)

Titanium(IV) in aqueous $HClO_4$ is predominantly TiO^{2+}, with possibly some $Ti(OH)_2^{2+}$ and Ti^{4+} present.[6,11] The approximate H_2O exchange rates for yl (double bonded) oxygen in TiO^{2+}, Ref. 11, is $1.6 \times 10^4 s^{-1}$ which is about nine orders of magnitude larger than that for VO^{2+}, suggesting a mechanistic difference. In trimeric and tetrameric species, which exist at high (>0.05 M) Ti(IV) concentrations and low $[H^+]$, exchange rate constants for oxo and hydroxo bridges are $\approx 10^2 s^{-1}$ and for terminal aqua or hydroxo groups $k_{exch} \approx 3 \times 10^3 s^{-1}$. (Ref. 11) Formation rate constants and activation parameters for reaction of Ti(IV) with a number of anions are similar ($k \approx 10^3 M^{-1} s^{-1}$; $\Delta H^{\ddagger} \approx 45 \ kJ mol^{-1}$ and $\Delta S^{\ddagger} \approx 17 \ J K^{-1} mol^{-1}$) and this suggests that substitution is I_d controlled.[12]

The reactions of Ti(III) and Ti(IV) species with dioxygen moieties have evoked interest. The Ti(III) complex $Ti(edta)H_2O^-$ reacts with O_2 to give $Ti(edta)(O_2)^{2-}$. A superoxo intermediate $Ti(edta)(O_2)^-$ is formed ($k = 1.0 \times 10^4 M^{-1} s^{-1}$) which is scavenged rapidly by the excess $Ti(edta)H_2O^-$ ($k > 10^6 M^{-1} s^{-1}$).[13] The second-order rate constant for reaction of TiO^{2+} with H_2O_2[14] has been confirmed[15,16] and reactions of the product $Ti(O_2)^{2+}$ have been examined.[15,17] Reductions of $Ti(O_2)^{2+}$ by Fe(II), Ti(III) and S(IV) are much slower than the corresponding reductions of H_2O_2. This resides partly in the reactions occuring only via rate determining dissociation of peroxide from $Ti(O_2)^{2+}$, Ref. 17. Oxidation of $Ti(O_2)^{2+}$ has also been examined.[15] The reaction with Ce(IV) produces the protonated superoxotitanium(IV), probably $TiO(HO_2^{\bullet})^{2+}$

$$Ti(O_2)^{2+} + Ce(IV) \xrightarrow[(^-H^+)]{H_2O} TiO(HO_2^{\bullet})^{2+} + Ce(III) \quad k = 1.1 \times 10^5 M^{-1} s^{-1} \quad (8.1)$$

Although the latter decomposes within a minute (1 M $HClO_4$, 25°C),

$$TiO(HO_2^{\bullet})^{2+} \underset{}{\overset{k}{\rightleftharpoons}} TiO^{2+} + HO_2^{\bullet} \qquad k = 0.10 \ s^{-1} \qquad (8.2)$$

$$TiO(HO_2^{\bullet})^{2+} + HO_2^{\bullet} \xrightarrow{fast} Ti(O_2)^{2+} + O_2 + H_2O \qquad (8.2(a))$$

its spectrum and kinetic reactivity can be determined by use of a multimixer (Sec. 3.3.2). Excess $Ti(O_2)^{2+}$ and Ce(IV) are mixed in mixer one, left for 0.23–0.31 s, the time for maximum production of $TiO(HO_2^{\bullet})^{2+}$ and the mixture then reacted in a second mixer with the examining reductant.[6,15]

Reactions of H_2O_2 with TiO^{2+}, $TiO(nta)H_2O^-$ and $TiO(tpypH_4)^{4+}$, 1 all leading to a 1:1 peroxo species, appear to be associatively activated, based on negative $\Delta V^{\ddagger}$ values determined by a high-pressure, stopped-flow technique.[16] It can be shown using ^{18}O-labelling and ir or nmr analysis that the O–O bond in the H_2O_2 remains intact (P = tpyp H_4^{4+}):[18]

$$Ti^{18}O(P)^{4+} + H_2O_2 \rightarrow Ti(O_2)P^{4+} + H_2^{18}O \qquad (8.3)$$

Ligand exchange and isomeric rearrangement of cis- and trans-isomers of ML_2Cl_4 (M = Ti, Zr and Hf; L = neutral base) have been examined by variable pressure 2D ^1H nmr spectroscopy.[19]

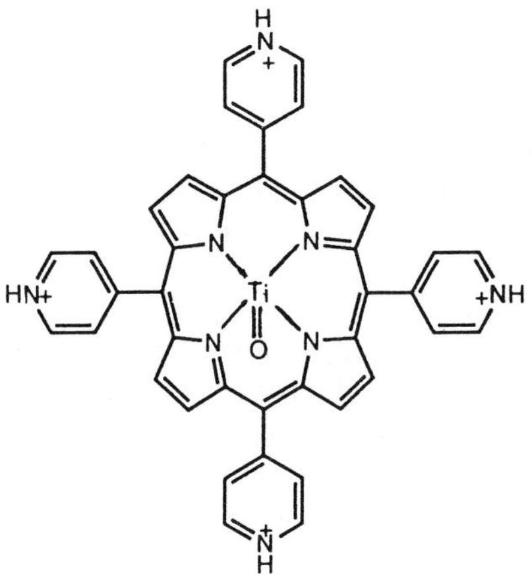

1

Vanadium

The lower oxidation states are stabilized by soft ligands e.g. CO (Prob. 3). The aquated vanadium ions represent an interesting series of oxidation states. They are all stable with respect to disproportionation and labile towards substitution. They undergo a number of redox reactions with one another, all of which have been studied kinetically. Many of the reactions are [H$^+$]-dependent. There has been recent interest in the biological aspects of vanadium since the discovery that vanadate can mimic phosphate and act as a potent inhibitor[20] (Prob. 4).

Vanadium(II)

The parameters for water exchange with $V(H_2O)_6^{2+}$ (Table 4.1) indicate an I_a mechanism.[21,22] The observed $\Delta V^{\ddagger}$ (-2.1 cm^{3}mol^{-1}) for anation of V^{2+} by SCN$^-$ includes a ΔV_{os} (estimated $+3.2$ cm^{3}mol^{-1}) and a $\Delta V_I^{\ddagger}$ for interchange (Sec. 4.2.2):

$$V(H_2O)_6^{2+} + SCN^- \underset{}{\overset{\Delta V_{os}}{\rightleftharpoons}} [V(H_2O)_6 \cdot SCN]^+ \underset{}{\overset{\Delta V_I^{\ddagger}}{\rightleftharpoons}} V(H_2O)_5SCN^+ + H_2O$$

$$(8.4)$$

The computed value for $\Delta V_I^{\ddagger}$ of -5.3 cm^{3}mol^{-1} is supportive also of an I_a mechanism.[23]

The reaction of V(II) with O$_2$ to give substantial amounts of V(IV) must represent a two electron oxidation step (Sec. 5.7.2) since V(III) reacts more slowly with O$_2$ than would account for the appearance of V(IV).[24] The suggested mechanism[25] resembles that for other bivalent metal ions,

$$V^{2+} + O_2 \rightarrow VO_2^{2+} \qquad k_1 = 2 \times 10^3 M^{-1}s^{-1} \qquad (8.5)$$

$$VO_2^{2+} + H_2O \rightarrow VO^{2+} + H_2O_2 \qquad k_2 \approx 10^2 s^{-1} \qquad (8.6)$$

The value of k_1 suggests an outer sphere process (Table 5.2) perhaps giving $V^{IV}(O_2^{2-})^{2+}$ via $V^{III}(O_2^-)^{2+}$.

If excess V^{2+} is present further reactions occur

$$VO_2^{2+} + V^{2+} \underset{k_r}{\overset{k_f}{\rightleftarrows}} VO_2V^{4+} \qquad k_f = 3.7 \times 10^3 M^{-1}s^{-1}, \quad k_r = 20 \ s^{-1} \qquad (8.7)$$

$$VO_2V^{4+} \longrightarrow 2 VO^{2+} \qquad k = 35 \ s^{-1} \qquad (8.8)$$

Two electron changes also occur with H_2O_2 [24], Tl(III) and Hg(II). Towards V(V), I_2 and Br_2 however, V(III) is the only product of oxidation of V(II) (Sec. 5.7.2).

The oxidation of V(II) by a large number of Co(III) complexes has been studied (Tables 5.2, and 5.7). Some oxidations are clearly outer-sphere and others inner-sphere (controlled by substitution in V(II)), and several are difficult to assign (Table 5.2). In general $\Delta H^{\ddagger}$ values are much lower for outer-sphere than inner-sphere redox reactions and outer-sphere processes usually give LFER in reactions with $Co(NH_3)_5X^+$, Ref. 26.

The ion $V(pic)_3^-$ is a useful highly-colored strong outer-sphere reductant ($E^0 = -0.41$ V at $\mu = 0.5$ M and 25 °C). [27]

Vanadium(III)

Vanadium(III) reacts with O_2 and ClO_4^- and is easily hydrolyzed ($pK_a = 3.0$), all important points to consider in studying its reaction kinetics. An I_a mechanism is favored for H_2O exchange (Table 4.5) and for other ligand substitutions. This is supported by the activation parameters [28] and the correlation of k_f with the basicity of the entering ligand (Table 8.2). [28,29]

Table 8.2 Rate Constants (k_f) for Interaction of Ligands (L) with $V(H_2O)_6^{3+}$ at 25 °C [a]

L	pK_a [b]	k_f $M^{-1}s^{-1}$
Cl^-	–	$\leqslant 3$
NCS^-	-1.8	1.1×10^2
H_2O	-1.7	54
$C_2O_4H^-$	1.2	1.3×10^3
$salH^-$ [c]	2.8	1.4×10^3
$4\text{-}NH_2salH^-$	3.6	7×10^3

[a] Selected from fuller compilation in Ref. 28.   [b] Dissociation constant of the conjugate acid of L.
[c] $salH_2$ = salicylic acid

The ion VOH^{2+} appears not to be more reactive than V^{3+} (Fig. 1.5) which is very unusual for M(III) ions. [28-30] The reaction of VOH^{2+} with HN_3 is favored however over the reaction between V^{3+} and N_3^- because the $\Delta S^{\ddagger}$ value for hydrolysis of VN_3^{2+} suggests that it incipiently produces VOH^{2+} and HN_3. [30]

Since V(III) is not easily reduced, only the reactions with Cr(II) and Eu(II) have been studied. Inverse $[H^+]$ terms in the rate law can be ascribed to reactions of VOH^{2+}, although this conclusion is equivocal (Sec. 2.1.7(c)). For the oxidation of V(III) by Cr(VI) the net activation process is

$$V^{3+} + HCrO_4^- \rightleftharpoons [VHCrO_4^{2+}]^{\ddagger} \xrightarrow{H_2O} VO^{2+} + H_3CrO_4 \qquad (8.9)$$

and thus in the three-step oxidation by Cr(VI), Eqs. (1.118–1.120), the first step is rate-determining. The second-order rate constant ($3.9 \times 10^2 M^{-1}s^{-1}$ at 25°C) may be too high to represent a V^{3+}-substitution controlled inner-sphere oxidation.

Vanadium(IV)

The labilities of the coordinated water in the $VO(H_2O)_5^{2+}$ (normally abbreviated VO^{2+}) and substituted VO^{2+} ions have been nicely studied by the nmr method.[31] There are three types of attached oxygen in **2**. Their increasing

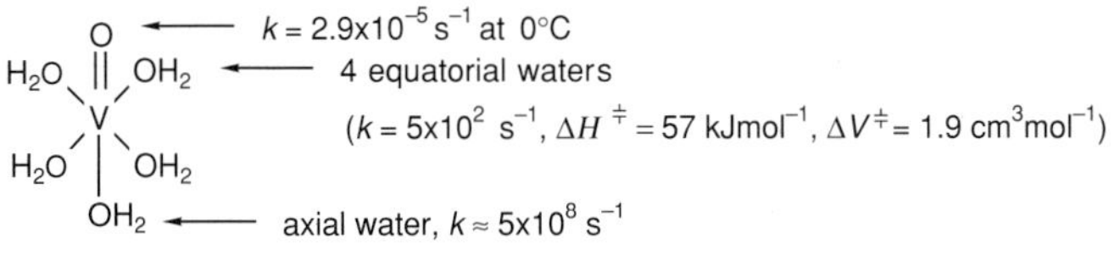

$$k = 2.9 \times 10^{-5} s^{-1} \text{ at } 0°C$$
4 equatorial waters
$$(k = 5 \times 10^2 \, s^{-1}, \Delta H^{\ddagger} = 57 \text{ kJmol}^{-1}, \Delta V^{\ddagger} = 1.9 \text{ cm}^3\text{mol}^{-1})$$
axial water, $k \approx 5 \times 10^8 \, s^{-1}$

2

labilities correlate with increased V–O bond distances: 1.60 Å (V=O), 2.00 Å (V–OH$_2$ equatorial) and 2.40 Å (V–OH$_2$ axial). For the V(IV), H$_2$O exchange involving yl oxygen, both $VO(H_2O)_5^{2+}$ (k_0) and $VO(OH)(H_2O)_4^+$ (k_{OH}) are involved.

$$VO(H_2O)_5^{2+} \rightleftharpoons VO(OH)(H_2O)_4^+ + H^+ \qquad K_a \qquad (8.10)$$

$$V = (k_0 + k_{OH}K_a[H^+]^{-1})[VO^{2+}] \qquad (8.11)$$

At 0°C and $\mu = 2.5$ M, $k_0 = 2.4 \times 10^{-5}s^{-1}$, $k_{OH} = 1.3 \, s^{-1}$ and $K_a = 4 \times 10^{-7}$M.[32] An internal electronic rearrangement, coupled with proton transfer, is suggested for the mechanism for both the k_0 and k_{OH} terms. In this a yl oxygen is converted into a labile coordinated water:[32,33]

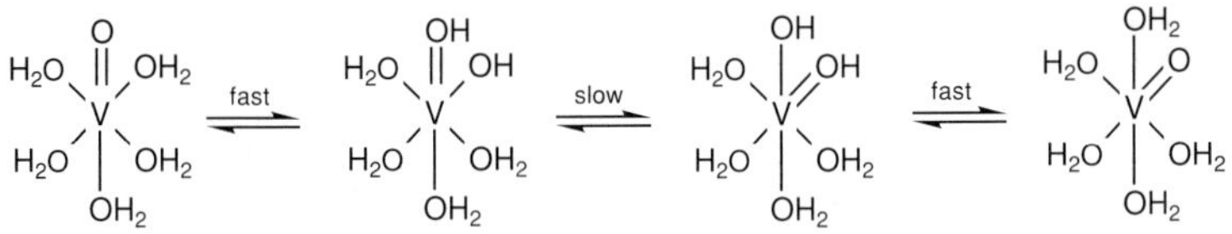

$$(8.12)$$

The V(IV)-H$_2$O exchange is catalyzed by V(V)[32] (Prob. 5(a)). The enhanced reactivity for the base form is observed in dimerization,[34] substitution and redox reactions (below). The mechanism of substitution of VO^{2+} remains uncertain. One of the problems is to assess the contribution of the highly reactive $VO(OH)^+$. Rate constants for complexing by VO^{2+} are all $\approx 10^3 M^{-1}s^{-1}$, Ref. 35, consistent with an I_d mechanism. By using chelating ligands to tie up various positions of the coordination sheath, one can isolate and distinguish between the types of coordinated oxygen.[31,32,36] Coordinated chelates in which the apical group is blocked by one of the donor atoms of the chelate group lead to slower anation of the remaining equatorial waters. An associative mechanism is favored from the low $\Delta H^{\ddagger}$, large negative $\Delta S^{\ddagger}$ and varying k_f with different nucleophiles.[37-39]

Being a d^1 one-electron reductant, VO^{2+} gives free radicals with two-electron oxidants e. g. with HSO_5^- Ref. 40 (and Cl_2 Ref. 41).

$$VO^{2+} + HSO_5^- \rightarrow VO_2^+ + H^+ + SO_4^{\bar{\cdot}} \tag{8.13}$$

$$VO^{2+} + SO_4^{\bar{\cdot}} + H_2O \rightarrow VO_2^+ + HSO_4^- + H^+ \tag{8.14}$$

The reduction of VO^{2+} by V(II), Cr(II), Eu(II) and Cu(I) implicates oxo or hydroxo binuclear species (Table 5.3). The oxidation of VO^{2+} by M(III)-bipyridine complexes and by $IrCl_6^{2-}$ has acid independent (k_0) and inverse acid (k_1) dependent terms, ascribable to reactions of $VO(H_2O)_5^{2+}$ and $VO(OH)(H_2O)_4^+$ respectively (Eqn. 8.10). Towards $Ru(bpy)_3^{3+}$, $k_0 = 1.0\ M^{-1}s^{-1}$ and $k_1 = 6.4 \times 10^5 M^{-1}s^{-1}$ at 25 °C and $\mu = 1.0$ M.[42,43] Enhanced reactivity of $VO(nta)OH^{2-}$ Ref. 44 has also been noted. A roughly linear relationship, originally developed by Rosseinsky, between $\log k_{Fe^{2+}}$ and $\log k_{VO(OH)^+}$ holds for a number of oxidants (Fig. 8.1).[42]

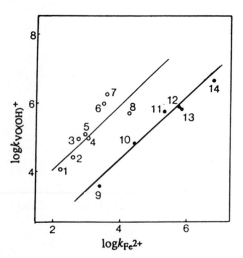

Fig. 8.1 Best linear plot of $\log k$ ($VO(OH)^+$) vs $\log k$ (Fe^{2+}) for reduction of eight Ni(III) macrocycles (upper line) and six tris(polypyridine) metal complexes. $Ni([9]aneN_3)_2^{3+}$ (1), $Ni(5,12\,diMe[14]4,11-dieneN_4)^{3+}$ (2), $Ni([14]aneN_4)^{3+}$ (3), $Ni([10]aneN_3)^{3+}$ (4), $Ni(2,9\,diMe[14]aneN_4)^{3+}$ (5), $Ni(rac-(5,14)-Me_6[14]aneN_4)^{3+}$ (6), $Ni(meso-(5,14)-Me_6[14]aneN_4)^{3+}$ (7), $Ni(Me_6[14]4,11-dieneN_4)^{3+}$ (8), $Os(bpy)_3^{3+}$ (9), $Fe(bpy)_3^{3+}$ (10), $Ru(Me_2bpy)_3^{3+}$ (11), $Ni(Me_2bpy)_3^{3+}$ (12), $Ru(bpy)_3^{3+}$ (13), and $Ni(bpy)_3^{3+}$ (14).[42] Reproduced with permission from D. H. Macartney, A. McAuley and D. A. Olubuyide, Inorg. Chem. **24**, 307 (1985) © (1985) American Chemical Society.

Vanadium(V)

Vanadium(V) exists as a large variety of monomeric and polymeric species over the pH zero-to-14 range. Vanadium-51 and ^{17}O nmr are important tools for their characterization.[45] The O:V ratio decreases in these species as increasing [H⁺] removes coordinated O^{2-} groups. The exchange of the coordinated oxo group in VO_2^+ (high acidity),[46] VO_4^{3-} (pH > 12)[47] and $V_{10}O_{28}^{6-}$ (>1 mM total vanadium at pH 4–10)[48-50(a)] with H_2O has been examined by mass spectroscopy and nmr methods with the results shown in Table 8.3. The decavanadate,

$V_{10}O_{28}^{6-}$, ion is especially interesting. It assembles with ease quickly when alkaline VO_4^{3-}-containing solutions are acidified. Despite this, the pK values of VO_4^{3-} can be determined by potentiometry in a flow apparatus (Sec. 3.10.1). Mechanisms for the formation and decomposition of $V_{10}O_{28}^{6-}$ have been suggested.[45,48,50(a)] Although there are seven structurally different oxygens in **3** these exchange with very similar (although probably not identical rates).[48-50(a)] Each type of oxygen can be identified by nmr[50(a)] (Sec. 3.9.5). A small change in half-life of exchange when the concentration of vanadium is changed ten-fold supports, as a likely mechanism for exchange, a breakdown to a more labile half-bonded intermediate, rather than complete scission to cyclic metavanadates.[48,50(a)] One or two "capping" vanadiums can be replaced in $V_{10}O_{28}^{6-}$ to give $MoV_9O_{28}^{5-}$ and $Mo_2V_8O_{28}^{4-}$. Using ^{17}O-nmr, it can be shown that the polyanion structure is substantially broken down during the substitution.[50(b)]

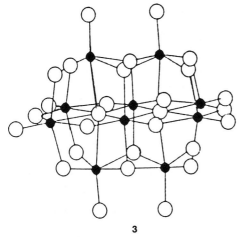

3

Table 8.3 Exchange Behavior of V(V) Ions Towards H_2O

Species	Exchange Rate	Comments	Refs.
VO_2^+	$t_{1/2} \approx 0.15$ s (0°C) by competition method	Unusually rapid for MO_2^{n+} may reside in *cis*–V$\overset{O^+}{\underset{O}{\diagdown}}$ structure	46
VO_4^{3-}	$t_{1/2} \approx 20$ s (0°C), OH$^-$ > 0.1 M $\Delta H^{\ddagger} = 92$ kJ mol^{-1}; $\Delta S^{\ddagger} = 60$ J K^{-1}mol^{-1} ($\mu = 1.4$ M)	4 oxygens equivalent. Solvent-assisted I_d mechanism favored.	47
$V_{10}O_{28}^{6-}$	$t_{1/2} = 7$-15 h (25°C), pH-dependent	Near equivalency of all oxygens mandates reversible, extensive breakage as exchange mechanism	48-50(a)

In the $VO_2(H_2O)_4^+$ ion, the two V=O bonds are *cis* to one another, k's are 10^4-10^6M^{-1}s^{-1} (Prob. 6) for water replacement by ligand and a dissociative mechanism is favored.[39] In ligand interchange, when there are available H_2O's on the V(V) complex, an intermediate with both departing and entering ligands attached to the metal, is involved. With V(V) complexes in which no bonded labile H_2O's remain (e. g. the edta or edda complex) a special mechanism for chelating ligand interchange is proposed. In this an outer-sphere preequilibrium complex is formed within which interchange occurs ($k = 0.1$–0.6 s^{-1}, fairly independent of ligand).[51]

A number of reductions of VO_2^+ show acid catalysis with no rate saturation at high $[H^+]$. This is consistent with a protonation equilibrium,

$$VO_2(H_2O)_4^+ + H^+ \rightleftharpoons VO(OH)(H_2O)_4^{2+} \qquad K \qquad (8.15)$$

separate reactivity for the two vanadium species and a relatively large value for K^{-1} (estimated as 30 M). Marcus treatment cannot be applied to the VO_2^+/VO^{2+} couple because two protons as well as one electron are involved. The same restriction does not apply to the couple $VO(OH)(H_2O)_4^{2+}/VO(OH)(H_2O)_4^+$ (the reactive V(IV) species) and extended Marcus treatment (Eqn. 5.37) shows a self-exchange rate constant for the $VO(OH)^{+/2+}$ couple of $\approx 10^{-3}$M^{-1}s^{-1} (Fig. 8.2).[43] Vanadium(V) reacts with H_2O_2 to give peroxo species. $VO(O_2)^+$ is sufficiently stable that its oxidation may be investigated.[52] The reaction with $S_2O_8^{2-}$ is Ag$^+$ catalyzed (Sec. Ag(I)).

$$S_2O_8^{2-} + Ag^+ \rightarrow SO_4^{2-} + SO_4^{-} + Ag^{2+} \qquad \text{rds} \qquad (8.16)$$

$$SO_4^{-} + Ag^+ \rightarrow SO_4^{2-} + Ag^{2+} \qquad (8.17)$$

$$Ag^{2+} + VO(O_2)^+ \rightarrow Ag^+ + VO_3^{2+} \cdot \qquad (8.18)$$

$$VO_3^{2+} \cdot \rightarrow VO^{2+} + O_2 \qquad (8.19)$$

Oxidation of VO^{2+} by $S_2O_8^{2-}$ to produce VO_2^+ is also Ag$^+$ catalyzed (Prob. 7). The formation and decay of complex radicals resulting from reactions of $VO(O_2)^+$ can be studied by double-mixing using epr and spectral monitoring.[53] Vanadium(V) oxidation of organic compounds often proceeds via complex formation.[54,55]

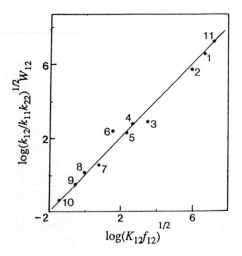

Fig. 8.2 Marcus plot of $\log (k_{12}/k_{11}k_{22})^{1/2} W_{12}$ vs $\log (K_{12}f_{12})^{1/2}$ for the oxidation of VO(OH)$^+$ and the reduction of VO(OH)$^{2+}$. The points are calculated by using $k_{11} = 1 \times 10^{-3}$M^{-1}s^{-1} for VO(OH)$^{+/2+}$ and the solid line represents the theoretical slope of unity. Oxidants: Ni(bpy)$_3^{3+}$ (1), Ni(4,4'-(CH$_3$)$_2$bpy)$_3^{3+}$ (2), Ru(bpy)$_3^{3+}$ (3), Ru(4,4'-(CH$_3$)$_2$bpy)$_3^{3+}$ (4), Fe(bpy)$_3^{3+}$ (5), IrCl$_6^{2-}$ (6). Reductants: Os(4,4'-(CH$_3$)$_2$bpy)$_3^{2+}$ (7), Os(4,7-(C$_6$H$_5$)$_2$phen)$_3^{2+}$ (8), Os(bpy)$_3^{2+}$ (9), Os(5-Clphen)$_3^{2+}$ (10), V(H$_2$O)$_6^{2+}$ (11).[43] Reproduced with permission from D. H. Macartney, Inorg. Chem. **25**, 2222 (1986). © (1986) American Chemical Society.

Chromium

Chromium produces some of the most interesting and varied chemistry of the transition elements. Chromium(O) and chromium(I) are stabilized in organometallics (Prob. 8). There have been extensive studies of the redox chemistry of Cr(II), Cr(III) and Cr(VI). Generally the Cr(IV) and Cr(V) oxidation states are unstable in solution (see below, however). These species play an important role in the mechanism of oxidation by Cr(VI) of inorganic and organic substrates and in certain oxidation reactions of Cr(II) and Cr(III). Examination of the substitution reactions of Cr(III) has provided important information on octahedral substitution (Chap. 4).

Chromium(II)

XAFS measurements on the beautiful blue ion $Cr(H_2O)_6^{2+}$ suggest tetragonal distortion with $Cr-O = 1.99\text{Å}$ (equatorial) and $Cr-O = 2.30\text{Å}$ (axial).[56,57] Solvent proton relaxation measurements on Cr(II) in CH_3OD indicate (a) two exchanging coordinated CH_3OD, k_{exch} per $CH_3OD = 1.2 \times 10^8 s^{-1}$ at 25 °C and (b) axial, equatorial interconversion ($k = <7 \times 10^4 s^{-1}$) in $Cr(CH_3OD)_6^{2+}$ at -80 °C.[58] Cr^{2+} is prepared in solution either by reduction of Cr(III) with Zn/Hg or by dissolution of Cr metal, usually in perchloric acid. These methods lead also to production of Zn^{2+} or Cl^- ions respectively, which do not interfere in most studies. The aqueous ion is extremely labile ($k_{exch} > 10^8 s^{-1}$)[21,56] and this, together with its weak complexing ability have limited the number of studies of its interaction with ligands.[57]

The rate constant for the rapid reaction of Cr^{2+} with O_2 ($k = 1.6 \times 10^8 M^{-1} s^{-1}$) can be measured by e_{aq}^- reduction of Cr^{3+} in the presence of O_2. The product, CrO_2^{2+} is long-lived in the absence of Cr^{2+} and O_2.[59,60] It reacts with Cr^{2+}

$$CrO_2^{2+} \xrightarrow{Cr^{2+}} CrO_2Cr^{4+} \xrightarrow[2H^+]{(2Cr^{2+})} Cr(OH)_2Cr^{4+} \tag{8.20}$$

to give the dihydroxy bridged product[61] which is the normally observed product of the reaction of Cr(II) with O_2. Oxidants such as Cu^{2+}, Fe^{3+} and Cl_2 give, with Cr(II), the mononuclear species Cr^{3+} and $CrCl^{2+}$ as the sole products.[62] CrO_2^{2+}, which is probably $Cr^{III}(O_2^-)^{2+}$ Ref. 63 can be easily prepared and is stable for hours or more.[64,65] Both inner- and outer-sphere redox reactions have been noted.[63] CrO_2^{2+} reacts with excess $N_2H_5^+$ in acid solution

$$CrO_2^{2+} + N_2H_5^+ \rightarrow Cr^{3+} + H_2O_2 + 1/2\,N_2 + NH_3 \tag{8.21}$$

with a rate law

$$-d[CrO_2^{2+}]/dt = k_0[CrO_2^{2+}] + k_1[N_2H_5^+][H^+][CrO_2^{2+}] \tag{8.22}$$

The first (smaller) term involves the spontaneous decomposition of CrO_2^{2+}. For the second term the following mechanism has been suggested[60]

$$CrO_2H^{3+} \rightleftharpoons CrO_2^{2+} + H^+ \tag{8.23}$$

$$CrO_2H^{3+} + N_2H_5^+ \rightleftharpoons CrO_2H^{2+} + {}^\bullet N_2H_5^{2+} \quad \text{rds} \tag{8.24}$$

$$\cdot N_2H_5^{2+} \rightarrow \cdot N_2H_4^+ + H^+ \tag{8.25}$$

$$2 \cdot N_2H_4^+ \rightarrow N_2 + 2\,NH_4^+ \tag{8.26}$$

Cr^{2+} reacts with H atoms, generated by pulse radiolysis[66] or uv flash photolysis,[67] to give $Cr(H_2O)_5H^{2+}$, $k = 1.5 \times 10^9\,M^{-1}s^{-1}$. The product may be regarded as the first member of a series of compounds of general formulae $Cr(H_2O)_5R^{2+}$ which arise from reaction of Cr^{2+} with RCl in anaerobic acid solution. It is the most reactive member of the series towards acidolysis (e.g. $k(CrH^{2+})/k(Cr(CH_3)^{2+}) = 2 \times 10^6$) and in other electrophilic reactions.[65,68]

Chromium(II) is a very effective and important reducing agent that has played a significant and historical role in the development of redox mechanisms (Chap. 5). It has a facile ability to take part in inner-sphere redox reactions (Prob. 9). The coordinated water of Cr(II) is easily replaced by the potential bridging group of the oxidant, and after intramolecular electron transfer, the Cr(III) carries the bridging group away with it; and as it is an inert product, it can be easily identified. There have been many studies of the interaction of Cr(II) with Co(III) complexes (Tables 2.6 and 5.7) and with Cr(III) complexes (Table 5.8). Only a few reductions by Cr(II) are outer-sphere (Table 5.7). By contrast, $Cr(edta)^{2-}$ Ref. 69 and $Cr(bpy)_3^{2+}$ are very effective outer-sphere reductants (Table 5.7).

The Cr(II)-acetate complex is a dimer and undergoes a number of substitution and redox reactions via the monomer with a common mechanism:

$$Cr_2^{II} \rightleftharpoons 2\,Cr^{II} \qquad k_1, k_{-1} \tag{8.27}$$

$$Cr^{II} + X \rightarrow \text{products} \qquad k_2 \tag{8.28}$$

The rate law

$$-d[Cr_2^{II}]/dt = -1/2\,(d[X]/dt) = k_1[Cr_2^{II}] \tag{8.29}$$

is observed for $X = edta^{4-}$, $Co(C_2O_4)_3^{3-}$ and $Co(NH_3)_5Cl^{2+}$ where $k_2(X) \gg k_{-1}$. The rate law:

$$-d[Cr_2^{II}]/dt = (k_1/k_{-1})^{1/2} k_2 [Cr_2^{II}]^{1/2} [X] \tag{8.30}$$

holds for $X = Co(edta)^-$, $Co(NH_3)_5OH^{2+}$ and $Cr(NH_3)_5X^{2+}$, for which, being less reactive oxidants, $k_2(X) \ll k_{-1}$[70] (See (2.55) and (2.56)). With I_3^-, a mixed rate law is observed.[71] With excess Cr(II) (Prob. 10), condition (8.29) leads to a zero-order loss of I_3^-. (Fig. 8.3)

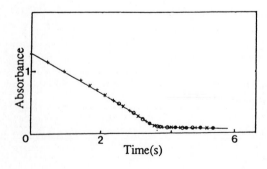

Fig. 8.3 Absorbance vs time traces for reaction (8.28), $X = I_3^-$ showing the zero-order character when $[Cr_2^{II}] = 4.25$ mM and $[I_2]_0$(mM) $= 0.6(+), 0.4(X), 0.2(\bigcirc)$ and $0.05(\bullet)$. In all reactions, NaI $= 25$ mM, $H_2O = 0.15$ M, $\lambda = 410$ nm (2 mm cell) and 25 °C. Ref. 71. Reproduced with permission from L. M. Wilson and R. D. Cannon, Inorg. Chem. **24**, 4366 (1985). © (1985) American Chemical Society.

Chromium(III)

The ion $Cr(H_2O)_6^{3+}$ is one of the few aqua species that is sufficiently inert that the solvent exchange rate and solvation number may be determined by conventional sampling or nmr analytical methods. Improved techniques enable lower concentrations of Cr(III) and H^+ to be used than in the early studies of the 50's. This allows the determination of an extended rate law for H_2O exchange with Cr(III)

$$k_{obs} = k_0 + k_1 [H^+]^{-1} \tag{8.31}$$

The first and second terms relate to exchange of $Cr(H_2O)_6^{3+}$ and $Cr(H_2O)_5OH^{2+}$ respectively, with results shown in Table 8.4.[72] These values reflect a strongly associative activation mode for $Cr(H_2O)_6^{3+}$ exchange and a predominantly dissociative one for $Cr(H_2O)_5OH^{2+}$. The lack of pressure dependence of $\Delta V^{\ddagger}$ for $Cr(H_2O)_6^{3+}$ also indicates that the solvation sheath is completely retained during interchange.[72] A large span in rate constants for anations of $Cr(H_2O)_6^{3+}$ (but not $Cr(H_2O)_5OH^{2+}$) also reflects an associative character,[72-74] Table 8.4.[72,73] The hydrolysis of Cr(III) yields a number of polynuclear species (Prob. 11) such as $Cr_2(OH)^{5+}$, $Cr_2(OH)_2^{4+}$, $Cr_3(OH)_4^{5+}$ and others, although there is disagreement as to their formulation.[76-78] Incisive recent reviews by Danish chemists[78,79] afford entry into all aspects of thermal and photochemical substitution processes involving Cr(III). In general, bridge cleavage, interconversions and formation reactions of trivalent metal ion polynuclear complexes occur through both spontaneous and acid-catalyzed paths, although for Cr(III) the former dominate. The decrease in stability as the number of hydroxo bridges increases, reflects increased bridge strain (Table 8.5).[78(a)]

Table 8.4 Rate Constants for the Reactions of $Cr(H_2O)_6^{3+}$ and $Cr(H_2O)_5OH^{2+}$ with X^{n-}, $\mu \approx 1.0$ M and 25 °C. From Ref. 72.

X^{n-}	$10^7 \times k_{Cr^{3+}}$ $M^{-1}s^{-1}$	$10^5 \times k_{CrOH^{2+}}$ $M^{-1}s^{-1}$
H_2O	24 (s^{-1})	18 (s^{-1})
$\underline{N}CS^-$	17	9.7
$\underline{S}CN^-$	0.1	2.1
NO_3^-	7.1	15
Cl^-	0.3	4.2
Br^-	0.09	2.7
I^-	0.008	0.46
SO_4^{2-}	110	61

Table 8.5 Stability of μ-Hydroxo Bridged Chromium(III) Complexes

Complex	$Cr-O-Cr$ Angle (degrees)	$t_{1/2}$ (hours)
$(NH_3)_5Cr(OH)Cr(NH_3)_5^{5+}$	166	19
$(NH_3)_4Cr(OH)_2Cr(NH_3)_4^{4+}$	102	1.6
$(NH_3)_3Cr(OH)_3Cr(NH_3)_3^{3+}$	83	0.008

The vast majority of substitution reactions of Cr(III) take place slowly and are easily measured. The complexes thus are a close second to those of Co(III) in importance as substrates for exploring all aspects of octahedron reactivity. They do moreover have the advantage of insensitivity to redox processes. The assignment of mechanism to anation and thus hydrolysis reactions of Cr(III) complexes, especially involving $Cr(NH_3)_5H_2O^{3+}$, has presented problems[80] but associative activation is now generally supported, Table 8.6[75] (Chapter 4). Acid and base hydrolyses, induced aquation and stereochemical change have been thoroughly investigated with Cr(III) complexes.[81] Thermal reactions of Cr(III) complexes are usually but, particularly with aqua ions not always, stereoretentive.[79] Rate parameters for water exchange with a number of Cr(III) amines may be used to explore kinetic cis- and trans-effects, although the rules have been less well established and with much fewer ligands than with Pt(II) chemistry.[79,82] Certain coordinated ligands will labilize Cr(III) (Prob. 12). These include polyaminocarboxylates,[83] porphyrins and a number of oxoions including OH^-. An I_d mechanism is favored.[84]

Chromium(III) organocations (see previous section) have attracted a good deal of attention.[65] The nature of the R group in $Cr(H_2O)_5R^{2+}$ controls the reactivity. When R is a primary group, the complex is stable in O_2. A chain mechanism holds for O_2 reaction with a complex containing a seondary or tertiary alkyl R group while reaction is indirect and via unimolecular homolysis with benzylchromium(III) (Sec. 2.1.6).

Table 8.6 Activation Parameters for Some Replacement Reactions of $Cr(NH_3)_5H_2O^{3+}$ at 50°C and $\mu = 1.0$ M. From Ref. 75

Entering Ligand	$10^4 k$ $M^{-1}s^{-1}$	$\Delta H^{\ddagger}$ $kJ\,mol^{-1}$	$\Delta S^{\ddagger}$ $JK^{-1}mol^{-1}$
H_2O	13.7 (s^{-1})	97	0
N_3^-	3.2	96	-15
NCS^-	4.2	102	$+13$
Cl^-	0.7	107	$+8$
Br^-	3.7	110	$+32$
$CCl_3CO_2^-$	1.8	105	$+9$
$HC_2O_4^-$	6.5	111	$+39$
$C_2O_4^{2-}$	29	104	$+32$

Chromium(IV) and (V)

Many more Cr(V) than Cr(IV) compounds have been characterized.[85,86] The Cr(V) complexes **4** are relatively stable in aqueous solution in air.[87] Often $R_1 = R_2 = Et$ in **4** has been

4

used to investigate this oxidation state.[86] At pH > 6, **4** decomposes (8.32) by second-order kinetics, suggesting a mechanism (8.33) and (8.34)

$$3 \, Cr(V) \rightarrow 2 \, Cr(VI) + Cr(III) \qquad\qquad (8.32)$$

$$2 \, Cr(V) \rightarrow Cr(VI) + Cr(IV) \qquad rds \qquad\qquad (8.33)$$

$$Cr(V) + Cr(IV) \rightarrow Cr(VI) + Cr(III) \qquad\qquad (8.34)$$

Oxidation of one-electron reductants Fe^{2+} and VO^{2+} by **4** produces a Cr(III) bischelate cation via an intermediate which is probably Cr(IV).[86] Two-electron reducing agents such as N_2H_4 produce a Cr(III) chelated monocarboxylate

$$Cr(V) + N_2H_4 \rightarrow Cr(III) + N_2H_2 + 2\,H^+ \qquad rds \qquad\qquad (8.35)$$

$$Cr(V) + N_2H_2 \rightarrow Cr(III) + N_2 + 2\,H^+ \qquad\qquad (8.36)$$

Finally, hydroxylamine produces a Cr(I) product:

$$OCr^VL_2 \rightleftharpoons OCr^VL + L \xrightarrow{\quad NH_3OH^+ \quad} OCr^VL(NH_2OH) + H^+ \qquad (8.37)$$

$$OCr^VL(NH_2OH) \rightarrow Cr^I(HL)NO \qquad\qquad (8.38)$$

Chromium(IV) and (V) are important intermediates in oxidation by Cr(VI), see next section. Chromium(V) is generally more reactive than Cr(VI). It is believed that Cr(V) is most effective for $C-H$ rupture whereas Cr(IV) best breaks $C-C$ bonds.[85]

Chromium(VI)

Some small doubt has been cast on the occurrence of $HCrO_4^-$ in Cr(VI) solutions in acid.[88] If this were confirmed, a large amount of kinetic data would need reinterpreting!

The rate of exchange of oxygen between Cr(VI) and water has been measured and the following rate law obtained:

$$V = k_1[CrO_4^{2-}] + k_2[HCrO_4^-] + k_3[H^+][HCrO_4^-]$$
$$+ k_4[HCrO_4^-][CrO_4^{2-}] + k_5[HCrO_4^-]^2 \qquad\qquad (8.39)$$

At 25°C, pH 7–12 and $\mu = 1.0$ M, $k_1 = 3.2 \times 10^{-7} s^{-1}$, $k_2 = 2.3 \times 10^{-3} s^{-1}$, $k_3 = 7.3 \times 10^5 M^{-1} s^{-1}$, $k_4 \approx 10^{-3} M^{-1} s^{-1}$ and $k_5 = 9.0 \, M^{-1} s^{-1}$ (for the transfer of one oxygen only). The proton evidently facilitates exchange by polarizing the Cr=O bond.[89]

The rate of attainment of the equilibria

$$Cr_2O_7^{2-} + H_2O \rightleftharpoons 2\,HCrO_4^- \qquad\qquad (8.40)$$

$$HCrO_4^- \rightleftharpoons H^+ + CrO_4^{2-} \qquad\qquad (8.41)$$

has been measured by flow and relaxation methods. The first step is generally base (Sec. 2.5.3) and acid-catalyzed. The highly negative values of $\Delta V^{\ddagger}$ (around -20 cm^{3}mol^{-1}) for OH$^-$, NH$_3$, H$_2$O and lutidine base-catalyzed hydrolysis of Cr$_2$O$_7^{2-}$ are in agreement with an I_a mechanism (Sec. 2.5.3).[90] The k_5 value in (8.39) is within an order of magnitude of that for the reverse of (8.40).[33]

Since the stable product of oxidation by Cr(VI) is Cr(III) we are necessarily involved with three-electron reactions, with their attendant interests and complications.[20,85] For inorganic reductants, R oxidized to O, the reaction scheme (1.118) holds

$$Cr(VI) + R \rightleftharpoons Cr(V) + O \qquad k_1, k_{-1} \tag{8.42}$$

$$Cr(V) + R \rightleftharpoons Cr(IV) + O \qquad k_2, k_{-2} \tag{8.43}$$

$$Cr(IV) + R \rightleftharpoons Cr(III) + O \qquad k_3, k_{-3} \tag{8.44}$$

With reducing agents inert to substitution, for example, Fe(phen)$_3^{2+}$, Fe(CN)$_6^{4-}$ or Ta$_6$Br$_{12}^{2+}$ the first step is the difficult one and straightforward second-order kinetics are observed:

$$V = k_1[Cr(VI)][R] \tag{8.45}$$

For a number of inorganic reductants, the second step is rate-limiting and

$$V = \left(\frac{k_1}{k_{-1}}\right) k_2[Cr(VI)][R]^2[O]^{-1} \tag{8.46}$$

Rarely is the third step rate-limiting but appears to occur with Ag(I)[B36]

$$V = \left(\frac{k_1}{k_{-1}}\right)\left(\frac{k_2}{k_{-2}}\right) k_3[Cr(VI)][R]^3[O]^{-2} \tag{8.47}$$

Support for the general scheme (8.42)–(8.44) also emerges from induced oxidation experiments (Prob. 4, Chap. 2) and from tracer experiments on the Cr(II)$-$Cr(VI) reaction.[91]

The reactions of Cr(VI) with H$_2$O$_2$ lead to a variety of peroxo species dependent on the conditions. The kinetics of formation or hydrolysis, or both, of CrO(O$_2$)$_2$ (Secs. 2.1 and 2.1.7(c)), Cr(O$_2$)$_4^{3-}$, Cr$_2$(O$_2$)$^{4+}$ and Cr$_3$(O$_2$)$_2^{5+}$ have been investigated.[92] The formation of CrO(O$_2$)$_2$ has been compared with those of a number of Cr(VI)-substrate complexes for which

$$V = k[HCrO_4^-][H^+][X^{n-}] \tag{8.48}$$

where X^{n-} = H$_2$PO$_4^-$, NCS$^-$, H$_2$O$_2$ and others.[93] All k's are approximately 10^5 M^{-2}s^{-1} and a dissociative mechanism is favored for the rds[93]

$$CrO_3(OH)^- + H^+ \rightleftharpoons CrO_3(OH_2) \quad \text{rapid} \tag{8.49}$$

$$CrO_3(OH_2) + X^{n-} \rightarrow CrO_3X^{n-} + H_2O \quad \text{rds} \tag{8.50}$$

The oxidation of organic substances by Cr(VI) has been reviewed and the importance of Cr(V) and Cr(IV) intermediates assessed. [85] The mechanisms appear to depend on the nature of the substrates and the medium. One favored mechanism for 2-propanol and other substrates in aqueous acid is [94]

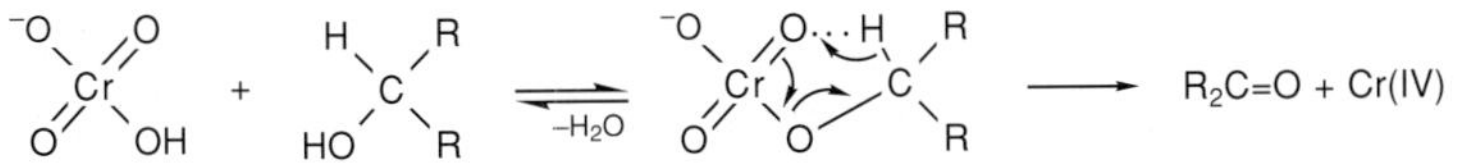

$$R_2C=O + Cr(IV) \qquad (8.51)$$

$$Cr(VI) + Cr(IV) \longrightarrow 2Cr(V) \text{ fast} \qquad (8.52)$$

$$Cr(V) + R_2CHOH \longrightarrow Cr(III) + R_2C=O \qquad (8.53)$$

A kinetic isotope effect supports $C-H$ bond breakage as a rds. [95] The intermediacy of Cr(V) is demonstrated by epr. The Cr(IV) species is epr-silent. [85] Free radical formation in some systems has also been demonstrated. [85]

Molybdenum

There has been considerable interest in the chemistry of molybdenum in recent years, in part due to its occurrence in nitrogenase which catalyzes the reaction:

$$N_2 + 8H^+ + 8e^- \rightarrow 2NH_3 + H_2 \qquad (8.54)$$

as well as in a number of other sulfhydryl enzymes, such as xanthine oxidase, nitrate reductase etc. [96] There is a wide range ($\approx 10^{10}$) of rate constants for substitution in molybdenum species. Generally, there is increased lability as the oxidation state becomes higher. [97]

Molybdenum(II)

$Mo_2(H_2O)_8^{4+}$ is one of only three dimeric aqua ions with no bridging ligands (Hg_2^{2+} and Rh_2^{2+} are the others). The structure **5** is eclipsed with δ-bond formation and quadruple metal-metal bonding. [97] A rapid pre-equilibrium involving substitution of anion into the axial (end) waters,

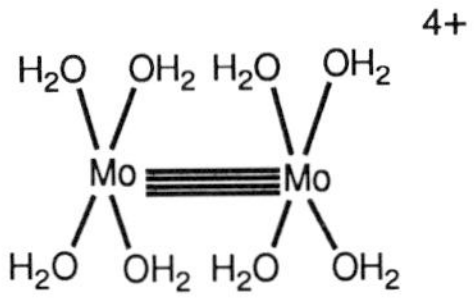

5

(not shown), is followed by a slower anion movement to the basal plane. [98] This mechanism resembles that proposed for substitution in VO^{2+} and TiO^{2+} (Table 8.7). [39,98]

Table 8.7 Comparison of Kinetic Parameters for the Formation of $1:1$ Complexes of Metal Ions with NCS^- at $25\,^\circ C$. From Ref. 98.

Metal Ion	$10^{-3}\,k$ $M^{-1}s^{-1}$	$\Delta H^{\ddagger}$ $kJ\,mol^{-1}$	$\Delta S^{\ddagger}$ $JK^{-1}mol^{-1}$
Mo_2^{4+}	0.59	58	$+3$
VO^{2+}	11.5	45	-16
TiO^{2+}	6.1	49	-17

Molybdenum(III)

The pale yellow $Mo(H_2O)_6^{3+}$ ion is obtained by reduction of Mo(VI) by Zn/Hg. Ligation rate constant values are highly dependent on the incoming ligand (Table 8.8) and this fact, together with the $-11.4\ cm^3mol^{-1}$ value for $\Delta V^{\ddagger}$ for the SCN^- reaction strongly supports an associative mechanism, possibly as extreme as A.[99] Surprisingly, $Mo(H_2O)_6^{3+}$ is more labile than $Cr(H_2O)_6^{3+}$ by a factor of 10^5. The reactions of $Mo(H_2O)_6^{3+}$ with O_2 and with ClO_4^- in acid are triphasic and give the $Mo^V{}_2O_4(H_2O)_6^{2+}$ ion. With O_2 the first step is believed to be

$$Mo^{3+} + O_2 \rightleftharpoons MoO_2^{3+} \qquad k_1, k_{-1} \tag{8.55}$$

and the value for k_1 ($180\ M^{-1}s^{-1}$) suggests an associative inner-sphere reaction.[100, 101]

Table 8.8 Second Order Rate Constants (k, $M^{-1}s^{-1}$) for Monocomplex Formation between L and $Mo(H_2O)_6^{3+}$ at $25\,^\circ C$. From Ref. 99.

L	k	L	k
Cl^-	4.6×10^{-3} [a]	$Co(C_2O_4)_3^{3-}$	0.34[b]
NCS^-	0.32[a]	$MoO_2(H_2O)_5^{3+}$	42[b]
$HC_2O_4^-$	0.49[a]	O_2	180[b]

[a] $\mu = 1.0\ M$ [b] $\mu = 2.0\ M$

Molybdenum(IV)

In low concentrations in 0.3 to 2.0 M H^+, Mo(IV) is present as MoO^{2+} or $Mo(OH)_2^{2+}$. At Mo(IV) concentrations exceeding 1 mM, the red trimeric species $Mo_3O_4(H_2O)_9^{4+}$ is present, the structure of which, **6**, is identical in solution and the solid state.[102] Both capping (a) and bridging (b) oxygens do not exchange with $H_2{}^{17}O$ in two years. The exchange of the H_2O's trans to bridging oxygen (c) is rapid and must be determined by nmr line broadening whereas the exchange of H_2O's trans to capping oxygen (d) can be determined by nmr signal loss. These water exchanges are via $Mo_3O_4(OH_2)_8OH^{3+}$, and it is believed that deprotonation occurs at a d water ($K_a = 0.30\ M$). Exchange rate constants are $1.6 \times 10^2 s^{-1}$ (c) and $1.5 \times 10^{-3}s^{-1}$ (d).[103] Anation of $Mo_3O_4^{4+}$ by SCN^- also occurs via the conjugate base and this substitution, as well as the water exchange are assigned an I_d mechanism (Prob. 13). There are two steps in the anation by $C_2O_4^{2-}$ and these are attributed to a fast (ring closure) and a slow (first bond formation) process i.e. the two steps are interposed (Sec. 1.6.2). [104(a)] Reduction of $Mo_3O_4(H_2O)_9^{4+}$ by Eu^{2+} produces Mo(III) via species containing Mo(III) Mo(III) Mo(IV). [104(b)]

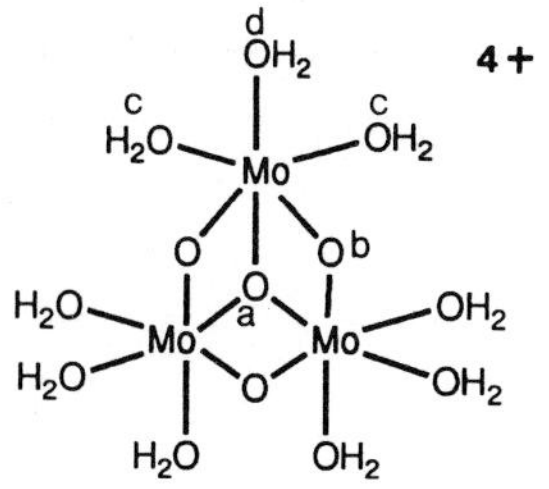

6

Molybdenum(V)

The diamagnetic binuclear species has the structure **7** (the number of coordinated H_2O's is uncertain) [105] and therefore three types of exchangable oxygen. The coordinated waters are labile. For the slower measurable oxygens, two well separated exchanges are observed, with $t_{1/2} = 4$ min (0°C) and 100 hours (40°C), each corresponding to two oxygens. Laser Raman spectra of solutions after the measurable faster exchange indicates these arise from yl O's. [106] Mo(V) is monomeric in organic solvents [39] (Prob. 14).

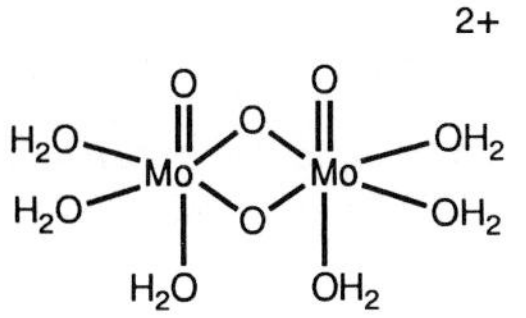

7

Molybdenum(VI)

Molybdenum(VI) and tungsten(VI) are stable oxidation states and only mild oxidants. Exchange with H_2O obeys the rate law:

$$V = k_0[XO_4^{2-}] + k_1[XO_4^{2-}][OH^-] \tag{8.56}$$

The values of k_0 (0.33 s^{-1} for X = Mo and 0.44 s^{-1} for X = W) are much larger than that for CrO_4^{2-} (3.2 × 10^{-7} s^{-1}), and this arises from smaller $\Delta H^{\ddagger}$ terms. [107] The tetrahedral MoO_4^{2-} ion reacts with chelating ligands X to form octahedral complexes. All kinetic data can be represented by

$$V = k[MoO_4^{2-}][X][H^+] \tag{8.57}$$

Plots of $\log k$ against $\log K_{HX}$ (K_{HX} is the constant for protonation of X) are linear (Fig. 8.4), and the greater reactivity of the more basic ligands supports an associative mechanism. [108] There is a similar behavior with WO_4^{2-} Ref. 39 and both contrast with that for $HCrO_4^-$ where an identical rate law (8.48) holds but a dissociative mechanism is favored. Tetrahedral coordination is maintained in reactions of Cr(VI) so that $Cr-O$ cleavage must occur. This probably accounts for the greater reactivity of $HMoO_4^-$ where adduct formation involves direct addition. [93]

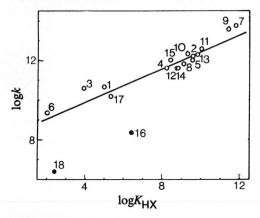

Fig. 8.4 Relationship between $\log k$ and $\log K_{HX}$ for Mo(VI) complex formation. L = Hox (1), ox$^-$ (2), Hoxs$^-$ (3), oxs^{2-} (4), Hcat$^-$ (5), H$_2$edta^{2-} (6), Hthb^{2-} (7), H$_2$thb$^-$ (8), Hpg^{2-} (9), H$_2$pg$^-$ (10), ep^{3-} (11), Hep^{2-} (12), dopa^{3-} (13), Hdopa^{2-} (14), H$_2$ga^{2-} (15), Co(NH$_3$)$_5$OH^{2+} (16), Has^{2-} (17), and Hnta^{2-} (18). [108] Hox = 8-hydroxyquinoline; H$_2$oxs = 8-hydroxyquinoline-5-sulfonic acid; H$_2$cat = catechol; H$_3$thb = 1,2,4-trihydroxybenzene; H$_3$pg = 1,2,3-trihydroxybenzene; H$_3$ep = (3,4-dihydroxy-phenyl)-2-(methylamino)ethanol; H$_3$dopa = (3,4-dihydroxyphenyl)alanine; H$_4$ga = 3,4,5-trihydroxy-benzoic acid; H$_3$as = 1,2-dihydroxyanthraquinone-3-sulfonic acid. Reproduced with permission from S. Funahashi, Y. Kato, M. Nakayama and M. Tanaka, Inorg. Chem. **20**, 1752 (1981).© (1981) American Chemical Society.

The substitution reaction

$$MoO_4^{2-} + 4\,H_2S \rightleftharpoons MoS_4^{2-} + 4\,H_2O \tag{8.58}$$

is surprisingly clean. The various steps have been delineated and a general rate law observed:

$$V = k\,[MoO_xS_{4-x}^{2-}]\,[HS^-]\,[H^+] \tag{8.59}$$

It is proposed that the proton is introduced into the activated complex with the $MoO_xS_{4-x}^{2-}$ species (rather than with HS$^-$), thereby allowing activation of the Mo=O bond [109]

$$\tag{8.60}$$

Manganese

Manganese represents the epitome of that characteristic property of the transition element namely the variable oxidation state. The aqueous solution chemistry includes all oxidation states from Mn(II) to Mn(VII), although these are of varying stability. Recently attention has been focused on polynuclear manganese complexes as models for the cluster of four manganese atoms which in conjunction with the donor side of Photosystem(II) is believed involved in plant photosynthetic oxidation of water. The Mn_4 aggregate cycles between 6 distinct oxidation levels involving Mn(II) to Mn(IV).[110]

Manganese(II)

The nmr lines for coordinated $H_2{}^{17}O$ are too broad for the accurate determination of the coordination number of Mn(II). The formula $Mn(H_2O)_6^{2+}$ is deduced from spectral considerations. The average parameters for water exchange are given in Table 4.1.[21,111 (a)] On the basis of these values an I_a mechanism for H_2O exchange, and for ligation by bpy and tpy[111 (b)] is suggested.

The aqua ion is not easily reduced nor oxidized. It is the slowest reacting of the bivalent transition metal ions with e_{aq}^- ($k = 7.7 \times 10^7 M^{-1}s^{-1}$) and the product Mn_{aq}^+ is very reactive.[112] However $Mn(CNR)_6^+$ (R = a variety of alkyl and aryl groups) is stable and the self-exchange in the Mn(I,II) hexakis(isocyanide) system has been studied by ^{55}Mn and ^1H nmr line broadening. The effects of solvent, temperature, pressure and ligand have been thoroughly explored.[113]

The $Mn(H_2O)_6^{2+/3+}$ exchange involves transfer of electrons from the high spin $d^5[(\pi d)^3(\sigma^*d)^2]$ +2 ion to the high spin $d^4[(\pi d)^3(\sigma^*d)^1]$ +3 ion. The transfer to a ligand-directed metal orbital of antibonding character should be accompanied by a substantial change in the Mn$-$O bond length, and therefore a reduced rate (Sec. 5.4). The electron transfer rate constant calculated from a semiclassical model is $10^{-4\pm1}M^{-1}s^{-1}$. The values estimated from Marcus treatment vary from 10^{-3} to $10^{-9}M^{-1}s^{-1}$, depending on the system considered (usually Mn^{3+} as an oxidant).[114] On the other hand a value of $3.2 \times 10^3 M^{-1}s^{-1}$ is estimated for the $Mn(tpps)^{4-/3-}$ exchange. The $>10^7$ enhanced reactivity compared with $Mn(H_2O)_6^{2+/3+}$ is ascribed to a change from a hard to a soft environment for Mn.[115] See also Ref. 116.

Manganese(III)

Manganese(III) is a powerful oxidant, with interesting mechanistic chemistry.[117] It can be generated *in situ* from MnO_4^- and Mn^{2+} in acid solution. By using excess Mn^{2+} ions and high acidity (3–5 M HClO$_4$) the marked disproportionation and hydrolytic tendencies of Mn(III) are suppressed and such solutions are stable for days at room temperature.

A number of second-order reactions with inorganic reductants have been studied. When the reductant has ligand properties, for example, $C_2O_4^{2-}$, Br$^-$ or HN$_3$, reaction is considered to occur via complex formation. Irradiation of $Mn(C_2O_4)_3^{3-}$ in aqueous solution leads to reduction and formation of CO_2 similar to that with the thermal reaction (Prob. 15):

$$Mn(C_2O_4)_3^{3-} \rightarrow Mn(C_2O_4)_2^{2-} + C_2O_4^{-} \tag{8.61}$$

$$Mn(C_2O_4)_3^{3-} + C_2O_4^{-} \rightarrow Mn(C_2O_4)_2^{2-} + C_2O_4^{2-} + 2\,CO_2 \tag{8.62}$$

Other first-row transition metal oxalate complexes behave similarly.[118]

Manganese(V)

MnO_4^{2-} is reduced by e_{aq}^{-} in base to yield MnO_4^{3-}. This ion is postulated as an intermediate in a number of studies.[119]

Manganese(VI)

The green tetrahedral ion MnO_4^{2-} is stable in basic solution. It can be prepared by reducing MnO_4^{-} with $Fe(CN)_6^{4-}$. There is uncertainty about the $MnO_4^{2-} - H_2O$ exchange rate.[33] The ion disproportionates in acid and the kinetics have been studied by stopped-flow.[120] At 610 nm where loss of MnO_4^{2-} is monitored, the reaction is first-order. At 520 nm where formation of MnO_4^{-} is observed, the reaction is second-order. These observations and the H^+ dependency suggest a mechanism

$$HMnO_4^{-} \rightleftharpoons MnO_4^{2-} + H^{+} \qquad \text{fast} \tag{8.63}$$

$$HMnO_4^{-} \rightarrow MnO_3 + OH^{-} \qquad \text{first-order} \tag{8.64}$$

$$MnO_3 + HMnO_4^{-} \rightarrow MnO_4^{-} + Mn(V) \qquad \text{second-order} \tag{8.65}$$

The MnO_4^{2-}, MnO_4^{-} electron transfer has been studied using radioisotopes and quenched-flow as well as by nmr, with good agreement between the results (Tables 1.3 and 3.3). The rate of outer-sphere electron transfer is given by

$$V = (k_0 + k_M [M^{+}]) [MnO_4^{-}] [MnO_4^{2-}] \tag{8.66}$$

The activation parameters for the cation-independent pathway (k_0) can be accounted for by a modified semiclassical Marcus-Hush theory. Lower enthalpies, and more positive volumes of activation are noted for the M^+-catalyzed pathway.[121]

Manganese(VII)

As with MnO_4^{2-} there is disagreement on the rate law for water exchange for MnO_4^{-}. One term in the rate law ($k_0[MnO_4^{-}]$) appears certain and for $M = Mn$, $k_0 = 1.9 \times 10^{-5}\,s^{-1}$ and

$M = Re$, $k_0 = 7.7 \times 10^{-7}s^{-1}$ at $25\,^{\circ}C$.[33] It is a useful oxidant for inorganic and organic substrates, often with complicated kinetics. In the stepwise mechanism

$$Mn(VII) + red \;\rightleftharpoons\; Mn(VI) + ox \tag{8.67}$$

$$Mn(VI) + red \;\rightleftharpoons\; Mn(V) + ox \tag{8.68}$$

$$Mn(V) + 3\,red \;\rightarrow\; Mn(II) + 3\,ox \quad rapid \tag{8.69}$$

the first step is rate-determining for red $= W(CN)_8^{4-}$, $Mo(CN)_8^{4-}$ and $Fe(CN)_6^{4-}$, whereas with $Fe(phen)_3^{2+}$, reduction of Mn(VI) is rate-determining.[B36] Two stages are discerned with V(IV)[122]

$$Mn(VII) + V(IV) \;\rightarrow\; Mn(VI) + V(V) \tag{8.70}$$

$$Mn(VI) + V(IV) \;\rightarrow\; Mn(V) - V(V) \;\rightarrow\; Mn(II) \tag{8.71}$$

The simple stoichiometry in (8.72) belies simple kinetics. The dynamics are

$$2\,MnO_4^- + 5\,H_2O_2 + 6\,H^+ \;\rightarrow\; 2\,Mn^{2+} + 8\,H_2O + 5\,O_2 \tag{8.72}$$

frighteningly complex, with induction, inhibition and autocatalytic features! For the first phase the rate law is

$$-d\,[MnO_4^-]/dt = \{a + b\,[H^+]\}\,[MnO_4^-]\,[H_2O_2] \tag{8.73}$$

which is interpreted as a nucleophilic inner-sphere attack of H_2O_2 on Mn(VII).[123]

It is a useful oxidant for hydrocarbons, alkenes, alcohols and aldehydes.[124] Permanganate reacts with carbon-carbon double bonds to form a cyclic manganate(V) diester. The nature of the products is determined by subsequent rapid processes.[124]

Iron

The complexes of Fe(II) and Fe(III), the important oxidation states in aqueous solution, have played major roles in our understanding of the mechanisms of substitution and redox processes.

Iron(II)

Data for water exchange with $Fe(H_2O)_6^{2+}$ are shown in Table 4.1.[111] The value for $\Delta V^{\ddagger}$ indicates an interchange dissociative mechanism, which is also reflected in data for the reaction of Fe^{2+} with tpy ($\Delta V^{\ddagger} = +3.5$ cm^{3}mol^{-1},[125]) and other ligands.[111] One of the earliest studies of substitution in a labile metal ion was of the reaction of Fe^{2+} with bpy and phen in acid solution (Sec. 2.1.4).

A classical D mechanism is favored for substitution in a number of low spin Fe(II) complexes of the form $trans$-Fe(N$_4$)XY where N$_4$ is a planar tetradentate ligand (porphine, phthalocyanine or macrocycle) or two bidentate glyoximates and X and Y are neutral ligands CO, py etc. (Structures 8).[126] These substitutions are characterized by a small range of on-rates for a particular (N$_4$) and X group, when Y is the replaced group. Equilibrium positions are dominated by the off-rates which can span many orders of magnitude. Thus $\Delta H^{\ddagger}_{-1}$ for the dissociation process is a useful measure of the coordinate bond energy.[126] There have been few studies of ligand replacement in non-octahedral complexes. (Table 4.8)

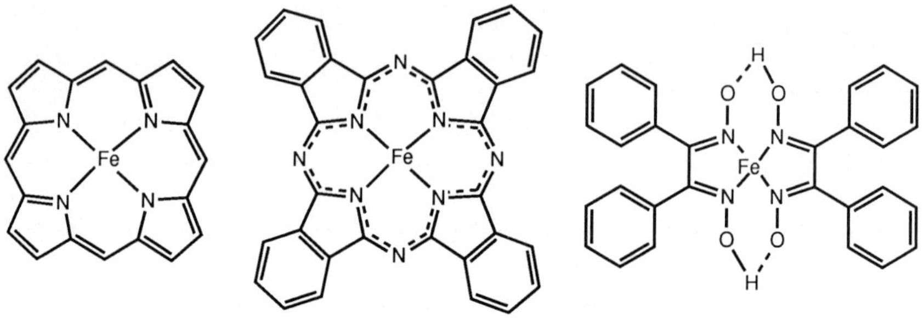

(A)   (B)   (C)

8 - Structures of four-coordinated Fe(II) porphyrin (A), phthalocyanine (B) and bis(diphenylglyoxime) (C) complexes.

A number of oxidations of Fe(H$_2$O)$_6^{2+}$ by one-electron outer-sphere reactants, IrCl$_6^{2-}$, Fe(phen)$_3^{3+}$, Mn(III) and Co(III) are rapid and second-order and give the expected LFER.[116,127] The measured rate constant for the Fe$^{2+/3+}$ self-exchange is 4 M^{-1}s^{-1}. The value computed from a number of cross reactions involving Fe^{2+} or Fe^{3+} is $\approx 10^{-3}$M^{-1}s^{-1}. This suggests that the directly measured self-exchange may be inner-sphere.[116] (Sec. 5.4 and Fig. 2.8) In the slow inner-sphere oxidations of Fe^{2+}, electron transfer is almost always rate-determining, (Fig. 5.1(b)) since substitution in the coordination sphere of Fe(II) is very rapid. Diagnosis of an inner-sphere process requires identifying quickly the Fe(III) complex formed (by flow methods, Sec. 5.3 a)). Rapid electron transfer between coordinately saturated Fe(II) and Fe(III) complexes must be outer-sphere. Oxidation of Fe(II) by two-electron oxidants must of necessity be more complex, since an unstable oxidation state of either iron or the oxidant must be produced. In some cases the production of a binuclear Fe(III) complex is evidence for the participation of Fe(IV), for example with HOCl and O$_3$ (but not O$_2$ nor H$_2$O$_2$[127]).

$$\text{Fe}^{II} + \text{ox} \rightarrow \text{Fe}^{IV} + \text{red} \tag{8.74}$$

$$\text{Fe}^{II} + \text{Fe}^{IV} \rightarrow [\text{Fe}^{III}]_2 \quad \text{fast} \tag{8.75}$$

The interaction of Fe(II) with small molecules has received much attention. The Fe(II)/oxygen system must be one of the most studied chemical interactions. Since the Fe-porphyrin complex forms the core of the naturally occurring iron respiratory proteins myoglobin and hemoglobin [128] the reaction of iron(II) porphyrin (PFe^{II}) with O_2 in noncoordinating solvents has been extensively studied. A generally accepted mechanism is

$$PFe^{II} + O_2 \rightleftharpoons PFeO_2 \tag{8.76}$$

$$PFeO_2 + PFe^{II} \rightarrow PFe^{III}-O-O-Fe^{III}P \tag{8.77}$$

$$PFe^{III}-O-O-Fe^{III}P \rightarrow 2\,PFe^{IV}O \tag{8.78}$$

$$PFe^{IV}O + PFe^{II} \rightarrow 2\,PFe^{III}-O-Fe^{III}P \tag{8.79}$$

By working at low temperatures ($-70\,°C$) it has been possible to identify the oxygen adduct in (8.76), the peroxo bridged complex in (8.77) and the Fe(IV) species in (8.78) as intermediates. [129] In aqueous solution, using $Fe(tmpyp)^{4+}$, a dimer product does not result from the O_2 reaction, which suggests that an Fe(IV) species is not implicated. [130] The first step (8.76) is isolated in the reactions of myoglobin and hemoglobin, hence imparting the unique oxygen-carrying and storage ability of these respiratory proteins. The simple formulation of (8.76) does however hide a wealth of detail (Sec. 2.1.2). The ion $Fe(edta)^{2-}$ is sensitive to O_2 but is a useful, gentle reductant (Prob. 17).

A combination of Fe^{2+} and H_2O_2 is termed Fenton's reagent and is an effective oxidizing mixture towards organic substrates. In aqueous solution, the $OH^{\bullet}$ radical appears to be the actual oxidant [131,132]

$$Fe(II) + H_2O_2 \rightarrow Fe^{III}OH^{2+} + OH^{\bullet} \tag{8.80}$$

From observations of widely different products in nonaqueous solvents, it is concluded that higher valence iron complexes are primary oxidants (as in biochemical oxidations [133-136]):

$$Fe(II) + H_2O_2 \rightarrow Fe(H_2O_2)^{2+} \quad \text{(which may be } FeO^{2+}) \tag{8.81}$$

$$Fe(H_2O_2)^{2+} + RH \rightarrow Fe(II) + ROH \tag{8.82}$$

$Fe(H_2O)_6^{2+}$ and other Fe(II) complexes react with NO to give brown-black complexes, e.g. $Fe^I(H_2O)_5NO^{2+}$ Ref. 137. There is a wide range of formation rate constants, $4 \times 10^2 M^{-1}s^{-1}$ to $>6 \times 10^7 M^{-1}s^{-1}$, and the interaction is conveniently studied by temperature-jump. [138]

Iron(III)

Nmr $H_2^{17}O$ line broadening gives the following rate law for exchange with Fe(III) solutions

$$k_{exch} = k_1 + k_2[H^+]^{-1} \tag{8.83}$$

k_1 is the rate constant for exchange of an aqua ligand on $Fe(H_2O)_6^{3+}$ with bulk water, and $k_2 = k_{OH}K_a$ in which k_{OH} is the rate constant for exchange on $Fe(H_2O)_5OH^{2+}$

$$Fe(H_2O)_6^{3+} \rightleftharpoons Fe(H_2O)_5OH^{2+} + H^+ \quad K_a \tag{8.84}$$

The data, including results from pressure effects, are shown in Tables 4.5 and 4.6. The complexing of Fe(III) with a large number of ligands has been studied (Table 2.1). The general consensus is that substitution in $Fe(H_2O)_6^{3+}$ is associative and that of $Fe(H_2O)_5OH^{2+}$ is dissociative. The apparent dependency on the nature of the ligand of the ligation rate constants for $Fe(H_2O)_6^{3+}$ but not for $Fe(H_2O)_5OH^{2+}$, strengthened by negative and positive values for $\Delta V^{\ddagger}$ respectively, [139-141] support I_a and I_d mechanisms respectively. Where proton ambiguity arises, the reaction of $FeOH^{2+}$ (with HA) is usually favored over reaction of Fe^{3+} (with A^-) Sec. 2.1.7(b). Porphyrins have the ability when coordinated to Fe(III) (and also Cr(III) and Co(III)) to labilize the remaining coordinated water(s) by a factor of 10^4–10^5 compared with that of the hexaaqua ion (Table 8.9). [142] This lability may be the explanation for the high rates of dimerization of the porphyrin complexes [130] e.g.

$$Fe(tmpyp)(H_2O)^{5+} + Fe(tmpyp)OH^{4+} \rightleftharpoons (tmpyp)Fe-\underset{H}{O}-Fe(tmpyp)^{9+} + H_2O \qquad (8.85)$$

Table 8.9 Water Exchange on Fe(III)-Porphyrin Complexes [142]

Complex	k_{exch} (per H_2O), s^{-1}	$\Delta H^{\ddagger}$, kJ mol^{-1}	$\Delta S^{\ddagger}$, J K^{-1} mol^{-1}
$Fe(tmpyp)(H_2O)_n^{5+}$	7.8×10^5	57	61
$Fe(tpps)(H_2O)_n^{3-}$	1.4×10^7	57	84

A number of Fe(III) complexes undergo such dimerization in solution.

The aquated iron(III) ion is an oxidant. Reaction with reducing ligands probably proceeds through complexing. Rapid scan spectrophotometry of the Fe(III)-cysteine system shows a transient blue Fe(III)-cysteine complex and formation of Fe(II) and cystine. [143] The reduction of Fe(III) by hydroquinone, in concentrated solution has been probed by stopped-flow linked to x-ray absorption spectrometry. The changing charge on the iron is thereby assessed. [144] In the reaction of Fe(III) with a number of reducing transition metal ions M in acid, the rate law

$$V = (a + b[H^+]^{-1})[Fe^{III}][M] \qquad (8.86)$$

is obeyed. [144] The inverse proton dependency is dominant and is ascribed to an inner-sphere reaction of a hydroxy form of either Fe(III) or M. [145] The absence of an inverse $[H^+]$ dependant term heralds an outer-sphere mechanism, e.g. in the reaction of Fe(III) with $Co(sep)^{2+}$ Ref. 146.

Some Iron Systems of Special Interest

Iron(II) and (III) cyano complexes provide very interesting mechanistic chemistry. The $Fe(CN)_5H_2O^{3-}$ ion is a popular substrate for study. It is easily prepared *in situ* by the dissolution of solid $Na_3[Fe(CN)_5NH_3]$ for 15 minutes at 25°C in aqueous solution. For a large variety of entering groups Y (unspecified charge) a D mechanism is favored (Prob. 18):

$$Fe(CN)_5H_2O^{3-} \rightleftharpoons Fe(CN)_5^{3-} + H_2O \qquad k_1, k_{-1}, K_1 \qquad (8.87)$$

$$Fe(CN)_5^{3-} + Y \rightleftharpoons Fe(CN)_5Y^{3-} \qquad k_2, k_{-2}, K_2 \qquad (8.88)$$

There is an excellent LFER between $\log k_{-2}$ and $\log K_1 K_2$ for reaction of neutral unidentates, and even for the binuclear-forming $Rh(NH_3)_5X^{3+}$ and $Co(CN)_5X^{2-}$ (Fig. 8.5[147]), also
Refs. 148, 149. Rate constants for the replacement of H_2O in $Fe(CN)_5H_2O^{3-}$ range from

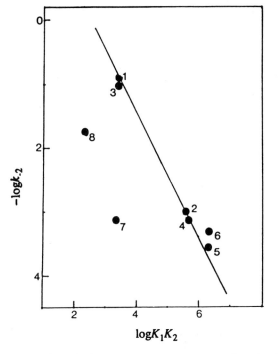

Fig. 8.5 LFER for $Fe^{II}(CN)_5L$ complexes.
Representative examples from a large
number of data include $C_6H_5CN(1)$, py(2),
4-CNpy in which attachment is through
NC(3), 4-CNpy in which attachment is
through the py N(4), 4-CH$_3$Npy(5),
4-bpy $Rh(NH_3)_5^{3+}$ (6),
4-CNpy $Co(CN)_5^{2-}$ (7), and NC-
$Co(CN)_5^{3-}$ (8). The line represents the
equation $\log k_{-2} = 2.45 - 0.98 \log K_1 K_2$
observed for a number of neutral ligands
L and interpreted in terms of a D
mechanism (A. D. James and R. S. Murray, J. Chem. Soc. Dalton Trans. 1530
(1975); A. P. Szecsy, S. S. Miller and A.
Haim, Inorg. Chim. Acta **28**, 189 (1978);
K. J. Pfenning, L. Lee, H. D. Wohlers
and J. D. Petersen, Inorg. Chem. **21**, 2477
(1982)).

$2 \times 10^2 M^{-1}s^{-1}$ (neutral ligands) to $6 \times 10^3 M^{-1}s^{-1}$ for $+3$ charged entering ions.[148] Dilute
solutions (<0.1 mM) of the aqua complex should be used to avoid complications of
polynuclear formation. In addition, O_2 oxidizes $Fe(CN)_5H_2O^{3-}$ fairly rapidly in acid solution:[150]

$$4\,Fe(CN)_5H_2O^{3-} + 4\,H^+ + O_2 \rightarrow 4\,Fe(CN)_5H_2O^{2-} + 2\,H_2O \qquad (8.89)$$

$$V = k\,[Fe(CN)_5H_2O^{3-}]\,[O_2]\,Fe^{2+}] \qquad (8.90)$$

Indeed, the oxidation of $Fe(CN)_6^{4-}$ by O_2 (as well as by H_2O_2 and BrO_3^-) proceeds via the
rds of dissociation of the hexa- to the penta-cyano complex.[151] The value of k in (8.90) is
$5.6 \times 10^6 M^{-2}s^{-1}$ at pH > 3.8. Traces of Fe^{2+} from decomposition of the cyano complex
promote catalytic oxidation (Prob. 19). A large number of complexes of the type
$Fe(CN)_5X^{n-}$ for both Fe(II) and Fe(III) have been studied and cross-reaction redox kinetics
abound.[151,152] Care has to be exercised in the use of $Fe(CN)_6^{3-}$. Daylight can induce changes
in the complex even within an hour[153] and catalytic effects (traces of Cu^{2+} Sec. 3.1.4) have
to be considered. In addition, the sensitivity of the values of E^0 and rate constants to
medium effects lessen the value of the iron-cyano complexes as reactant partners for the
demonstration of Marcus relationships.[154] Nevertheless, they, with other inorganic complexes, have been extensively employed to probe the peripheral characteristics of metalloproteins.[155]

Nitric oxide coordinated to iron modifies, in a striking manner, the properties and reactivity of free NO (Sec. 6.2). Probably the most famous such coordinated entity is the nitroprusside ion, $Fe(CN)_5NO^{2-}$. An incisive review of its reactions particularly related to its hypertensive action (it reduces blood pressure of severely hypertensive patients) is available.[156] Nitroprusside ion reacts with a variety of bases

$$Fe(CN)_5NO^{2-} + X^{n-} \rightleftharpoons Fe(CN)_5NO(X)^{(2+n)-} \tag{8.91}$$

with attack at the O or the N atom of the coordinated nitrosyl group.[157] A full study of the primary reaction of $Fe(CN)_5NO^{2-}$ with a variety of aliphatic thiols has been carried out by a temperature-jump/stopped-flow combination. This is necessary because (8.91) $X^{n-} = RS^-$ is followed within seconds by redox reactions.[158] The primary products of reduction of nitroprusside by dithionite are $Fe(CN)_4NO^{2-}$ and CN^-. Using nitroprusside enriched to 90% in ^{13}C, the epr of the iron product is best interpreted in terms of coupling to four (and not five) ^{13}C nuclei and a single ^{14}N of NO.[159] This supports the formulation of the product as $Fe(CN)_4NO^{2-}$, containing the $Fe^I - NO^+$ rather than the $Fe^{II} - NO$ entity.[156]

A number of Fe(II) and Fe(III) chelates exist in low spin, high spin equilibrium in solution. They therefore afford an excellent opportunity to study the dynamics of a relatively "simple" electron-transfer and a number of very rapid reaction techniques have been applied to these systems (Chapters 3 and 7). Spin state interconversions are slightly more rapid in Fe(III) complexes.[160]

Iron(IV) and Iron(V)

These oxidation states attract attention because Fe(IV) and Fe(V) cation radical porphyrins are active intermediates in biological hydroxylation.[161] In strong base, Fe(IV) is produced from Fe(III) by powerful oxidants and Fe(V) arises from radical reduction of Fe(VI)

$$Fe(OH)_4^- + OH^\bullet \rightarrow Fe^{IV} = O(OH)_n^{(2-n)+} + (3 - n)OH^- + H_2O \tag{8.92}$$

$$FeO_4^{2-} + e_{aq}^- \rightarrow FeO_4^{3-} \tag{8.93}$$

Both products are unstable and decompose to Fe(III) with rate constants $2 \ s^{-1}$ (Fe(IV) in $1M \ OH^-$) and $4 \ s^{-1}$ (Fe(V) in $5M \ OH^-$).[162,163]

Iron(VI)

Iron(VI) ferrates are easily prepared, stable solids. They are strong oxidizing agents ($E^0 = +0.72V$ in alkaline solution) and show a high degree of selectivity.[163] In aqueous basic solution the ion is FeO_4^{2-}; all O's are equivalent towards H_2O exchange. At pH 9.6–14,

$$V = k \, [FeO_4^{2-}] \tag{8.94}$$

where $k = 1.6 \times 10^{-2}s^{-1}$. The oxidation states Fe(IV) and Fe(V) can be shown not to play an important role in the exchange.[164]

Ruthenium

Increasing attention is being given to the reactivity of ruthenium species which show unusual behavior compared with their Co analogs. Aspects of current interest are mixed valence states,[165] ruthenated proteins to probe electron transfer in them (Chap. 5) and the photochemistry and photophysics of Ru(II) polypyridine complexes.[166]

Ruthenium(II)

The exchange data for $Ru(H_2O)_6^{2+}$ indicate an I mechanism.[167] The larger activation volume for $Fe(H_2O)_6^{2+}$ than $Ru(H_2O)_6^{2+}$ exchange (i.e. more dissociative character) arises from the presence of two e_g electrons in the former. This apparently more than offsets the larger ionic radius for $Fe(H_2O)_6^{2+}$ ($0.78 Å$ vs $0.73 Å$) which would favor an associative behavior.[167] There is a strong *trans* effect (which is probably operational in the transition state) in the 10^3 enhanced rate of water exchange in $Ru(\eta^6\text{-}C_6H_6)(H_2O)_3^{2+}$ compared with $Ru(H_2O)_6^{2+}$. An interchange mechanism for both is proposed.[168] Substitution reactions of ruthenium(II) ammines have been widely studied and are predominantly dissociative even in nonaqueous solution.[169,170] The $Ru(NH_3)_5H_2O^{2+}$ ion reacts with a number of unidentate ligands, including nitrogen, with similar $((0.3\text{-}30) \times 10^{-2}M^{-1}s^{-1})$ rate constants suggesting a dissociative activation mode.[171-173] Because of the relatively slow rate for water substitution in $Ru(NH_3)_5H_2O^{2+}$ ($\approx 0.1 M^{-1}s^{-1}$) and the relatively large rate constant for self-exchange ($\approx 3 \times 10^3 M^{-1}s^{-1}$), the ruthenium(II) ion tends to react by outer-sphere redox mechanisms.[174] Ruthenium(II) ammines react with O_2 in acidic solution:

$$2\,Ru(II) + 2\,H^+ + O_2 \rightarrow 2\,Ru(III) + H_2O_2 \tag{8.95}$$

$$-d\,[Ru(II)]/dt = 2k_1\,[Ru(II)]\,[O_2] \tag{8.96}$$

for which a mechanism invoking the intermediacy of O_2^- is proposed:

$$Ru(II) + O_2 \rightleftharpoons Ru(III) + O_2^- \qquad k_1, k_{-1} \tag{8.97}$$

$$O_2^- + H^+ \rightleftharpoons HO_2^{\cdot} \tag{8.98}$$

$$Ru(II) + HO_2^{\cdot} \rightarrow Ru(III) + HO_2^- \tag{8.99}$$

Although autoxidation of $Ru(sar)^{2+}$ has similar characteristics in acidic solution, in base hydrogen atom transfer from $Ru(sar)^{2+}$ to O_2^- leads to a deprotonated Ru(III) species which is oxidized to relatively stable $Ru^{IV}(sar\text{-}2\,H^+)^{2+}$ Ref. 175. The strong deviation from linearity for semi-log plots, with a large excess of O_2, is removed when Fe(II) is added. This suppresses the k_{-1} step and doubles the rate. Compare Sec. 2.2.1 (b). The value of k_{-1} can be assessed as $1.3 \times 10^8 M^{-1}s^{-1}$ Ref. 176. The behavior of pentacyanoruthenium complexes has been compared with the iron analogs. Substitution in $M^{II}(CN)_5L^{n-}$ with both $M = Fe$ and Ru is dissociative, with decreased lability for the Ru(II) species, Table 8.10.[177]

Table 8.10 Rate Constants for Reactions of $Ru(CN)_5H_2O^{3-}$ and $Fe(CN)_5H_2O^{3-}$ with Ligands at 25 °C. From Ref. 177.

Ligand	k_1 (Ru)[a] $M^{-1}s^{-1}$	$10^5 \times k_{-1}$ (Ru)[b] s^{-1}	k_1 (Fe)[a] $M^{-1}s^{-1}$	$10^3 \times k_{-1}$ (Fe)[b] s^{-1}
$bpyH^+$	44	4.2	2050	2.6
bpy	14	6.8	365	0.62
pyz	10	1.8	380	0.42
py	5.4	3.3	365	1.1
imid	5.1	11	240	1.3
dmso	13	0.85	240	0.075

[a] Formation of $M(CN)_5L^{n-}$ [b] Hydrolysis of $M(CN)_5L^{n-}$

$Ru(CN)_5NO^{2-}$ reactions with OH^-, SH^- and SO_3^{2-} resemble those of the nitroprusside ion, with attack at the coordinated nitrosyl to give analogous transients and similar second-order rate constants.[178] Ruthenium(II) complexes of the general type $Ru(N_2)_3^{2+}$, $N_2 = $ bidentate ligands, are important reactants. The relative inertness of $Ru(NH_3)_6^{2+}$ and $Ru(diimine)_3^{2+}$ towards substitution makes these complexes definite, although weak, outer-sphere reductants (Tables 5.4, 5.5, 5.6 and 5.7).[179] Ruthenium(II) complexes of the general type $Ru(diimine)_3^{2+}$, and particularly the complex $Ru(bpy)_3^{2+}$, have unique excited state properties.[166] They can be used as photosensitizers in the photochemical conversion of solar energy, Scheme 8.1 [180]

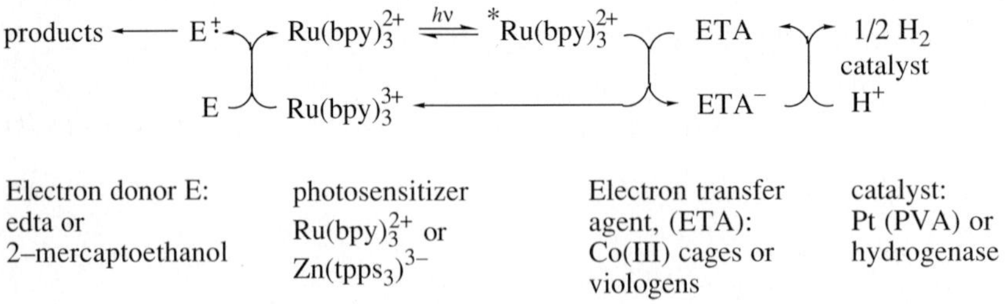

products ← E^+ ⟍ $Ru(bpy)_3^{2+}$ ⇌^{hv} $^*Ru(bpy)_3^{2+}$ ⟍ ETA ⟍ 1/2 H_2 catalyst

E ⟋ $Ru(bpy)_3^{3+}$ ← ⟋ ETA^- ⟋ H^+

$$(8.100)$$

Electron donor E:	photosensitizer	Electron transfer	catalyst:
edta or	$Ru(bpy)_3^{2+}$ or	agent, (ETA):	Pt (PVA) or
2–mercaptoethanol	$Zn(tpps_3)^{3-}$	Co(III) cages or	hydrogenase
		viologens	

It is essential that the quenching of $^*Ru(bpy)_3^{2+}$ by the ETA is via electron and not energy transfer. Either or both pathways have been observed with cobalt(III) complexes[180,181] (Prob. 22).

Ruthenium(III)

Substitution reactions of Ru(III) are very slow.[170,182] An I_a mechanism is assigned to water exchange for $Ru(H_2O)_6^{3+}$ and an I mechanism for $Ru(H_2O)_5OH^{2+}$ (Tables 4.5 and 4.6).[167] Coordinated edta labilizes the remaining H_2O coordinated to Ru(III). Activation parameters for:

$$Ru(edta)H_2O^- + L \rightleftharpoons Ru(edta)L^- + H_2O \qquad (8.101)$$

indicate an associative mechanism.[183,184] The bell-shaped k/pH profile (Fig. 1.13) resembles that for the Co(III) analog with a similar assignment namely to reaction of Ru(edta)OH^{2-}, $\text{Ru(edta)H}_2\text{O}^-$ and $\text{Ru(Hedta)H}_2\text{O}$.[183,184] Data for substitution in Ru(II)edta complexes are more equivocal because the more labile Ru(III)-edta complexes catalyze substitution in Ru(II) (Compare Sec. 5.7.3)[183] Associative mechanisms are common with Ru(III) complexes.[185] Complete stereoretention in substitution and induced hydrolyses is observed.[186] Overall, Ru(III) behaves more like Ru(II) than like the obvious analog, Co(III).

$\text{Ru(H}_2\text{O)}_6^{2+/3+}$ is the only known low spin d^6, d^5 pair of metal aqua ions. Exchange can be measured by broadening of the ^{99}Ru nmr signals of $\text{Ru(H}_2\text{O)}_6^{2+}$ in the presence of $\text{Ru(H}_2\text{O)}_6^{3+}$. The results check well with those measured directly at lowered temperatures using a fast injection technique. The measured rate constant 20 $\text{M}^{-1}\text{s}^{-1}$ at 25 °C in 2.5 M H^+ agrees well with that estimated from cross-reactions data and the Marcus expression.[167,187] $\text{Ru(H}_2\text{O)}_6^{2+/3+}$ is the only unambiguous outer-sphere self-exchange among the hexaaqua ions. This arises because H_2O exchange on both ions is much slower than electron transfer.[187] The higher value (3.3 × $10^3\text{M}^{-1}\text{s}^{-1}$ at 4 °C Ref. 188) for the $\text{Ru(NH}_3\text{)}_6^{2+/3+}$ couple is ascribed to lower reorganisational energy (in Eqn. (5.24)) for the ammines ($\Delta d = 0.04\,\text{Å}$, $\text{Ru}-\text{N}$ and $\Delta d = 0.09\,\text{Å}$, $\text{Ru}-\text{O}$)[189] (Prob. 23). The differences in the $\text{Ru}-\text{N}$ bonds in $\text{Ru(sar)}^{2+/3+}$ complexes must be even smaller since k (self-exch) = 1.2 × $10^5\text{M}^{-1}\text{s}^{-1}$ Ref. 116.

Ruthenium(VI) and (VII)

The rate constant for electron exchange between RuO_4^- and RuO_4^{2-} is greater than $10^4\text{M}^{-1}\text{s}^{-1}$ in 0.1 M OH^- at 25 °C. This is faster than the electron exchange between MnO_4^{2-} and MnO_4^-, although the standard potentials for the two couples are close (0.60 and 0.56 V for Ru and Mn species respectively).[190]

Cobalt

It would be hardly possible to do full justice to the kinetic behavior of cobalt even in a book devoted to that subject. Only some important features will be emphasized. The stable oxidation states in aqueous solution are Co(II) and Co(III).

Cobalt(I)

Kinetic information on the lower oxidation states (Prob. 24) is sparse for Werner-type complexes.[112] Co^{2+} and Co(II)-bpy complexes are reduced by e_{aq}^- to give Co(I) complexes except that the mono species yields $\text{Co}^{\text{II}}(\text{bpy}^{\bullet-})(\text{H}_2\text{O})_4^+$. The rate constants are in the range 3.5–7.4 × $10^{10}\text{M}^{-1}\text{s}^{-1}$ for the mono, bis and tris bipyridine complexes and 3.0 × $10^9\text{M}^{-1}\text{s}^{-1}$ for the hexaaquacobalt(II) ion. The radicals $\text{CO}_2^{\bullet-}$ and $(\text{CH}_3)_2\text{COH}^{\bullet}$ react more slowly ($k \approx 10^7 M^{-1}s^{-1}$) and concurrently produce Co(I) and radical addition products. The examination of a number of cross-reactions of the type:

$$Co(bpy)_2^+ + Co(bpy)_3^{2+} \rightleftharpoons Co(bpy)_2^{2+} + Co(bpy)_3^+ \qquad (k = 2 \times 10^9 M^{-1}s^{-1})$$

$$(8.102)$$

leads to self-exchange rate constant estimates of $>10^8 M^{-1}s^{-1}$ for $Co(bpy)_3^{+/2+}$ couples.[191] These Co(I) species react very rapidly with acid forming Co(III) hydrides e.g.[192]

$$Co(bpy)_3^+ + H_3O^+ \rightleftharpoons Co(bpy)_2(H_2O)H^{2+} + bpy \qquad (8.103)$$

Cobalt(I) oximes related to vitamin B_{12} have been extensively investigated.[193,194]

Cobalt(II)

A variety of geometries have been established with Co(II). The interconversion of tetrahedral and octahedral species has been studied in nonaqueous solution (Sec. 7.2.4). The low spin, high spin equilibrium observed in a small number of cobalt(II) complexes is rapidly attained (relaxation times < ns) (Sec. 7.3). The six-coordinated solvated cobalt(II) species has been established in a number of solvents and kinetic parameters for solvent(S) exchange with $Co(S)_6^{2+}$ indicate an I_d mechanism (Tables 4.1–4.4). The volumes of activation for Co^{2+} complexing with a variety of neutral ligands in aqueous solution are in the range +4 to +7 cm^{3}mol^{-1}, reemphasizing an I_d mechanism.[125]

Substitution in 5- and 4-coordinated cobalt(II) complexes is associative. The macrocycle Me$_4$cyclam imposes a 5-coordinated structure in the complex $Co(Me_4cyclam)CH_3CN^{2+}$ which exchanges with solvent CH$_3$CN by an I_a mechanism ($\Delta V^{\ddagger} = -9.6$ cm^{3}mol^{-1}).[195] A strongly associative mechanism for (tetrahedral) $Co(Ph_3P)_2Br_2/Ph_3P$ exchange in CDCl$_3$ is supported by a $\Delta V^{\ddagger} = -12$ cm^{3}mol^{-1} value. (Sec. 4.8)

Exchange reactions involving complexes of Co(II) and Co(III) have proved very interesting because of the generally high-spin and low-spin characteristics, respectively, of these oxidation states. The differences in self-exchange rate constants have been rationalised in terms of reorganizational energy differences (Chap. 5). An interesting correlation of log k (self-exchange) with the number of amine protons in twenty-four $Co(N)_6^{2+/3+}$ systems has been noted.[196]

The interaction of cobalt(II) complexes with O$_2$ has been intensively studied[197,198] (Prob. 25). In general, octahedral complexes have been examined in water and square planar complexes in nonaqueous solution. The types of ligands which promote O$_2$ affinity for Co(II) are porphyrins,[199] salicylidenamines, aliphatic polyamines, amino acids or peptides, and saturated and unsaturated macrocyclic tetraamines.[200] Three or more strong donor groups attached to cobalt(II) appear to be necessary to promote its O$_2$ binding.[198] A general mechanism (charges omitted) for O$_2$ uptake by octahedral complexes of Co(II), deduced by kinetics[201] is

$$Co(L) + O_2 \rightleftharpoons Co(L)O_2 \qquad k_1, k_{-1} \qquad (8.104)$$

$$Co(L)O_2 + Co(L) \rightleftharpoons LCoO_2CoL \qquad k_2, k_{-2} \qquad (8.105)$$

Square planar Co(II) complexes with tetradentate ligands tend to stop at (8.104) in nona-queous and aqueous solution. [200,202] Otherwise, reaction results in the μ-peroxo complex and this can undergo further bridging to form a variety of double bridged complexes. [198] Even-tually mononuclear Co(III) complexes result. Recently ^{59}Co nmr has been used to help to disentengle the various species which arise in the overall oxygenation process. [203] The values of k_1 range from 10^4 to $5 \times 10^5 \mathrm{M}^{-1}\mathrm{s}^{-1}$. [201] Oxidation and reduction reactions of the μ-peroxo and μ-superoxo cobalt complexes have been kinetically investigated. [198] The k_{-2} path can be studied by scavenging with strong ligand or H^+ (to remove Co(L)) or with $S_2O_4^{2-}$ (to remove O_2). If the scavenging is rapid, the loss of $LCoO_2CoL$ reflects the value of k_{-2} [198] (Prob. 26). The enhanced reactivity of $Co([14]aneN_4)(H_2O)O_2^{2+}$ compared with aqueous O_2 towards a number of outer-sphere and inner-sphere reductants resides mainly in a greater driv-ing force and little radical or special kinetic properties need be assigned to the 1:1 adduct. [202]

The yellow Co(II)-cyanide solid complex contains square-pyramidal five-coordinated $Co(CN)_5^{3-}$. The green ion in solution is possibly $Co(CN)_5H_2O^{3-}$ with very weak H_2O coor-dination in the sixth axial position. [204] It is an important inner-sphere reductant (Prob. 8 of Chap. 5, and Tables 5.3 and 5.7). [205] Generally

$$Co(CN)_5^{3-} + XY \rightarrow Co(CN)_5X^{3-} + Y^\bullet \quad rds \qquad (8.106)$$

$$Co(CN)_5^{3-} + Y^\bullet \rightarrow Co(CN)_5Y^{3-} \qquad (8.107)$$

where XY may be H_2O, Br_2, I_2, ICN, NH_2OH, H_2O_2 and RX. [205] As would be expected with this mechanism, there is an inverse correlation of second-order rate constant with the bond energy of the XY additive. [205] The reaction of $Co(CN)_5^{3-}$ with H_2 to give $Co(CN)_5H^{3-}$ differs from the reactions mentioned above in being a third-order process (Secs. 1.12 and 2.1.3). Homolytic (as opposed to heterolytic) splitting of H_2 is supported from the results of ex-amining the reaction in D_2O. The Co-D/Co-H ratio in the product (estimated from infra-red stretching data) changes from approximately zero to one, the increasing value arising from known exchange of the product. [206]

$$Co(CN)_5H^{3-} + OD^- \rightleftharpoons Co(CN)_5^{4-} + HOD \qquad (8.108)$$

Cobalt(III)

The aquated Co(III) ion is a powerful oxidant. The value of $E^0 = 1.88$ V ($\mu = 0$) is indepen-dent of Co(III) concentration over a wide range suggesting little dimer formation. [207] It is stable for some hours in solution especially in the presence of Co(II) ions. This permits ex-amination of its reactions. The $CoOH^{2+}$ species is believed to be much more reactive than Co_{aq}^{3+} Ref. 208. Both outer sphere and substitution-controlled inner sphere mechanisms are displayed. As water in the $Co(H_2O)_6^{3+}$ ion is replaced by NH_3 the lability of the coordinated water is reduced. The cobalt(III) complexes which have been so well characterized by Werner are thus the most widely chosen substrates for investigating substitution behavior. This in-cludes proton exchange in coordinated ammines, and all types of substitution reactions [209] (Chap. 4) as well as stereochemical change (Table 7.8). The CoN_5X^{n+} entity has featured widely in substitution investigations. [210] There are extensive data for anation reactions of

$Co(NH_3)_5H_2O^{3+}$, $Co(NH_3)_5dmso^{3+}$ and other aqua and diaqua cobalt(III) complexes.[211] I_d mechanisms prevail, supported by large positive $\Delta V^{\ddagger}$'s and isokinetic plots. The inert Co(III) complexes are important for studying the reactivity of coordinated ligands (Sec. 6.3). Cobalt(III) complexes are often chosen as the oxidant partner in the study of redox reactions (see many of the tables in Chap. 5) (Prob. 28). They can present a large variety of bridging groups, both inorganic and organic, to the reducing agents that take part in inner-sphere redox reactions (Sec. 5.6).

Although Co(III) is often considered the classical representative of inert behavior, there are a number of cobalt(III) complexes that react rapidly enough to require that the rates be determined by flow methods. Table 8.11 shows a representative selection of such labile complexes.

Table 8.11 Some Labile Cobalt(III) Systems at 25 °C

Category	Complex System	k, $M^{-1}s^{-1}$	Ref.
(a)	$Co(tpps)(H_2O)_2^{3+} + SCN^-$	1.0×10^2	a
(b)	$Co(NH_3)_5OSO_2CF_3^{2+} + H_2O$	$0.027\,(s^{-1})$	212
	$+ OH^-$	$> 10^6$	
(c)	$Co(NH_3)_5H_2O^{3+} + HCrO_4^-$	2.3	b
	$Co(NH_3)_5OH^{2+} + HCrO_4^-$	2.2×10^{-2}	b
(d)	$Co(en)_2(SO_3)H_2O^+ - H_2O$ exch.	$13.2\,(s^{-1})$	214

a $\Delta H^{\ddagger} = 77$ kJ mol^{-1}; $\Delta S^{\ddagger} = +60$ JK^{-1}mol^{-1}; $\Delta V^{\ddagger} = +15.4$ cm^{3}mol^{-1}; J. G. Leipoldt, R. van Eldik and H. Kelm, Inorg. Chem. **22**, 4147 (1983). Previous work: K. R. Ashley and S. Au-Young, Inorg. Chem. **15**, 1937 (1976). D or I_d mechanism is proposed.
b A. Okumura, N. Takeuchi and N. Okazaki, Inorg. Chim. Acta **213**, 127 (1985).

The lability may arise in the following circumstances category (a)–(d), Table 8.11:
(a) from a strong in-plane chelating ligand, such as a porphyrin, labilizing axial unidentate ligands (*cis*-effect). A neat suggestion is that the electron-rich porphyrin donates electron density to the Co(III) imparting partial labile Co(II) character.
(b) when an excellent leaving group is involved e. g. as in $Co(NH_3)_5OSO_2CF_3^{2+}$ (Table 8.11) or in $Co(NH_3)_5OClO_3^{2+}$ ($k_{H_2O} = 0.1$ s^{-1}).[212,213] Unlike (c) below, Co–O bond cleavage occurs completely in base and in acid.[212] These then are good precursors ($CF_3SO_3^-$ much safer than ClO_4^-!) for synthesis, by solvolysis, of a large number of complexes of the type $Co(NH_3)_5L^{n+}$ including $Co(NH_3)_5H_2{}^{18}O^{3+}$ (Refs. 212, 213). This lability also applies to Rh, Ir, Cr, Ru and Os complexes of the type $M(NH_3)_5OSO_2CF_3^{2+}$.
(c) because a Co-ligand bond is not broken in the substitution e. g. in the interchange of H_2O and certain oxoanions in Co(III) ammine complexes[212,215] as well as the classical reactions involving CO_2 uptake[212,216]

$$Co(NH_3)_5OH_2^{3+} \underset{}{\overset{-H^+}{\rightleftharpoons}} Co(NH_3)_5OH^{2+} \underset{k_{-2}}{\overset{CO_2, k_2}{\rightleftharpoons}} Co(NH_3)_5OCO_2H^{2+}$$

$$\updownarrow \qquad (8.109)$$

$$Co(NH_3)_5OCO_2^+ + H^+$$

The values of k_2 and k_{-2} vary little with different cobalt(III) complexes and a transition state for CO_2 uptake can be represented as **9**

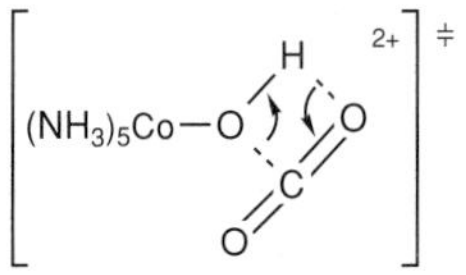

9

^{18}O experiments show no $Co-O$ bond cleavage in the interchange (Sec. 2.2.2). A similar situation is encountered with SO_2 uptake (Prob. 27).

(d) from the presence of strong *trans*-labilizing groups such as alkyl, SO_3^{2-} and NH_2^- groups.[211,214] The *trans*-effect sequence for Co(III) and Pt(II) bears similarities. Larger effects for CN^- and NO_2^- in Pt chemistry probably reside in the increased electron density associated with the d^8 configuration.

There are many binuclear complexes involving peroxo, amido, and hydroxo single and multibridged combinations. The studies of the kinetics of their substitution, disproportionation and redox reactions have been well summarized.[78,198] Cobalt(III) bridged superoxo complexes react with strong reducing agents to give bridged peroxo complexes without participation of the Co(III) center. The formation and cleavage of a $Co-C$ bond is an important feature in vitamin B_{12} chemistry.[217]

Cobalt(IV)

Little kinetic data exist for this oxidation state. Cobalt(IV) has been proposed as an intermediate in the Co(II)-catalyzed reaction of H_2O with $Ru(bpy)_3^{3+}$ at pH > 5

$$Ru(bpy)_3^{3+} + 1/2\,H_2O \rightarrow Ru(bpy)_3^{2+} + H^+ + 1/4\,O_2 \tag{8.110}$$

The Co(IV) species (perhaps CoO^{2+}) reacts with H_2O/OH^- to regenerate Co(II) and produce H_2O_2. The kinetics are quite involved.[218]

Rhodium and Iridium

Rhodium(I) and Iridium(I)

Both oxidation states undergo the very important oxidative addition reactions in which molecules (XY) such as O_2, H_2, CO, SO_2, C_2H_4 and CH_3I add directly in solution[219] e.g. with Vaska's compound,

$$Ir(Ph_3P)_2(CO)Cl + XY \rightleftharpoons Ir(Ph_3P)_2(CO)ClXY \qquad k_1, k_{-1} \tag{8.111}$$

LFER and negative $\Delta S^{\ddagger}$ and $\Delta V^{\ddagger}$ values indicate an associative mechanism with strongly polar character to the activated complex for MeI oxidations of Rh(I) β-diketonates.[220]

Ligand replacement reactions, which have been little studied,[221] are strongly associative, as with other square-planar complexes.[220] Replacement, as the first step in oxidative addition, has been detected by ir and conductivity monitoring in the reaction:[222]

$$Ir(OR)(CO)(Ph_3P)_2 + 2\,CO \rightarrow Ir(CO)_3(Ph_3P)_2^+ + OR^- \rightarrow Ir(COOR)(CO)_2(Ph_3P)_2$$

$$(8.112)$$

Reactions of Wilkinson's catalyst $Rh(Ph_3P)_3Cl$ involve the three-coordinate intermediate $Rh(Ph_3P)_2Cl$. This intermediate can be generated and its reactivity probed by flash photolytic methods

$$Rh(CO)(Ph_3P)_2Cl \underset{h\nu}{\rightleftharpoons} Rh(Ph_3P)_2Cl + CO \qquad (8.113)$$

The second-order decay of the three-coordinate transient (to form a dimer, $k = 2 \times 10^7 M^{-1}s^{-1}$) can be accelerated in the presence of substrates, Ph_3P, CO etc. The second-order rate constant for reaction with CO, $1.0 \times 10^5 M^{-1}s^{-1}$ is consistent with an early estimate ($>7 \times 10^4 M^{-1}s^{-1}$) from thermal studies.[223]

Rhodium(II)

Well characterized in the solid state, the aqua ion probably exists as $Rh_2(H_2O)_{10}^{4+}$. It results from the reduction of mononuclear Rh(III) complexes:[224]

$$2\,Rh(H_2O)_5Cl^{2+} + 2\,Cr^{2+} \rightarrow Rh_2(H_2O)_{10}^{4+} + 2\,CrCl^{2+} \qquad (8.114)$$

It is slowly oxidized by air. Tetra-μ-carboxylate-dirhodium(II) complexes undergo facile substitution with rate constants $\approx 10^6 M^{-1}s^{-1}$ [225,226]

$$Rh_2X_4(H_2O)_2 + L \rightarrow Rh_2X_4(H_2O)L + H_2O \qquad (8.115)$$

$Rh_2(OCOR)_4X_2$ undergo oxidation to $Rh_2(OCOR)_4X_2^+$ by Ce(IV), Fe(III) and V(V). For $Rh_2(OAc)_4X_2$ there is a linear $\log k_f$ vs $\log K_{12}$ relationship, slope 0.5. This simple approach works because k_{11} in Eqn. (5.35) is similar for the three oxidants.[227]

$Rh(bpy)_3^{2+}$ results from e_{aq}^- or CO_2^- reduction of $Rh(bpy)_3^{3+}$. The reactivity of this very labile Rh(II) complex towards O_2, dissociation and disproportionation has been extensively studied.[228,229] Reduction of $Rh(bpy)_3^{3+}$ by e_{aq}^- yields $Rh(bpy^\bullet)(bpy)_2^+$.[228]

Rhodium(III)

The ion $Rh(H_2O)_6^{3+}$ is well characterized in aqueous solution from $H_2^{18}O$ exchange studies.[230] The hydroxy species is more labile. There is an absence of pressure dependency measurements, but an I_d mechanism is favored.[33,230] The formation and cleavage of some hydroxo-bridged rhodium(III) and iridium(III) complexes have been studied.[78] Kinetic studies of oxidation by Rh(III) indicate that $RhOH^{2+}$ is the sole oxidant.[224] Kinetic data for substitution in $Rh(N)_5H_2O^{3+}$ are sparse. An I_a mechanism is favored for water exchange[33,231] and there is a definite, although not substantial, degree of ligand assistance in net substitution reactions.[232] Substitution reactions of Rh(III) and Ir(III) complexes invariably occur without stereochange.

Iridium(III)

$Ir(H_2O)_6^{3+}$ (pK 4.4) was first characterized in 1976.[233] Only rate studies of interaction with $H_2C_2O_4$ and $HC_2O_4^-$ have been reported.[234] Substitution reactions of Ir(III) are extremely slow and stereoretentive. Replacement reactions must often be studied above 100 °C.[235] The inertness of Ir(III) has been exploited in order to study the base hydrolysis of coordinated trimethylphosphate in $Ir(NH_3)_5(OP(OMe)_3)^{3+}$

$$Ir(NH_3)_5(O=P(OMe)_3)^{3+} + OH^- \rightarrow Ir(NH_3)_5(OP=O(OMe)_2)^{2+} + MeOH \qquad (8.116)$$

The rate is 400-fold faster than for the free ligand. In both cases intermolecular attack of OH^- at the phosphorus center is involved. Hydrolysis of the corresponding Co(III) complex is through 100% $Co-O$ bond cleavage, thus preventing a test of the effect of metal coordination.[236] Oxidation of $Ir(H_2O)_6^{3+}$ yields binuclear Ir(IV) and Ir(V) species although little is known of their chemistry.[237] Electrochemical reduction in nonaqueous solution of $Ir(bpy)_3^{3+}$ produces Ir(III) species in which 1, 2 or 3 coordinated bpy's are reduced. The situation resembles $Ru(bpy)_3^{2+}$ reduction to 1+, 0 and 1− species.[238]

Nickel

Although Ni(II) is usually the stable oxidation state in solution, a considerable number of Ni(I), Ni(III) and Ni(IV) complexes have been prepared.[239,240] These are strong reducing or oxidizing agents but are often stable in aqueous solution for sufficient time for examination. Discovery of Ni(I) and Ni(III) in methanogenic bacteria and other bioinorganic material in the past 15 years has heightened interest in these oxidation states.[239-241] Epr is useful for characterizing intermediates or products of reactions involving these oxidation states. It can distinguish Ni(III) from a Ni(II)-stabilized radical since the latter has only one signal at $g \approx 2.00$.[242]

Nickel(0)

The tetrahedral complexes of the d^{10} Ni(0) system undergo dissociative substitution (Table 4.15). Kinetic data are shown in Table 8.12.[243] Infrared monitoring methods feature prominently in these studies (Sec. 3.9.2).

Table 8.12 Activation Parameters for Substitution Reactions of Ni^0L_4 in Toluene at 25 °C[243]

NiL_4	k, s^{-1}	$\Delta H^{\ddagger}$ kJ mol^{-1}	$\Delta S^{\ddagger}$ J K^{-1} mol^{-1}
$Ni(CO)_4$	2.0×10^{-2}	93	33
$Ni(PF_3)_4$	2.1×10^{-6}	119	46
$Ni[P(OEt)_3]_4$	9.9×10^{-7}	110	8

Nickel(I)

This oxidation state which resembles Cu(II)[240] may be prepared by electrochemical,[244,245] photochemical[246,247] or pulse radiolytic[244] reduction of nickel(II). Nickel(I) macrocycles are powerful reductants and their spectra and redox potentials have been measured.[244] The reactions of the Ni(I) complexes Ni(tmc)[+] **10** and **11** with RX are similar.[246,247]

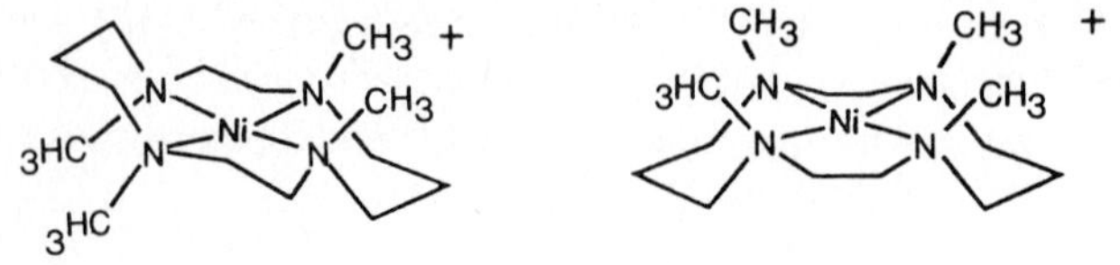

RRSS   RSRS
10   11

$$Ni(tmc)^+ + RX \rightarrow Ni(tmc)X^+ + R^{\bullet} \tag{8.117}$$

$$Ni(tmc)^+ + R^{\bullet} \rightarrow Ni(tmc)R^+ \tag{8.118}$$

$$Ni(tmc)R^+ + H_2O \rightarrow Ni(tmc)^{2+} + RH + OH^- \tag{8.119}$$

$$-d[Ni(tmc)^+]/dt = k[Ni(tmc)^+][RX] \tag{8.120}$$

When RX = propyl iodide, the rates of formation and hydrolysis of Ni(tmc)R[+] are comparable and can be disentangled using a program for numerical integration of kinetic equations.[246] One-electron reduction electrochemically of Ni(II) porphyrins in DMF produces the Ni(I) complexes and not the Ni(II)-porphyrin anion radicals as had been supposed.[248] Factor F_{430} found in methane-forming bacteria is a nickel complex of an unsaturated tetraaza macrocycle.[249]

Nickel(II)

This ion continues to be the most studied of the labile metal ions for probing the phenomena of substitution,[250] rearrangements (Chap. 7) and ligand reactivity (Chap. 6). The number of coordinated waters in the hexaaqua ion $Ni(H_2O)_6^{2+}$ has been established by both the nmr peak area method (Chap. 4) and by neutron scattering techniques in concentrated nickel salt solution.[21,251] Solvent exchange rates for a large variety of solvated nickel ions over temperature and pressure ranges have been determined by nmr methods (Table 8.13). An I_d mechanism for exchange is strongly supported. With the hindered $Ni(dmf)_6^{2+}$, the dissociative mode is favored and an extreme D mechanism has been proposed.[252] There have been many studies of ligand substitution in Ni(II).[250] This has been possible because of the wide variety of stable complexes that Ni(II) forms and the ease with which rates of exchange and formation may be measured by nmr line broadening, flow and conventional methods. Again, in general, rate parameters favor the I_d mechanism for interchange.[253] The kinetics of chelation have been probed mainly using Ni(II) as the species of choice (Sec. 4.4.1). Substituting water in the nickel(II) ion by other ligands (particularly containing nitrogen donors) affects the rates and mechanisms of further substitution (Prob. 29). Table 4.9 shows water exchange rate constants of a number of nickel(II) complexes. [250,254–256] With the rules for substitution behavior in Ni(II) well established, we can use these to assess the mode and sequence of binding of metal ions to complex molecules. The interaction of Ni^{2+} with the coenzyme flavin adenine dinucleotide has been studied by temperature-jump either directly spectrally or via pH indicators (Sec. 1.8.2). Four relaxation times ranging from 100 μs to 35 ms are observed and may be rationalized in terms of the sequence (8.121) (A-P-F representing linked adenine, phosphate and flavin moieties). For the first step, a characteristic rate constant $2 \times 10^5 M^{-1} s^{-1}$ is observed. [257]

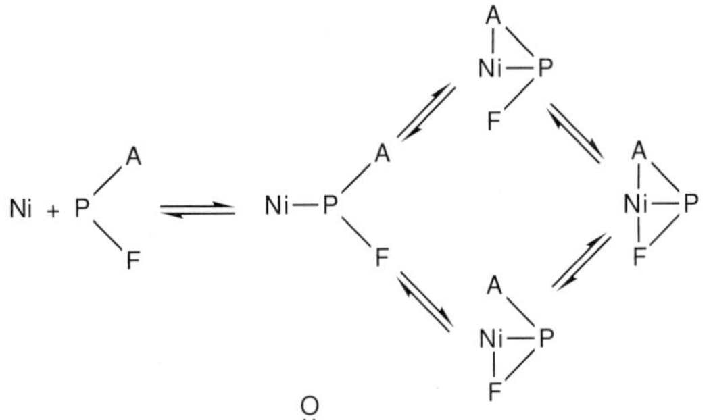

$$(8.121)$$

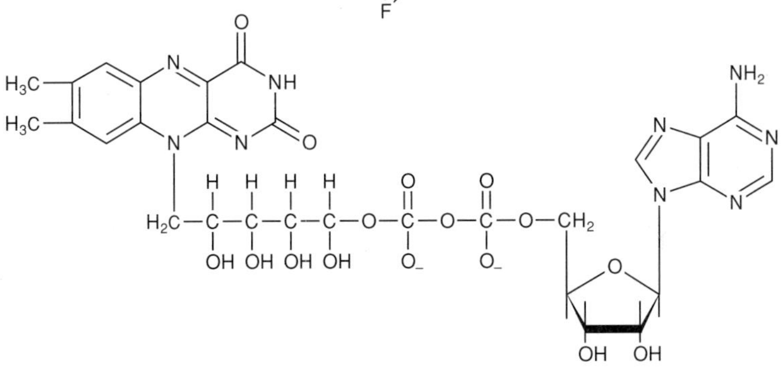

F—P—A

Table 8.13 Rate Parameters for Solvent Exchange on $Ni(S)_6^{2+}$, B17, Table 2.8.

Species	$k\ (s^{-1})$	$\Delta H^{\ddagger}\ kJ\,mol^{-1}$	$\Delta S^{\ddagger}\ JK^{-1}mol^{-1}$	$\Delta V^{\ddagger}\ cm^3\,mol^{-1}$
$Ni(H_2O)_6^{2+}$	3.2×10^4	57	$+32$	$+ 7$
$Ni(CH_3OH)_6^{2+}$	1.0×10^3	66	$+34$	$+11$
$Ni(CH_3CN)_6^{2+}$	2.8×10^3	64	$+37$	$+10$
$Ni(dmf)_6^{2+}$	3.8×10^3	63	$+34$	$+ 9$
$Ni(dmso)_6^{2+}$	1.7×10^4	49	$+ 1$	$-$
$Ni(NH_3)_6^{2+}$	7.0×10^4	57	$+41$	$+ 6$

The rapid interconversion between tetrahedral and planar geometries, between *cis*- and *trans*-planar forms and between planar and octahedral species has been studied using Ni(II) complexes (Secs. 7.2.1–7.2.3). Chelates can be resolved into optical forms and their racemization studied. (Sec. 7.6.1).

There have been few studies of substitution in complexes of nickel(II) of stereochemistries other than octahedral. Substitution in 5-coordinated and tetrahedral complexes is discussed in Secs. 4.9 and 4.8 respectively. The enhanced lability of the nickel(II) compared with the cobalt(II) tetrahedral complex is expected from consideration of crystal field activation energies. The reverse holds with octahedral complexes (Sec. 4.8).

Nickel(II) ion and complexes are often included in the study of the catalytic properties of metal ions (Chap. 6). Nickel(II) (and copper(II)) have a marked ability to promote ionization from coordinated amide and peptide likages (Sec. 6.3.2). Nickel(II) can also help assemble reactants in a specific fashion to produce macrocycles.[258]

Nickel(III)

Relatively plentiful in both solid and solution states,[239,240,259,260] Ni(III) is generated from Ni(II) by strong oxidants, such as $OH^{\bullet}$, $Br_2^{\bar{\ }}$ and $(SCN)_2^{\bar{\ }}$, produced by pulse radiolysis and flash photolysis. Rate constants are $\approx 10^9 M^{-1} s^{-1}$ for oxidation by $OH^{\bullet}$ and $Br_2^{\bar{\ }}$ and $\approx 10^8 M^{-1} s^{-1}$ for $(SCN)_2^{\bar{\ }}$ Ref. 259. The most popular means of production in both aqueous and nonaqueous solution is electrolytic.[261,262] The ligands which stabilize Ni(III) are cyanide, deprotonated peptides, amines and aminocarboxylates, α-diimines and tetraaza macrocycles, including porphyrins.[259,260] Low spin d^7 Ni(III) resembles low spin Co(II). The kinetics of the following types of reactions have been studied:

a) substitution − The formation rates for (8.122) (L, variety of macrocycles, peptides) are insensitive to L and X^- and an I_d mechanism is proposed, consistent with the small Ni^{3+} ion.[261,263,264]

$$NiL(H_2O)_2^{3+} + X^- \rightleftharpoons NiL(H_2O)X^{2+} + H_2O \qquad (8.122)$$

b) redox − The kinetics of reduction of Ni(III) complexes by outer-sphere processes enables an estimation of $Ni(bpy)_3^{3+/2+}$, $Ni(oxime)^{3+/2+}$ and $Ni(sar)^{3+/2+}$ self-exchanges, all of which are $(1-3) \times 10^3 M^{-1} s^{-1}$. This relatively low rate constant can be rationalized in terms of a transfer of an e_g electron.[115,259,265,266]

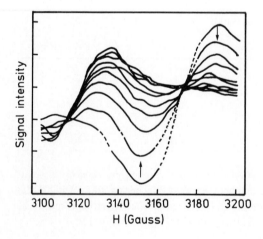

Fig. 8.6 Conversion of a violet-black to a yellow Ni(III) species, Ni(H$_{-1}$AlaAla)$_2^-$ monitored by epr and showing isosbestic points. There is 0.35 s between scans which are run in 0.125 M H$^+$ and at 25 °C. Ref. 242. Reproduced with permission from S. A. Jacobs and D. W. Margerum, Inorg. Chem. **23**, 1195 (1984). © (1984) American Chemical Society.

c) rearrangement — A black to yellow Ni(III) rearrangement has been observed with Ni(III) bis(dipeptides). It is accompanied by isosbestic points (from spectral and epr monitoring) (Fig. 8.6) and is acid catalyzed. The two forms are considered tetragonally compressed ($g_{II} > g_\perp$, electron in $d_{x^2-y^2}$) and tetragonally elongated ($g_\perp > g_{II}$, electron in d_{z^2}) octahedral complexes respectively. The yellow form undergoes an acid-independent internal redox reaction to give Ni(II). The black $\rightarrow$ yellow $\rightarrow$ Ni(II) sequence has been fully analyzed (Sec. 1.6.2).[242]

Nickel(IV)

This oxidation state[239,240] is particularly stabilized by dioximes e. g. **12** (R_2 is normally H).

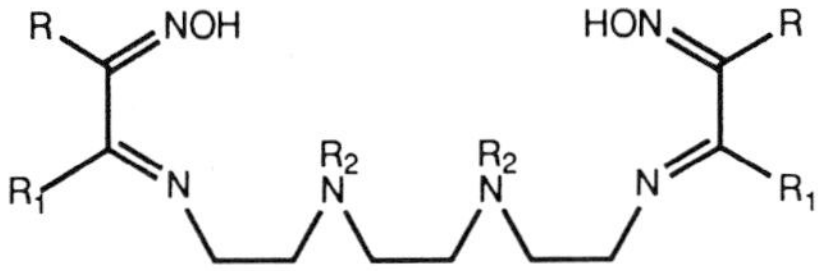

12

Most of the studies of the reactivity of this oxidation state have been carried out with these types of complexes.[267] The low spin d^6 Ni(IV) resembles Co(III) in being diamagnetic, substitution inert and capable of resolution into optical isomers.[268] Reductions are outer-sphere; when Co(edta)$^{2-}$,[268] ascorbate[269] and Co(phen)$_3^{2+}$ Ref. 269 are used, intermediate Ni(III) species are detected and characterized. Oxidation of Co(edta)$^{2-}$ by Ni^{IV} (S-Me$_2$L)$^{2+}$ or Ni^{III} (S-Me$_2$L)$^+$, L = **12**, R=R$_1$=CH$_3$; R$_2$=H, produces a $\approx10\%$ excess of (+)-Co(edta)$^-$ over the ($-$) form Sec. 5.7.4.[269] The comproportionation reaction (protons omitted)

$$\text{Ni}(\mathbf{12})^{2+} + \text{Ni}(\mathbf{12}) \rightleftharpoons 2\,\text{Ni}(\mathbf{12})^+ \qquad k_1, k_{-1} \qquad (8.123)$$

is important at pH > 5 and values for k_1 and k_{-1} have been determined[267] (Prob. 30).

Palladium

Palladium(0)

Substitution is dissociative (Sec. 4.8).

Palladium(II)

Water exchange of Pd(H$_2$O)$_4^{2+}$ has been studied by the ^{17}O-nmr line width of coordinated water in 0.8–1.7 M HClO$_4$. The parameters (Table 4.12) support an associative mechanism. The exchange rate is over 10^6 times faster than for Pt(H$_2$O)$_4^{2+}$ (Ref. 270). The ligation of

$Pd(H_2O)_4^{2+}$ has been discussed.[270] For the reaction with I^- an I_a mechanism is proposed with bond breaking less important for Pd(II) than Pt(II).[271] Many of the characteristics established for substitution in Pt(II) are also seen in Pd(II).[221]

The reaction system (8.124) illustrates a number of features of Pd(II)

$$Pd(H_2O)_4^{2+} + MeCN \underset{k_{-1}}{\overset{k_1}{\rightleftharpoons}} Pd(MeCN)(H_2O)_3^{2+} \underset{k_{-2}}{\overset{k_2}{\rightleftharpoons}} cis\text{- and } trans\text{-}Pd(MeCN)_2(H_2O)_2^{2+}$$
$$+ H_2O \qquad\qquad\qquad + H_2O \qquad (8.124)$$

square-planar substitution. Step 1 is monitored at 353 nm, which is an isosbestic point for the mono/bis mixture. Step 2 is monitored at 296 nm. The second step is complicated by the production of a *cis, trans* mixture of unknown composition. Because of this $k_2/k_{-2} \neq K_2$ (the experimentally determined equilibrium constant for the second step). The first step is straightforward and $k_1/k_{-1} = K_1$. Comparison of step 1 with a number of analogous substitutions involving *mono* complexes of Pd(II) is shown in Table 8.14.[272]

There is the usual strong dependence of the rate constant on the nature of the entering ligand (k_1) and a smaller dependence on the leaving group (k_{-1}). Although $\Delta V_{\pm 1}^{\ddagger}$ is only slightly negative, (Table 8.14) the value is not changed when H_2O is the incoming ligand and there are three different leaving groups. Their release into the medium would be expected to produce a large expansion. This suggests that the leaving group is still tightly bound in the transition state and that an A rather than an I_a mechanism pertains.[272]

Table 8.14 Rate Constants for Formation (k_1) and Aquation (k_{-1}) of $PdL(H_2O)_3^{(2-n)+}$ in Aqueous Solution at 25°C in 1 M $HClO_4$ [272]

L^{n-}	k_1, $M^{-1}s^{-1}$	k_{-1}, s^{-1}	$\Delta V_1^{\ddagger}$, $cm^3 mol^{-1}$	ΔV_{-1}, $cm^3 mol^{-1}$
H_2O	41^a	560^b	-2.2	-2.2
Me_2SO	2.45	0.24	-9.2	-1.7
$MeCN$	309	16	-4.0	-1.5
Cl^-	1.8×10^4	0.83	$-$	$-$
Br^-	9.2×10^4	0.83	$-$	$-$
I^-	1.1×10^6	0.92	$-$	$-$

^{a} Water exchange rate $k_{exch} = 560$ s^{-1} per coordination site recalculated to second-order units $4k_{exch}/55$
^{b} Water exchange rate k_{exch} for a particular coordination site.

Palladium(IV)

Substitution in Pd(IV) is generally fast.

Platinum

Platinum(0)

The compounds of zero valent platinum $Pt(Ph_3P)_4$ and $Pt(Ph_3P)_3$ were first discovered in 1958.[273] Tetrahedral complexes $Pt(PF_3)_4$ and $Pt(P(OEt)_3)_4$ undergo nucleophilic substitution by a dissociative mechanism (Sec. 4.8).

Platinum(I)

The binuclear complex **13** undergoes substitution of Br^- by Cl^- in two steps. The assignment $k_1 = 93\ M^{-1}s^{-1}$ and $k_2 = 19\ M^{-1}s^{-1}$ for the successive replacements in CH_2Cl_2 has been made on the basis of the expected absorbance of the intermediate (Sec. 1.6.2). The Pt(I) center may be regarded as square planar. The rate enhancement over analogous Pt(II) complexes may be ascribed to a large *trans* effect from the $Pt-Pt$ bond.[274]

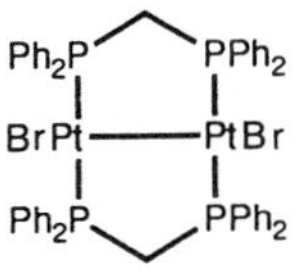

13

A variety of small molecules can be inserted between the two metals in these complexes. The reaction with CH_2N_2 in CH_2Cl_2 to insert $-CH_2-$ is second order and kinetic parameters for different X's suggest a rate-limiting transfer of an electron pair from the $Pt-Pt$ bond to the methylene group of CH_2N_2[275]

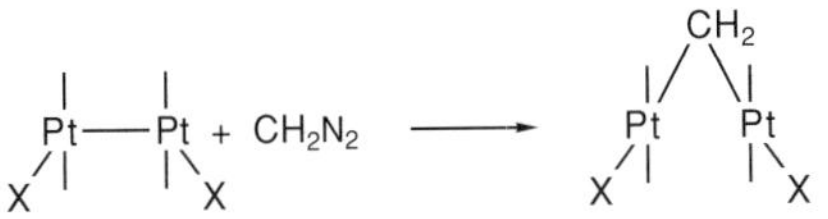

$$(8.125)$$

Platinum(II)

Up to 0.5 M solutions of $Pt(H_2O)_4^{2+}$ can be prepared. The slow exchange of $Pt(H_2O)_4^{2+}$ with solvent H_2O can be followed from the increase in height of the ^{17}O signal of $Pt(H_2O)_4^{2+}$ after mixing with ^{17}O-enriched H_2O, see Fig. 1.11. The kinetic data for the exchange (Table 4.12) support an I_a or A mechanism.[276] The values are more accurate than those obtained by using ^{195}Pt nmr, since a wider temperature range could be employed. Nearly all the mechanistic studies of nucleophilic substitutions in square-planar complexes have been made on compounds of platinum(II).

Platinum(III)

This oxidation state has usually a transitory existence. It may be generated by electrochemical, photolytic and thermal non-complementary reactions using a 1e redox agent[277,278]

$$Pt(IV) + red \rightleftharpoons Pt(III) + ox \qquad (8.126)$$

$$Pt(III) + red \rightarrow Pt(II) + ox \qquad (8.127)$$

Reduction of Pt(IV) by e_{aq}^- and oxidation of Pt(II) by $OH^\bullet$ produces reactive Pt(III) species which can be detected by conductivity, polarographic and uv/visible spectral means.[278,279]

Platinum(IV)

Normally Pt(IV) is an extremely inert oxidation state. Substitution in Pt(IV) is catalyzed by Pt(II)[280]

$$\textit{trans-}Pt^{IV}L_4XY + Z + Pt^{II}L_4 \rightarrow \textit{trans-}Pt^{IV}L_4YZ + X + Pt^{II}L_4 \tag{8.128}$$

and is associated with a mechanism:

$$PtL_4 + Z \rightleftharpoons PtL_4Z \tag{8.129}$$

$$PtL_4Z + PtL_4XY \rightarrow PtL_4YZ + PtL_4X \tag{8.130}$$

$$PtL_4X \rightleftharpoons PtL_4 + X \tag{8.131}$$

Modifications of this mechanism have been made recently (Sec. 2.1.3).

Copper

Copper(I)

The ion Cu^+ is extremely labile. Rate constants for the formation of maleate or fumarate complexes are $\approx 10^9 M^{-1}s^{-1}$ Ref. 281. It can be prepared in an acid perchlorate solution by reaction of Cu^{2+} with a one-electron reducing agent such as Cr^{2+}, V^{2+} or Eu^{2+} Ref. 282. Although there is a marked tendency for disproportionation, solutions of Cu^+ are metastable for hours in the absence of oxygen, particularly when concentrations of Cu(I) are low and the acidity is high. Espenson[282] has capitalized on this to study the rates of reduction by Cu^+ of some oxidants, particularly those of Co(III), Table 5.7 (see Prob. 6(c) Chap. 5).

Copper(I) complexes can be generated by radical reduction of the corresponding Cu(II) complex. A few copper(I) macrocycles are remarkably stable[283] but usually they are strong reducing agents reactive in O_2, and disproportionate to copper(II) and metallic copper.[284]

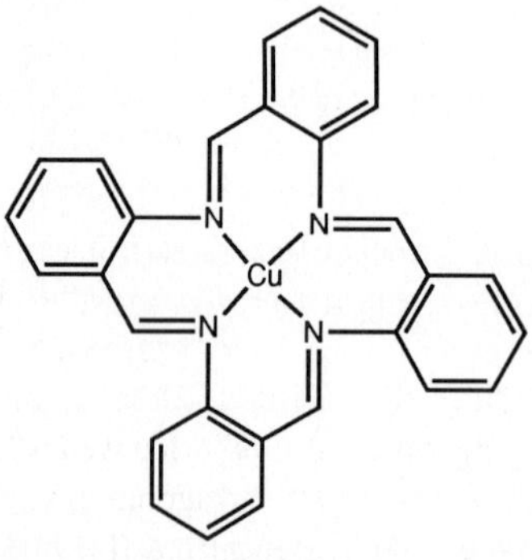

14

Under extreme conditions some Cu(II) or Cu(I) macrocycles undergo controlled potential reductive electrolysis to Cu(0) complexes.[284] The Cu(I)-Cu(II) exchange has not been investigated in a perchloric acid medium. In 12 M HCl, fast exchange has been noted by nmr line broadening (Sec. 3.9.6(a)). One of the few self-exchanges which has been directly measured (by nmr line broadening) is $Cu(taab)^{+/2+}$, **14**, for which a value of $5 \times 10^5 M^{-1}s^{-1}$ in CD_3OD has been determined.[285] The calculation of the self-exchange for Cu(I)-Cu(II) complexes is difficult because there are large structural differences between the two oxidation states and the Marcus approach needs modification. Estimated self-exchange rate constants vary from $\approx 10^{-5} M^{-1}s^{-1}$ (aqua complex) to $5 \times 10^7 M^{-1}s^{-1}$ (chloro complex).[285] Cleavage of DNA is efficiently induced by treatment with a mixture of Cu(II) ion, phen, H_2O_2 and a reducing agent. This has stimulated interest in the $Cu(phen)_2^+ -H_2O_2$ reaction. The latest data using $OH^\bullet$ scavengers (e. g. MeOH) (Prob. 31) indicate that $OH^\bullet$ does not result from the reaction $(Cu^I = Cu(phen)_2^+)$:

$$Cu^I + H_2O_2 \longrightarrow Cu^{II} + OH^- + OH^\bullet \tag{8.132}$$

but that the mechanism may be

$$Cu^I + H_2O_2 \longrightarrow Cu^I(H_2O_2) \quad k_1 \tag{8.133}$$

$$Cu^I(H_2O_2) \xrightarrow{Cu^I} 2\,Cu^{II} + 2\,OH^- \tag{8.134}$$

$$\Big\downarrow_{RH} Cu(II) + OH^- + H_2O + R^\bullet \tag{8.135}$$

Assuming the rate of step (8.133) is less than that associated with step (8.134), $k_{obs} = 2\,k_1 = 4.6 \times 10^3 M^{-1}s^{-1}$ (Ref. 286). Cu(I) is important in the binding and activation of oxygen in biological systems.[287]

Copper(II)

The six water molecules are arranged around Cu(II) ion in a tetragonally distorted octahedron.[21,288] Substitution can easily occur at the very labile axial positions. Rapid inversion of the axial and equatorial ligands leads then to apparent easy substitution in the equatorial position.[289] The water exchange is too rapid $(k \approx 10^{10} s^{-1})$ to be determined accurately by ^{17}O nmr.[21] The Cu(II)-methanol system is more amenable to study. The mean lifetime at 25°C of a particular tetragonal distortion in $Cu(MeOH)_6^{2+}$ is 1.2×10^{-11} s while that for a particular CH_3OH is 1.4×10^{-8} s.[290] Thus the six coordinated methanols are indistinguishable. For these $k_{exch} = 3.1 \times 10^7 s^{-1}$, $\Delta H^\ddagger = 17.2$ kJmol^{-1}, $\Delta S^\ddagger = -44.0$ JK^{-1}mol^{-1} and $\Delta V^\ddagger = +8.3$ cm^{3}mol^{-1}, consistent with a dissociative activation mode.[291] From a comparison of CH_3OH exchange with $M(CH_3OH)_6^{2+}$, M = Mn, Fe, Co, Ni and Cu (see Table 4.2) it is concluded that electronic occupancy of d-orbitals and effective radius dictates the dissociative mode, as much as tetragonal distortion. The tetragonal distortion does however enhance the lability of Cu(II)[291] (Prob. 32). It is technically difficult to measure rates of formation of Cu(II) complexes. A temperature jump study gives for the formation of $Cu(NH_3)(H_2O)_5^{2+}$, $k_f = 2.3 \times 10^8 M^{-1} s^{-1}$, $\Delta H^{\ddagger} = 18.8$ kJ mol^{-1} and $\Delta S^{\ddagger} = -21$ JK^{-1} mol^{-1}, supporting a dissociative mechanism.[292] The fairly constant ratio for $k_{Cu^{2+}}/k_{Ni^{2+}}$ for reactions of Cu^{2+} and Ni^{2+} with a wide variety of ligands suggests a common mechanism, i.e. dissociative for both.[293] The rate constants for formation of bis and tris-bipyridine and -phenanthroline complexes of Cu(II) are appreciably higher than for the mono and this is unusual. It is attributed to a "stacking" phenomenon.[294] The rate-determining step for Cu(II) complexing by polyamines or nitrogen macrocycles varies with the nature of the ligand entering (Sec. 4.5).[295] The interchange of coordinated and free ligand in Cu complexes has been studied. When the Cu complex has one or more coordinated water e.g.

$$CuL(H_2O)_m^{(2-n)+} + cyclamH^+ \xrightarrow{k} Cu(cyclam)^{2+} + H^+ + L^{n-} + mH_2O \qquad (8.136)$$

the relationship of $\log k$ with $\log K_{CuL}$ is shown in Figure 8.7.[296] The expression:

$$\log k = \log [k_0/(1 + K_{CuL} K_0^{-1})] \qquad (8.137)$$

is obeyed, with $k_0 = 1.1 \times 10^7 M^{-1} s^{-1}$ and $K_0 = 2.3 \times 10^{10} M^{-1}$ at 25°C and $\mu = 1.0$ M (KNO$_3$). This means that for the weaker Cu(II) complexes, when $pK_{CuL} < 10.4$, k is independent of the nature of L ($= 1.1 \times 10^7 M^{-1} s^{-1}$) and unwrapping of the coordinated L is not a prerequisite for interchange. For very stable Cu complexes, $pK_{CuL} \gg 10.4$, there is an inverse relationship between the stability of CuL and its reactivity in ligand exchange reactions. A similar behavior occurs with the reaction of Ni(II) complexes with cyclamH$^+$ ($\log K_0 \approx 11.5$) but now only the most stable complex (with edda) deviates from the ligand-independent rate region.[296]

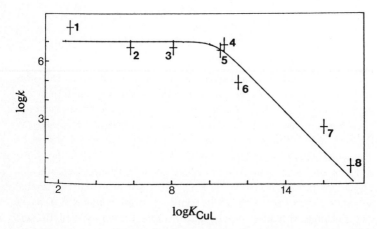

Fig. 8.7 Plot of $\log k$ vs $\log K_{CuL}$ for the reaction with cyclamH$^+$ of Cu complexes with various ligands L = succinate(1), malonate(2), glycine(3), en(4), ida(5), nta(6), dien(7), and Hedta(8). The curve is calculated using (8.137) with $k_0 = 1.1 \times 10^7 M^{-1} s^{-1}$ and $K_0 = 2.3 \times 10^{10} M^{-1}$. Ref. 296. Reproduced with permission from Y. Wu and T. A. Kaden, Helv. Chim. Acta **68**, 1611 (1985).

In the replacement of one ligand by another in a bis chelated Cu(II) species not containing coordinated water however, e. g.

$$Cu(L)_2 + L_1 \rightleftharpoons Cu(LL_1) + L \qquad (8.138)$$

the rate constants depend on the nature of L and L_1 and an associative mechanism is particularly attractive for rationalizing this behavior.[297] Copper(II) reacts with a number of reducing ligands such as sulfite and dithizone, and eventually gives a Cu(I) product. The reaction probably proceeds through a Cu(II) complex within which the redox process occurs. The reaction of Cu(II) with CN^- has been thoroughly assessed (Fig. 8.8)[298]

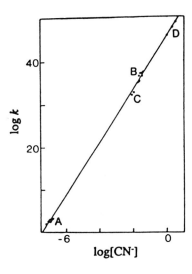

Fig. 8.8 Plot of $\log k\,(M^{-1}s^{-1})$ vs $\log[CN^-]$ (M) for data by J. H. Baxendale and D. T. Westcott, J. Chem. Soc. 2347 (1959) (A); N. Tanaka, M. Kamada and T. Murayana, Bull. Chem. Soc. Japan **31**, 895 (1959); G. Nord and H. Matthes, Acta Chem. Scand. **A28**, 13 (1974) (C); and S. Yoshimura, A. Katagiri, Y. Deguchi and S. Yoshizawa, Bull. Chem. Soc. Japan **53**, 2437 (1980) (D). The line corresponds to $\log k = 6.17 \log[CN^-] + 46$. Ref. 298. Reproduced with permission from A. Katagiri, S. Yoshimura and S. Yoshizawa, Inorg. Chem. **20**, 4143 (1981). © (1981) American Chemical Society.

$$Cu(II) + CN^- \rightarrow Cu(I) + 1/2\,(CN)_2 \qquad (8.139)$$

$$V = k\,[Cu^{2+}]^2 \text{ (excess CN)} \quad \therefore \quad V = a\,[Cu^{2+}]^2[CN^-]^6 \qquad (8.140)$$

and the reaction is believed to proceed via $Cu_2^{II}(CN)_6^{2-}$ Ref. 298. The reduction of transition metal salts by H_2 has been investigated in detail.[299] The reaction with Cu(II) is autocatalytic (due to Cu(I)). The suggested mechanism is

$$Cu^{2+} + H_2 \rightleftharpoons CuH^+ + H^+ \qquad (8.141)$$

$$CuH^+ + Cu^{2+} \rightleftharpoons 2\,Cu^+ + H^+ \qquad (8.142)$$

mainly

$$Cu^+ + H_2 \rightleftharpoons CuH + H^+ \qquad (8.143)$$

$$CuH + Cu^{2+} \rightarrow CuH^+ + Cu^+ \qquad (8.144)$$

$$2\,Cu^+ \rightarrow Cu^0 + Cu^{2+} \qquad (8.145)$$

Copper(III)

Interest is mounting in this state,[300,301] promoted once again by its possible implication in biological systems. Galactose oxidase, for example, is a copper enzyme which catalyses the oxidation of galactose to the corresponding aldehyde.[301] The tervalent oxidation state may be prepared from Cu(II) by chemical, anodic[302] and radical oxidation. Cu(III) complexes of peptides[303] and macrocycles[302] have been most studied, particularly from a mechanistic viewpoint. The oxidation of I^- by Cu(III)-deprotonated peptide complexes[304] and by imine-oxime complexes[305] have a similar rate law

$$V = (k_1[I^-] + k_2[I^-]^2)[Cu(III)] \tag{8.146}$$

There is a LFER between the values of $\log k_1$ (and $\log k_2$) and $E^0/0.059$ for the Cu(III), Cu(II) couple. The slope for the k_1 path is 0.56 and this suggests that the rate limiting step is electron transfer to form an $I^\bullet$ radical. The slope for the k_2 path is 0.95, indicating that the activated complex occurs after electron transfer.[304] A number of redox reactions involving Cu(III) are reported.[301] The Cu(II)-(III) peptide exchange rate constant is $5 \times 10^4 M^{-1} s^{-1}$ for **15** (coordinated waters omitted) determined directly from the broadening of the nmr signal of Cu(III) by Cu(II). This value is in good agreement with that estimated from cross-reactions.[306]

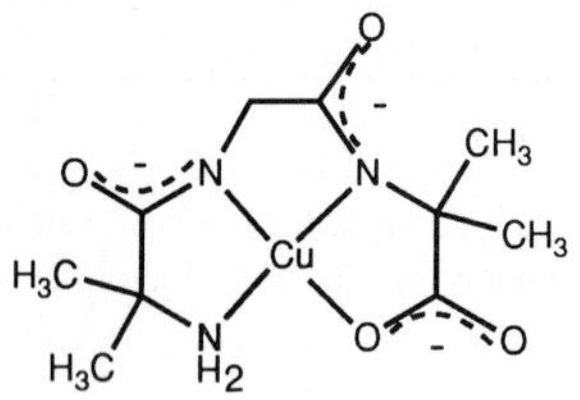

15

Silver

Silver(I)

The Ag^+ ion is labile.[307] Even with cryptands, which react sluggishly with most labile metal ions, Ag^+ reacts with a rate constant around $10^6 M^{-1} s^{-1}$ (in dmso).[308] The higher stability of Ag(I) complexes compared with those of the main groups I and II resides in much reduced dissociation rate constants.[308] Dissociation tends to control the stability of most metal cryptand complexes.[309] Silver(I) is a useful electron mediator for redox reactions since Ag(I) and Ag(II) are relatively rapid reducers and oxidizers, respectively. Silver(I) thus promotes oxidation by sluggish, if strong, oxidants and catalyses a number of oxidations by $S_2O_8^{2-}$ in which the rate-determining step is

$$Ag^+ + S_2O_8^{2-} \rightarrow Ag^{2+} + SO_4^{2-} + SO_4^- \tag{8.147}$$

for which

$$V = k[Ag^+][S_2O_8^{2-}]$$

The oxidized species does not feature in the rate law. (Table 8.15.[52,310,311]) With carboxylic acids, after step (8.147)

$$Ag^+ + SO_4^- \rightarrow Ag^{2+} + SO_4^{2-} \tag{8.148}$$

$$Ag^{2+} + RCH_2COOH \rightarrow Ag^+ + H^+ + RCH_2CO_2^{\bullet} \tag{8.149}$$

$$RCH_2CO_2^{\bullet} \rightarrow RCH_2^{\bullet} + CO_2 \tag{8.150}$$

$$SO_4^- + RCH_2COOH \nearrow RCH_2^{\bullet} + CO_2 + H^+ + SO_4^{2-} \tag{8.151}$$

$$\searrow RCHCOOH^{\bullet} + H^+ + SO_4^{2-} \tag{8.152}$$

The competition amongst (8.148), (8.151) and (8.152) can be assessed.[312] Ag(II) is also the active oxidizer in the Ag(I) catalysis of the Cr(III)-Co(III) and Fe(II)-Co(III) reactions[313] (Prob. 33). In the former case, care has to be exercised to consider a stoichiometric factor which arises, since the rate in terms of the change of concentration of reactant (Co(III)) differs from that in terms of the product (Cr(VI)).[313]

Table 8.15 Kinetic Data for Ag^+ Ion Catalyzed Oxidations by $S_2O_8^{2-}$ (25°C)

Reductant	μ, M	$10^3 \times k, M^{-1}s^{-1}$	Ref.
Cr^{3+}	0.23	5.7	310
	0.48	5.3	310
Mn^{2+}	0.10	8.2	310
	0.57	4.4	310
N_2H_4	0.20	9.2	310
$VO(O_2)^+$	1.0	3.0 (20°C)	52
VO^{2+}	0.65	5.0	52

Silver(I) has a marked ability to catalyze rearrangement and degradation reactions of highly strained polycyclics (Sec. 6.8).

Silver(II)

As well as an important kinetic transient (see above) aqueous Ag(II) ($E^0 \approx +2.0V$) can be obtained by oxidation of Ag(I) by $OH^{\bullet}$ radicals. The pK's of Ag^{2+} remain controversial. The kinetics of oxidation of a variety of substrates have been discussed.[314]

Ag(II) can be stabilized by nitrogen-containing heterocycles such as py, polypyridyls and macrocycles. In fact, Ag(I) disproportionates in the presence of the 14-membered tetraaza macrocycle meso-Me$_6$[14]aneN$_4$(M)

$$2\,Ag(I) + M \rightleftharpoons Ag(II)M^{2+} + Ag(mirror) \tag{8.153}$$

The Ag(II) complex has a characteristic d^9 epr spectrum and is stable in solution. It is oxidized reversibly to Ag(III) complexes.[315]

Silver(III)

A review on this oxidation state includes some kinetic data.[301] One of the simplest examples is the diamagnetic square planar ion $Ag(OH)_4^-$, easily prepared in less than millimolar concentrations by electrochemical oxidation of a silver anode in strong base. It is stable for a few hours in 1 M NaOH, and forms orange complexes with tri- and tetrapeptides which resemble those of Cu(III)[316] e.g. ($H_{-3}G_4^{4-}$ and $H_{-4}G_4^{5-}$ are deprotonated tetraglycine):

$$Ag(OH)_4^- + H_{-3}G_4^{4-} \rightarrow Ag(H_{-4}G_4)^{2-} + 3\,OH^- + H_2O \tag{8.154}$$

An associative mechanism is supported, consistent with a low-spin d^8 configuration. Other ligands such as arsenite reduce $Ag(OH)_4^-$ in a rapid second-order reaction. It is uncertain whether it occurs via complex formation.[317] Silver(III) macrocycles including porphyrin complexes have been characterized.[301,315]

Gold

Gold(I)

Only limited data are available for substitution in this state. An associative mechanism is favored and Au(I) appears to be less reactive than Ag(I) or Hg(II).[318]

Gold(III)

The aqua ion $Au(H_2O)_4^{3+}$ has not been characterized either in solution or in the solid state. Most of the substitution studies have involved the halide complexes AuX_4^- and $Au(NH_3)_4^{3+}$ (Ref. 319). A number of earlier generalizations have been confirmed.[221] Rates are very sensitive to the nature of both entering and leaving ligands and bond formation and breaking are nearly synchronous. The double-humped energy profiles witnessed with Pd(II) and Pt(II) are not invoked; the five-coordinate species resulting from an associative mechanism is the transition state:

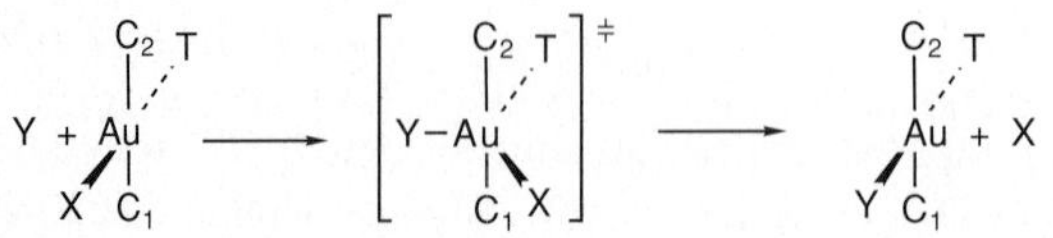

$$Y + \underset{X \quad C_1}{\overset{C_2 \quad T}{Au}} \longrightarrow \left[\underset{C_1 \quad X}{\overset{C_2 \quad T}{Y-Au}} \right]^{\ddagger} \longrightarrow \underset{Y \quad C_1}{\overset{C_2 \quad T}{Au}} + X \tag{8.155}$$

Substitution in Au(III) is characterized by small $\Delta H^{\ddagger}$'s, negative $\Delta S^{\ddagger}$'s, rapidity greater than Pd(II), and an insignificant k_1 in the usual two term rate law (Eqn. (4.94)), i.e. H_2O is a poor entering group. The effects of entering, leaving and non-reacting ligands have been reasonably well assessed.[319] They are as important as with other square planar complexes. Linear $\Delta G^{\ddagger}$ vs. ΔG^0 plots (slope ≈ 1.0) for a common substrate with various nucleophiles, generally,

$$AuL_4 + Y \rightarrow AuL_3Y + L \tag{8.156}$$

indicate associative character (Fig. 8.9).[319] The leaving group has less influence as the reactivity of the entering group increases, again an indication of a lesser role for bond breaking with the entry of efficient nucleophiles. Both *cis*- and *trans*-effects of nonreacting ligands are noted and appear to be more important than for Pd or Pt. The orders for the *trans*-effect $NH_3 < Cl^- < Br^- < SCN^- \approx R_3P$ and *cis*-effect $Cl^- \leqslant Br^- < NH_3 < CN^-$ are usual.[319,320]

Substitution and redox processes are often mixed up when reducing nucleophiles are used. Direct reduction ($AuCl_4^- + I^-$), reduction via substitution ($AuCl_4^- + Ph_3P$) and reduction plus substitution ($AuBr_4^- + SCN^-$) have all been noted (Prob. 34). Reduction by compounds of the main groups V, VI and VII has been extensively examined.[319,320]

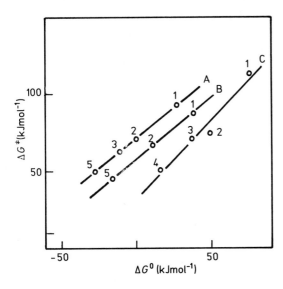

Fig. 8.9 Relationship between $\Delta G^{\ddagger}$ and ΔG^0 for ligand substitution in $AuCl_4^-$ (A), $AuBr_4^-$ (B) and $Au(NH_3)_4^{3+}$ (C) at 25 °C. For $L = H_2O(1)$, $Cl^-(2)$, $Br^-(3)$, $I^-(4)$ and $SCN^-(5)$. Ref. 319. Reproduced with permission from L. H. Skibsted, Adv. Inorg. Bioinorg. Mechs. **4**, 137 (1986).

Zinc and Cadmium

The chemistry of Zn and Cd is predominantly substitutive and catalytic in nature and dominated by the +2 oxidation state. The Zn^+ and Cd^+ ions are formed, albeit slowly ($k < 10^5 M^{-1}s^{-1}$) by e_{aq}^- reduction of Zn^{2+} and Cd^{2+}. Only Cd^{2+} is reduced by $CO_2^{\bar{\cdot}}$.[112] The Zn^+ ion is the stronger reductant.[321] From Marcus treatment of the reductions of $Ru(NH_3)_6^{3+}$, it is estimated that $E^0(Cd^{2+/+}) \approx -0.5 V$ and $E^0(Zn^{2+/+}) \approx -1.0 V$.[112] There are relatively few data for substitution in Zn^{2+} and Cd^{2+}. The comparison of the complexing of Zn^{2+} and Cd^{2+} by bpy (using high-pressure, stopped-flow) is very interesting (Fig. 8.10).[322] The positive ($+7.1 cm^3 mol^{-1}$) and negative ($-5.5 cm^3 mol^{-1}$) values for $\Delta V^{\ddagger}$, respectively, are strong evidence for I_d and I_a character in the substitutions, a formulation which was previously suspected.[322] There is a wider range of rate constants ($10^5 - 5 \times 10^9 M^{-1}s^{-1}$) for ligation of Cd(II). For recent systems studied by ultrasonics, $Cd(H_2O)_6^{2+}$-SCN^- and $Zn(dmso)_6^{2+}$-NO_3^-, Cl^- in dmso, see Refs. 323 and 324. The complexing of a number of Zn^{2+} complexes, ZnL^+, (L a series of anions) by phen has been studied

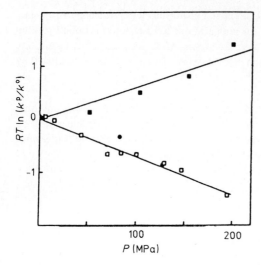

Fig. 8.10 Effect of pressure on the reactions of Zn^{2+} ($\square$) and Cd^{2+} ($\blacksquare$) with bpy at 0°C. Concentrations of Zn^{2+} or Cd^{2+} were 50–200 μM in excess of bpy (5 μM).[322] Reproduced with permission from Y. Ducommun, G. Laurenczy and A. E. Merbach Inorg. Chem. **27**, 1148 (1988). © (1988) American Chemical Society.

by stopped-flow. The values of the second-order formation rate constants for ZnL^{2+} and Zn^{2+} are linearly related to the electron-donating ability of L (8.157) where

$$\log k_{ZnL^{2+}(-H_2O)} = \log k_{Zn^{2+}(-H_2O)} + \gamma E \qquad (8.157)$$

E = electron donor constant of L; γ = a constant, characteristic of the metal and related to a softness parameter[325]

$$\gamma = -5.8\sigma + 5.7 \qquad (8.158)$$

Zinc porphyrins find use in catalytic cycles promoting the $H^+ \rightarrow 1/2\,H_2$ transformation. The Zn excited state (tpps$_3$ represents the tri(sulfonated phenyl) derivative of porphyrin):

$$Zn(tpps_3)^{3-} \xrightarrow{h\nu} {}^*Zn(tpps_3)^{3-} \qquad (8.159)$$

replaces ${}^*Ru(bpy)_3^{2+}$ in scheme 8.100.[326]

Zinc is the metal constituent of a number of very important enzymes including carbonic anhydrase, carboxypeptidase, thermolysin and alcohol dehydrogenase.[327-330] In a number of these enzymes, the zinc ion and groups important in the activity are structurally superimposable.[330] There is a distorted tetrahedral geometry about the Zn and invariably a coordinated H_2O is present. The ionization of this is promoted by the metal. The preferred mechanisms involve nucleophilic attack by the $Zn-OH^-$ entity on the C=O moiety in the substrate. This action is invariably assisted by a nearby amino acid e. g. His-64 in carbonic anhydrase II, Glu-270 in carboxypeptidase A and Glu-143 in thermolysin.[327-330] A number of important concepts are invoked to attempt an explanation of the mechanism of these complex catalyses. They include promoted nucleophilic attack, inter- and intramolecular hydrogen transfer, rate-limiting step, isotope effects and others.[327] A well-supported mechanism for the action of the efficient enzyme carbonic anhydrase is shown in (8.160); reactants in circles, products in squares:

$$\text{EZnOH}^- + \boxed{\text{CO}_2}\ \rightleftharpoons\ \text{EZnOH}^- \cdot \text{CO}_2\ \rightleftharpoons\ \text{EZnOCO}_2\text{H}^- + \boxed{\text{H}_2\text{O}} \rightleftharpoons$$

$$\text{EZnH}_2\text{O} + \boxed{\text{HCO}_3^-}\ \rightleftharpoons\ \text{EZnH}_2\text{O}\ \rightleftharpoons\ \text{EZnOH}^- + \boxed{\text{H}^+}$$

(8.160)

The ionization is the rds. The spontaneous rate is considered to be accelerated by both intermolecular (from the buffer) and intramolecular (from His-64) proton transfer.[328] Cadmium can replace the native zinc from metalloenzymes and ^{113}Cd nmr is being used increasingly to probe metal ligation sites in metalloproteins.[331] The cadmium product often does not show enzyme activity however.

Mercury

Although Hg has two oxidation states, there is a relatively small amount of redox chemistry associated with the group. Many reactions of Hg(I) and Hg(II) appear to involve the disproportionation equilibrium:

$$\text{Hg}_2^{I} \rightleftharpoons \text{Hg}^{II} + \text{Hg}^0$$

(8.161)

Mercury atoms, Hg^0, are sufficiently soluble in water to remain as part of a homogeneous equilibrium.[332]

Mercury(II)

Mercury(II) complexes are very labile. The dynamics of equilibria involving Hg^{2+}, phen, bpy and OH^- have been recently examined by temperature-jump.[333] The rate constants for phen and bpy complexing correspond to diffusion-controlled values, but those of phenH$^+$ and bpyH$^+$ are 2–3 orders of magnitude lower. The species CH_3Hg^+ has the ability to form strong complexes in aqueous solution with a wide variety of N, P, O and S donor atoms.[334,335] Temperature-jump and stopped-flow studies of $\text{CH}_3\text{Hg}(\text{II})$ transfer between ntps(4-nitro-2-sulfonatothiophenolate) and a variety of unidentate ligands have been reported:

$$\text{CH}_3\text{Hg}(\text{ntps}) + \text{X}^{n-} \rightleftharpoons \text{CH}_3\text{HgX}^{(2-n)+} + \text{ntps}^{2-} \qquad k_1, k_{-1}, K_1$$

(8.162)

The $\log k_1/\log K_1$ or $\log k_{-1}/\log K_1$ plots are shown in Fig. 8.11. The curve resembles that for other atom transfers[336] including acid-base interactions.[337] See also Ref. 335.

For strongly endergonic reactions the slope of $\log k_1$ vs $\log K_1$ is 1.0. There is a diffusion-controlled limit for exergonic reactions, the value dependent on the charge types involved.[334,335] This behavior favors an I_a mechanism. It is probably the major pathway for $\text{CH}_3\text{Hg}(\text{II})$ exchange among thiol ligands in biological systems.[335]

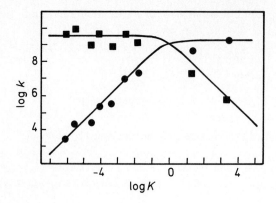

Fig. 8.11 Log k vs log K for $CH_3Hg(II)$ transfer reaction (8.161). $X^{n-} = Br^-$, imid, SO_3^{2-}, I^-, OH^-, phos$^-$, $S_2O_3^{2-}$, CN^-, RS^- (from left to right).[334]

References

1. A. D. Hugi, L. Helm and A. E. Merbach, Inorg. Chem. **26**, 1763 (1987).
2. T. W. Swaddle, Inorg. Chem. **22**, 2663 (1983).
3. P. Chaudhuri and H. Diebler, J. Chem. Soc. Dalton Trans. 1693 (1986); Z. Phys. Chem. (Munich) **139**, 191 (1984).
4. H. Ogino, E. Kikkawa, M. Shimura and N. Tanaka, J. Chem. Soc. Dalton Trans. 894 (1981).
5. A. McAuley, O. Olubuyide, L. Spencer and P. R. West, Inorg. Chem. **23**, 2594 (1984).
6. F. P. Rotzinger and M. Grätzel, Inorg. Chem. **26**, 3704 (1987).
7. B. H. Berrie and J. E. Earley, Inorg. Chem. **23**, 774 (1984).
8. R. Marćec, M. Orhanović, J. A. Wray and R. D. Cannon, J. Chem. Soc. Dalton Trans. 663 (1984).
9. M. S. Ram, A. H. Martin and E. S. Gould, Inorg. Chem. **22**, 1103 (1983).
10. R. N. Bose, R. D. Cornelius and A. C. Mullen, Inorg. Chem. **26**, 1414 (1987).
11. P. Comba and A. Merbach, Inorg. Chem. **26**, 1315 (1987).
12. G. A. K. Thompson, R. S. Taylor and A. G. Sykes, Inorg. Chem. **16**, 2881 (1977).
13. F. J. Kristine, R. E. Shepherd and S. Siddiqui, Inorg. Chem. **20**, 2571 (1981).
14. M. Orhanović and R. G. Wilkins, J. Amer. Chem. Soc. **89**, 278 (1967).
15. R. C. Thompson, Inorg. Chem. **23**, 1794 (1984).
16. M. Inamo, S. Funahashi and M. Tanaka, Inorg. Chem. **22**, 3734 (1983); **24**, 2475 (1985).
17. R. C. Thompson, Inorg. Chem. **25**, 184 (1986).
18. J.-M. Latour, B. Galland and J.-C. Marchon, J. Chem. Soc. Chem. Communs. 570 (1979); M. Inamo, S. Funashi and M. Tanaka, Bull. Chem. Soc. Japan, **59**, 2629 (1986).
19. M. T.-Rossier, D. H.,-Cleary, U. Frey and A. E. Merbach, Inorg. Chem. **29**, 1374 (1990).
20. R. Wever and K. Kustin, Adv. Inorg. Chem. **35**, 81 (1990); D. C. Crans, R. L. Bunch and L. A. Theisen, J. Amer. Chem. Soc. **111**, 7597 (1989).
21. J. P. Hunt and H. L. Friedman, Prog. Inorg. Chem. **30**, 359 (1983).
22. Y. Ducommun, D. Zbinden and A. E. Merbach, Helv. Chim. Acta. **65**, 1385 (1982).
23. P. J. Nichols, Y. Ducommun and A. E. Merbach, Inorg. Chem. **22**, 3993 (1983).
24. J. H. Swinehart, Inorg. Chem. **4**, 1069 (1965).
25. J. D. Rush and B. H. J. Bielski, Inorg. Chem. **24**, 4282 (1985).
26. B27, p. 438.
27. A. M. Lannon, A. G. Lappin and M. G. Segal, Inorg. Chem. **23**, 4167 (1984).
28. A. D. Hugi, L. Helm and A. E. Merbach, Helv. Chim. Acta **68**, 508 (1985); P. Y. Sauvageat, Y. Ducommun and A. E. Merbach, Helv. Chim. Acta **72**, 1801 (1989).

29. B. Perlmutter-Hayman and E. Tapuhi, Inorg. Chem. **18**, 2872 (1979).

30. J. H. Espenson and J. R. Pladziewicz, Inorg. Chem. **9**, 1380 (1970).

31. K. Wuthrich and R. E. Connick, Inorg. Chem. **7**, 1377 (1968); Y. Kuroiwa, M. Harada, M. Tomiyasu and H. Fukotomi, Inorg. Chim. Acta **146**, 7 (1988).

32. M. D. Johnson and R. K. Murmann, Inorg. Chem. **22**, 1068 (1983).

33. H. Gamsjäger and R. K. Murmann, Adv. Inorg. Bioinorg. Mechs. **2**, 317 (1983).

34. H. Wendt, Inorg. Chem. **8**, 1527 (1969).

35. T. M. Che and K. Kustin, Inorg. Chem. **19**, 2275 (1980).

36. S. Funahashi, S. Funada, M. Inamo, R. Kurita and M. Tanaka, Inorg. Chem. **21**, 2202 (1982).

37. M. Nishizawa and K. Saito, Inorg. Chem. **19**, 2284 (1980).

38. O. Yokoyama, H. Tomiyasu and G. Gordon, Inorg. Chem. **21**, 1136 (1982).

39. K. Saito and Y. Sasaki, Adv. Inorg. Bioinorg. Mechs. **1**, 179 (1982).

40. R. C. Thompson, Inorg. Chem. **20**, 3745 (1981).

41. A. Adegite, J. Chem. Soc. Dalton Trans. 1199 (1975).

42. D. H. Macartney, A. McAuley and O. A. Olubuyide, Inorg. Chem. **24**, 307 (1985).

43. D. H. Macartney, Inorg. Chem. **25**, 2222 (1986).

44. M. Nishizawa, Y. Sasaki and K. Saito, Inorg. Chem. **24**, 767 (1985).

45. E. Heath and O. W. Howarth, J. Chem. Soc. Dalton Trans. 1105 (1981); D. C. Crans, C. D. Rithner and L. A. Theisen, J. Amer. Chem. Soc. **112**, 2901 (1990).

46. K. M. Rahmoeller and R. K. Murmann, Inorg. Chem. **22**, 1072 (1983).

47. R. K. Murmann, Inorg. Chem. **16**, 46 (1977).

48. R. K. Murmann and K. C. Giese, Inorg. Chem. **17**, 1160 (1978).

49. H. Gamsjäger and R. K. Murmann, Adv. Inorg. Bioinorg. Mechs. **2**, 317 (1983).

50. (a) P. Comba and L. Helm, Helv. Chim. Acta **71**, 1406 (1988); (b) O. W. Howarth, L. Pettersson and I. Andersson, J. Chem. Soc. Dalton Trans. 1915 (1989).

51. J. Lagrange, K. Aka and P. Lagrange, Inorg. Chem. **21**, 130 (1982).

52. R. C. Thompson, Inorg. Chem. **21**, 859 (1982); **22**, 584 (1983).

53. J. D. Rush and B. H. J. Bielski, J. Phys. Chem. **89**, 1524 (1985).

54. J. H. Ferguson and K. Kustin, Inorg. Chem. **18**, 3349 (1979).

55. P. O. I. Virtanen, S. Kurkisuo, H. Nevala and S. Pohjola, Acta Chem. Scand. **A40**, 200 (1986).

56. T. K. Sham, J. B. Hastings and M. L. Perlman, Chem. Phys. Letters **83**, 391 (1981).

57. K. Micskei and I. Nagypál, J. Chem. Soc. Dalton Trans. 1301 (1990).

58. L. Chong-de and R. B. Jordan, Inorg. Chem. **26**, 3855 (1987).

59. R. M. Sellers and M. G. Simic, J. Amer. Chem. Soc. **98**, 6145 (1976).

60. S. L. Bruhn, A. Bakač and J. H. Espenson, Inorg. Chem. **25**, 535 (1986).

61. R. W. Kolaczkowski and R. A. Plane, Inorg. Chem. **3**, 322 (1964).

62. M. Ardon and R. A. Plane, J. Amer. Chem. Soc. **81**, 3197 (1959).

63. M. E. Brynildson, A. Bakač and J. H. Espenson, Inorg. Chem. **27**, 2592 (1988).

64. M. E. Brynildson, A. Bakač and J. H. Espenson, J. Amer. Chem. Soc. **109**, 4579 (1987).

65. J. H. Espenson, Prog. Inorg. Chem. **30**, 189 (1983); Adv. Inorg. Bioinorg. Mechs. **1**, 1 (1982).

66. H. Cohen and D. Meyerstein, J. Chem. Soc. Dalton Trans. 2449 (1974).

67. D. A. Ryan and J. H. Espenson, Inorg. Chem. **20**, 4401 (1981).

68. A. Petrou, E. V-Astra, J. Konstantatos, N. Katsaros and D. Katakis, Inorg. Chem. **20**, 1091 (1981).

69. R. A. Henderson and A. G. Sykes, Inorg. Chem. **19**, 3103 (1980).

70. R. D. Cannon and J. S. Stillman, Inorg. Chem. **14**, 2202; 2207 (1975).

71. L. M. Wilson and R. D. Cannon, Inorg. Chem. **24**, 4366 (1985).

72. F.-C. Xu, H. R. Krouse and T. W. Swaddle, Inorg. Chem. **24**, 267 (1985).

73. C. Arnau, M. Ferrer, M. Martinez and A. Sanchez, J. Chem. Soc. Dalton Trans. 1839 (1986).

74. M. A. Abdullah, J. Barrett and P. O'Brien, J. Chem. Soc. Dalton Trans. 1647 (1984).

75. S. Castillo-Blum and A. G. Sykes, Inorg. Chem. **23**, 1049 (1984).

76. H. Stünzi, F. P. Rotzinger and W. Marty, Inorg. Chem. **23**, 2160 (1984); **25**, 489 (1986).

77. L. Mønsted, O. Mønsted and J. Springborg, Inorg. Chem. **24**, 3496 (1985).

78. (a) J. Springborg, Adv. Inorg. Chem. **32**, 55 (1988); (b) P. Andersen, Coordn. Chem. Revs. **94**, 47 (1989).

79. L. Mønsted and O. Mønsted, Coordn. Chem. Revs. **94**, 109 (1989).

80. Table I in Ref. 79 lists 14 references since 1972, showing varying confidence in an I_a or I_d mechanism!

81. D. A. House, Coordn. Chem. Revs. **23**, 223 (1977).

82. L. Mønsted and O. Mønsted, Acta Chem. Scand. **A34**, 259 (1980).

83. H. Ogino and M. Shimura, Adv. Inorg. Bioinorg. Mechs. **4**, 107 (1986).

84. D. R. Prasad, T. Ramasami, D. Ramaswamy and M. Santappa, Inorg. Chem. **19**, 3181 (1980).

85. M. Mitewa and P. R. Bontchev, Coordn. Chem. Revs. **61**, 241 (1985).

86. E. S. Gould, Acc. Chem. Res. **19**, 66 (1986).

87. M. Krumpolc, B. G. DeBoer and J. Roček, J. Amer. Chem. Soc. **100**, 145 (1978).

88. D. A. House in Mechanisms of Inorganic and Organometallic Reactions, Vols. 5 and 6, Chap. 6.

89. A. Okumara, M. Kitani, Y. Toyomi and N. Okazaki, Bull. Chem. Soc. Japan, **53**, 3143 (1980).

90. P. Moore, Y. Ducommun, P. J. Nichols and A. E. Merbach, Helv. Chim. Acta **66**, 2445 (1983).

91. L. S. Hegedus and A. Haim, Inorg. Chem. **6**, 664 (1967).

92. S. B. Brown, P. Jones and A. Suggett, in Inorganic Reaction Mechanisms, J. O. Edwards, ed., Wiley-Interscience, NY, 1970, p. 159.

93. A. Haim, Inorg. Chem. **11**, 3147 (1972); S. Funahashi, F. Uchida and M. Tanaka, Inorg. Chem. **17**, 2784 (1978).

94. W. Watanabe and F. H. Westheimer, J. Chem. Phys. **17**, 61 (1949).

95. H. Kwart and J. H. Nickle, J. Amer. Chem. Soc. **98**, 2881 (1976).

96. T. G. Spiro, ed., Molybdenum Enzymes, Wiley-Interscience, NY, 1985.

97. D. T. Richens and A. G. Sykes, Comments Inorg. Chem. **1**, 141 (1981).

98. J. E. Finholt, P. Leupin and A. G. Sykes, Inorg. Chem. **22**, 3315 (1983).

99. D. T. Richens, Y. Ducommun and A. E. Merbach, J. Amer. Chem. Soc. **109**, 603 (1987).

100. E. F. Hills, P. R. Norman, T. Ramasami, D. T. Richens and A. G. Sykes, J. Chem. Soc. Dalton Trans. 157 (1986).

101. E. F. Hills, C. Sharp and A. G. Sykes, Inorg. Chem. **25**, 2566 (1986).

102. R. K. Murmann and M. E. Shelton, J. Amer. Chem. Soc. **102**, 3984 (1980).

103. D. T. Richens, L. Helm, P.-A. Pittet, A. E. Merbach, F. Nicolo and G. Chapuis, Inorg. Chem. **28**, 1394 (1989).

104. (a) B.-L. Ooi and A. G. Sykes, Inorg. Chem. **27**, 310 (1988); (b) D. T. Richens and C. G. Photin, J. Chem. Soc. Dalton Trans. 407 (1990).

105. M. Ardon and A. Pernick, J. Amer. Chem. Soc. **95**, 6871 (1973); **96**, 1643 (1974); Inorg. Chem. **12**, 2484 (1973).

106. R. K. Murmann, Inorg. Chem. **19**, 1765 (1980).

107. H. von Felten, B. Wernli, H. Gamsjäger and P. Baertschi, J. Chem. Soc. Dalton Trans. 496 (1978).

108. S. Funahashi, Y. Kato, M. Nakayama and M. Tanaka, Inorg. Chem. **20**, 1752 (1981).

109. M. A. Harmer and A. G. Sykes, Inorg. Chem. **19**, 2881 (1980).

110. J. B. Vincent, C. Christmas, H.-R. Chang, Q. Li, P. D. W. Boyd, J. C. Huffman, D. N. Henrickson and G. Christou, J. Amer. Chem. Soc. **111**, 2086 (1989); G. W. Brudvig and R. H. Crabtree, Prog. Inorg. Chem. **37**, 99 (1989).

111. (a) Y. Ducommun, K. E. Newman and A. E. Merbach, Inorg. Chem. **19**, 3696 (1980); (b) R. Mohr, L. A: Mietta, Y. Ducommun and R. van Eldik, Inorg. Chem. **24**, 757 (1985).

112. G. V. Buxton and R. M. Sellers, Coordn. Chem. Revs. **22**, 195 (1977).

113. M. Stebler, R. M. Nielson, W. F. Siems, J. P. Hunt, H. W. Dodgen and S. Wherland, Inorg. Chem. **27**, 2893 (1988).

114. D. H. Macartney and N. Sutin, Inorg. Chem. **24**, 3403 (1985).
115. R. Langley and P. Hambright, Inorg. Chem. **24**, 1267 (1985).
116. (a) P. Bernhard and A. M. Sargeson, Inorg. Chem. **26**, 4123 (1987); (b) J. T. Hupp and M. J. Weaver, Inorg. Chem. **22**, 2557 (1983).
117. G. Davies, Coordn. Chem. Revs. **4**, 199 (1969).
118. A. Harriman, Coordn. Chem. Revs. **28**, 147 (1979).
119. L. J. Kirschenbaum and D. Meyerstein, Inorg. Chim. Acta **53**, L99 (1981).
120. J. H. Sutter, K. Colquitt and J. R. Sutter, Inorg. Chem. **13**, 1444 (1974). See also D. G. Lee and T. Chen, J. Amer. Chem. Soc. **111**, 7534 (1989) for the second-order disproportionation of MnO_4^{2-} at pH 9–11 using initial rates.
121. L. Spiccia and T. W. Swaddle, Inorg. Chem. **26**, 2265 (1987).
122. F. W. Moore and K. W. Hicks, Inorg. Chem. **16**, 716 (1977).
123. R. H. Simoyi, P. DeKepper, I. R. Epstein and K. Kustin, Inorg. Chem. **25**, 538 (1986).
124. D. G. Lee and T. Chen, J. Amer. Chem. Soc. **111**, 7534 (1989); T. Ogino and N. Kikuiri, J. Amer. Chem. Soc. **111**, 6174 (1989) and references therein.
125. R. Mohr and R. van Eldik, Inorg. Chem. **24**, 3396 (1985).
126. X. Chen and D. V. Stynes, Inorg. Chem. **25**, 1173 (1986) and references therein, D. V. Stynes, Pure Appl. Chem. **60**, 561 (1988).
127. T. J. Conocchioli, E. J. Hamilton, Jr. and N. Sutin, J. Amer. Chem. Soc. **87**, 926 (1965).
128. E. Antonini and M. Brunori, Hemoglobin and Myoglobin in Their Reactions with Ligands, North-Holland Publishing, Amsterdam, 1971.
129. A. L. Balch, Y.-W. Chan, R.-J. Cheng, G. N. La Mar, L. Latos-Grazynski and M. W. Renner, J. Amer. Chem. Soc. **106**, 7779 (1984) and references therein.
130. G. A. Tondreau and R. G. Wilkins, Inorg. Chem. **25**, 2745 (1986).
131. C. Walling, Acc. Chem. Res. **8**, 125 (1975).
132. C. Walling and K. Amarnath, J. Amer. Chem. Soc. **104**, 1185 (1982).
133. P. Jones and I. Wilson, Metal Ions in Biological Systems, H. Sigel, ed., Marcel Dekker, NY, 1978, Vol 7, pp 185–240 (catalase) FeO^{2+} acting as Compound I.
134. H. B. Dunford and J. S. Stillman, Coordn. Chem. Revs. **19**, 187 (1976) (peroxidase) FeO^{2+} acting as compound II; J. E. Frew and P. Jones, Adv. Inorg. Bioinorg. Mechs. **3**, 175 (1984).
135. P. G. Debrunner, I. C. Gonzales, S. G. Sligar and G. C. Wagner, in Metal Ions in Biological Systems Vol 7, pp 241–275 (cytochrome P 450); J. H. Dawson and K. S. Eble, Adv. Inorg. Bioinorg. Mechs. **4**, 1 (1986).
136. D. Dolphin, B. R. James and H. C. Welborn, Adv. Chem. Ser. **201**, 563 (1982) (cytochrome P 450 shows only some parallelisms).
137. F. T. Bonner and K. A. Pearsall, Inorg. Chem. **12**, 21 (1973); **21**, 1978 (1982).
138. D. Littlejohn and S. G. Chang, J. Phys. Chem. **86**, 537 (1982).
139. M. Grant and R. B. Jordan, Inorg. Chem. **20**, 55 (1981).
140. T. W. Swaddle and A. E. Merbach, Inorg. Chem. **20**, 4212 (1981).
141. A. L. Crumbliss, Comments Inorg. Chem. **8**, 1 (1988).
142. I. T. Ostrich, G. Liu, H. W. Dodgen and J. P. Hunt, Inorg. Chem. **19**, 619 (1980).
143. R. F. Jameson, W. Linert, A. Tschinkowitz and V. Gutmann, J. Chem. Soc. Dalton Trans. 943 (1988).
144. N. Yoshida, T. Matsushita, S. Saigo, H. Oyanagi, H. Hashimoto and M. Fujimoto, J. Chem. Soc. Chem. Communs. 354 (1990).
145. A. Bakač, M. E. Brynildson and J. H. Espenson, Inorg. Chem. **25**, 4108 (1986).
146. N. Rudgewick-Brown and R. D. Cannon, Inorg. Chem. **24**, 2463 (1985).
147. K. J. Pfenning, L. Lee, H. D. Wohlers and J. D. Petersen, Inorg. Chem. **21**, 2477 (1982).
148. A. Yeh and A. Haim, J. Amer. Chem. Soc. **107**, 369 (1985).
149. Inorganic Reaction Mechanisms, The Chemical Society, Vol 6, 1979, p. 221.
150. H. E. Toma, Inorg. Chim. Acta **15**, 205 (1975).

151. G. Rabai and I. R. Epstein, Inorg. Chem. **28**, 732 (1987) for the BrO_3^- oxidation and references.

152. H. E. Toma and A. A. Batista, J. Inorg. Biochem. **20, 53** (1984).

153. M. W. Fuller, K.-M. F. Le Brocq, E. Leslie and I. R. Wilson, Aust. J. Chem. **39**, 1411 (1986).

154. R. A. Marcus and N. Sutin, Biochem. Biophys. Acta **811**, 265 (1985).

155. See, for example, A. G. Sykes, Chem. Soc. Revs. **14**, 283 (1985).

156. A. R. Butler and C. Glidewell, Chem. Soc. Revs. **16**, 361 (1987).

157. A. G. Sharpe, The Chemistry of Cyano Complexes of the Transition Metals, Academic Press, NY, 1976, p. 134.

158. M. D. Johnson and R. G. Wilkins, Inorg. Chem. **23**, 231 (1984); P. J. Morando, E. B. Borghi, L. M. de Schteingart and M. A. Blesa, J. Chem. Soc. Dalton Trans. 435 (1981); A. R. Butler, A. M. Calsy-Harrison and C. Glidewell, Polyhedron, **7**, 1197 (1988).

159. C. Glidewell and I. L. Johnson, Inorg. Chim. Acta **132**, 145 (1987).

160. J. K. Beattie, Adv. Inorg. Chem. **32**, 1 (1988).

161. R. E. White and M. J. Coon, Ann. Revs. Biochem. **49**, 315 (1980); I. Tabushi, Coordn. Chem. Revs. **86**, 1 (1988).

162. J. D. Rush and B. H. J. Bielski, J. Amer. Chem. Soc. **108**, 523 (1986).

163. B. H. J. Bielski and M. J. Thomas, J. Amer. Chem. Soc. **109**, 7761 (1987) and references therein.

164. H. Goff and R. K. Murmann, J. Amer. Chem. Soc. **93**, 6058 (1971).

165. D. E. Richardson and H. Taube, Coordn. Chem. Revs. **60**, 107 (1984).

166. A. Juris, V. Balzani, F. Barigelletti, S. Campugna, P. Belser and A. von Zelewsky, Coordn. Chem. Revs. **84**, 85 (1988). T. J. Meyer, Pure Appl. Chem. **62**, 1003 (1990).

167. P. Bernhard, L. Helm, I. Rapaport, A. Ludi and A. E. Merbach, J. Chem. Soc. Chem. Commun. 302 (1984); Inorg. Chem. **27**, 873 (1988).

168. M. S. Rothlisberger, W. Hummel, P. A. Pittet, H. B. Burgi, A. Ludi and A. E. Merbach, Inorg. Chem. **27**, 1358 (1988).

169. R. A. Leising, J. S. Ohman and K. J. Takeuchi, Inorg. Chem. **27**, 3804 (1988); extensive references to Ru substitution.

170. H. Taube, Comments Inorg. Chem. **1**, 17 (1981).

171. R. E. Shepherd and H. Taube, Inorg. Chem. **12**, 1392 (1973).

172. C. M. Elson, I. J. Itzkovitch and J. A. Page, Can. J. Chem. **48**, 1639 (1970).

173. J. F. Ojo, O. Olubuyide and O. Oyetunji, J. Chem. Soc. Dalton Trans. 957 (1987).

174. A. Haim, Prog. Inorg. Chem. **30**, 273 (1983).

175. P. Bernhard, A. M. Sargeson and F. C. Anson, Inorg. Chem. **27**, 2754 (1988).

176. D. M. Stanbury, O. Haas and H. Taube, Inorg. Chem. **19**, 519 (1980).

177. J. M. A. Hoddenbagh and D. H. Macartney, Inorg. Chem. **25**, 2099 (1986).

178. J. A. Olabe, L. A. Gentil, G. Rigotti and A. Navaza, Inorg. Chem. **23**, 4297 (1984).

179. P. C. Ford, Coordn. Chem. Revs. **5**, 75 (1970), for a review of the reactivity of Ru(II)-amine complexes.

180. I. I. Creaser, L. R. Gahan, R. J. Geue, A. Launikonis, P. A. Lay, J. D. Lydon, M. G. McCarthy, A. W.-H. Mau, A. M. Sargeson and W. H. F. Sasse, Inorg. Chem. **24**, 2671 (1985).

181. K. Zahir, W. Bottcher and A. Haim, Inorg. Chem. **24**, 1966 (1985).

182. B. Anderes, S. T. Collins and D. K. Lavallee, Inorg. Chem. **23**, 2201 (1984).

183. T. Matsubara and C. Creutz, Inorg. Chem. **18**, 1956 (1979).

184. H. C. Bajaj and R. van Eldik, Inorg. Chem. **27**, 4052 (1988).

185. M. T. Fairhurst and T. W. Swaddle, Inorg. Chem. **18**, 3241 (1979).

186. J. A. Broomhead, L. A. P. Kane-Maguire and D. Wilson, Inorg. Chem. **14**, 2575; 2579 (1975).

187. P. Bernhard, L. Helm, A. Ludi and A. E. Merbach, J. Amer. Chem. Soc. **107**, 312 (1985).

188. P. J. Smolenaers and J. K. Beattie, Inorg. Chem. **25**, 2259 (1986).

189. P. Bernhard, H.-B. Burgi, J. Hauser, H. Lehmann and A. Ludi, Inorg. Chem. **21**, 3936 (1982).

190. E. V. Luoma and C. H. Brubaker, Jr., Inorg. Chem. **5**, 1618 (1966).

191. H. A. Schwarz, C. Creutz and N. Sutin, Inorg. Chem. **24**, 433 (1985).
192. C. Creutz, H. A. Schwarz and N. Sutin, J. Amer. Chem. Soc. **106**, 3036 (1984).
193. B12 (2 volumes), D. Dolphin, ed. Wiley, NY, 1982.
194. J. M. Pratt, Chem. Soc. Revs. **14**, 161 (1985).
195. L. Helm, P. Meier, A. E. Merbach and P. A. Tregloan, Inorg. Chim. Acta **73**, 1 (1983).
196. P. Hendry and A. Ludi, Adv. Inorg. Chem. **35**, 117 (1990).
197. E. C. Niederhoffer, J. H. Timmons and A. E. Martell, Chem. Revs. **84**, 137 (1984).
198. S. Fallab and P. R. Mitchell, Adv. Inorg. Bioinorg. Mechs. **3**, 311 (1984).
199. I. Tabushi and T. Sasaki, J. Amer. Chem. Soc. **105**, 2901 (1983) report cooperative O_2 binding by a dimeric cobalt porphyrin.
200. R. Machida, E. Kimura and M. Kodama, Inorg. Chem. **22**, 2055 (1983) includes other refs.
201. J. Simplicio and R. G. Wilkins, J. Amer. Chem. Soc. **91**, 1325 (1969).
202. K. Kumar and J. F. Endicott, Inorg. Chem. **23**, 2447 (1984).
203. D. R. Eaton and A. O'Reilly, Inorg. Chem. **26**, 4185 (1987).
204. L. D. Brown and K. N. Raymond, Inorg. Chem. **14**, 2591 (1975).
205. L. G. Marzilli, P. A. Marzilli and J. Halpern, J. Amer. Chem. Soc. **93**, 1374 (1971); J. Halpern, Acc. Chem. Res. **3**, 386 (1970).
206. J. Halpern, Inorg. Chim. Acta, **77**, L105 (1983).
207. G. Biedermann, S. Drecchio, V. Romano and R. Zingales, Acta Chem. Scand. **A40**, 161 (1986).
208. J. C. Brodovitch and A. McAuley, Inorg. Chem. **20**, 1667 (1981), extensive table of oxidations by $CoOH^{2+}$ in $HClO_4/ClO_4^-$ medium.
209. J. O. Edwards, F. Monacelli and G. Ortaggi, Inorg. Chim. Acta **11**, 47 (1974) comprehensive tables of reactions of $+3$ metal ions covering older literature.
210. D. A. House, Coordn. Chem. Revs. **23**, 223 (1977) comprehensive review and tables of data.
211. M. C. Ghosh, P. Bhattacharya and P. Banerjee, Coordn. Chem. Revs. **91**, 1 (1988).
212. N. E. Dixon, W. G. Jackson, M. J. Lancaster, G. A. Lawrance and A. M. Sargeson, Inorg. Chem. **20**, 470 (1981); D. A. Buckingham, P. J. Cresswell, A. M. Sargeson and W. G. Jackson, Inorg. Chem. **20**, 1647 (1981), contains a useful table of acid and base hydrolysis rate constants.
213. G. A. Lawrance, Adv. Inorg. Chem. **34**, 145 (1989).
214. D. R. Stranks and J. K. Yandell, Inorg. Chem. **9**, 751 (1970).
215. R. van Eldik, Adv. Inorg. Bioinorg. Mechs. **3**, 275 (1984).
216. J. A. Palmer and R. van Eldik, Chem. Revs. **83**, 651 (1983).
217. P. J. Toscano and L. G. Marzilli, Prog. Inorg. Chem. **31**, 105 (1984); L. Randaccio, N. B. Pahor, E. Zangrando and L. G. Marzilli, Chem. Soc. Revs. **18**, 225 (1989).
218. B. S. Brunschwig, M. H. Chou, C. Creutz, P. Ghosh and N. Sutin, J. Amer. Chem. Soc. **105**, 4832 (1983).
219. B1, Chap. 5; P. P. Deutsch and R. Eisenberg, Chem. Revs. **88**, 1147 (1988) review the stereochemistry of H_2 oxidative addition to Ir(I).
220. J. G. Leipoldt, E. C. Steynberg and R. van Eldik, Inorg. Chem. **26**, 3068 (1987); G. J. van Zyl, G. J. Lamprecht, J. G. Leipoldt and T. W. Swaddle, Inorg. Chim. Acta **143**, 223 (1988).
221. L. Cattalini, in Inorganic Reaction Mechanisms, Part I, J. O. Edwards, ed. Wiley-Interscience, NY, 1970, p. 263; R. J. Cross, Adv. Inorg. Chem. **34**, 219 (1989).
222. W. M. Rees, M. R. Churchill, J. C. Fettinger and J. D. Atwood, Organometallics, **4**, 2179 (1985).
223. D. A. Wink and P. C. Ford, J. Amer. Chem. Soc. **109**, 436 (1987).
224. E. F. Hills, M. Moszner and A. G. Sykes, Inorg. Chem. **25**, 339 (1986) and references therein.
225. K. Das, E. L. Simmons and J. L. Bear, Inorg. Chem. **16**, 1268 (1977).
226. C. R. Wilson and H. Taube, Inorg. Chem. **14**, 2276 (1975).
227. R. B. Ali, K. Sarawek, A. Wright and R. D. Cannon, Inorg. Chem. **22**, 351 (1983).
228. Q. G. Mulazzani, S. Emmi, M. Z. Hoffman and M. Venturi, J. Amer. Chem. Soc. **103**, 3362 (1981).
229. H. A. Schwarz and C. Creutz, Inorg. Chem. **22**, 707 (1983).

230. M. J. Pavelich and G. M. Harris, Inorg. Chem. **12**, 423 (1973).

231. T. W. Swaddle, Adv. Inorg. Bioinorg. Mechs. **2**, 95 (1983).

232. M. Martinez and M. Ferrer, Inorg. Chem. **24**, 792 (1985).

233. P. Beutler and H. Gamsjäger, J. Chem. Soc. Chem. Communs. 554 (1976); J. Chem. Soc. Dalton Trans. 1415 (1979).

234. M. R. McMahon, A. McKenzie and D. T. Richens, J. Chem. Soc. Dalton Trans. 711 (1988).

235. R. A. Bauer and F. Basolo, Inorg. Chem. **8**, 2237 (1969).

236. P. Hendry and A. M. Sargeson, J. Chem. Soc. Chem. Communs. 164 (1984).

237. S. E. Castillo-Blum, D. T. Richens and A. G. Sykes, Inorg. Chem. **28**, 954 (1989).

238. V. T. Coombe, G. A. Heath, A. J. MacKenzie and L. J. Yellowlees, Inorg. Chem. **23**, 3423 (1984).

239. K. Nag and A. Chakravorty, Coordn. Chem. Revs. **33**, 87 (1980). Comprehensive account of Nickel(I), (III) and (IV).

240. A. G. Lappin and A. McAuley, Adv. Inorg. Chem. **32**, 241 (1988).

241. R. Cammack, Adv. Inorg. Chem. **32**, 297 (1988).

242. S. A. Jacobs and D. W. Margerum, Inorg. Chem. **23**, 1195 (1984).

243. R. D. Johnston, F. Basolo and R. G. Pearson, Inorg. Chem. **10**, 247 (1971).

244. N. Jubran, G. Ginzburg, H. Cohen, Y. Koresh and D. Meyerstein, Inorg. Chem. **24**, 251 (1985).

245. F. V. Lovecchio, E. S. Gore and D. H. Busch, J. Amer. Chem. Soc. **96**, 3109 (1974) − epr shows that the reduced products are either d^9 nickel(I) or d^8 nickel(II) stabilized anion radicals depending on the nature of the macrocycle unsaturation.

246. M. S. Ram, A. Bakač and J. H. Espenson, Inorg. Chem. **25**, 3267 (1986).

247. A. Bakač and J. H. Espenson, J. Amer. Chem. Soc. **108**, 713 (1986).

248. D. Lexa, M. Momenteau, J. Mispelter and J.-M. Saveant, Inorg. Chem. **28**, 30 (1989).

249. P. E. Rouvière and R. S. Wolfe, J. Biol. Chem. **263**, 7913 (1988); B. Jaun, Helv.Chim. Acta, **73**, 2209 (1990).

250. R. G. Wilkins, Comments Inorg. Chem. **2**, 187 (1983).

251. J. R. Newsome, G. W. Neilson, J. E. Enderby and M. Sandstrom, J. Chem. Phys. Lett. **82**, 399 (1981); G. W. Neilson and J. E. Enderby, Adv. Inorg. Chem. **34**, 195 (1989).

252. P. J. Nichols, Y. Fresard, Y. Ducommun and A. E. Merbach, Inorg. Chem. **23**, 4341 (1984).

253. B. Mohr and R. van Eldik, Inorg. Chem. **24**, 3396 (1985) summarizes data.

254. R. J. Pell, H. W. Dodgen and J. P. Hunt, Inorg. Chem. **22**, 529 (1983).

255. P. Moore, J. Sachinidis and G. R. Willey, J. Chem. Soc. Dalton Trans. 1323 (1984).

256. P. Moore, Pure Appl. Chem. **57**, 347 (1985).

257. J. Bidwell, J. Thomas and J. Stuehr, J. Amer. Chem. Soc. **108**, 820 (1986).

258. A recent example, with references to previous literature, is in nickel(II) template synthesis of macrotricyclic complexes from formaldehyde and triamines, M. P. Suh, W. Shin, S.-G. Kang, M. S. Lah and T.-M. Chung, Inorg. Chem. **28**, 1602 (1989).

259. R. I. Haines and A. McAuley, Coordn. Chem. Revs. **39**, 77 (1981).

260. S. Bhattacharya, R. Mukherjee and A. Chakravorty, Inorg. Chem. **25**, 3448 (1986), extensive list of references.

261. M. Jacobi, D. Meyerstein and J. Lilie, Inorg. Chem. **18**, 429 (1979).

262. P. Morliere and L. K. Patterson, Inorg. Chem. **21**, 1833; 1837 (1982).

263. C. K. Murray and D. W. Margerum, Inorg. Chem. **21**, 3501 (1982).

264. M. G. Fairbank and A. McAuley, Inorg. Chem. **25**, 1233 (1986).

265. D. H. Macartney and N. Sutin, Inorg. Chem. **22**, 3530 (1983).

266. M. G. Fairbank and A. McAuley, Inorg. Chem. **26**, 2844 (1987).

267. A. G. Lappin, D. P. Martone and P. Osvath, Inorg. Chem. **24**, 4187 (1985).

268. A. G. Lappin, M. C. M. Laranjeira and R. D. Peacock, Inorg. Chem. **22**, 786 (1983) and references therein.

269. A. G. Lappin and M. C. M. Laranjeira, J. Chem. Soc. Dalton Trans. 1861 (1982).

270. L. Helm, L. I. Elding and A. E. Merbach, Helv. Chim. Acta **67**, 1453 (1984).

271. L. I. Elding and L. F. Olsson, Inorg. Chim. Acta **117**, 9 (1986).

272. B. Hellquist, L. I. Elding and Y. Ducommun, Inorg. Chem. **27**, 3620 (1988).

273. L. Malatesta and C. Cariello, J. Chem. Soc. 2323 (1958).

274. M. Shimura and J. H. Espenson, Inorg. Chem. **23**, 4069 (1984).

275. S. Muralidharan and J. H. Espenson, Inorg. Chem. **22**, 2786 (1983).

276. L. Helm, L. I. Elding and A. E. Merbach, Inorg. Chem. **24**, 1719 (1985).

277. J. Halpern and P. Pribanić, J. Amer. Chem. Soc. **90**, 5942 (1968).

278. W. L. Waltz, J. Lilie, A. Goursot and H. Chermette, Inorg. Chem. **28**, 2247 (1989); E. Bothe and R. K. Broszkiewicz, Inorg. Chem. **28**, 2988 (1989) and references therein.

279. H. A. Boucher, G. A. Lawrance, P. A. Lay, A. M. Sargeson, A. M. Bond, D. F. Sangster and J. C. Sullivan, J. Amer. Chem. Soc. **105**, 4652 (1983).

280. W. R. Mason, Coordn. Chem. Revs. **7**, 241 (1972).

281. D. Meyerstein, Inorg. Chem. **14**, 1716 (1975).

282. O. J. Parker and J. H. Espenson, J. Amer. Chem. Soc. **91**, 1968 (1969).

283. N. Jubran, H. Cohen, Y. Koresh and D. Meyerstein, J. Chem. Soc. Chem. Communs. 1683 (1984).

284. A. M. Bond and M. A. Khalifa, Inorg. Chem. **26**, 413 (1987) and references therein.

285. E. J. Pulliam and D. R. McMillin, Inorg. Chem. **23**, 1172 (1984); M. J. Martin, J. F. Endicott, L. A. Ochrymowycs and D. B. Rorabacher, Inorg. Chem. **26**, 3012 (1987); H. Doine, Y. Yano and T. W. Swaddle, Inorg. Chem. **28**, 2319 (1989).

286. G. R. A. Johnson and N. B. Nazhat, J. Amer. Chem. Soc. **109**, 1990 (1987).

287. K. D. Karlin and Y. Gultneh, Prog. Inorg. Chem. **35**, 219 (1987); R. G. Wilkins, in Oxygen Complexes and Oxygen Activation by Transition Metals, A. E. Martell, ed., Plenum, 1988, p. 49.

288. M. Magini, Inorg. Chem. **21**, 1535 (1982).

289. M. Eigen, Ber. Bunsenges. Phys. Chem. **67**, 753 (1963).

290. R. Poupko and Z. Luz, J. Chem. Phys. **57**, 3311 (1972).

291. L. Helm, S. F. Lincoln, A. E. Merbach and D. Zbinden, Inorg. Chem. **25**, 2550 (1986).

292. L. S. W. L. Sokol, T. D. Fink and D. B. Rorabacher, Inorg. Chem. **19**, 1263 (1980).

293. R. G. Wilkins, Pure Appl. Chem. **33**, 583 (1973).

294. I. Fábian and H. Diebler, Inorg. Chem. **26**, 925 (1987) and references therein.

295. J. A. Drumhiller, F. Montavon, J.-M. Lehn and R. W. Taylor, Inorg. Chem. **25**, 3751 (1986).

296. Y. Wu and T. A. Kaden, Helv. Chim. Acta **67**, 1868 (1984); **68**, 1611 (1985).

297. R. G. Pearson and R. D. Lanier, J. Amer. Chem. Soc. **86**, 765 (1964).

298. A. Katagiri, S. Yoshimura and S. Yoshizawa, Inorg. Chem. **20**, 4143 (1981).

299. E. A. Van Hahn and E. Peters, J. Phys. Chem. **69**, 547 (1965); J. Halpern, Ann. Rev. Phys. Chem. **16**, 103 (1965).

300. F. C. Anson, T. J. Collins, T. G. Richmond, B. D. Santarsiero, J. E. Toth and B. G. R. T. Treco, J. Amer. Chem. Soc. **109**, 2974 (1987).

301. W. Levason and M. D. Spicer, Coordn. Chem. Revs. **76**, 45 (1987).

302. L. Fabbrizzi, T. A. Kaden, A. Perotti, B. Seghi and L. Siegfried, Inorg. Chem. **25**, 321 (1986) – di- and trivalent Ni and Cu with rigid and flexible dioxotetraaza macrocycles.

303. D. W. Margerum, Pure Appl. Chem. **55**, 23 (1983).

304. J. M. T. Raycheba and D. W. Margerum, Inorg. Chem. **20**, 45 (1981).

305. N. I. Al-Shattti, M. A. Hussein and Y. Sulfab, Trans. Metal. Chem. **9**, 31 (1984).

306. C. A. Koval and D. W. Margerum, Inorg. Chem. **20**, 2311 (1981).

307. M. M. Farrow, N. Purdie and E. M. Eyring, Inorg. Chem. **14**, 1584 (1975).

308. B. G. Cox, J. Garcia-Rosas, H. Schneider and N. van Truong, Inorg. Chem. **25**, 1165 (1986).

309. B. G. Cox, N. van Truong and H. Schneider, J. Amer. Chem. Soc. **106**, 1273 (1984).

310. D. M. Yost and H. Russell, Jr., Systematic Inorganic Chemistry, Prentice-Hall, NY, 1946.

311. M. Cyfert, Inorg. Chim. Acta **73**, 135 (1983).

312. A. Kumar and P. Neta, J. Amer. Chem. Soc. **102**, 7284 (1980).

313. D. H. Huchital, N. Sutin and B. Warnquist, Inorg. Chem. **6**, 838 (1967).

314. E. Mentasti, C. Baiocchi and J. S. Coe, Coordn. Chem. Rev. **54**, 131 (1984).

315. M. Pesavento, A. Profumo, T. Soldi and L. Fabbrizzi, Inorg. Chem. **24**, 3873 (1985) and earlier refs.

316. L. J. Kirschenbaum and J. D. Rush, J. Amer. Chem. Soc. **106**, 1003 (1984).

317. L. J. Kirschenbaum and J. D. Rush, Inorg. Chem. **22**, 3304 (1983).

318. P. N. Dickson, A. Wehrli and G. Geier, Inorg. Chem. **27**, 2921 (1988).

319. L. H. Skibsted, Adv. Inorg. Bioinorg. Mechs. **4**, 137 (1986).

320. L. I. Elding and L. H. Skibsted, Inorg. Chem. **25**, 4084 (1986).

321. G. V. Buxton and R. M. Sellers, Compilation of Rate Constants for the Reactions of Metal Ions in Unusual Valency States, NSRDS-NBS 62, 1978; D. M. Stanbury, Adv. Inorg. Chem. **33**, 69 (1989).

322. Y. Ducommun, G. Laurenczy and A. E. Merbach, Inorg. Chem. **27**, 1148 (1988).

323. K. Tamura, J. Phys. Chem. **91**, 4596 (1987).

324. H. B. Silber, L. U. Kromer and F. Gaizer, Inorg. Chem. **20**, 3323 (1981).

325. S. Yamada, K. Ohsumi and M. Tanaka, Inorg. Chem. **17**, 2790 (1978).

326. I. Okura, N. Kaji, S. Aono, T. Kita and Y. Yamada, Inorg. Chem. **24**, 453 (1985).

327. T. G. Spiro, ed. Zinc Enzymes, Wiley-Interscience, New York 1983.

328. D. N. Silverman and S. Lindskog, Accs. Chem. Res. **21**, 30 (1988).

329. D. W. Christianson and W. N. Lipscomb, Accs. Chem. Res. **22**, 62 (1989).

330. B. W. Matthews, Accs. Chem. Res. **21**, 333 (1988).

331. M. F. Summers, J. van Rijn, J. Reedijk and L. G. Marzilli, J. Amer. Chem. Soc. **108**, 4254 (1986) for refs.

332. R. Davies, B. Kipling and A. G. Sykes, J. Amer. Chem. Soc. **95**, 7250 (1973).

333. H. Gross and G. Geier, Inorg. Chem. **26**, 3044 (1987).

334. I. W. Erni and G. Geier, Helv. Chim. Acta **62**, 1007 (1979); G. Geier and H. Gross, Inorg. Chim. Acta **156**, 91 (1989).

335. D. L. Rabenstein and R. S. Reid, Inorg. Chem. **23**, 1246 (1984).

336. R. A. Marcus, Ann. Revs. Phys. Chem. **15**, 155 (1964).

337. M. Eigen, Angew. Chem. Int. Ed. Engl. **75**, 489 (1963).

Selected Bibliography and Sources of Information

Hydrolysis and Polymerization

C. F. Baes and R. E. Mesmer, The Hydrolysis of Cations, Wiley, New York, 1976.

Reduction Potentials

G. Milazzo and S. Caroli, Tables of Standard Electrode Potentials, Wiley, 1978.

A. J. Bard, R. Parsons and J. Jordan, Standard Potentials in Aqueous Solution, Dekker, New York, 1985.

D. M. Stanbury, Reduction Potentials of Free Radicals, Advances Inorganic Chemistry, Academic, New York, 1989, Vol. 33, p. 69.

Kinetics

Specialist Periodical Reports, Vols. 1-7, Chemical Society, London. Covers the literature from 1969 to 1978 on all aspects of inorganic reaction mechanisms.

M. V. Twigg, ed., Mechanisms of Inorganic and Organometallic Reactions, Vols. 1-7, Plenum, Ongoing series, 1979 to 1989 coverage.

R. van Eldik, J. Asano and W. J. LeNoble, Chem. Rev. **89**, 549 (1989) – Compilation of $\Delta V^{\ddagger}$ values. Collections of rate constants for reactions of radicals and metal complexes excited states are cited in Refs. 143–148 and 358 in Chap. 3.

General information on the coordination chemistry of the transition elements is contained in G. Wilkinson, R. D. Gillard and J. A. McCleverty, eds. Comprehensive Coordination Chemistry, Vols. 3 (Early Transition Elements), 4 (Middle Transition Elements and 5 (Later Transition Elements). These are detailed and authoritative accounts. In addition, periodical issues of Coordination Chemistry Reviews contain reviews of the current chemistry of the transition (and main group) elements.

Problems

1. The following rate constants ($M^{-1}s^{-1}$) were obtained for the redox reaction partners listed:

	Cr^{2+}	Fe^{2+}	$Fe(CN)_6^{4-}$	$Ti(III)^a$
Co(edta)$^-$	6.6×10^3	6.0×10^{-4}	0.21	0.29
Co(edtaH)H$_2$O or Co(edta)H$_2$O$^-$	7.6×10^3	8.5×10^{-3}	3.4	3.0
Co(edtaH)Cl$^-$ or Co(edta)Cl^{2-}	$>2 \times 10^6$	1.4	7.0	9.0

^{a} value of a (s^{-1}) in $-d[Co(III)]/dt = a[[Co(III)][Ti(III)]]/b + [H^+]$

Differentiate between inner- or outer-sphere mechanisms for the four reductants.
R. Marćec and M. Orhanovic, Inorg. Chem. **17**, 3672 (1978); H. Ogino, E. Kikkawa, M. Shimura and N. Tanaka, J. Chem. Soc. Dalton Trans. 894 (1981).

2. Zr(IV) forms a peroxocomplex formulated $Zr_4(O_2)_2(OH)_4^{8+}$. This dissociates in 2 M $HClO_4$ acid with $k = 1.3 \times 10^{-4}s^{-1}$ at 25°C. Both the rate of oxidation of $Zr_4(O_2)_2(OH)_4^{8+}$ by Ce(IV) and the reduction by S(IV) in 2 M $HClO_4$:

$$Zr_4(O_2)_2(OH)_4^{8+} + 4\,Ce(IV) \rightarrow 4\,Zr(IV) + 4\,Ce(III) + 2\,O_2$$

$$Zr_4(O_2)_2(OH)_4^{8+} + 2\,SO_2 \rightarrow 4\,Zr(IV) + 2\,HSO_4^-$$

are independent of Ce(IV) or S(IV) concentrations and at 25°C

$$V = 2.4 \times 10^{-3} [Zr_4(O_2)_2(OH)_4^{8+}]$$

Suggest a mechanism for the redox processes.
R. C. Thompson, Inorg. Chem. **24**, 3542 (1985).
(For the kinetics of conversion of $Zr_4(OH)_8^{8+}$ to monomeric Zr(IV), $-d[Zr_4(OH)_8^{8+}]/dt = k[H^+][Zr_4(OH)_8^{8+}]$)
D. H. Devia and A. G. Sykes, Inorg. Chem. **20**, 910 (1981).

3. $V(CO)_6$ is the only homoleptic metal carbonyl radical, comparable with $Re(CO)_5$ and $Mn(CO)_3L_2$ radicals, which however are more difficult to study. For the substitution reaction in hexane solution:

$$V(CO)_6 + R_3X \rightarrow V(CO)_5R_3X + CO$$

$$-d\,[V(CO)_6]/dt = k\,[V(CO)_6]\,[R_3X]$$

The values of k decrease for R_3X in the order:

$(n\text{-}C_4H_9)_3P > Ph_3P > Ph_3As$.   For $R_3X = Ph_3P$, at 25 °C ,

$k = 0.25\ M^{-1}s^{-1}$, $\Delta H^{\ddagger} = 41.8\ kJmol^{-1}$ and $\Delta S^{\ddagger} = -116\ JK^{-1}mol^{-1}$.

Suggest a mechanism (and rationalize it) for these substitutions.
Q.-Z. Shi, T. G. Richmond, W. C. Trogler and F. Basolo, J. Amer. Chem. Soc. **104**, 4032 (1982); see also D. J. Darensbourg, Adv. Organomet. Chem. **21**, 113 (1982).

4. Orthovanadate(V) is a potent inhibitor of many phosphate metabolizing enzymes. In what ways might VO_4^{3-}-mediated enzymes resemble or differ from PO_4^{3-}-mediated enzymes?
N. D. Chasteen, J. K. Grady and C. E. Holloway, Inorg. Chem. **25**, 2754 (1986).

5. (a) Suggest a reason why VO_2^+ catalyzes the exchange of VO^{2+} with H_2O with a rate law:

$$V = \{k_0 + k_1[VO_2^+]\}\,[VO^{2+}]$$

M. D. Johnson and R. K. Murmann, Inorg. Chem. **22**, 1068 (1983).

(b) The ^{51}V nmr signal of VO_2^+ is broadened by the addition of VO^{2+}. This is ascribed to an exchange process, the rate law for which can be expressed by

$$V = \frac{a\,[VO^{2+}]_0[VO_2^+]_0^2}{1 + b\,[VO_2^+]_0}$$

at 50 °C in 5.4 M $HClO_4$. Suggest a reason for the rate law.
K. Okamoto, W.-S. Jung, H. Tomiyasu and H. Fukutomi, Inorg. Chim. Acta **143**, 217 (1988).
Information for (a) and (b): There is evidence for the existence of a 1:1 VO_2^+/VO^{2+} mixed valence dimer ($K = 0.67\ M^{-1}$ with the conditions of the nmr experiment), but a VO_2^+ dimer in only much higher acid. C. Madic, G. M. Begun, R. L. Hahn, J. P. Launay and W. E. Thiessen, Inorg. Chem. **23**, 469 (1984).

6. Suggest how the rapid exchange of VO_2^+ oxygens with H_2O may arise from the cis structure shown. [Coordinated H_2O's exchange rapidly]

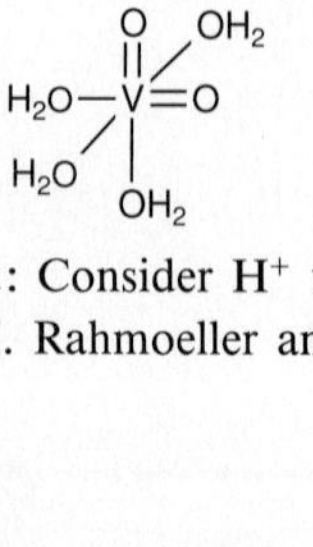

(Hint: Consider H^+ transfer)
K. M. Rahmoeller and R. K. Murmann, Inorg. Chem. **22**, 1072 (1983).

7. For the Ag^+ catalyzed oxidation of VO_3^+ by $S_2O_8^{2-}$, using VO_3^+ in deficiency, the absorbance at 356 nm (which measures the total concentration of VO_2^+ and VO_3^+ since it is an isosbestic point for these species) vs time is represented as below

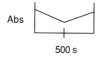

Abs

500 s

What is happening?
R. C. Thompson, Inorg. Chem. **22**, 584 (1983).

8. How might one measure the kinetic parameters for the self-exchange ($k \approx 10^8 M^{-1}s^{-1}$) of $(\eta^6\text{-arene})_2Cr^I/(\eta^6\text{-arene})_2Cr^0$ in solution? What would be the advantages of such a system for testing outer-sphere electron transfer theories? How might the rate constant vary with the optical and static dielectric constants for various solvents?
T. T.-T. Li, M. J. Weaver and C. H. Brubaker, Jr., J. Amer. Chem. Soc. **104**, 2381 (1982). See also M. S. Chan and A. C. Wahl, J. Phys. Chem. **86**, 126 (1982).

9. $Cr(H_2O)_6^{2+}$ reacts with iodoacetamide to give $(H_2O)_5CrI^{2+}$ (50%), $(H_2O)_5CrOC(NH_2)CH_3^{3+}$ (18%) and $(H_2O)_5CrCH_2CO(NH_2)^{2+}$ (30%). The reaction is first order in each reactant and independent of $[H^+]$. Suggest a mechanism.
P. Kita and R. B. Jordan, Inorg. Chem. **25**, 4791 (1986).

10. In the reaction of the Cr(II)-acetate dimer (8.27)–(8.30), what kinetics will be observed if $[Cr(II)]_T \gg X$ for $X = Co(C_2O_4)_3^{3-}$, Co(edta)$^-$ and I^-?
L. M. Wilson and R. D. Cannon, Inorg. Chem. **24**, 4366 (1985).

11. $Cr_4(NH_3)_{12}(OH)_6^{6+}$ does not exhibit acid-base properties. Acid cleavage gives $Cr(H_2O)_6^{3+}$ and *cis*-$Cr(NH_3)_4(H_2O)_2^{3+}$ in a 1:3 ratio. What is the likely structure for the tetramer?
P. Andersen, T. Damhus, E. Pedersen and A. Petersen, Acta Chem. Scand. **A38**, 359 (1984); J. Springborg, Adv. Inorg. Chem. **32**, 55 (1988).

12. The rate constants for thiocyanate anation of a number of complexes of the type $CrL(H_2O)^{n-}$, where L is an N-substituted ethylenediamine-triacetate (L) are shown in the Table.

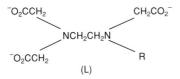

(L)

Table. Rate Constants for the Thiocyanate Anations of $CrL(H_2O)^{n-}$ (k_1) and the Aquation Reactions (k_{-1}) of $CrL(NCS)^{(n+1)-}$

R	k_1 $M^{-1}s^{-1}$	k_{-1} s^{-1}
$CH_2CO_2^-$	14	27
CH_2CO_2H	0.8	0.03
CH_2CH_2OH	3.3	0.24
$CH_2CO_2Co(NH_3)_5^{2+}$	13.3	4.2
H	0.03	0.0024
L = $5H_2O$ [a]	1.8×10^{-6}	9.2×10^{-9}

[a] $Cr(H_2O)_6^{3+}$ + SCN^- reaction

What do the data suggest as to the cause of the unusually high rate constants for these Cr(III) reactions?
H. Ogino and M. Shimura, Adv. Inorg. Bioinorg. Mechs. **4**, 107 (1986).

13. Examine the structure of $Mo_3O_4(H_2O)_9^{4+}$ **6** and draw the possible isomers for $Mo_3O_4(NCS)_2^{2+}$. It is observed that the aquation of $Mo_3O_4(NCS)_2^{2+}$ is biphasic. Given that the Mo's would be expected to behave independently and that rapid inter-conversion of isomers is unlikely, indicate which is the likely structure of $Mo_3O_4(NCS)_2^{2+}$.
P. Kathirgamanathan, A. B. Soares, D. T. Richens and A. G. Sykes, Inorg. Chem. **24**, 2950 (1985).

14. The rate constants for the reaction

$$MoOCl_3(OPPh_3)_2 + Cl^- \rightarrow MoOCl_4(OPPh_3)^- + Ph_3PO$$

in dichloromethane were measured using the Mo(V) compound in deficiency (0.15 mM). The reactions were generally first-order to at least 85% completion. The dependence of the first-order rate constant k_{obs} on the concentrations of Cl^- (added as Et_4NCl) and Ph_3PO is tabulated (25°C)

$[Cl^-]$ [a] mM	$[Ph_3PO]$ mM	k_{obs} s^{-1}
0.13	0.00	44
0.26	0.00	45
0.36	0.00	49
0.15 [b]	0.00	42
0.13	1.0	43
0.13	2.0	31
0.13	3.0	25
0.13	6.8	17
0.13	12	11.5

[a] Estimated from measured dissociation constant of Et_4NCl in CH_2Cl_2.
[b] 0.15 mM Br^- instead of Cl^-

Suggest a mechanism for the reaction and determine any relevant rate constants.
C. D. Garner, M. R. Hyde, F. E. Mabbs and V. I. Routledge, J. Chem. Soc. Dalton Trans. 1175 (1975); also 1198 (1977).

15. $Mn(C_2O_4)_2^-$ decomposes in acid solution:

$$2\,Mn(C_2O_4)_2^- \rightarrow 2\,Mn^{2+} + 3\,C_2O_4^{2-} + 2\,CO_2$$

In the absence of O_2,

$$-d\,[Mn(C_2O_4)_2^-]/dt = k\,[Mn(C_2O_4)_2^-]\,[H^+]$$

Addition of Co(III) complexes (which can be good radical scavengers) does not change the rate law, but decreases the rate. With increasing concentration of Co(III), a limiting rate is reached which is about 50% of that in the absence of Co(III).

In the presence of O_2 the rate is *markedly* inhibited, the rate law changes and considerable amounts of H_2O_2 are formed.

Suggest a reasonable mechanism for the decomposition.
M. Kimura, M. Ohota and K. Tsukahara, Bull. Chem. Soc. Japan, **63**, 151 (1990).

16. N_2-saturated saline (0.56 M) solutions of 20 mM Fe^{2+} at pH 4.8–6.0 were flash photolyzed at 265 nm. A broad transient absorption was observed with a peak between 700–800 nm and an intensity which was roughly proportional to the laser intensity. The transient signal decayed with $t_{1/2} \approx 400$ ns. It was scavenged by Fe^{3+} and Fe^{2+} ions and did not appear when N_2O-saturated solutions were used. Explain.
P. S. Braterman, A. G. Cairns-Smith, R. W. Sloper, T. G. Truscott and M. Claw, J. Chem. Soc. Dalton Trans. 1441 (1984).

17. $Fe(edta)^{2-}$ is sensitive to O_2 but is a useful gentle reductant which, for example, unlike many reductants, does not attack the porphyrin ring in metal-porphyrin complexes. For the reactions of $Co(tpps)^{3-}$ and $Co(tap)^{5+}$ with (excess) $Fe(edta)^{2-}$, predict the rate vs $[Fe(edta)^{2-}]$ profile, and the effect of ionic strength on any second-order rate constants for the two reactions.
R. Lagley, P. Hambright and R. F. X. Williams, Inorg. Chem. **24**, 3716 (1985).

18. The formation of the iron(II) product in the reaction:

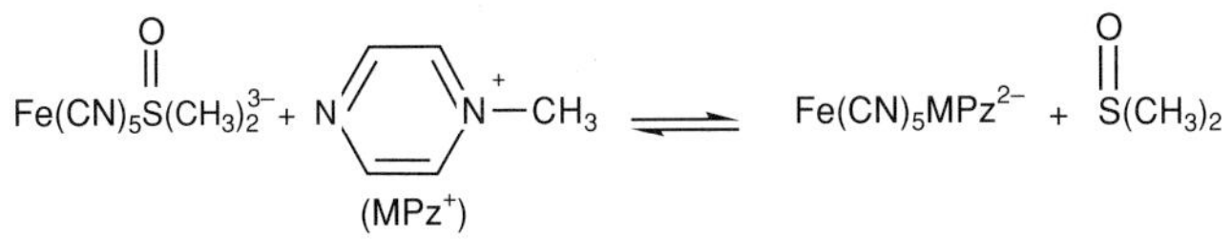

in aqueous solution, can be monitored at 655 nm. Quite surprisingly, the specific rate of approach to equilibrium (k_{obs}) decreases as the concentrations of MPz^+ increases (small excess of dmso). Rationalize this behavior on the basis of a D mechanism for the interchange.
J. M. Malin, H. E. Toma and E. Giesbrecht, J. Chem. Educ. **54**, 385 (1977).

19. Give a plausible reason for the irreproducibility of substitution reactions of $Fe^{III}(CN)_5H_2O^{2-}$. The ion had been prepared in solution from a sample of solid $Na_2Fe(CN)_5H_2O$ which had not been overly purified. Addition of a small amount of Br_2 slowed down the substitution reaction and gave more consistent data.
A. D. James, R. S. Murray and W. C. E. Higginson, J. Chem. Soc. Dalton Trans. 1273 (1974).

20. The effect of redox mediators on the rate of reduction of the Fe(III) myoglobin-fluoride adduct by dithionite is shown in the figure. Suggest reasons for these behaviors. The net reaction is $MbFe(III). \ F^- \xrightarrow{S_2O_4^{2-}} MbFe(II) + F^-$ (very slow)

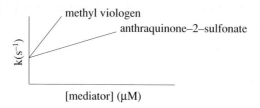

R. P. Cox, Biochem. J. **167**, 493 (1977).

21. $Fe^{II}(CN)_5X^{3-}$ species, pulse-irradiated at 530 nm, display bleaching within the 20 ns pulse. After bleaching, the original absorbance is redeveloped at rates linearly depending on the concentration of free X in solution. For **1** (but not **2**) a second small absorbance change ($\approx 10\%$) accompanied the first. Explain this behavior.

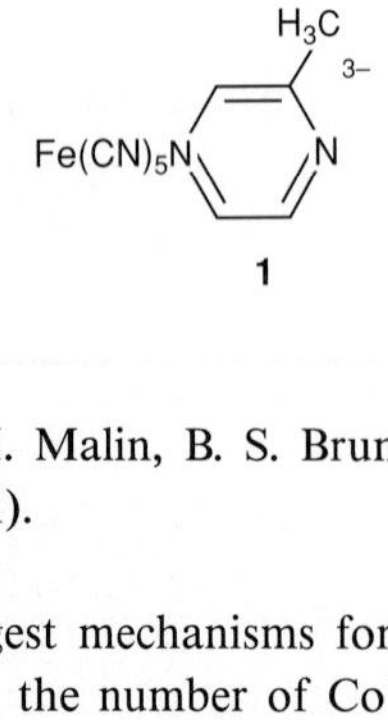

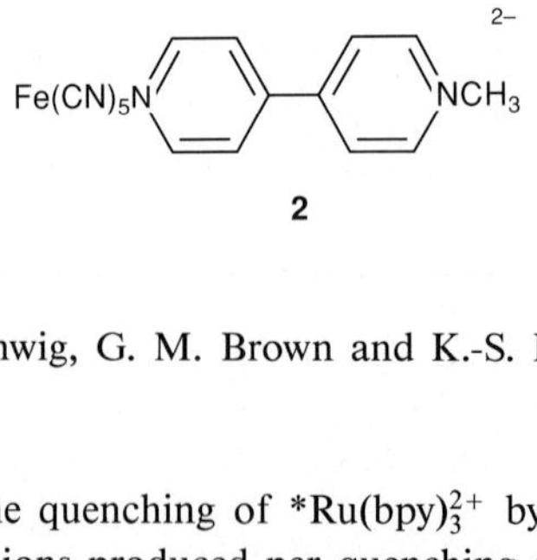

J. M. Malin, B. S. Brunschwig, G. M. Brown and K.-S. Kwan, Inorg. Chem. **20**, 1438 (1981).

22. Suggest mechanisms for the quenching of $*Ru(bpy)_3^{2+}$ by $Co(en)_3^{3+}$ and therefore how N_{Co}, the number of Co^{2+} ions produced per quenching event, might equal values between zero and one. How might some mechanisms be eliminated?
K. Zahir, W. Bottcher and A. Haim. Inorg. Chem. **24**, 1966 (1985).

23. The rate of the electron exchange of the $Ru(en)_3^{3+/2+}$ couple is slightly faster than that of the $Ru(NH_3)_6^{3+/2+}$ couple. Show that this is expected, by examining Eqn. (5.23) and in particular considering the electrostatic work terms and the outer sphere reorganizational energies (λ_o) contributions to the activation free energy.
J. K. Beattie and P. J. Smolenaers, J. Phys. Chem. **90**, 3684 (1986).

24. $Co_2(CO)_8$ reacts with Lewis bases to give various products. With $AsPh_3$ the monosubstituted intermediate $Co_2(CO)_7AsPh_3$ first forms. The first-order rate constant k is independent of $[AsPh_3]$ but reduced by added CO. At 15 °C, $k = 5.1 \times 10^{-3}s^{-1}$, $\Delta H^{\ddagger} = 93$ kJmol^{-1}, and $\Delta S^{\ddagger} = +42$ JK^{-1}mol^{-1} (no added CO). The exchange rate of $Co_2(CO)_8$ with ^{13}CO is $3 \times 10^{-4}s^{-1}$ at 0 °C. Outline the possible mechanisms for substitution in $Co_2(CO)_8$ and indicate the likely one for $AsPh_3$.
M. Absi-Halabi, J. D. Atwood, N. P. Forbus and T. L. Brown, J. Amer. Chem. Soc. **102**, 6248 (1980).

25. The interaction of Co(II) complexes (CoL^{2+}) with O_2 can be studied by mixing in a stopped-flow apparatus Co^{2+} (syringe 1) with a ligand L/O_2 mixture (syringe 2). What are the advantages of this procedure and what are likely to be the problems?
S. Nemeth and L. I. Simandi, Inorg. Chem. **22**, 3151 (1983); M. Kodama and E. Kimura, J. Chem. Soc. Dalton Trans. 327 (1980).

26. All three reactions shown below (μ-peroxo complex in deficiency)

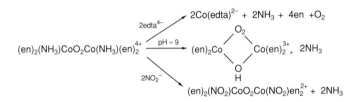

proceed by first-order kinetics, independent of $[edta^{4-}]$, pH or $[NO_2^-]$, with a very similar value for k, $4.9 \times 10^{-3}s^{-1}$ at 25 °C. Explain.
Y. Sasaki, K. Z. Suzuki, A. Matsumoto and K. Saito, Inorg. Chem. **21**, 1825 (1982).

27. How would you expect the uptake by $Co(NH_3)_5H_2O^{3+}$ of SO_2 to compare with that of CO_2 regarding rates and products?
R. van Eldik and G. M. Harris, Inorg. Chem. **19**, 880 (1980); R. van Eldik, J. von Jouanne and H. Kelm, Inorg. Chem. **21**, 2818 (1982).

28. The reduction of $Co(bpy)_3^{3+}$ by reducing radicals, e_{aq}^-, etc. has rate constants $\sim 10^8 - 10^{10}M^{-1}s^{-1}$. Transient species $(\lambda = 300$ nm, $\varepsilon = 4.2 \times 10^4M^{-1}s^{-1})$ are observed which undergo further spectral changes, see Figure, $k \approx 3.5$ s^{-1}. No such change accompanies $Co(phen)_3^{3+}$ reduction, at least within a few seconds. Give a reasonable explanation for this difference of behavior.
M. G. Simic, M. Z. Hoffman, R. P. Cheney and Q. G. Mulazzani, J. Phys. Chem. **83**, 439 (1979).
[Hint: Data in P. Ellis and R. G. Wilkins, J. Chem. Soc. 299 (1950) are relevant.]

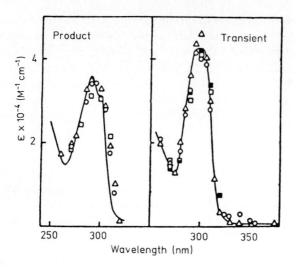

Problem 28. Spectra of transient (a) and final product (b) of one-electron reduction of 40 μM Co(bpy)$_3^{3+}$ in aqueous solution by e_{aq}^- (○) and $CO_2^{\bar{\cdot}}$ (△) at pH 6.9 and $(CH_3)_2COH^\bullet$ at pH 7.0 (□) and 0.5 (■). The solid line in (b) is the spectrum of a 3:1 mixture of bpy and Co^{2+} ion. Reproduced with permission from M. G. Simic, M. Z. Hoffman, R. P. Cheney and Q. G. Mulazzani, J. Phys. Chem. **83**, 439 (1979). © (1979) American Chemical Society.

29. The Ni complex Ni(tmc)L^{2+} (*Trans* I) in which the conformation at the nitrogens is *RSRS* is 5-coordinated. Exchange parameters are (L = D$_2$O):

$$k = 1.6 \times 10^7 s^{-1}; \quad \Delta H^{\ddagger} = 24.7 \text{ kJmol}^{-1} \quad \text{and} \quad \Delta S^{\ddagger} = -24 \text{ JK}^{-1}\text{mol}^{-1}$$

Compare these with the corresponding values (L = D$_2$O) for the solvent exchange of the 6-coordinated species Ni(tmc)L$_2^{2+}$ in the *Trans* III, *RRSS* configuration:

$$k = 1.6 \times 10^8 s^{-1}; \quad \Delta H^{\ddagger} = 37.4 \text{ kJmol}^{-1} \quad \text{and} \quad \Delta S^{\ddagger} = +38 \text{ JK}^{-1}\text{mol}^{-1}$$

Suggest mechanisms for the two exchanges. [Similar behavior is observed when L = CH$_3$CN and dmf].
P. Moore, J. Sachinidis and G. R. Willey, J. Chem. Soc. Dalton Trans. 1323 (1984).

30. The disproportionation of the Ni(III) complex of **12**, R=R$_1$=CH$_3$; R$_2$=H, can be followed by the initial rate method. The dependence of the initial rate on [Ni(III)] and pH is shown below. Suggest a mechanism for the disproportionation (protonation of the Ni(III) complex is known to occur with a pK_a = 4.0. Disproportionation at pH > 6 is very slow).

Initial Rate Data for the Disproportionation of Nickel(III) at 25 °C and $\mu = 0.1$ M

pH	$[Ni(III)]_0$ $10^4 \times M$	Initial Rate $10^7 \times Ms^{-1}$
1.13	0.23	0.12
1.16	0.93	1.8
1.17	2.3	9.8
1.68	2.1	19
2.17	2.3	47
2.67	2.3	155
3.10	1.7	211
3.41	1.7	340
3.58	1.7	380

A. G. Lappin, D. P. Martone and P. Osvath, Inorg. Chem. **24**, 4187 (1985).

31. The reaction of Cu(I) with H_2O_2 has been usually assumed to produce OH$^\bullet$ radicals as the reactive intermediate, e. g. with $Cu(phen)_2^+$,

$$Cu(phen)_2^+ + H_2O_2 \rightarrow Cu(phen)_2^{2+} + OH^\bullet + OH^-$$

In the presence of scavengers (alcohols, RH) the OH$^\bullet$ radical reacts as follows:

$$OH^\bullet + RH \rightarrow R^\bullet + H_2O$$

How would you go about using the scavenging reaction to prove (or disprove!) the production of OH$^\bullet$ in the Cu(I)-H_2O_2 interaction?
G. R. A. Johnson and N. B. Nazhat, J. Amer. Chem. Soc. **109**, 1990 (1987).

32. The rate constants for the $Cu_2(OAc)_4$, HOAc exchange determined by ^{1}H and ^{17}O nmr line broadening methods are in good agreement. Suggest a reason why the exchange rate constants ($\approx 10^4 s^{-1}$) are so much smaller than those normally encountered for solvent exchange with Cu(II).
A. Hioki, S. Funahashi and M. Tanaka, Inorg. Chem. **25**, 2904 (1986).

33. The oxidation of the ions Cr^{3+}, Co^{2+} and VO^{2+} by SO_4F^- is catalyzed by Ag^+. For all these reactions

$$V = -d[SO_4F^-]/dt = k[Ag^+][SO_4F^-]$$

with $k = 1.3 \times 10^3 M^{-1}s^{-1}$ at 17 °C. What mechanism do these data suggest?
R. C. Thompson and E. H. Appleman, Inorg. Chem. **20**, 2114 (1981).

34. The rate of reaction of $AuCl_4^-$ with Fe(II) to give $AuCl_2^-$ is retarded by Fe(III) but accelerated by Cl^- ions. Suggest a mechanism.
K. Moodley and M. J. Nicol, J. Chem. Soc. Dalton Trans. 993 (1977).

Problems − Hints to Solutions

Solutions to most problems are contained in the literature cited. Therefore only hints are given; in a few cases the complete solution or numerical answer is provided.

Chapter 1

1. The dimensions of kT/h are (erg deg $^{-1}$)(deg)(erg s $^{-1}$) $^{-1}$, i.e. s $^{-1}$. The equation is dimensionally correct therefore only for a unimolecular reaction.

2. Using the approach of Sec. 1.2.1 it is apparent that $V = d[N_2]/dt = k[Cu^{2+}][H_2O_2]$. The value of V must be changed from units of ml N_2/min into moles/liter per second. For first entry $V = 18.0$ μM s $^{-1}$ and $k = 2.3 \times 10^2$ M^{-1}s^{-1}. The amount of N_2, 7.3 ml per min represents 2-3% of total production of N_2 and thus initial rate conditions are used. At N_2H_4 concentrations used, Cu(II) must be coordinated to N_2H_4 and reaction is between $Cu(N_2H_4)_n^{2+}$ and H_2O_2.

3. It is apparent that $V = k$ [A] [Enzyme] and thus for first entry, converting ΔD into M, $k = (0.8)(5.5 \times 10^3)^{-1}(20)^{-1}(1.1 \times 10^{-3})^{-1}(21 \times 10^{-6})^{-1} = 3.2 \times 10^2$M^{-1}s^{-1}. The amount of A transformed in 20 s represents 13% of complete transformation, i.e. close to an initial rate condition.

4. Division of the zero-order rate constant (which represents a rate) by the product $[S_2O_8^{2-}]_0$ $[Ag^+]_0$ leads to a constant (second-order) rate constant, $(1.35 \times 10^{-6})(9.5 \times 10^{-3})^{-1}$ $(23 \times 10^{-3})^{-1} = 6.2 \times 10^{-3}$ M^{-1}s^{-1} for 1st entry. The rds involves $S_2O_8^{2-} + Ag^+ \rightarrow$ $SO_4^- + SO_4^{2-} + Ag^{2+}$.

5. A log k/log $[BrO_3^-][H^+]^2$ plot is curved and corresponds to $k = k_1 + k_2[BrO_3^-][H^+]^2$ with $k_1 = (1.2 \pm 0.3) \times 10^{-3}$s^{-1} and $k_2 = 50.8 \pm 0.4$ M^{-3}s^{-1} from all data.

6. Plots of $[CH_3^\cdot]$ vs time (zero-order) and of ln $[CH_3^\cdot]$ vs time (first order) are non linear. Only the plot of $[CH_3^\cdot]^{-1}$ vs time is linear so that the reaction is second order with $2k$ = slope = 2.4×10^{10}M^{-1}s^{-1}.

7. Assume the general rate law (Sec. 2.1.6):

$$V = \{a [S_2O_4^{2-}] + b [S_2O_4^{2-}]^{1/2}\} [azurin] = k_{obs} [azurin]$$

From a plot of $k_{obs} [S_2O_4^{2-}]^{-1/2}$ vs $[S_2O_4^{2-}]^{1/2}$, the slope is a (6.6 × 10^2M^{-1}s^{-1}) and the intercept b (= 170M$^{-1/2}$s^{-1} = $K^{1/2}k_2$, whence $k_2 = 4.6 \times 10^6$M^{-1}s^{-1} for SO_2^- reaction).

8. Use Eqn. (1.22). For (a), $a = 1/2$ and for (b), $a = 3/2$ and these values of a represent the order of the reaction with respect to the reagent in deficiency.

9. Must be clear on definitions.
$-d(a - x)/dt = dx/dt = 2k(a - x)(b - x/2)$ and integration leads to
$(2b - a)^{-1}$ ln $[a(2b - x)][2b(a - x)]^{-1} = kt$.

10. Integrate equations of the form shown after algebraic manipulation.

$$d\,[MA_2]/dt = k_{-1}\,[MAB]^2 - k_1\,[MA_2]\,[MB_2]$$

11. From (1.74)

$$d\,[B]/dt = A_0 k_1 (k_2 - k_1)^{-1}[-k_1 \exp(-k_1 t) + k_2 \exp(-k_2 t)]$$

$[B] = [B]_{max}$ when $d\,[B]/dt = 0$, whence $[B]_{max}$ and t_{max} can be derived.

12. Treat the rate law

$$-\frac{1}{a}\frac{d\,[A]}{dt} = k_1\,[A]^a\,[B]^b - k_{-1}\,[D]^d\,[E]^e$$

as described in Sec. 1.8.1. The perturbation of the concentration of A is $+a\Delta x$ and using $[A] = [A]_e + a\Delta x = [A]_e (1 + (a\Delta x/[A]_e))$ aids in the general derivation.

13. From Sec. 1.8.1, for $C \rightleftharpoons A + B$, k_1, k_{-1}, and $[A] = [B]$, $k = 2k_{-1}[A] + k_1$. It is easy to show

$$k^2 = 4k_1 k_{-1}\,([A] + [C]) + k_1^2$$

Treatment of Figure shows for $L = Et_2NCS_2$, $k_1 = 2.9\,s^{-1}$ and $k_{-1} = 1.5 \times 10^3 M^{-1}s^{-1}$; $L = Et_2NCSSe$, $k_1 = 3.8\,s^{-1}$ and $k_{-1} = 9.0 \times 10^2 M^{-1}s^{-1}$ and $L = Et_2NCSe_2$, $k_1 = 10.8\,s^{-1}$ and $k_{-1} = 8.3 \times 10^2 M^{-1}s^{-1}$.

14. Adopt approach of (1.153). In general,

$$k_I, k_{II} = 1/2\,(\alpha_{11} + \alpha_{22}) \pm \{[1/2\,(\alpha_{11} + \alpha_{22})]^2 + \alpha_{12}\alpha_{21} - \alpha_{11}\alpha_{22}\}^{1/2}$$

where α_{11}, α_{22}, α_{12} and α_{21} are appropriate functions of k_1, k_{-1}, k_2 and k_{-2} (see (1.170)–(1.173)).

15. Use $-d\,[X]/dt = 0$, $[S]_0\,(= [S] + [P]) \gg [E]_0\,(= [E] + [X])$ to derive equation. For initial rate measurements, we can set $[P] = 0$ and obtain an equation which conforms to that of (1.104) since as $[S] \to \infty$, rate $\to V_s$. The same approach to the extended mechanism yields a similar equation with V_s etc. more complex.

16. From Eqns. (1.125) and (1.126) it is clear that

$$k_I + k_{II} = (k_{-1} + k_2)\,[H^+] + (k_1 + k_{-2})\,[Cu^{2+}]$$

$$k_I k_{II} = (k_1 k_2 + k_{-1}k_{-2})\,[Cu^{2+}]\,[H^+] + k_1 k_{-2}\,[Cu^{2+}]^2$$

Computer treatment of the data yields best fit to these equations when $k_1 + k_{-2} = 2.08 \times 10^4 M^{-1}s^{-1}$; $(k_{-1} + k_2)\,[H^+] = 6.5\,s^{-1}$; $k_1 k_{-2} = 1.96 \times 10^7 M^{-2}s^{-2}$ and $(k_1 k_2 + k_{-1}k_{-2}) = 1.79 \times 10^7 M^{-2}s^{-2}$ whence individual k's.

17. V_{exch} must be calculated for each entry using Eqn. (1.191). The value of $V_{exch}/[Ag(II)]^2$ is reasonably constant and the second-order exchange rate constant is $1.1 \times 10^3 M^{-1}s^{-1}$ for the 1st entry. A plausible mechanism is $2Ag(II) \rightleftharpoons Ag(III) + Ag(I)$.

18. A plot of k vs pH conforms to Eqn (1.207) with $k_{AH} = 190\ M^{-1/2}s^{-1}$, $k_A = 0.48 M^{-1/2}s^{-1}$ and $K_{AH} = 2.5 \times 10^{-5}M$. The limiting (acid) k_{AH} and K_{AH} are difficult to estimate.

19. Because k decreases in the acid and base regions, only the 2nd and 3rd terms in (1.226) are important. For carbonic anhydrase-B, $k_1 = 3.5 \times 10^6 M^{-1}s^{-1}$, $k_2 = 2 \times 10^8 M^{-1}s^{-1}$, $pK_S = 9.3$ and $pK_E = 7.5$ and for the carboxymethylated derivative,

$k_1 = 3.0 \times 10^6 M^{-1} s^{-1}$, $k_2 = 4.5 \times 10^6 M^{-1} s^{-1}$, $pK_S = 9.3$ and $pK_E = 9.1$. The two possible routes for only isoenzyme $-B$ can be distinguished on the basis of the values of k_1 or k_2. The latter appears too large (exceeding a diffusion-controlled value) and reaction of the basic form of the enzyme with the acidic (neutral) sulfonamide is favored.

20. For the ionization of $OH^\bullet \rightleftharpoons O^{\overline{\cdot}} + H^+$, the reactivity of the basic form is negligible and Eqn (1.207) can be used with $k_{OH^\bullet} = 1.2 \times 10^{10} M^{-1} s^{-1}$ and $pK_{OH^\bullet} = 11.9 \pm 0.2$.

21. The value of $K (= k_1/k_{-1})$ for the equilibrium:

$$H^\bullet + OH^- \rightleftharpoons H_2O + e^-_{aq} \qquad k_1, k_{-1}$$

in conjunction with that for $H_2O \rightleftharpoons H^+ + OH^-$, will yield the desired K_A for $H^\bullet \rightleftharpoons H^+ + e^-_{aq}$ ($2.3 \times 10^{-10} M$).

Chapter 2

1. See literature references.

2. The reaction of A involves an hydrolytic path as well as one which is first-order dependent on $[H^+]$. Reactions of B and C involve only two or three protons in the respective activated complexes. Proton-assisted dissociation of macrocycle complexes are discussed in Sec. 4.5.1 (a). Preequilibria involving H^+ are favored.

3. The formation of the dimer $Fe_2(OH)_2^{4+}$ suggests an Fe(IV) intermediate which then reacts with Fe(II). It is most likely to arise with 2-electron oxidants.

4. The induction factor of 2 arises from Cr(V) oxidation of I^- as rds, Eqn. (1.121).

5. Use $^{15}NH_2 {}^{18}OH$ and examine product. This is $N^{15}N^{18}O$ which shows the O and central N of NNO arise from NH_2OH. Possibly an intermediate $(NC)_5FeN(=O)^{15}NH^{18}OH^{2-}$ results from addition of $^{15}NH_2 {}^{18}OH$ to NO moiety of $Fe(CN)_5NO^{2-}$. The intermediate then loses H^+ to OH^- in rds and cleaves to $(NC)_5Fe(H_2O)^{3-}$ and $N^{15}N^{18}O$.

6. The isotope effect suggests abstraction of H(D) from the NH(ND) group of $Co(sep)^{2+}$ is necessary for the $O_2^{\overline{\cdot}}$ reaction to form the (thermodynamically stable) HO_2^- product. Direct outer sphere oxidation of $Co(sep)^{2+}$ by O_2 will not produce an isotope effect.

7. The expression

$$\ln \frac{k}{T^n} = \ln C - \frac{U}{RT}$$

with $n = 0$, 1/2 and 1 is fully discussed in B8, p. 116–118.

8. In Eqn. (4.48), when $k_{-1} \gg k_2$, proton transfer is a pre-equilibrium and when $k_{-1} \ll k_2$, the act of deprotonation becomes rate-limiting. Since these processes are likely to have different heats of activation, in the intermediate region, $k_{-1} \approx k_2$, the heat of activation is changing and the Eyring plot will be curved. See Sec. 2.6.

9. A change in the rds is suggested over the temperature range examined. A near zero value of $\Delta H^{\ddagger}$ is usually associated with an exothermic preequilibrium (Sec. 2.6). Formation of a $PCu(I) \cdot Fe(III)$ adduct is suggested within which electron transfer occurs. At lower temperature a higher $\Delta H^{\ddagger}$ obtains and formation of the adduct may involve a reorganisational step.

10. The approach outlined in Sec. 2.5 indicates that a dissociative mechanism applies.

11. With a dissociative mechanism expected for

$$Ni^{2+} + L \rightleftharpoons NiL^{2+} \qquad k_1, k_{-1}, K$$

since k_1 is fairly constant, k_{-1} parallels K. In addition, K often is related to the stability of HL^+ (K_a^{-1}), and thus $\log k_{-1}$ is linearly related to pK_a.

12. (a) Small negative value suggest an electrophilic reaction with small amount of formal carbanion transfer from Cr to Hg.

(b) The low value of ρ indicates a nonpolar transition state.

(c) Electron-donating groups increase basicity of porphyrin and weaken Ru py interaction. This would be expected to lead to an increased rate if dissociative mechanism.

13. (a) When the activated complex resembles the reactants, an "early" transition state forms and a peak in the reaction profile occurs earlier. The value of $|\Delta V^{\ddagger}|$ is smaller than with situation (b).

14. (a) The considerable solvation of OH^- influences the value of $\Delta V^{\ddagger}$. See Ref. 163.

(b) Negative and positive $\Delta V^{\ddagger}$ values are consistent with associative and dissociative mechanisms. These differences might be related to the structures of the Co(II) and Cu(II) species respectively.

15. Examine the volume profile for the base hydrolysis on the basis of a conjugate base mechanism:

$Co(NH_3)_5X^{(3-n)+} + OH^-$

(a) $\Delta V^{\ddagger}_{\text{exp}} - \Delta V_0 \equiv \Delta V^{\ddagger}$ is a constant involving the common 5-coordinate intermediate $+ H_2O$. The value $\approx 20 \text{ cm}^3\text{mol}^{-1}$ suggests that an H_2O molecule is completely "absorbed" during the final step.

(b) If $V^{\ddagger} = V(Co(NH_3)_4NH_2^{2+}) + V(X^{n-})$ [product-like activated complex], $V(H_2O) = 18.1 \text{ cm}^3\text{mol}^{-1}$ and $V(OH^-) = 0.5 \text{ cm}^3\text{mol}^{-1}$, then $V(Co(NH_3)_4NH_2^{2+}) = 66, 77, 75$ and $76 \text{ cm}^3\text{mol}^{-1}$ for the four entries.

(c) This is a graphical representation of (b). Slope $= 0.90$; intercept $= 92 \pm 3 \text{ cm}^3\text{mol}^{-1}$. $V(Co(NH_3)_4NH_2^{2+}) = 92 - 17.6 = 74 \text{ cm}^3\text{mol}^{-1}$.

(d) $\Delta V^{\ddagger}_{\text{exp}} = \Delta V_K + \Delta V^{\ddagger}_K$.

16. Increasing concentrations of Na^+ give rise to $NaFe(CN)_6^{3-}$ and $Na_2Fe(CN)_6^{2-}$ and changing slope of plot may arise in terms of a reactive $NaFe(CN)_6^{3-}$ and a non-reactive $Na_2Fe(CN)_6^{2-}$.

17. The rate law is

$$-d\,[Cr(H_2O)_5ONO^{2+}]/dt = \{k_1\,[H^+] + k_2\,[H^+]^2\}\,[Cr(H_2O)_5ONO^{2+}]$$

with $k_1 = 0.32 \text{ M}^{-1}\text{s}^{-1}$ and $k_2 = 1.2 \text{ M}^{-2}\text{s}^{-1}$. The relative magnitudes of k_1 and k_2 preclude a medium effect. Reactive mono- and di-protonated species are considered the effective reactants.

Chapter 3

1. Examine the effect of CN^- concentrations on the rate constant for reaction of Hg(II) species with the Cys-212. Expect saturation kinetics from which rate constants for reaction of ArHgOH and ArHgCN may be assessed as well as equilibrium constant K for

$$ArHgOH + CN^- \rightleftharpoons ArHgCN + OH^- \qquad K$$

Show that the value of K agrees with that independently found spectrally (in absence of enzyme). Other cysteine-containing proteins should give a similar pattern and value for K.

2. Reduce the resolved Co(III) complex very rapidly in mixer 1. Allow the reduced complex to racemize for a known time before oxidizing in a second mixer. The change of optical rotation in the Co(III) product reflects the amount of racemization of Co(II) complex in the time between its generation and its destruction.

3. At lower pressure $t_{1/2} = 2.50$ ms, $k_1 = 2.4 \times 10^4 \, M^{-1}s^{-1}$ and $k_{-1} = 1.1 \times 10^2 \, s^{-1}$. At high pressure, $t_{1/2} = 3.16$ ms, $k_1 = 1.2 \times 10^4 \, M^{-1}s^{1-}$ and $k_{-1} = 1.2 \times 10^2 \, s^{-1}$. Use equation for $A + B \rightleftharpoons C$ in Table 1.2, and calculate equilibrium concentrations from values of K. These data yield $\Delta V_1^{\ddagger} = 17.5 \, cm^3 mol^{-1}$ and $\Delta V_{-1}^{\ddagger} \approx 0 \, cm^3 \, mol^{-1}$ using Eqn. 2.125 and 1000 kg cm^{-2} = 98 MPa.

4. (a) For geminate recombination see Sec. 2.1.2.
   (b) The $2.8 \, s^{-1}$ step is determined by scavenging of L by excess CO. Subpicosecond laser photolysis of $B-Fe-L$ and picosecond monitoring allow determination of other rate constants and overall equilibrium, as well as detection of $B-Fe$.

5. Very fast change corresponds to $Co(NH_3)_5Cl^{2+} \xrightarrow{h\nu} Co(NH_3)_5^{2+} + Cl\cdot$; the slower changes correspond to consumption of H^+. The equivalent conductance of H^+ ($350 \, \Omega^{-1}mol^{-1}cm^2$) is much higher than those of other species.

6. The e_{aq}^- is most reactive ($k \sim 10^{10} \, M^{-1}s^{-1}$), $CO_2^{\bar{}}$ and $(CH_3)_2\dot{C}OH$ less reactive ($k \sim 10^8 - 10^9 \, M^{-1}s^{-1}$) and H· least reactive ($k \sim 10^7 - 10^8 \, M^{-1}s^{-1}$). Since the intermediate A arises from all radicals it is likely the one-electron reduced, $Fe(CN)_5NO^{3-}$. This will be labile and likely to lose CN^- reversibly.

7. Conductivity changes observed arise from the formation or loss of H^+ (Question 5).

8. If the spontaneous disproportionation of $Br_2^{\bar{}}$ is accelerated when Co(II) is present (in concentrations comparable to those used in the B_{12r}/Br_2 experiments) then (2) is the favored fast step. This step is inner-sphere if $Co(III)-Br$ is detected and its aquation to $Co(III)-H_2O$ noted.

9. Compare the properties (spectral and hydrolysis rate) of the intermediate with those of $Fe(C_2O_4)^+$ generated from Fe^{3+} (excess) and $C_2O_4^{2-}$. Consider the role of ion pairs and triplets in understanding the rate law.

10. Use $A_o = a/R_e - a/R_o$ and $A_o - x = a/R_e - a/R_t$ in the expression for a second-order reaction.

$$t = \frac{1}{A_o k} \left[\frac{A_o}{A_o - x} \right] - \frac{1}{A_o k}$$

to derive the desired equation. Concentration = a/resistance (R) and x, the number of moles of A that react in time t. See B7 or B16.

11. The broadening of the line A equals k_1/π. That of line B will give a linear plot vs [A]/[B] of slope k_1/π.

12. At coalescence temperature 40°, $2k = \sqrt{2}.\pi.48 = 214$ s^{-1}. As the temperature is raised intermolecular exchange occurs. At $\sim 105°$, $2k = \sqrt{2}.\pi.21 = 94$ s^{-1}.

13. A plot of $k_{obs} (=\pi\Delta W)$ vs [CN$^-$] is linear with slope k in the rate law $V = k$ [Pd(CN)$_4^{2-}$] [CN$^-$]. From all data, $k = 1.2 \times 10^2$ M^{-1}s^{-1}.

14. Direct oxygen transfer would lead to a uniphasic release of H$^+$, whereas Cl$^+$ transfer results in two steps, first a release of base and then a larger release of H$^+$.

15. Heterolytic splitting would lead to an initial Co-D/Co-H ratio in the product of 1.0. (R = 0.53). It is apparent that this is not the case. Since D$_2$O does not enter into the homolytic splitting, R should equal 0. At early times this appears to be so. Why might the value slowly increase?

16. When P−SH reacts with a mixture of ESSE (k_1) and RSSR (k_2) (both in excess) the appearance of colored ES$^-$ will result from ESSE directly and from RSSR indirectly. A plot of the observed rate constant (k, s^{-1}) vs [RSSR], [ESSE] constant, will be linear with slope k_2.

17. Use

$$\frac{[Cr^2]_0}{[Co^{2+}]_e} = 1 + \left[\frac{k_1}{k_2}\right]\left[\frac{[VO^{2+}]_{aver}}{[Co(NH_3)_5F^{2+}]_{aver}}\right]$$

to compute $k_1/k_2 = 1.52 \pm 0.04$.

18. a) pH-stat (Sec. 3.10.1) b) conductivity c) resolved pdta^{4-} (Sec. 3.9.4) d) infrared (Sec. 3.9.2) e) H$^+$ change (indicator) or appearance of enzyme activity f) HPLC (Sec. 3.11 (a)) g) conductivity h) fast flow/rapid freeze esr (Table 3.3).

Chapter 4

1. d [SCN$^-$]/$dt = (k_{-1} + k_{-2}$ [H$^+$]$^{-1} + k_{-3}$ [H$^+$]$^{-2}$) [Cr(H$_2$O)$_5$NCS^{2+}]

   For example, $k_1/k_{-1} = K_1$

2. The equation indicates that the value of $V^{\ddagger}$ is close to the mean of the initial and final states. This gives a clue to the mechanism.

3. Only one tautomeric form will give the chelate directly. The other might form the chelate via the unidentate metal complex. A biphasic reaction might be anticipated.

4. (a) Assuming that the rate constant for the open form $\sim 10^4$ M^{-1}s^{-1}, its contribution to observed rate constant will be only $10^4 \times 10^{-3}$ i.e. 10 M^{-1}. The observed value 380 M^{-1}s^{-1} must therefore be associated with H-bonded form, which is ~ 30 fold lower than for normal Ni^{2+} complexing.

   (b) Internal hydrogen bonding in tsa$^-$, but not dhba$^-$, must be weak and ring closure and/or proton loss cannot be rate limiting.

5. For (a) $k_{obs} = k_1 k_2 k_{-1}^{-1}$ [L$_1$] when ring closure is fast. For (b) $k_{obs} = k_3$[L$_1$]. A large positive value for $\Delta V^{\ddagger}$ rules out the associative mechanism (b).

6. (a) Cu([12]ane N$_4$)$^{2+}$ is square pyramidal with Cu^{2+} 0.5 Å out of the plane.

   (b) This behaviour is consistent with Fig. 4.6 with the activated complexes forming at different stages, [Cu$^{2+} \cdot 3$OH$^-$] and [Cu$^{2+} \cdot 2$OH$^-$] respectively.

7. Ni^{2+} displays normal (I_d) behavior. With $Ni(trien)(H_2O)_2^{2+}$ ring closure of $Ni(trien)(H_2O)XY^{2+}$ is rate determining at high [XY]. A first order reaction of $Ni([12]aneN_4)^{2+}$ controls the addition of XY.

8. (a) Stage II must involve a preequilibrium containing one CN^- ion and/or HCN species. The k_5 term arises from spontaneous dissociation of $NiA(CN)_2$.

   (b) Ni(edda) species containing different number of CN^- ligands arise as $[CN^-]$ increases. The rigidity of cydta prevents unwrapping and CN^- binding, with a consequent slower rate.

   (c) The two terms arise from dimer dissociation and CN^- attack on the dimer.

9. (a) See Sec. 4.5.3(a). A preequilibrium step is involved in (a) and the first (deformation) step is rate determining in (b).

   (b) Deformation of porphyrin ring is unnecessary with N-substituted porphyrins.

10. See Secs. 7.31 and 7.32.

11. (a) The first order (solvolytic) path will be missing and $V = k [Pt]_{total} [phen]$.

    (b) Meaningful OH^- dependence for Me_3dien and Me_4dien complexes expected since they show pseudo-octahedral complex characteristics (why not Me_5dien complexes?) (Sec. 4.7.3). Values of k_1 (no$[I^-]$term) and k_2 decrease with increasing steric hindrance, with little $[I^-]$ dependence for Me_4dien and Me_5dien complexes.

12. Use Jahn-Teller considerations for $Cu(II)$ $(Cu(Me_6tren)dmf^{2+}$ is trigonal bipyramidal). Consider the environment of the dmf in the two Co(II) complexes.

Chapter 5

1. For the Ce(IV) reaction, $\log K_{12} (= 16.9 \, \Delta E) = 18.2$ and $k_{12} = 2.9 \times 10^8 \, M^{-1}s^{-1}$. For the MnO_4^- reaction, $\log K_{12} = 3.4$ and $k_{12} = 4.2 \times 10^4 \, M^{-1}s^{-1}$.

2. Much larger inner-shell rearrangement energy terms are involved in the CO_2^-; CO_2 couple relating to linear CO_2 but bent CO_2^- species.

3. The incursion of inner-sphere pathways (for 1st two entries), solvation differences (O_2^-, O_2) and stereochemical changes are the bases for the differences in theoretical and experimental rate constants.

4. See the approach used in Table 5.5. The transfer of σ^* d electrons will lead to relatively large $M-N$ differences in the two oxidation states and therefore slower rates. There is a low-spin $\rightleftharpoons$ high-spin equilibrium for $Fe(sar)^{2+}$.

5. All couples involve moderate bond length changes and consequent slow rates.

6. (a) Formation of precursor complex is rate determining with Cr^{2+} reactions.

   (b) Compare the likelihood of H^+ ionization with the two complexes and the consequent kinetic behavior. Also consider inner-sphere and outer-sphere paths for the two isomers.

   (c) Use $k (ROH^{2+})/k (ROH_2^{3+})$ and $k (RN_3^{2+})/k (RNCS^{2+})$ ratios (R = $Co(NH_3)_5$), the widespread range of k's and similarities in patterns between Cu(I) and Cr(II) reductions to support inner-sphere processes for Cu(I).

   (d) Resonance transfer mechanism favored for V^{2+}.

7. Very large ratios for $k (RN_3^{2+})/k (RNCS^{2+})$ suggest inner-sphere. Large ratios of $k (RSCN^{2+})/k (RN_3^{2+})$ indicate outer-sphere and small ratios (<10) are probably inner-sphere. (Sec. 5.6(c)).

8. The first group must be inner-sphere. The second group must involve a preequilibrium containing a CN^- ion, followed by an outer-sphere redox reaction.

9. Consider what might occur between Fe^{2+} and the product of the reduction, $Co(bamap)(H_2O)^{2-}$, and how Zn^{2+} might interfere.

10. The Ru(II) product is inert and ligand dissociation does not readily occur (compare Eqn. 5.82).

11. The colors correspond to a) an ion pair b) $Fe(CN)_6^{3-}$ and finally c) $Co(edta)^-$.

12. Consider the effects of conjugation on electron transfer and of free rotation around a $C-C$ bond modifying the closeness of metal centers.

13. Examine the rate law arising from the scheme:

$$ L \underset{Eu^{3+}}{\overset{Eu^{2+}}{\rightleftharpoons}} L^{\cdot} \xrightarrow{Co(III)} L + Co^{2+} $$

(L = 4-pyridinecarboxylic acid) and rationalise the ineffectiveness of the 3-derivative.

14. In general, the larger k_H/k_D value corresponds to an outer-sphere process or an inner-sphere process which is not substitution controlled.

15. The very fast change relates to direct reduction of the Fe(III) center by the radical. The amount of absorbance change of this compared to the slow change can be understood if the hydrophobic nature of the heme site is considered. The rate of the slow change is similar for all systems since it involves a (common) intramolecular electron transfer. See (5.8.4).

16. The k_1 and k_2 terms represent intramolecular and intermolecular conversion of $PCu^{II}Ru^{II}$ into $PCu^{I}Ru^{III}$.

Chapter 6

1. The product is $Co(NH_3)_4(NH_2CH_2CONH)^{2+}$ so that two protons are lost from the Co(III) reactant. A series of preequilibria involving either one or two OH^- ions and a rds involving either one or no OH^- ions, respectively, must be operative.

2. The key to understanding the observations is the potential for binding by OH^-, SCN^- or CH_2CONH^-, in the axial position of the Cu(II) complex. This binding can either aid (OH^-) or inhibit hydrolysis.

3. The k_{obs}/pH profile for hydrolysis of the free ligand can be rationalized in terms of mononanionic and neutral species. Saturating effect occurs with Cu^{2+} ion present, due to complete complex formation (consider the species formed and the most reactive likely).

4. Consider the effect of hydrogen bonding between $Cr-OH_2$ and $Cr-OH$ groups in the bridge on the stability of the aquahydroxo bridged species.

5. For the ionization

$$ VO(H_2O)_5^{2+} \rightleftharpoons VO(H_2O)_4OH^+ + H^+ \quad k_1, k_{-1}, K_1 \,(= 4 \times 10^{-7}M) $$

and

$$ Ni(H_2O)_6^{2+} \rightleftharpoons Ni(H_2O)_5OH^+ + H^+ \quad k_2, k_{-2}, K_2 \,(= 10^{-10}M) $$

calculate k_1 and k_2 on the basis that k_{-1} and k_{-2} are diffusion-controlled rate constants $\sim 10^{10}$ $M^{-1}s^{-1}$. Compare these values with k and determine whether exchange is controlled by ionization.

6. No significant dissociation of $Os(bpy)_3^{2+}$ occurs during the observations. There must be four different types of CH groups. The fastest (which?) must correspond to OD^- attack.

7. Convince yourself that the three glycine ligands in III are stereochemically different. Assume NH_2 is a stronger ligand than CO_2^- and predict which chelate ring is the strongest bound (*trans*-effect) and therefore has the highest CH exchange rate (why?).

8. Consider the effect of internal hydrogen bonding in IV on the rate of the OH^- reaction, and the effect of charge on the reaction of IV, $X = CN^-$ compared with that of $X = NH_3$.

9. (a) Consider the influence of R on the likelihood of $C-OR$ or $CO-R$ cleavage and in turn the influence of the cleavage position on the rate.

   (b) The first step is probably the formation of $(NH_3)_5CoO=C(NR_2)_2^{3+}$ and $(NH_3)_5CoNRC(=O)NR_2^{2+}$. Only reaction of the latter leads to the product $(NH_3)_5CoNCO^{2+}$ ($R = H$ or CH_3; consider steric effects).

   (c) Consider the position of proton attachment to the coordinated $C_2O_4^{2-}$ and its likely effect on the rate.

   (d) Structural differences in reactant and product present an additional energy barrier to a normally rapid acid-base reaction.

   (e) Metal ion catalysis involves metal chelation with the enol and keto forms of acac and their isomerization.

10. The product of one-electron oxidation is $Co(NH_3)_5C_2O_4^{2+}$ which contains the powerful reducing radical $C_2O_4^{\bar{\cdot}}$. See Sec. 5.8.3.

Chapter 7

1. Use Le Chatelier's principle or Eqn. 3.7 to show ΔV is slightly negative. No charge change means only intrinsic ΔV observed. Consider (a) the number of water molecules transferred and (b) the effect of spin state change (size of complex) on value of ΔV.

2. $\Delta V_1^{\ddagger} = +5.2$ and $\Delta V_{-1}^{\ddagger} = -2.9$ cm^3 mol^{-1}

3. (a) Camphor is a natural substrate for P-450 and when bound promotes (largely) a high spin state. See Sec. 7.3.

   (b) This conforms to $A + B \rightleftharpoons C \rightleftharpoons D$ with $A + B \rightleftharpoons C$ rapid compared with $C \rightleftharpoons D$ (Sec. 1.8.2) and almost complete.

   (c) $K_d = ([P\text{-}450_{ls} \cdot S] + [P\text{-}450_{hs} \cdot S])/[P\text{-}450_{ls}] [S]$

   and

   $K_s = [P\text{-}450_{hs} \cdot S]/[P\text{-}450_{ls} \cdot S] = k_2/k_{-2}$

   combined give the desired equation.

   (d) Fe(III) is 6-coordinated in the low spin state (extra H_2O) and 5-coordinated in high spin state. Correlate access of Fe in high spin state with value of k_{-2}.

4. (a) Use $[Co(NH_3)_5 {}^{17}OH_2^{3+}$ and $HNO_2]$ and $[Co(NH_3)_5OH_2^{3+}$ and $HN^{17}O_2]$ respectively.

   (b) Demonstrate that the O of $Co(NH_3)_5H_2O^{3+}$ originates from O's of $Co(NH_3)_5NO_2^{2+}$. This would mean $Co-NO_2^{2+} \rightarrow CoO_3SCF_3^{2+} \rightarrow CoH_2O^{3+}$ cannot occur. Also show that $Co(NH_3)_5ONO^{2+}$ reacts more rapidly than $Co(NH_3)_5NO_2^{2+}$ with CF_3SO_3H, otherwise there would be no isosbestic points in the $CoNO_2^{2+} \rightarrow Co-H_2O^{3+}$ transformation.

5. Draw the structures of the ligands, acetone, ascorbate and 4-Meimid and show that there are 2 distinct sites in each for coordination. Nmr is an effective method for monitoring the rearrangements in (b) and (c). That in (a) can be only measured by an indirect method.

6. Show that the scheme

$$(NH_3)_5RuNH_2CH_2CO_2Et^{3+} \underset{k_{-1}}{\overset{k_1}{\rightleftharpoons}} (NH_3)_5RuOC(OEt)CH_2NH_2^{3+} \overset{K(+H^+)}{\rightleftharpoons}$$

$$(NH_3)_5RuOC(OEt)CH_2NH_3^{4+} \overset{k_2}{\longrightarrow} products$$

conforms to the equation with $c/a = 1/k_1$. Thus k_1 is consistent with the value for the analogous rearrangement of the glycine analog.

7. First should observe deuteration at the sarcosinato-N atom (collapse of Me doublet signal around 2 ppm to a singlet). Later a peak (at 2.25 ppm) ascribed to formation of the $\Lambda(R)$ isomer should appear.

8. A plausible intermediate in the rearrangement and in the reaction with Ph_3P is a four-coordinated Ir compound. The similarity in k values suggests a dissociative mechanism for the reactions.

9. Two isomers result from the *meso* ligand (Δ-*cis-β*- and *trans*-) and three isomers from the *racemic* ligand (Λ-*cis-α*-, Δ-*cis-β*- and *trans*-).

10. Require value of k_2. From (7.44), this equals $k_R k_{-1}/2k_1$; $k_1 = k_E$ and $k_{-1} \approx 10^9\,s^{-1}$. Calculated values of k_2 are $1.4 \times 10^5\,s^{-1}$, $2.4 \times 10^6\,s^{-1}$ and $3 \times 10^4\,s^{-1}$ respectively. Value decreases as oxidation state increases.

Chapter 8

1. Reductions by Cr^{2+} are usually inner sphere and have a much higher rate constant with oxidants which can present potential chloride bridge. On this basis reductions by Fe^{2+} are also inner sphere, whereas those of Fe $(CN)_6^{4-}$ and Ti(III) are outer sphere.

2. There appears to be a common intermediate whose formation from $Zr_4(O_2)_2(OH)_4^{8+}$ is rate-determining in the oxidation and reduction reactions. A slow ring-opening process may expose an $-O_2H$ arm, which would be expected to react rapidly with Ce(IV) or S(IV). This same intermediate may be involved in the acid hydrolysis.

3. Nucleophilicity order for R_3X, sign of $\Delta S^{\ddagger}$ and existence of 7-coordinated vanadium carbonyls, are consistent with an A mechanism.

4. Dominant forms of V(V) are $H_2VO_4^-$ and HVO_4^{2-} at pH $\sim$ 7, analogous to those of P(V). However, V(V) species are more labile and relatively easily interconvertible. Vanadium also has a rich redox chemistry which is missing with phosphorus.

5. (a) and (b) The forms of the rate law suggest that a mixed valence oxy-bridged dimer, $V_2O_3^{3+}$, formed reversibly from VO^{2+} and VO_2^+, will lead to (a) enhanced H_2O exchange of VO^{2+} (k_1 term) and (b) easier O exchange with VO_2^+ (develop equations). An alternative explanation for (b), $2VO_2^+ \rightleftharpoons V_2O_4^{2+}$; $V_2O_4^{2+} + VO^{2+} \rightleftharpoons$ exchange is unlikely.

6. A similar rationalisation of the rapid exchange as that in (8.12) can be invoked.

7. Both segments of the biphasic plot are zero-order. The intermediate is obviously VO^{2+}. Thus VO_3^+ and VO^{2+} do not feature in the rate law for the oxidations to VO^{2+} and VO_2^+ respectively. The rds probably must therefore involve only Ag^+ and $S_2O_8^{2-}$ in each case (see Table 8.15).

8. Esr line broadening. When one of the couples if uncharged, the electrostatic work term is approximately zero. Structural differences between the two species are also likely to be small (λ_i small). The effect of solvent is contained in λ_o. Using Eqns. (5.23) and (5.25)–(5.27), a plot of log k_{exch} vs $(D_{op}^{-1} - D_s^{-1})$ should be linear with negative slope.

9. Cr(II) reacts with a number of organic halides by hydrogen atom abstraction in the first (rds). One of the products ($\cdot CH_2CONH_2$) reacts further with Cr^{2+}.

10. The rate laws (8.29) and (8.30), with $[Cr(II)]_{total}$ in excess, lead to oxidation by $Co(C_2O_4)_3^{3-}$ and I_3^- showing initially pseudo zero-order kinetics. As the concentration of the oxidant decreases however, $k_2[oxid] \leqslant k_{-1}$ and some deviation from linearity for the plot occurs and eventually becomes first-order, although this may be near to the completion of reaction. With Co(edta)$^-$ in deficiency, the reaction is pseudo first-order.

11. All 6OH groups must be involved in bridges. Products and their ratio suggests that tetramer is $Cr((OH)_2Cr(NH_3)_4))_3^{6+}$.

12. Consider the effect of pendent arm R on the water lability of $CrL(H_2O)^{n-}$ and how this might operate. Compare the rate constants with those involving $Cr(H_2O)_6^{3+}$ and SCN^-.

13. There are five distinct isomers for $Mo_3O_4(NCS)_2^{2+}$, considering attachment to the same or different Mo's and whether *trans* to bridging or capped O's. Symmetrical isomers might be expected to give uniphasic kinetics.

14. Examination of data shows an independence of k_{obs} on both concentration and nature of halide, and an inverse dependence of k_{obs} on [Ph$_3$PO]. Try a D mechanism.

15. The radical anion CO_2^- often features in reactions involving oxalate. The reduction of the rate by Co(III) complexes might be understandable in terms of Sec. 2.2.1(a). With O_2, scavenging of CO_2^- also occurs. Now another radical O_2^- is formed and a chain reaction is set up, thus modifying the rate law.

16. Rationalise the spectral, kinetic and scavenging effects in terms of the production of e_{aq}^- Table 3.5).

17. Plots of rate vs [Fe(edta)$^{2-}$] are linear for reduction of Co(tpps)$^{3-}$ and curved for Co(tap)$^{5+}$ (ion-pairing). Read Sec. 2.9.1 for an understanding of the ionic strength effects.

18. For a D mechanism, use an equation analogous to (1.143). Derive k_{obs} at low and at high [MPz$^+$] and specify the conditions for the observed behaviour of k_{obs}.

19. Consider traces of Fe^{II}(CN)$_5$H$_2$O^{3-} as the impurity, and the effect of this on the rate of substitution in the Fe(III) analog (Sec. 5.7.3). Br$_2$ would oxidise this impurity.

20. The mediators are accelerating the reduction which is very slow otherwise. The reduction of methyl viologen and anthraquinone-2-sulfonate by $S_2O_4^{2-}$ is very rapid and leads to strongly reducing radicals.

21. The effect of X on the regeneration of Fe^{II}(CN)$_5$X^{3-} after bleaching supports the loss of X in the irradiation process. Consider the two ways in which 1 but not 2 may bind to Fe(CN)$_5^{3-}$ after bleaching. (This is probably the basis for the small change following binding). Sec. 7.4.1 provides a clue as to what is probably happening.

22. Set up a reaction scheme in which *Ru(bpy)$_3^{2+}$ forms an outer-sphere complex with Co(en)$_3^{3+}$. Both electron transfer and energy transfer can occur within this complex to produce Ru(bpy)$_3^{3+}$|Co(en)$_3^{2+}$ and/or Ru(bpy)$_3^{2+}$|*Co(en)$_3^{3+}$. The breakdown of these lead to Co(II) and/or Co(III) respectively. The relative values of the various rate constants linking these species will determine the value of N_{Co}.

23. There is a decreased electrostatic work term due to a large distance of closest approach with $Ru(en)_3^{3+/2+}$. This is small at larger μ and the important factor is a reduced outer-sphere reorganization energy for $Ru(en)_3^{3+/2+}$. Eqns. (5.23) and (5.25) lead to $\Delta G_o^* = 45/r$ kcal mol^{-1} at 25° in H_2O. Use $r = 6.6$ Å for $Co(NH_3)_6^{3+/2+}$ and $r = 10.2$ Å for $Co(en)_3^{3+/2+}$ ions and show that differences in ΔG_o^* will lead to $\sim 10^2$ higher rate constant for $Co(en)_3^{3+/2+}$ than for $Co(NH_3)_6^{3+/2+}$. See N. Sutin, Ann. Revs. Nucl. Sci. **12**, 285 (1962).

24. The kinetic data are consistent with either $Co-CO$ or $Co-Co$ bond cleavage in $Co_2(CO)_8$ as the rds is substitution. Comparison with CO exchange data suggests that both involve $Co-CO$ fragmentation in a first step. Compare Eqn. (2.109).

25. Advantage: Anaerobic manipulation of CoL^{2+} unnecessary. Disadvantage: Must show that formation of CoL is much faster than the subsequent reaction with O_2, otherwise kinetics could be complicated.

26. A common rate-determining process must attend each reaction and involve only the μ-peroxo complex. The products arising with edta^{4-} suggest fragmentation $(\rightarrow 2Co(en)_2(NH_3)H_2O^{2+} + O_2)$ is followed by more rapid reactions.

27. Consider the relative shapes of SO_2 and the sulfito product compared with those of CO_2 and the carbonate complex. In both cases, $Co-OH_2$ bond cleavage does not occur.

28. The spectra of the transient and final products are similar. The immediate product of the reduction will be $Co(bpy)_3^{2+}$. Consider the possible species in a dilute (μM) solution containing a 3:1 ligand/Co^{2+} mixture. $Co(phen)_3^{2+}$ is more inert than $Co(bpy)_3^{2+}$.

29. Use values of $\Delta S^\ddagger$ (and $\Delta H^\ddagger$) to assign I_a or I_d mechanisms. However, read L. Helm, P. Meier, A. E. Merbach and P. A. Tregloan, Inorg. Chem. Acta, **73**, 1 (1983) for a different interpretation for *Trans* I exchange behavior based on a $\Delta V^\ddagger$ value.

30. Would suspect disproportionation to be second-order. Confirm by examining first three entries. Calculate 2nd order rate constant for each entry. Since acid and base forms, $NiLH^{2+}$ and NiL^+, are unreactive, then maximum rate close to pH ~ 4 arises from reaction of $NiLH^{2+}$ with NiL^+ (k). Use equation analogous to (1.231) and confirm $k = (3.4 \pm 0.5) \times 10^3 M^{-1}s^{-1}$.

31. The radical R· would be capable of reducing $Cu(phen)_2^{2+}$ to $Cu(phen)_2^+$ and initiate a chain reaction (compare Prob. 15).

32. Consider whether the Jahn-Teller effect (Sec. 4.2.1 (a) and Chap. 8 Cu(II)) which is believed to be the reason for the lability of $Cu(H_2O)_6^{2+}$, can operate with $Cu_2(OAc)_4$.

33. Interaction of Ag^+ with SO_4F^- can lead to Ag^{2+} and SO_4^- both of which may oxidise the substrates. Examine Table 8.15 to determine whether the k value cited is feasible in the rate law provided.

34. Retardation effect means Fe(III) (and Au(II)) are products in a preequilibrium. Develop the rate law. It is difficult to assess the Cl$^-$ ion effect. Consider the possible role of Fe(II)- and Fe(III)-chloro species.

Subject Index

Horst Friebolin

Basic One- and Two-Dimensional NMR Spectroscopy

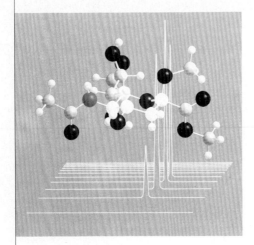

1991. XXI, 344 pages with 144 figures and
42 tables. Softcover. DM 58.00.
ISBN 3-527-28108-8

Translated by Jack K. Becconsall

**With a foreword by J.D. Roberts,
CALTECH.**

*"The book will be of interest and help to both
those needing to learn and those needing a
reference book to refresh their memories, or
extend their capabilities in NMR Spectroscopy. ...Go read about it in Friebolin, then we
can talk."*
J.D. Roberts

Written by an NMR expert with longstanding teaching experience, this basic textbook is a compact
and easily understandable introduction to the modern world of one- and two-dimensional NMR
spectroscopy.

Lucid and information-packed, the book is highlighted by its

• clear and descriptive illustrations
• numerous examples constructed especially for the material presented
• comprehensive treatment of the most important areas of NMR spectroscopy
• excellent didactic approach.

It is suitable both as a textbook for students as well as a selfstudy guide for professionals.

To order please contact your bookseller or:
VCH, P.O. Box 10 11 61, D-6940 Weinheim · VCH, Hardstrasse 10, P.O. Box, CH-4020 Basel ·VCH, 8 Wellington Court, Cambridge CB1 1HZ, UK
VCH, Suite 909, 220 East 23rd Street, New York, NY 10010-4606, USA

VCH